Structure and Function of Sarcoplasmic Reticulum

Academic Press Rapid Manuscript Reproduction

Proceedings of a Symposium on Structure and Function of
Sacroplasmic Reticulum
Held at Kansai Seminar House
Ikuno, Dojo-Cho, Kita-ku, Kobe, Japan
November, 1982

Structure and Function of Sarcoplasmic Reticulum

Edited by

Sidney Fleischer

Department of Molecular Biology
Vanderbilt University
Nashville, Tennessee

Yuji Tonomura

Department of Biology
Faculty of Science
Osaka University
Osaka, Japan

1985

ACADEMIC PRESS, INC.

(*Harcourt Brace Jovanovich, Publishers*)

Orlando San Diego New York London
Toronto Montreal Sydney Tokyo

ACADEMIC PRESS, INC.
Orlando, Florida 32887

United Kingdom Edition published by
ACADEMIC PRESS INC. (LONDON) LTD.
24–28 Oval Road, London NW1 7DX

LIBRARY OF CONGRESS CATALOGING IN PUBLICATION DATA

Symposium on Structure and Function of Sarcoplasmic Reticulum (1982 : Kobe, Japan)
Structure and function of sarcoplasmic reticulum.

Organized as part of the United States-Japan Cooperative Science Program.
Includes index.
1. Sarcoplasmic reticulum—Congresses. I. Fleischer, Sidney. II. Tonomura, Yuji. III. United States-Japan Cooperative Science Program. IV. Title.
QH603.S27S96 1985 591.1'852 83-21556
ISBN 0-12-260380-X (alk. paper)

PRINTED IN THE UNITED STATES OF AMERICA

85 86 87 88 9 8 7 6 5 4 3 2 1

Contents

B. Kinetic Studies of the Calcium Pumping Cycle

Section IV
The Ca^{2+} Release Process

Section V

Regulation of Ca^{2+} Uptake and Release

Contributors

Numbers in parentheses indicate the pages on which the authors' contributions begin.

Hiroshi Abe (561), *First Department of Medicine, Osaka University School of Medicine, Osaka, Japan*

Bruno Agostini (19), *Max-Planck-Institute of Medical Research, Heidelberg, Federal Republic of Germany*

Indu S. Ambudkar (577), *Department of Biological Chemistry, University of Maryland School of Medicine, Baltimore, Maryland 21201*

Nelly Bennett (225, 389), *Laboratoire de Biologie Moléculaire et Cellulaire, Centre d'Etudes Nucléaires de Grenoble, Grenoble/Cedex, France*

Charles Berenski (155), *Biophysics Laboratory, Veterans Administration Medical Center, State University of New York at Buffalo, Buffalo, New York 14214*

J. K. Blasie (51), *Departments of Chemistry and Biochemisty/Biophysics, University of Pennsylvania, Philadelphia, Pennsylvania 19104 and Department of Biology, Brookhaven National Laboratory, Upton, New York*

Werner Boll (305), *Department of Physiology, Max-Planck-Institute for Medical Research, Heidelberg, Federal Republic of Germany*

Mildred Cohn (207), *Department of Biochemistry and Biophysics, University of Pennsylvania School of Medicine, Philadelphia, Pennsylvania 19104*

Leopoldo de Meis (353), *Instituto Ciencias Biomedicas, Departmento Bioquímica, Universidade Federal Rio de Janeiro, Cidade Univerntaria-Ilha Fundao 21910, Rio de Janeiro, Brazil*

Yves Dupont (225, 389), *Laboratoire de Biologie Moléculaire et Cellulaire, Centre d'Etudes Nucléaires de Grenoble, Grenoble/Cedex, France*

Setsuro Ebashi (1), *Department of Pharmacology, Faculty of Medicine, University of Tokyo, Tokyo 113, Japan*

Makoto Endo (521), *Department of Pharmacology, University of Tokyo, Tokyo, Japan*

Alexandre Fabiato (479), *Department of Physiology, Medical College of Virginia, Richmond, Virginia 23284*

Sidney S. Fleischer (51, 119, 155), *Department of Molecular Biology, Vanderbilt University, Nashville, Tennessee 37203*

Jeffrey P. Froehlich (277, 297), *National Institute on Aging, National Institutes of Health, Gerontology Research Center, Baltimore City Hospitals, Baltimore, Maryland 21224*

Minoru Hamada (297), *Department of Medical Biochemistry, Ehime University, Ehime 791-02 Japan*

Hikaru Harafuji (411), *Department of Pharmacology, Juntendo University School of Medicine, Tokyo, Japan*

Wilhelm Hasselbach (19), *Max-Planck-Institute of Medical Research, Heidelberg, Federal Republic of Germany*

Phillip F. Heller (277), *National Institute on Aging, National Institutes of Health, Gerontology Research Center, Baltimore City Hospitals, Baltimore, Maryland 21224*

Leo Herbette (51), *Departments of Chemistry and Biochemistry/Biophysics, University of Pennsylvania, Philadelphia, Pennsylvania 19104, and Department of Biology, Brookhaven National Laboratory, Upton, New York, and Division of Cardiology, University of Connecticut Health Center, Farmington, Connecticut 06032*

Ken-ichiro Higashi (101), *Faculty of Pharmaceutical Sciences, University of Tokyo, Tokyo, Japan*

Lin Hymel (155), *Department of Molecular Biology Vanderbilt University, Nashville, Tennessee 37203*

Noriaki Ikemoto (395), *Department of Muscle Research, Boston Biomedical Research Institute, and Department of Neurology, Harvard Medical School, Boston, Massachusetts 02138*

Yasutada Imamura (63), *Department of Pure and Applied Sciences, College of General Education, The University of Tokyo, Tokyo, Japan*

Giuseppe Inesi (191, 249), *Department of Biological Chemistry, University of Maryland School of Medicine, Baltimore, Maryland 21201*

Makoto Inui (561), *First Department of Medicine, Osaka University School of Medicine, Osaka, Japan*

Chan Jung (155), *Biophysics Laboratory, Veterans Administration Medical Center, State University of New York at Buffalo, Buffalo, New York 14214*

Masaaki Kadoma (561), *First Department of Medicine, Osaka University School of Medicine, Osaka, Japan*

Tohru Kanazawa (325), *Department of Biochemistry, Asahikawa Medical College, Asahikawa, Japan*

Michiki Kasai (147, 537), *Department of Biophysical Engineering, Faculty of Engineering Science, Osaka University, Osaka, Japan*

Masao Kawakita (63), *Department of Pure and Applied Sciences, College of General Education, The University of Tokyo, Tokyo, Japan*

Do Han Kim (395), *Department of Muscle Research, Boston Biomedical Research Institute and Department of Neurology, Harvard Medical School, Boston, Massachusetts 02115*

Yutaka Kirino (101, 453), *Faculty of Pharmaceutical Sciences, University of Tokyo, Tokyo, Japan*

Y. Komatsu (461), *Faculty of Pharmaceutical Sciences, University of Tokyo, Tokyo 113, Japan*

Nagomi Kurebayashi (411, 429), *Department of Pharmacology, Juntendo University School of Medicine, Tokyo, Japan*

Mark Kurzmack (191), *Department of Biological Chemistry, University of Maryland School of Medicine, Baltimore, Maryland 21201*

Jean-Jacques Lacapere (225), *Département de Biologie, Centre d'Etudes Nucléaires, Saclay/Cedex, France*

David Lewis (191), *Department of Biological Chemistry, University of Maryland School of Medicine, Baltimore, Maryland 21201*

David H. MacLennan (91), *Banting and Best Department of Medical Research, University of Toronto, C. H. Best Institute, Toronto, Ontario, Canada*

Madoka Makinose (305), *Department of Physiology, Max-Planck-Institute for Medical Research, Heidelberg, Federal Republic of Germany*

Andreas Maurer (155), *Department of Molecular Biology, Vanderbilt University, Nashville, Tennessee 37240*

Pankaj Medda (19), *Max-Planck-Institute of Medical Research, Heidelberg, Federal Republic of Germany*

Andrea Migala (19), *Max-Planck-Institute of Medical Research, Heidelberg, Federal Republic of Germany*

Hiroshi Morii (439), *Department of Biology, Faculty of Science, Osaka University, Osaka, Japan*

Kazuo Nagasaki (537), *Department of Biophysical Engineering, Faculty of Engineering Science, Osaka University, Osaka, Japan*

Hiroshi Nakamura[1] (337), *Department of Etiology and Pathophysiology and Department of Molecular Physiology, National Cardiovascular Center Research Institute, Osaka, Japan*

Yoichi Nakamura (179), *Department of Biology, Faculty of Science, Osaka University, Osaka, Japan*

Kazuki Nunogaki (537), *Department of Biophysical Engineering, Faculty of Engineering Science, Osaka University, Osaka, Japan*

Yasuo Ogawa (411, 429), *Department of Pharmacology, Juntendo University School of Medicine, Tokyo, Japan*

Hiromichi Okuda (297), *Department of Medical Biochemistry, Ehime University, Ehime 791-02, Japan*

Masanori Osakabe (453), *Faculty of Pharmaceutical Sciences, University of Tokyo, Tokyo, Japan*

[1]Present address: Department of Biological Chemistry, University of Maryland, Baltimore, Maryland 21201.

D. Pascolini (51), *Departments of Chemistry and Biochemistry/Biophysics, University of Pennsylvania, Philadelphia, Pennsylvania 19104*

D. H. Pierce (51), *Departments of Chemistry and Biochemistry/Biophysics, University of Pennsylvania, Philadelphia, Pennsylvania 19104*

Elizabeth Pintado (207), *Department of Biochemistry and Biophysics, University of Pennsylvania School of Medicine, Philadelphia, Pennsylvania 19104*

Richard Pougeois (225), *Départment de Recherches Fondamentales, Centre d'Etudes Nucléaires, Grenoble/Cedex, France*

Reinhart A. F. Reithmeier (91), *Department of Biochemistry, University of Alberta, Edmonton, Alberta, Canada*

Kimiko Saito (63), *Department of Pure and Applied Sciences, College of General Education, The University of Tokyo, Tokyo, Japan*

J. George Sarmiento (297), *Department of Medicine, University of Connecticut, Farmington, Connecticut 06032*

Antonio Scarpa (51, 207), *Department of Biochemistry and Biophysics, University of Pennsylvania School of Medicine, Philadelphia, Pennsylvania 19104*

Adil E. Shamoo (577), *Department of Biological Chemistry, University of Maryland School of Medicine, Baltimore, Maryland 21201*

Munekazu Shigekawa (337), *Department of Etiology and Pathophysiology and Department of Molecular Physiology, National Cardiovascular Center Research Institute, Osaka, Japan*

Hiroshi Shimizu (461, 453), *Faculty of Pharmaceutical Sciences, University of Tokyo, Tokyo, Japan*

V. Skita (51), *Departments of Chemistry and Biochemistry/Biophysics, University of Pennsylvania, Philadelphia, Pennsylvania 19104*

Masahiro Sokabe (537), *Department of Behaviorology, Faculty of Human Science, Osaka University, Osaka, Japan*

Michihiro Sumida (297), *Department of Medical Biochemistry, Ehime University, Ehime 791-02 Japan*

Michihiko Tada (561), *First Department of Medicine, Osaka University School of Medicine, Osaka, Japan*

Takahisa Taguchi (147), *Department of Biophysical Engineering, Faculty of Engineering Science, Osaka University, Osaka, Japan*

Yuichi Takakuwa (325), *Department of Biochemistry, Asahikawa Medical College, Asahikawa, Japan*

Hitoshi Takenaka (297), *Department of Medicine, University of Connecticut, Farmington, Connecticut 06032*

Charles Tanford (259), *Department of Physiology, Duke University Medical Center, Durham, North Carolina 27706*

Manabu Tanifuji (537), *Department of Biophysical Engineering, Faculty of Engineering Science, Osaka University, Osaka, Japan*

Yuji Tonomura (163, 179), *Department of Biology, Faculty of Science, Osaka University, Osaka, Japan*

Wofgang Waas (19), *Max-Planck-Institute of Medical Research, Heidelberg, Federal Republic of Germany*

Shigeo Wakabayashi (337), *Department of Etiology and Pathophysiology, and Department of Molecular Physiology, National Cardiovascular Center Research Institute, Osaka, Japan*

Takahide Watanabe[2] (249), *Department of Biological Chemistry, University of Maryland, School of Medicine, Baltimore, Maryland 21201*

James M. Watras (297), *Department of Medicine, University of Connecticut, Farmington, Connecticut 06032*

Y. Yamaguchi (461), *Faculty of Pharmaceutical Sciences, University of Tokyo, Tokyo 113 Japan*

Taibo Yamamoto (163), *Department of Biology, Faculty of Science, Osaka University Osaka, Japan*

Robert E. Yantorno[3] (163), *Department of Biology, Faculty of Science, Osaka University, Osaka, Japan*

Kimiko Yasuoka-Yabe[4] (63), *Department of Pure and Applied Sciences, College of General Education, The University of Tokyo, Tokyo, Japan*

[2]Present address: Department of Molecular Physiology, National Cardiovascular Center Research Institute, Osaka, Japan.

[3]Present address: Upper Darby, Pennsylvania 19082.

[4]Present address: The Institute of Medical Science, The University of Tokyo, Tokyo, Japan

Preface

This volume on "The Structure and Function of Sarcoplasmic Reticulum" is based on an International Conference held at Kansai Seminar House in Kobe, Japan, November 1–4, 1982. To our knowledge, it was the first symposium devoted largely to sarcoplasmic reticulum. In this volume we have assembled what is known about sarcoplasmic reticulum, i.e., historical development, current knowledge and state of the art, and future perspectives.

The study of sarcoplasmic reticulum is important because of its central role in muscle physiology and as a prototype for the understanding of the molecular biology of a membrane pump. Among the many biomembranes serving to transport cations, sarcoplasmic reticulum attracts special interest since active calcium transport by this membrane system provides one of the simplest and most effective ways to gain insight into the molecular mode of energy transduction in which chemical energy is converted to osmotic work and *vice versa*. The maintenance of a large calcium ion concentration gradient across the cell membrane, i.e., low intracellular free calcium ion concentration, is essential for the viability of *all* cells. This is because the fine tuning of the intracellular free calcium ion concentration modulates many cell functions. A variety of transporters are present in different cell membranes, including the plasmalemma, endoplasmic reticulum, and the mitochondrion, which contribute to the regulation and maintenance of such gradients. The most important type is the ATP-driven calcium pump which translocates calcium from the cytoplasm either out of the cell or into the intracellular compartments.

Muscle contraction and relaxation are regulated by the intracellular free calcium concentration which acts as a "second messenger." In this regard, sarcoplasmic reticulum performs three functions: (1) it is an intracellular compartment for the storage of calcium; (2) it is involved in the calcium release process which triggers muscle contraction; and (3) it contains the calcium pumping machinery for the energized uptake of calcium enabling muscle to relax. Highly purified sarcoplasmic reticulum from fast skeletal muscle is a specialized membrane which consists mainly of the calcium pump protein (approximately 90% of the membrane protein), also referred to as the Ca^{2+}-ATPase, Ca^{2+}, Mg^{2+}-ATPase, or Ca^{2+}, Mg^{2+}-dependent ATPase. The calcium pump protein, of approximately 115 kilodaltons, has been shown by reconstitution studies to be the pumping molecule. Similar calcium pumping machinery is present in sarcoplasmic reticulum from many muscle types such as fast skeletal, slow skeletal, smooth and cardiac muscle, and from diverse sources, i.e., from insects to mammals and birds. Calcium pumps are also present in membranes from various types of nonmuscle cells. There are important differences in calcium pump proteins yet to be fully detailed in terms of primary protein sequence, structure, size. immunological cross-reactivity, and in regulation. Since isolated sarcoplasmic reticulum vesicles, or "fragmented sarcoplasmic reticulum," are capable of energized calcium uptake, the pumping cycle has been described in some detail. Less is known about the calcium release process although extensive research activity is ongoing in this direction. The calcium pump protein from heart sarcoplasmic reticulum has been shown to be modulated by both calcium (and calmodulin) and cyclic AMP by way of two different types of protein kinases. The study of this modulation furthers our understanding of how biological pumps can be regulated.

The book is organized in five sections. The first section consists of a historical background and critical overview by Professors Ebashi and Hasselbach, respectively, who participated in the conference and in the discovery of sarcoplasmic reticulum which occurred scarcely 25 years ago. The remaining four sections deal with the chemistry and ultrastructure of the calcium pump protein in sarcoplasmic reticulum (Section II), kinetics and thermodynamics of the calcium pumping mechanism (Section III), the calcium release process (Section IV), and regulation of calcium uptake and release (Section V). The contributions are from many of the key laboratories in the field. In essence, the various dimensions in the study of sarcoplasmic reticulum as a pump and its role in muscle contraction and relaxation have been included.

The study of sarcoplasmic reticulum is truly multidisciplinary in scope and international in effort. We are grateful to the participants of the conference for sharing their knowledge and enthusiasm and for contributing their chapters.

The symposium was organized by Sidney Fleischer and Yuji Tonomura as part of the United States–Japan Cooperative Science Program. In addition to scientists from the United States and Japan we were fortunate to have a number of participants from other countries as well. The conference was funded by the National Science Foundation of the United States (Grant number INT-8121353) and the Japan Society for the Promotion of Science. Further support was provided by Osaka University First Department of Medicine by Professors Hiroshi Abe and Michihiko Tada and by Osaka University Department of Biology, Faculty of Science by Professor Yuji Tonomura. The Advisory Committee included Professors Hiroshi Abe, Setsuro Ebashi, Makoto Endo, Giuseppe Inesi, and Michihiko Tada. The friendly atmosphere and the efficient operation of the conference was due to the considerable efforts of the host committee from the University of Osaka, including Y. Tonomura, T. Yamamoto, H. Abe, H. Takisawa, T. Arata, S. Yamashita, T. Watanabe, S. Wakabayashi, M. Kadoma, M. Yamada, M. Inui, and F. Ishikura.

Shortly after the conference was over we were informed that Professor Yuji Tonomura had died. His passing is a considerable loss to science, especially in the fields of muscle research and biological transport. Each of us who had the privilege of knowing him share a void with the loss of a dear friend, a gentle and wise colleague. This book is dedicated to him. The eulogy was written by Professors Manuel Morales and Giuseppe Inesi.

I am especially grateful to Professor Taibo Yamamoto for his help with coordinating matters associated with publication.

SIDNEY FLEISCHER

1. N. Kurebayashi
2. J. Sakamoto
3. S. Yamashita
4. J. Taniguchi
5. S. Yamada
6. F. Ohmori
7. K. Konishi
8. M. Tanifuji
9. T. Takagi
10. M. Makinose
11. M. Kasai
12. H. Shimizu
13. S. Wakabayashi
14. T. Kanazawa
15. G. Kawamura
16. M. Kawakita
17. M. Endo
18. T. Watanabe
19. M. Yamaguchi
20. H. Miyamoto
21. T. Taguchi
22. H. Takisawa
23. T. Kanemasa
24. Y. Ogawa
25. M. Sumida
26. A. Scarpa
27. S. Ebashi
28. J. Froehlich
29. G. Inesi
30. W. Hasselbach
31. D. H. MacLennan
32. Y. Tonomura
33. K. Nunogaki
34. A. Shamoo
35. S. Fleischer
36. C. Tanford
37. S. Yagi
38. M. Tada
39. M. Kadoma
40. T. Iwadare
41. K. Blasie
42. H. Nakamura
43. N. Ikemoto
44. J. Fujii
45. M. Shigekawa
46. Y. Dupont
47. T. Yamamoto
48. Y. Kirino
49. L. de Meis
50. M. Inui
51. A. Fabiato
52. H. Morii
53. T. Daiho
54. H. Harafuji

INTERNATIONAL CONFERENCE ON THE STRUCTURE AND FUNCTION
OF SARCOPLASMIC RETICULUM
Kobe, Japan 1982
Organizers: Sidney Fleischer and Yuji Tonomura

YUJI TONOMURA
1923–1982

YUJI TONOMURA

1923—1982

At the height of his career (when he died), Yuji Tonomura belonged to two worlds: that of contractility and that of ion transport, but his first love was contractility. It all began in the early postwar years. Though bedridden with tuberculosis that was nearly fatal, he collaborated with his lifelong friend Watanabe to produce the first analytic treatment of actomyosin ATPase and turbidity. Their paper appeared in *Nature*, but went unnoticed (it was much too early for enzyme kinetics studies). Thus a pattern that was thereafter to plague Yuji and distress his friends was set: delayed recognition. Tonomura did not suffer mathematical illiterates, and was not fond of popularizing his work. His thoroughly honest criticisms were seldom "softened" to mollify their targets. His relentless logic, complex graphs, and mathematical reasoning made him difficult to read and to listen to. All these attributes contributed to his delayed recognition. But in time we learned that it was unwise and self-defeating to ignore him. The importance of one of his greatest contributions—the discovery of the ADP·P_i intermediate—escaped the grasp of Western scientists for at least five years. Collectively, we were unable (or unwilling to study his work. Appreciation of his other major contributions was delayed because his experiments were ahead of their time. In 1964 he seemed to be saying, "If I can do with organic solvents what you can do with thiol reagents, then we must both be producing a change in conformation." The statement seemed a little mystical at the time, but twenty years and countless papers later, most of us now think that activation results from conformation distortion, not from blocking thiols.

Yuji was an accomplished physical biochemist, but his associations with organic synthesists were exceptionally fruitful. It was with Kubo that he discovered the reactive lysine of myosin and with Ikehara that he produced a wonderful family of ATP analogs tailored to track substrate movement. The ramifications of these discoveries continue to this day. Measurements of fluorescence energy transfer in muscle proteins are now rather fashionable, but one of the cleverest examples remains the Tonomura–Onishi study of the proximity of the tryptophan active site. In the last decade of his life, Tonomura was persuaded to visit America on short sabbaticals, each of which had beneficial results. First, his health invariably improved during each visit (a fact he attributed as much to being able to escape the Osaka tensions as the Osaka weather). Second, a number of beautiful contributions resulted [the first (and only) number average measurement of myosin and the EPR method for tracing the effects across oligomeric systems, experiments arguing for the importance of rotation in free energy transfer, etc.]. Third, a canard, widespread among his students, was repeatedly refuted. Although Yuji often shared a *Kirin* or a *Suntory* with his students, they never forgot that he was the Sensei. As is common in such a relationship, legends about him passed from one generation of students to the next. One of these was that although the *Sensei* was brilliant in direction, he was absolutely hopeless at the bench. It was a constant pleasure to be able to prove to these young students that in fact Yuji was a highly skilled experimenter who produced many neatly filled books with precise data. Tonomura's irreverence for symposia was unaffected by his mounting international recognition. In 1981 he was invited to a meeting in Paris, attended only by the best of the world's muscle researchers. "Well, it's my long-awaited chance to see the Louvre," he told his congratulating friends. He himself provided the research community with two notable symposia: the 1980 Myosin Conference in Sapporo, and the 1982 meeting on "The Structure and Function of Sarcoplasmic Reticulum," the proceedings of which appear in this volume.

The life of a man who is both admired and loved by his friends always seems to end too soon, particularly in terms of tasks left undone. For several years prior to his death, Tonomura had been grappling with a difficult problem: Are the two heads of a myosin molecule unsymmetrical? Settling the issue would have far-reaching consequences for contractility. As usual, it was Tonomura's group against a disbelieving world. Progressively, those of opposite beliefs began to relent under the pressure of experimental results. Researchers began to concede that, yes, there are different heavy chains in a myosin preparation. And, having failed to find more than one kind of heavy chain in smooth muscle, Yuji quickly conceded that the nonequivalence, could not be fundamental. At this time it seems that the resolution may lie in isozymes, but it is such a pity that Yuji did not live to settle the issue himself.

Tonomura's second productive love was ion transport. Early studies in collaboration with Yamamoto described the formation of a phosphorylated enzyme intermediate in the mechanism of ATP utilization by the calcium-dependent ATPase. Thereafter, in collaboration with Yamada, he reported that enzyme phosphorylation with P_i in the presence of a transmembrane calcium gradient is an early event in the reversal of the calcium pump. He then proceeded with Kanazawa to describe the kinetics of formation and breakdown of phosphorylated enzyme intermediate at low temperature, and in collaboration with Sumida noted the phenomenon of calcium translocation accompanying enzyme phosphorylation. His contributions extend to the Na^+, K^+-ATPase for which he defined, in collaboration with Yamauchi, cation binding parameters.

Tonomura's work in the field of ion transport and coupled enzyme reactions retained the complexity and sophistication which was inherent in his work. Nevertheless, the tangible and accessible nature of the phosphorylated enzyme intermediate soon led to the acceptance of his work. Many of us have been inspired by his ideas and experiments.

In addition to his direct scientific contributions, Tonomura's training of illustrious researchers is an example of the extraordinary influence a university professor can have not only in his own nation but internationally. Those of us who were fortunate to work with him in the same laboratory and to discuss ideas, methodology, and experimental results, know well the pervasive extent of his influence, and understand how it was possible for Professor Tonomura to inspire such a large school of excellent students.

Finally we like to recall Yuji, the human being and friend, his keen appreciation for Eastern and Western art, the benevolent pride in his own accomplishments, his serene conversation late at night over a beer, his humor and sense of values. Yuji's memory will continue to inspire and encourage us in the future.

MANUEL MORALES
GUISEPPE INESI

Section I

Historical Background and Overview

RELAXING FACTOR, SARCOPLASMIC RETICULUM AND TROPONIN --- A HISTORICAL SURVEY ---

Setsuro Ebashi

Department of Pharmacology, Faculty of Medicine,
University of Tokyo, Bunkyo-ku, Tokyo 113

INTRODUCTION

When we look back on the history of biological sciences, there can be seen several cases where important new biological concepts were first developed from studies on muscle and then generalized. Ca ion, which is now accepted as the most fundamental mediator of cellular functions in general, is a typical example.

In 1950's almost all muscle biochemists were enthusiastically engaged in the research on the myosin-actin-ATP interaction *in vitro*. The studies on the relaxing factor were not considered to be on the "royal road". This factor, however, identified with the sarcoplasmic reticulum and related to Ca ion. Subsequent effort to clarify the mode of action of Ca ion was led to the discovery of the third factor other than myosin and actin, i.e., the troponin-tropomyosin system, the regulatory system of muscle contraction motivated by Ca ion.

Nearly twenty years have passed since I left the field of sarcoplasmic reticulum research. During this time remarkable progress has been made and now I feel myself quite a stranger. I am thus not qualified to give a true historical survey.* Instead, I will describe my personal

---

* For the pioneering work of Heilbrunn (1) and Kamada (2) concerning the crucial role of Ca ion in muscle contraction, see ref. (3).

ISBN 0-12-260380-X

experiences on how I became interested in studying the 'relaxing factor' and how I left the field of sarcoplasmic reticulum. Emphasis will be placed on my failures rather than my success through good lack.

"DISCOVERY" OF THE RELAXING FACTOR

In 1947, Professor H. Kumagai, my teacher, suggested that I should work on smooth muscle pharmacology with an electrophysiological approach, and I spent several vain years on this. In the meantime, I had a chance to read a humble mimeographed copy of the book, "Chemistry of Muscular Contraction, 1947", by Szent-Györgyi (4), which excited me greatly. Naturally, I wished to start immediately working on the actomyosin system, but my lack of biochemical training compelled me to postpone beginning until early in 1952. At that time I was able to get the second edition of

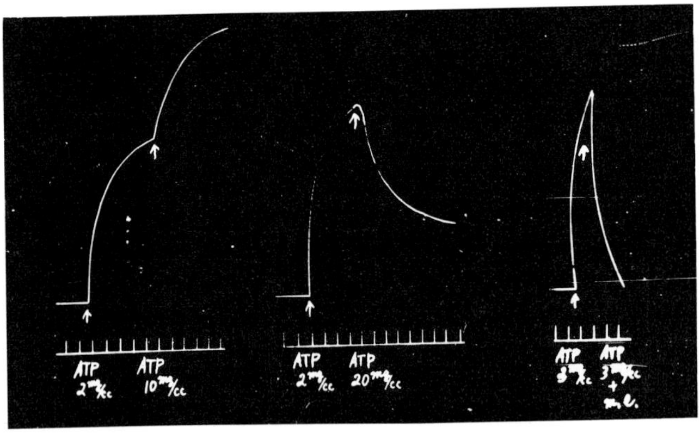

Fig. 1: Relaxation of glycerinated rabbit psoas by ATP in the presence of muscle extract (m.e.)

A piece of muscle was homogenized, centrifuged at a moderate speed, and the resulting supernatant was used as m.e. Glycerinated psoas was soaked in a solution containing 0.15 M KCl and 10 mM $MgCl_2$, and the contraction was started by adding ATP to the solution. Shortening of the fibers was recorded by a light lever on a smoked drum (time in minutes). Note that the relaxation by a moderate ATP concentration in the presence of m.e. is far more pronounced than that by a high ATP concentration without m.e.

the book (4), in which I found the preparation method of 'glycerol-extracted psoas'. This experimental material was much more familiar to a pharmacologist than actomyosin itself. So I began preliminary experiments in collaboration with K. Fujita.

For a short while we enjoyed the beautiful contraction of glycerinated rabbit psoas induced by ATP. Soon we noticed that muscle fibers, shortened by ATP, never relaxed even after ATP removal. This was not consistent with the traditional concept of muscle contraction, one of its outstanding characteristics being its reversibility. As a pharmacologist acquainted with the drug action on living muscle, I could not overlook this puzzling fact. Incidentally, there appeared then a very interesting paper of Bozler (5), which showed that the ATP-contracted glycerinated fibers could be relaxed by a higher concentration of ATP.

However, there was no possibility for such a prompt increase of ATP concentration in muscle cells and, therefore, this could not be a physiological mechanism of relaxation. After many failures I became suddenly aware of our daily experiences with fresh glycerinated fibers which tended to relax even with a relatively low concentration of ATP. This might indicate that a soluble intrinsic factor to induce relaxation should have been retained in fresh fibers (we know today that this interpretation was not right; gradual decrease of relaxing tendency was mainly due to the deterioration of the sarcoplasmic reticulum with elapsed time). So we prepared a simple muscle extract and applied it to ATP-contracted glycerinated fibers. As a result, clear relaxation was induced --- it was a dramatical success (Figs. 1 - 4).*

A few months after our observation, however, Prof. Kumagai notified me of a brief communication in Nature, in which Bendall (6) demonstrated an experiment very similar to ours; furthermore, we learned that the factor itself had been discovered already in 1951 by Marsh in Bailey's laboratory (7).

If it were nowadays, I should have abandoned the whole project, greatly disappointed, regretting my lack in getting up-to-date information in advance. But the situation in

---

* A part of this work was later published in Japanese as the doctoral thesis of Fujita (8).

which Japanese scientists were placed was terribly depressed at that time, so I instead took some degree of satisfaction even with this "now meaningless" result and continued the effort to isolate the active principle from the extract with the collaboration of F. Takeda (9, 10).

Our work had thus been almost in vain, but one little fruit was that we have deepened our understanding of the actin-myosin interaction in the presence of ATP as expressed in Fig. 5, which was indebted to original profound thoughts of Szent-Györgyi (4) and Bozler (5).

In the meantime, there appeared several papers asserting that the factor should be an ATP-regenerating system such as creatine kinase-phosphocreatine system (11) or myokinase (12). We could not agree with this idea; for instance, the active fraction obtained by ammonium sulfate fractionation was different from the fraction of such an ATP-regenerating system (Fig. 6) (10, 14).

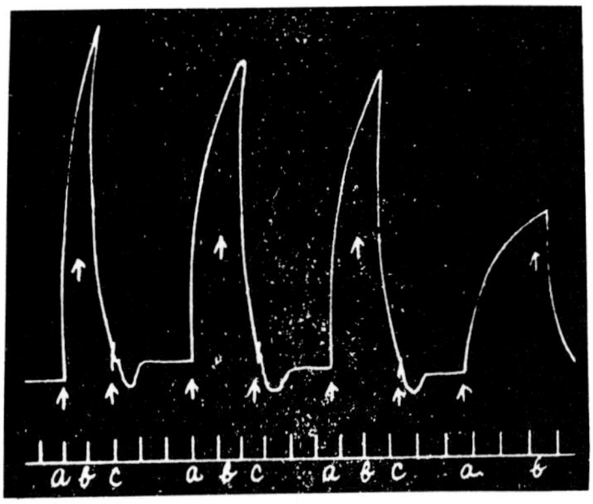

Fig. 2: Reversible contraction-relaxation cycle of glycerinated muscle fibers by ATP with the aid of 'muscle extract'.

a: ATP, 5 mM in final concentration, was added. b: ATP, 5 mM, plus 'muscle extract' were added. c: the bath solution was exchanged with the solution containing no ATP and 'muscle extract'. For others see the legend to Fig. 1.

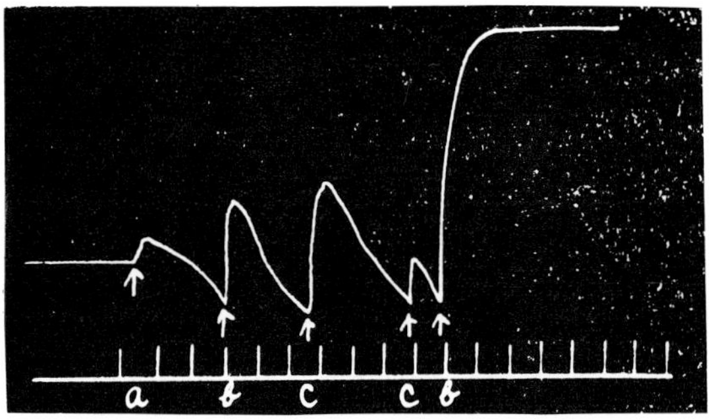

Fig. 3: Reversal of relaxation into contraction by Ca ion.

a: ATP, 5 mM in final concentration, plus 'muscle extract' were added to the solution. b: $CaCl_2$, 1 mM in final concentration, was further added. c: $CaCl_2$, 0.5 mM, was further added. For others see the legend to Fig. 1.

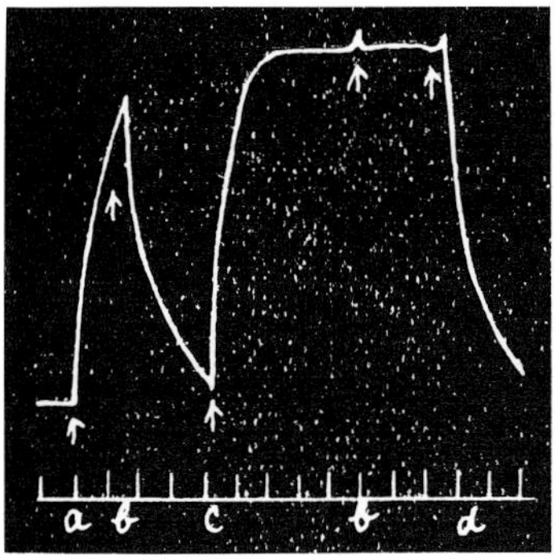

Fig. 4: Enhancement of relaxing action of 'muscle extract' by phosphate.

a, b: see the legend to Fig. 2. c: $CaCl_2$, 2 mM in final concentration, was added. d: Sodium phosphate, 30 mM in final concentration, was further added. Oxalate showed a stronger effect, but emphasis was laid on phosphate because we thought at that time that this might be somehow related to physiological mechanism. It is worthy of note that Kamada was interested in phosphate, ascribing some physiological role to it (2). For others see the legend to Fig. 1.

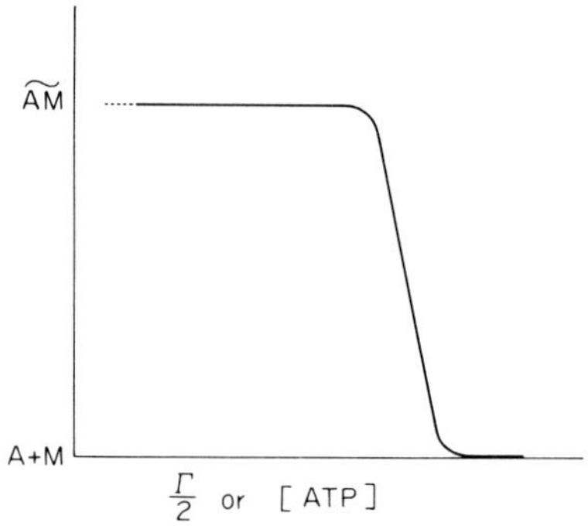

Fig. 5: Schematic illustration of the effects of ionic strengths and ATP concentrations on the response of actomyosin system to ATP in the presence of enough Mg ion.

AM: contracted state. A + M: relaxed state (myosin and actin are dissociated). At that time we did not know that MgATP, not ATP, was the substrate for the actomyosin system; the presence of Mg ion was considered to be necessary for maintaining the actomyosin system in the active state. So we used 'ATP', which was later replaced by 'MgATP'.

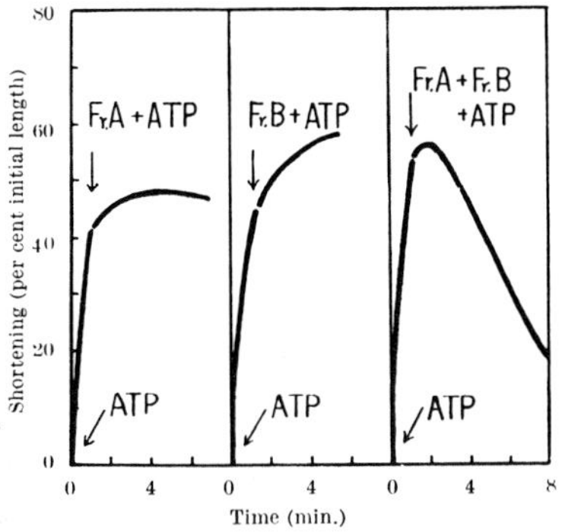

Fig. 6: Cooperation of two fractions derived from 'muscle extract' in inducing relaxation of glycerinated psoas.

Experiments were carried out in the solution containing 0.15M KCl and 10mM $MgCl_2$. Fr. A was the fraction precipitated by ammonium sulfate between 10 and 20 g per 100 ml, the main factor of 'muscle extract', being identifical with Kielley-Meyerhof enzyme. Fr. B, precipitated between 30 and 40 g per 100 ml contained myokinase, which explained the effect of Fr. B to some extent, but in the case of fairly well washed glycerinated fibers, myokinase could not replace Fr. B (see Fig. 2 in ref. 13). This puzzling problem has not yet been clarified. (quoted from ref. 13).

KIELLEY-MEYERHOF ENZYME

One day, I happened to come across the papers by Kielley and Meyerhof published in 1948 (14, 15), entitled "A new magnesium-activated adenosinephosphatase", and found that the properties of our fraction and the 'new' enzyme were very much alike. Both were precipitated by 20 grams per 100 ml of ammonium sulfate (0.35 saturation) and looked opalescent. The preparation obtained by Kielley and Meyerhof was certainly active as the relaxing factor. We prepared the factor only by high speed centrifugation, another important step of purification process of Kielley-Meyerhof enzyme, omitting the step of ammonium sulfate fractionation; the enzyme thus prepared was active as the routine preparation, or sometimes better than the latter, and could retain its activity for a much longer time. We were now convinced of the identity of the two preparations, the relaxing factor and Kielley-Meyerhof's enzyme --- both were the same (Fig. 6) (13, 16).

Indeed, the papers of Kielley and Meyerhof described nearly all the properties of the microsomal enzyme which is now called Ca-transport ATPase:

i) The enzyme is a particulate fraction (14, 15).

ii) It is activated by Mg ion, its optimum being 4mM, and is activated by Mn to a lesser extent (15).

iii) It contains a considerable amount of lipid, particularly lecithin (17).

iv) Clostridium welchii lecithinase abolishes its activity in parallel with the disappearance of lecithin (17) (Fig. 7).

The Ca-transport enzyme or Ca-Mg-ATPase was thus already discovered in 1948 (14, 17). What we did later was only to relate the enzyme to its physiological roles such as the relaxation (13, 16) or Ca uptake (18 - 20). I am puzzled why most research workers in this field have ignored this monumental work by Kielley and Meyerhof.

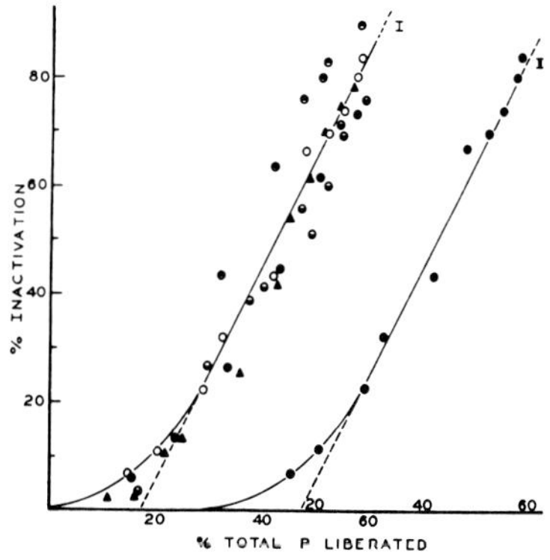

Fig. 7: Relationship between inactivation of ATPase activity of Kielley-Meyerhof enzyme and liberation of phosphate from the enzyme by Clostridium welchii lecithinase. 'I' contained the results of five series of experiments and 'II' showed those of a single experiment (the curve was shifted to the right side to indicate the uniformity of the results). This was the first experiment to show the involvement of phospholipid in enzymatic action. Later it was shown that Clostridium welchii lecithinase could abolish relaxing activities (16). (Kielley and Meyerhof; quoted from ref. 17).

SOLUBLE RELAXING FACTOR AND EDTA RELAXATION

The next step of the research was naturally to inquire into the mechanism of how the particulate enzyme would exert its physiological activity. Most researchers thought that a soluble 'true relaxing factor' of a small molecular weight must have been produced from the particulate fraction (cf. 21). The ambition of muscle biochemists was to be the first to isolate such a fascinating substance. I, however, was for no particular reason not happy with this idea.

In the meantime, Bozler (22) in 1954 and Watanabe (23) in 1955 independently reported that EDTA could induce the relaxation of contracted glycerinated fibers. This should have given a chance to Japanese scientists to reveal the secret of 'relaxation', because a pharmaceutical firm in Japan (Dojin Pharmaceutical Company) had been producing various kinds of chelating agents, originally invented by G. Schwarzenbach; GEDTA (glycoletherdiaminetetraacetic acid, EGTA) was naturally included among them. However, my lack of thorough consideration and some erroneous information caused me to miss this favorable opportunity (24) (Table I). I also tried to determine Ca uptake by the Kielley-Meyerhof enzyme in the presence of ATP, but colorimetric analysis by my poor hand did not allow me to detect any positive result (unpublished data).

Table I: Relative affinities for Ca and Mg ions and relaxing activities of various chelating agents.

	キレート作用		弛緩作用
	Ca	Mg	
Ethylenediaminetetraacetic acid (EDTA)	(100)	(100)	(100)
Cyclohexylethylenediaminetetraacetic acid (CyDTA)	82	151	10
Glycolether ethylenediaminetetraacetic acid (GEDTA)	13	23*	820
Diethylenetriaminepentaacetic acid(DTPA)	3.0	2.4	12
Ethylether ethylenediaminetetraacetic acid (EEDTA)	1.2	2.7	37
N-Hydroxyethylethylenediaminetriacetic acid (EDTA-OH)	0.7	—	25
Nitrylotriacetic acid (NTA)	0.02	0.2	5

キレート作用 : chelating action
弛緩作用 : relaxing action

* this value was derived from erroneous information and should read 0.03. Considering this point and the Mg interference with Ca binding capacities, we then found the linear relationship between Ca binding capacities and relaxing activities as shown in Fig. 8. (quoted from ref. 24).

This seemingly hopeless situation caused me to become very depressed and I wished to make a change, so early in 1958 I wrote to Prof. Fritz Lipmann in the hope that I could learn up-to-date biochemistry. His answer was that I should continue my work.

One midnight in Lipmann's laboratory, I was wondering why GEDTA was so effective as a relaxing agent (Table I). Suddenly I felt something was wrong. So I dashed to the library to consult the original book (English version) of Schwarzenbach (25) and found that a previous information about the binding constant of GEDTA for Mg was erroneous. Corrected results now revealed beautiful relationship between Ca binding capacities and relaxing activities (18).

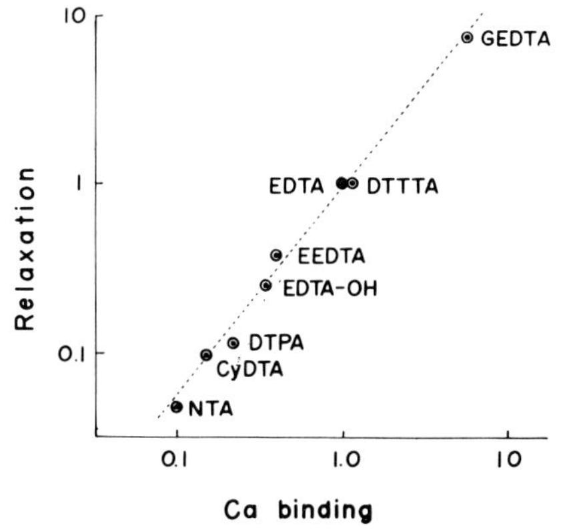

Fig. 8: Linear relationship between Ca binding capacities and relaxing activities on glycerinated fibers of various chelating agents. (quoted from ref. 18).

For abbreviations for chelating agents see Table 1. The experiments were carried out in the solution containing 0.15M KCl 0.01M $MgCl_2$ and 0.02M Tris-maleate buffer (pH 6.8). Ca binding capacities were calculated taking the interference by Mg ion into consideration. DTTTA, N',N"-Dimethyltriethylene tetramine-N,N'"-tetraacetic acid, was kindly supplied by Prof. Schwarzenbach, which was not included in the previous experiment shown in Table I (18). DTTTA has virtually no affinity for Mg, weaker than GEDTA, so that, in spite of its rather weak affinity for Ca, it shows a relatively strong Ca binding in the presence of a high concentration of Mg ion (26).

CALCIUM BINDING OF RELAXING FACTOR OR SARCOPLASMIC RETICULUM

There was then no doubt, I thought, that the relaxing factor, the Kielley-Meyerhof enzyme, should bind Ca ion utilizing the energy of ATP. Prof. Lipmann generously gave me permission to test my bold idea. I worked hard to carry through within a rather short term the project suggested by him, i.e., the ATP-ADP exchange reaction of the enzyme (Fig. 9), which reflects its most fundamental nature. In the meantime, I had the pleasure of becoming acquainted with A. Weber, who had detailed experimental evidence which made her sure that the inhibitory action of EDTA on the ATPase of myofibrils was due to its Ca binding (26). Eventually the day came to carry out the experiment, cherished in mind for months, early in June, 1959. The result was just expected;

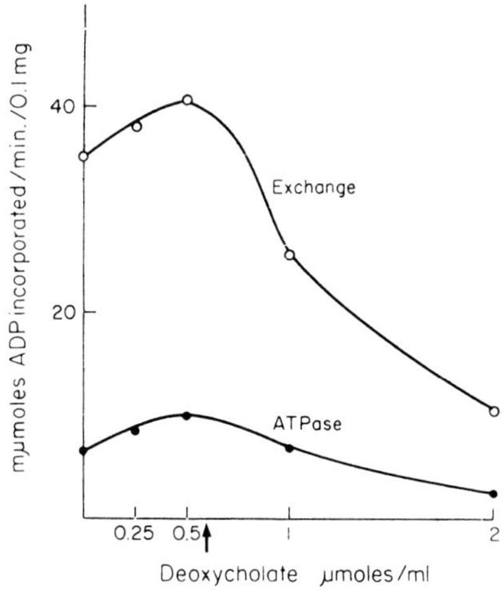

Fig. 9: Effect of deoxycholate on the ATP-ADP exchange reaction of the relaxing factor (Kielley-Meyerhof enzyme). Prof. Lipmann had predicted the association of rapid ATP-ADP exchange reaction with this membraneous ATPase, which clearly indicated the presence of phosphorylated intermediate of the enzyme. He was interested in the question whether or not deoxycholate, which could abolish the relaxing action and Ca uptake of the enzyme, would uncouple the exchange reaction from the ATPase reaction. Arrow indicates the deoxycholate concentration that abolished the Ca uptake. Since this concentration of deoxycholate partially solubilizes the membrane ATPase, the above result implies many suggestions even nowadays. (quoted from ref. 19).

this was only experience in my life in which the result completely lived up to my expectation (Fig. 10).

Everything then went very well. Every datum supported the Ca concept (27). Even such a low Ca ion concentration as 0.2 μM could exert a definite effect on the superprecipitation of well-washed natural actomyosin (myosin B) a few μM of Ca ion completely restored its contractility (Fig. 11). It was quite an easy task to construct the whole picture of the physiological contraction-relaxation cycle (27a). I thought that everything would have been solved and everyone would agree with me, but the situation was not so simple.

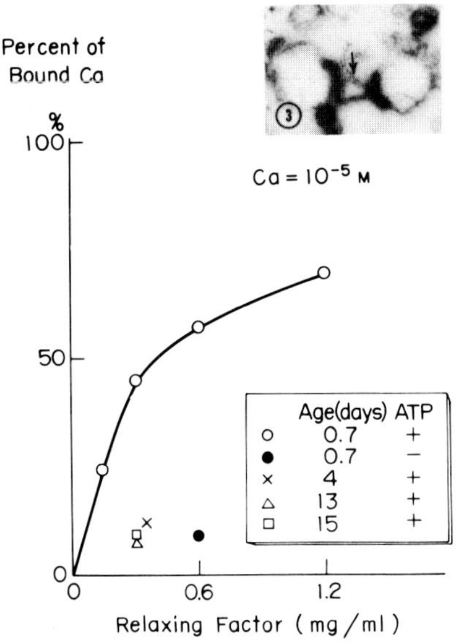

Fig. 10: Ca uptake of the relaxing factor.

The suspension of the relaxing factor in the solution containing 0.01 mM $CaCl_2$, mixed with ^{42}Ca, was spun down. The radioactivity in the pellet relative to the total was considered to indicate the amount of Ca contained in the relaxing factor, or fragmented sarcoplasmic reticulum. Since I was very anxious to know the results as soon as possible, I was bold enough to carry out this experiment without any preliminary tests. For further detail, see the original paper (19). Inlet: Triad-like structure found in the relaxing factor. The electron micrograph of the relaxing factor was composed almost exclusively of vesicular structures, in which we could occasionally see triad-like structures like the above. (quoted from ref. 19, but modified for this article).

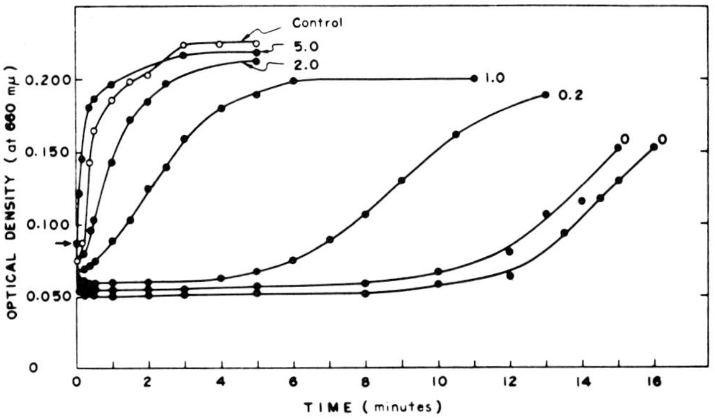

Fig. 11: Superprecipitation of EDTA-washed natural actomyosin at varied concentrations of Ca ion.

Natural actomyosin prepared by ordinary method was carefully washed with EDTA-containing solutions to remove bound Ca and then with EDTA-free solutions to remove EDTA. Superprecipitation was carried out in the solution containing 0.06 M KCl, 1 mM $MgCl_2$ 0.02 M $MgCl_2$, 0.02 M Tris-maleate buffer (pH 6.74) and a certain concentration of Ca ion indicated on the curve in μM. (quoted from ref. 27).

DISCOVERY OF TROPONIN

In 1962, a big symposium was organized by Dr. John Gergely in Dedham (21). All the data (cf. 21) further convinced me as well as A. Weber of the Ca concept, but contrary to our expectation the idea was very unpopular at the Symposium (cf. 21). One of the criticisms against the Ca concept was that Ca ion could certainly exert its effect on crude systems such as glycerinated fibers or natural actomyosin, but not on pure actomyosin; such an agent as being ineffective on the pure system could not be considered to be the true factor. One implication of this criticism might be that such a crude system would be contaminated by fragmented sarcoplasmic reticulum, thereby producing the "soluble relaxing factor".

In 1956, Perry (28) showed that EDTA could repress the ATPase of natural actomyosin, but not that of reconstituted actomyosin. After establishment of the Ca concept A. Weber worked out this problem (29) and showed that some reconstituted actomyosin preparations were somehow sensitive to GEDTA, and the actin side was responsible for whether or not the reconstituted system would be sensitive to Ca ion.

I felt that we must solve this enigmatic problem to convince those who would not believe the Ca concept and therefore started this task immediately after coming back from Dedham. My idea at first was that the routine procedure to prepare actin, i.e., acetone treatment, would be too drastic to retain the protein's subtle nature. I thought once that we had succeeded in obtaining such an actin preparation. Soon we realized, however, that this was not due to the success of preparing 'native actin', but the presence of the 'third factor' in that actin preparation (30). The factor was rather easily separated from actin and found to resemble tropomyosin, which had been discovered by Bailey long ago but of which the function had long been unknown. Since classical tropomyosin did not exhibit the physiological function of 'native tropomyosin' (30), there was no doubt that 'native tropomyosin' should contain another factor. Indeed, a new globular protein, the first Ca binding protein of biological importance, was isolated from 'native tropomyosin' and named troponin (31 - 32).

The story of the further work since that time has been described in previous review articles (33 - 36), so I will avoid overlapping. However, I would like to refer to two points. First, we were very lucky that troponin was very immunogenic; this enabled us to determine its localization in the myofibril and consequently formulate the thin filament model (Fig. 12). Second, the troponin system represents the most advanced or differentiated mechanism of Ca regulation, not only of muscle but of all cells, of which the secret resides in the function of troponin T, the most enigmatic subunit among three troponin subunits. The work along this line (cf. 37) will eventually solve the mechanism of how Ca ion can control the myosin-actin interaction.

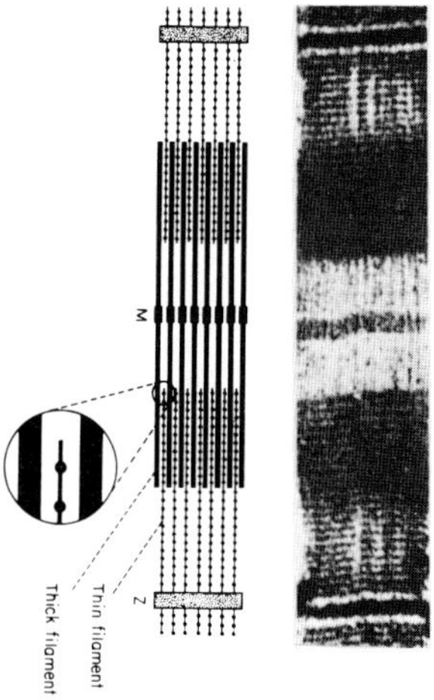

Fig. 12: A model for the fine structure of the thin filament, i.e. the troponin-tropomyosin-F-actin complex.

Lower figure shows the staining of sarcomere by antitroponin, which provided most important information in formulating the above model. (quoted from ref. 34, but modified for this article).

CALCIUM ERA

Establishment of the Ca concept in the contractile mechanism naturally tempted us to look for a Ca-dependent metabolic system. Indeed, Ozawa et al. (38, 39) have found in 1967 that phosphorylase b kinase is as sensitive to Ca ion as the contractile system. It was very impressive to realize that the well-known activation by cyclic AMP of this enzyme also required Ca ion (Fig. 12) (39).

This work inspired Kakiuchi, who then found Ca dependence of phosphodiesterase, the key enzyme for cyclic AMP metabolism (40), and eventually discovered 'modulator

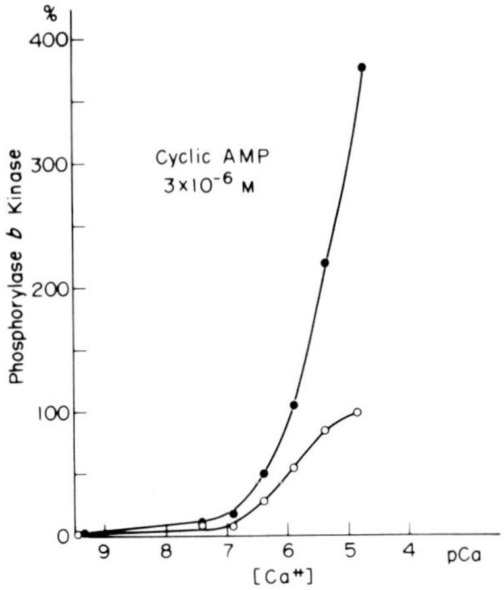

Fig. 13: Dependence of cardiac phosphorylate b kinase on Ca ion.

Open circle: without cyclic AMP. Closed circle: with cyclic AMP. Note that cyclic AMP-activated phosphorylase b kinase activated also requires Ca ion absolutely (39).

protein' in 1970 (41). Discovery of this protein, called calmodulin by Cheung (42), another discoverer of this protein, has brought the calcium era into the biochemical field. Numerous Ca binding proteins and Ca dependent processes have been appearing day by day, which I can hardly follow. I again begin to feel myself a stranger.

Acknowledgement: Reviewing my personal history, I realized how greatly indebted I am to the encouragement and advice, personal as well as scientific, of Professor Hiroshi Kumagai, Professor Fritz Lipmann and Professor Albert Szent-Gyorgyi. I have no words to express my deep sorrow over the death of Professor Yuji Tonomoura, who was the true leader of muscle science and bioenergetics in Japan.

REFERENCES

1. Heilbrunn, L.V. (1940). Physiol. Zool. 13:88
2. Kamada, T. and Kinosita, H. (1943). Japan J. Zool. 10: 469.
3. Ebashi, S. In "Calmodulin and Intracellular Ca^{++} Receptors", (eds. Kakiuchi, S. et al.), p. 421, Plenum Press, New York, 1982.
4. Szent-Györgyi, A. (1947 and 1951). Chemistry of Muscular Contraction, Academic Press, New York
5. Bozler, E. (1951). Amer. J. Physiol. 167:276.
6. Bendall, J.R. (1952). Nature 170:1058.
7. Marsh, B.B. (1951). Nature 167:1065
8. Fujita, K. (1954). Folia Pharmacol. Japan 50:183 (in Japanese).
9. Takeda, F. (1955). Folia Pharmacol. Japan 51:111 (in Japanese).
10. Ebashi, S., Takeda, F. Kumagai, H. (1955). Folia Pharmacol. Japan 51:107 (in Japanese).
11. Goodall, M.C., and Szent-Gyorgyi, A.G. (1953). Nature 172:84.
12. Bendall, J.R. (1954). Proc. Roy. Soc. (London), B, 142:409.
13. Kumagai, H., Ebashi, S., and Takeda, F. (1955), Nature 176:166.
14. Kielley, W.W., and Meyerhof, O. (1948). J. Biol. Chem. 174:387.
15. Kielley, W.W., and Meyerhof, O. (1948). J. Biol. Chem. 176:591.
16. Ebashi, S. (1957) In Conference on the Chemistry of Muscular Contraction, p. 89, Igaku-Shoin, Tokyo.
17. Kielley, W.W., and Meyerhof, O. (1950). J. Biol. Chem. 183:391.
18. Ebashi, S. (1960). J. Biochem. 48:150.
19. Ebashi, S., and Lipmann, F. (1962). J. Cell Biol. 14: 389.
20. Hasselbach, W., and Makinose, M. (1961). Biochem. Z. 333:518.
21. Gergely, J. (ed.) (1964). Biochemistry of Muscle Contraction, Little, Brown & Co., Boston.
22. Bozler, E. (1954). J. Gen. Physiol. 38:149.
23. Watanabe, S. (1955). Arch. Biochem. Biophys. 54:559.
24. Ebashi, S. (1959) In IVth Symposium on physicochemistry of biomacromolecules, 1958 (ed. R. Natori), p. 25, Tokyo, Nanko-do (in Japanese).
25. Schwarzenbach, G. (1955). Complexometric Titration, Methuen & Co, London.

26. Weber, A. (1959). J. Biol. Chem. 234:2764.
27. Ebashi, S. (1961). J. Biochem. 50:236.
27a. Ebashi, S. (1961). Progr. Theoret. Phys. Suppl. 17, 35.
28. Perry, V., and Grey, T.C. (1956). Biochem. J. 5 p.
29. Weber, A., and Winicur, S. (1961). J. Biol. Chem. 236:3198.
30. Ebashi, S. (1963). Nature 22:1010.
31. Ebashi, S., and Kodama, A. (1965). J. Biochem. 58:107.
32. Ebashi, S., Ebashi, F., and Kodama, A. (1967). J. Biochem. 62:137.
33. Ebashi, S., and Endo, M. (1968). Progr. Biophys. Molec. Biol. 18:123.
34. Ebashi, S., Endo, M., and Ohtsuki, I. (1969). Q. Rev. Biophys. 2:351.
35. Ebashi, S. (1974). Essays in biochemistry (ed. P.N. Campbell & F. Dickens), Vol. 10, p. 1, Academic Press, London.
36. Ebashi, S. (1980). Proc. R. Soc. Lond. B 207:259.
37. Ohtsuki, I. (1980). In 'Muscle Contraction, Its Regulatory Mechanism', (S. Ebashi, K. Maruyama, and M. Endo, eds.), p. 273 Japan Scientific Societies Press, Tokyo, and Springer-Verlag, Heidelberg.
38. Ozawa, E., and Ebashi, S. (1969). J. Biochem. 62:285.
39. Ozawa, E., and Hosoi, K., and Ebashi, S. (1967). J. Biochem. 61:531.
40. Kakiuchi, S., and Yamazaki, R. (1970). Proc. Japan Acad, 46:387.
41. Kakiuchi, S. Yamazaki, R., and Nakajima, H. (1970). Proc. Japan Acad. 46:587.
42. Cheung, W.Y. (1970). Biochem. Biophys. Res. Commun. 38:533.

THE SARCOPLASMIC RETICULUM CALCIUM PUMP: EARLY AND RECENT DEVELOPMENTS CRITICALLY OVERVIEWED

Wilhelm Hasselbach, Bruno Agostini, Pankaj Medda, Andrea Migala, Wolfgang Waas

Max-Planck-Institute of Medical Research
Heidelberg, FR Germany

I. INTRODUCTION

In the last two decades the sarcoplasmic membrane system has become a central element in our concept of excitating contraction coupling (1, 2). The complex membrane system serves as a sink for calcium ions during muscle relaxation and as a calcium source during excitation. As a muscle specific structure, it first only gained limited interest among physiologists and met skepticism among biochemists. However, after a decade it became apparent that these membranes are an ideal preparation for studying many aspects of membrane biology (3 - 7). In the following our attention will be focused on a few developments which led to the establishment of a concept concerning the energetics and the mechanism of the membranes' calcium transporting system, the sarcoplasmic reticulum calcium pump.

II. CALCIUM TRANSPORT ENERGETICS

A. The Application of Calcium Precipitating Agents

Our way to the calcium pump in the sarcoplasmic membranes started with attempts to stabilize the labile relaxing factor of Marsh present in aqueous muscle extracts (8). Relaxing factor activity meant that aqueous muscle extracts or suspensions of muscle microsomes can suppress the ATP induced contraction of muscle fibres or the ATPase of myofibrils and actomyosin particles (8 - 12). Marsh, who was a student of K. Bailey at Cambridge, had further shown that the relaxing activity of the factor can

ISBN 0-12-260380-X

be abolished by the addition of small quantities of calcium ions. Most astonishingly these calcium quantities were much smaller than the calcium contaminations present in the preparations, in aqueous muscle extracts as well as in suspensions of isolated microsomes. Therefore, we decided to diminish the contaminations of ionized calcium by adding calcium binding agents. Fortunately, at that time, agents like EGTA (ethylene glycol bis(2-aminoethylether) N,N'-tetraacetic acid) and EDTA (ethylene diamino-N,N'-tetraacetic acid) were not yet available to us. Therefore we used the classical calcium binding anions oxalate, phosphate and citrate. We found that the inhibitory activity of the preparations enormously increased when oxalate was added to the assay media, while citrate and phosphate

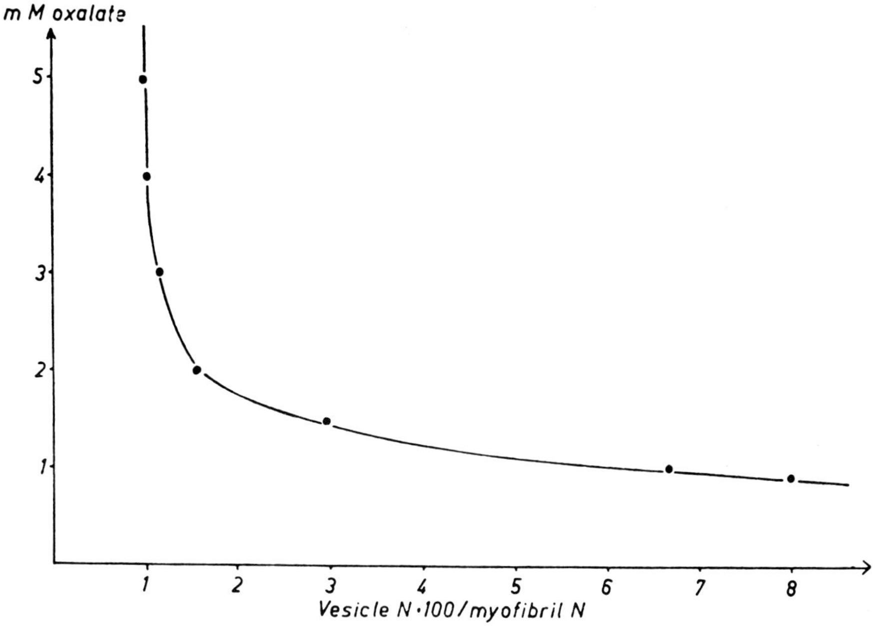

FIGURE 1. The effect of oxalate on the inhibitory activity of rabbit muscle vesicles. The curve represents the vesicle N/myofibril N ratio necessary to produce 50 per cent inhibition of myofibrillar ATPase activity. (13)

proved to be relatively little effective (Fig. 1). Self-evidently, we excluded that this enhancement of the factor's activity was brought about by a direct inhibiting effect of the agents on the actomyosin ATPase. In the following, in many laboratories interest became focused on the identification and characterization of the relaxing factor. The notion that its action was related to the presence of ATP regenerating systems was disproven by Portzehl 1957 who showed that the active principle was confined to muscle microsomes (14, 15). Hence, we assumed that in the presence of ATP the microsomes might produce a soluble relaxing factor and we therefore searched for an exhaustible precursor as a constituent of the microsomes (16).

In the course of these experiments we found that in the presence of oxalate, the addition of small quantities of calcium ions produced a transient abolition of the factor's inhibiting effect. The calcium induced activity transient appeared to model a contraction relaxation cycle. In fact, calcium addition to relaxed glycerol extracted muscle fibres could induce such a cycle. These findings stimulated us to study how the added calcium interacts with the different components present in the complex reaction system containing the contractile proteins, microsomes, ATP, magnesium, calcium and oxalate. In pursuing these studies which were performed with radioactive calcium we made the following observations (17 - 20): 1. The calcium concentration in the ATP containing media was reduced by the microsomes to less than 10 nM. At this low calcium level the calcium binding components of the contractile protein are completely deprived of bound calcium as it had simultaneously been shown by A. Weber (21). 2. Calcium was stored inside the vesicles as calcium oxalate which keeps the concentration of ionized calcium in the internal space low. The maximum uptake of 10 μ mol calcium/mg protein is hundred times greater than the maximal amount of calcium which is taken up in the absence of oxalate. Massive calcium deposits could be seen in the electronmicroscope as electron dense material inside vesicular structures (13). In glycerol extracted muscle fibres calcium oxalate was found to be deposited mainly in the terminal cisternae of the sarcoplasmic reticulum (22, 23). Thus the calcium storing microsomes were identified as structural elements

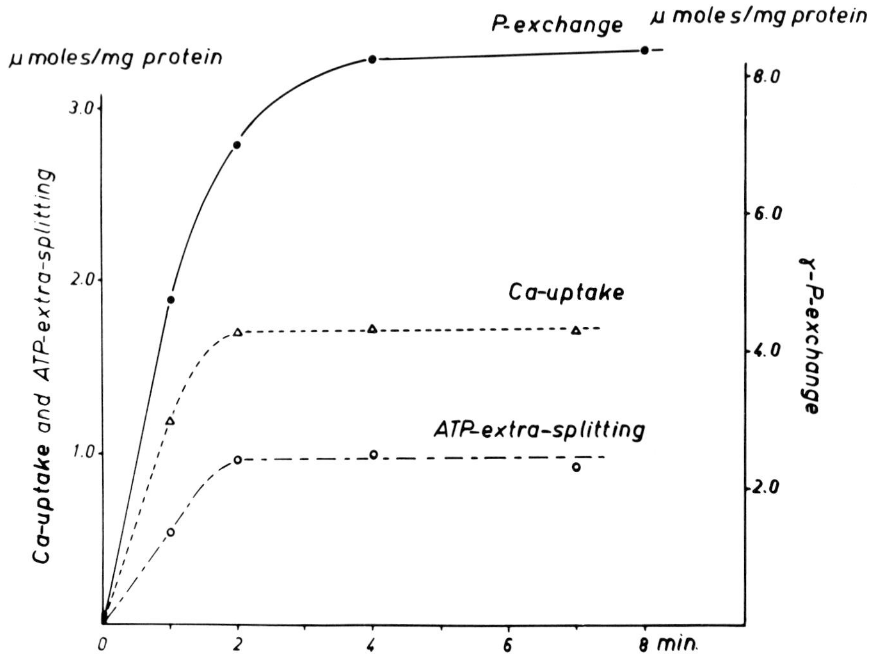

FIGURE 2. Time course of calcium uptake ATP-extra-splitting and phosphate exchange between ATP and ADP in the presence of 5 mM oxalate (19).

of the sarcoplasmic reticulum. 3. In the presence of oxalate the process of loading lasted for several minutes and is causally linked to an enhanced ATP consumption - the extra ATPase activity (Fig. 2). (17, 19)

The two events could be resolved in time and a stoichiometry of two was established as coupling ratio. The initial difficulties to simultaneously measure ATP splitting and calcium uptake were overcome by using neutral mercury nitrate for precipitating the calcium oxalate filled vesicles so that they could be removed from the solution by conventional filtration. This convenient but hazardous method was later replaced by the expensive Millipore filtration procedure which Martonosi (24) inaugurated 1964 after our results had become known in the USA 1962. The coupling ratio of two that

had been found was consistent with the dependence of calcium transport and ATP splitting on the calcium concentration exhibiting a Hill coefficient of two. 4. In the absence of oxalate calcium transport leads to an immediate rise of the internal calcium concentration which largely turns off calcium transport and ATP splitting. This interpretation was deduced from the activation of the vesicular ATPase by calcium in the presence of oxalate. The activating effect of calcium in oxalate containing assays was in contrast to Kielley and Meyerhof's original finding that the ATPase present in aqueous muscle extracts was inhibited by calcium ions (25). That this inhibition was caused by the increasing calcium concentration inside the vesicles was derived from the fact that oxalate could relieve inhibition even if the external calcium concentration was kept constant (26).
5. In the presence of oxalate, a second enzymatic activity residing in the microsomes, a nucleoside diphosphokinase, which had previously been found by M. Ulbrecht (27), was activated by small quantities of calcium like the ATPase and proceeded as long as calcium uptake took place. This transient calcium-dependent phosphate exchange between ATP and ADP was taken as evidence showing that a phosphorylated intermediate was formed during calcium transport. Its direct demonstration succeeded in 1966 (28) and was subsequently verified and extended by Tonomura and his co-workers (29 - 31). They especially concentrated on phosphorylation kinetics on the basis of which first reaction schemes were developed.

B. The Efficiency of the Calcium Transport System

The reported findings let us postulate the existence of an ATP driven active calcium transport in the muscle's microsomal membranes. This thesis was supported by the following criteria:
1. The observed calcium accumulation only occurs if ATP or other energy yielding substrates are split by the membranes (17, 32 - 34).
2. Calcium transport results in an enormous decrease of the product of the ion concentrations of calcium and oxalate or phosphate in the external solution as compared to the internal solution. In the latter the values of the ion concentration products are

TABLE 1. Ca^{++} concentration, ATP potential and pump efficiency η

	Frog satorius muscle	Rat cardiac muscle	Isolated SR vesicles
Reticular Ca_r^{++} mM	5-10*(38)	5-10*(39)	5-10*(35) 1** 0.25** 0.02***
External Ca_o^{++} μM	0,05(41)	0,1(41)	0.02(35) 0.25 0,14 0.005
Ca_r^2/Ca_o^2	$10^{10}-4\cdot 10^{10}$	$2.5\cdot 10^{9}-10^{10}$	$6\cdot 10^{10}-2\cdot 10^{11}$ $1.6\cdot 10^{7}$ $3.2\cdot 10^{6}$ $1.6\cdot 10^{7}$
$\frac{K\cdot ATP}{ADP\cdot P}$	10^{10}	$10^{9}-10^{10}$	10^{10} $2.5\cdot 10^{7}$ $6.2\cdot 10^{6}$ $2\cdot 10^{9}$
η	1-1.06	0,94-1.1	- 0,98 0,80 0.77

The reticular concentrations are calculated from the total amount of calcium and the internal Ca binding parameters (*) or from the solubility product of Ca phosphate (**) or Ca oxalate (***) (40, 42). The ATP potentials are calculated from the standard free energy of ATP and the prevailing concentrations of ATP, ADP and P: K • ATP/ADP • P (43).

determined by the solubility products of the respective calcium salts. Hence, in the presence of calcium precipitating anions it is possible to determine ion concentration ratios what is made difficult otherwise due to the uncertainty concerning the ion concentrations in the intravesicular space. The ion concentration ratios which are established across the membranes depend on the available free energy of ATP hydrolysis. Values up to 100 000 were reported (35). Evidently the calcium pump is a transport system with an extremely high concentrating power. The measured concentration ratios yield the concentration terms for the energy requirement of the pump. Since no membrane potential across the sarcoplasmic reticulum membranes has to be taken into account [cf. ref. 2 and Sec. IIC] concentration term alone determines the total energy requirement. In addition to the concentration ratio the overall economy of the system crucially depends on the molar ratio of calcium taken up to ATP hydrolysed. Most recent studies conducted in the presence as well as in the absence of calcium precipitating agents support the value of two as coupling ratio (6). This ratio remains unchanged even when the calcium transport rate considerably declines (32). Furthermore, a coupling ratio of two is observed not only during calcium uptake but also when the cycle of the pump is reserved, i.e. when ATP is synthesized from ADP and inorganic phosphate during calcium efflux from calcium loaded sarcoplasmic reticulum vesicles. The latter discovery resulted from experiments conducted by Barlogie with the intention to find conditions under which a rapid calcium release could be induced (36, 37). The efficiency of the pump is therefore given by the following expression

$$\eta = 2 \cdot \ln Ca_i/Ca_o / \ln Keq \cdot ATP/ADP \cdot P,$$

which relates the osmotic work per mole of ATP hydrolyzed to the available free energy. As shown in the following table 1, the efficiency of the pump in the isolated vesicles approaches values near one. Efficiencies near one are also obtained for the pump in vivo if published values for Ca_i, Ca_o and the ATP potential are used (38 - 42). The high efficiency of the pump indicates that the coupling between calcium movement and the reaction steps which lead to phosphate liberation must be very tight, i. e. the pump only consumes or synthesizes ATP in conjunction with net calcium movement and that there is no calcium movement without ATP participation (44). This is possible only when passive calcium leakage can practically be neglected. If leaks would be present, the system would permanently dissipate energy at the cessation of net calcium uptake because ATP must be split in order to compensate the leak. This, however, is not the case. Energy dissipation which is obtained from the ratio of the forward and the reverse reaction rates remains minimal (c.f. 45).

C. The Membranes' Ion Channels

As mentioned above, our considerations concerning the energetics of the calcium transport system are based on the assumption that no energy is required to overcome an electrical potential during calcium transport. The data compiled in table 2 show that the sarcoplasmic reticulum vesicles in fact contain channels for small cations and anions. Their conductivity is quite high and electrically shunts the membrane. Hence, the build up of an electrical potential in the usual KCl containing assay media is effectively prevented (2). These studies further confirm that the membranes are not only little permeable for calcium but also for magnesium ions. The uptake of oxalate and phosphate in conjunction with calcium transport - mentioned before - showed that the membranes possess in addition to the channels for small ions pathways for the bulkier calcium precipitating anions oxalate and phosphate.

Evidence has been put forward that both ions can compete for the same pathway (50). It is the very existence of this anion pathway which enabled us to elucidate stoichiometry and energetics of the calcium transport system. On the other hand, however, the uncritical use of these agents has caused many misunderstandings because the driving force for calcium oxalate or phosphate precipitation has often been underestimated.

TABLE 2. Ion permeability of SR membranes

Ions	Measurements of	Permeability 10^{-7}cm s^{-1}	Membrane resistance spec. Ω cm^2
K^+	Turbidity(46	0.6	28000
K^+	Fluorescence(47)	10 - 500	2600 - 500
K^+	Conductivity(48)	-	17
Na^+	"	-	68
Cl^-	"	-	3
Ca^{++}	Efflux(5,49)	0.02 - 0.1	10^7
Mg^{++}	Turbidity(46)	0.02	$3\cdot10^7$
Oxalate	Turbidity(46)	0.3	
Oxalate	Influx	6	
Phosphate	Turbidity(46)	1	

III. CALCIUM BINDING TO THE TRANSPORT PROTEIN AND ITS SUBUNIT STRUCTURE

The high concentration ratio which the pump can establish between external and internal space of the sarcoplasmic reticulum vesicles requires that during the pump cycle the energy furnished by ATP is used to cyclically produce true or apparent affinity changes for calcium. As illustrated by the most simple transport scheme the calcium concentration ratio is determined by the ratio of the products of the rate constants describing the four reaction steps. (Fig. 3).

In order to make $Ca_i/Ca_o > 1$, energy has to be fed into the cycle which can be accomplished by changing any of the eight constants or any combination of them (51). Therefore, the energy yielding substrate ATP could transfer its energy in any of the four reaction sequences. At first we assumed that the membranes were equipped with low affinity calcium binding sites to which ATP transiently conveys high affinity. After calcium translocation the calcium binding sites were supposed to return

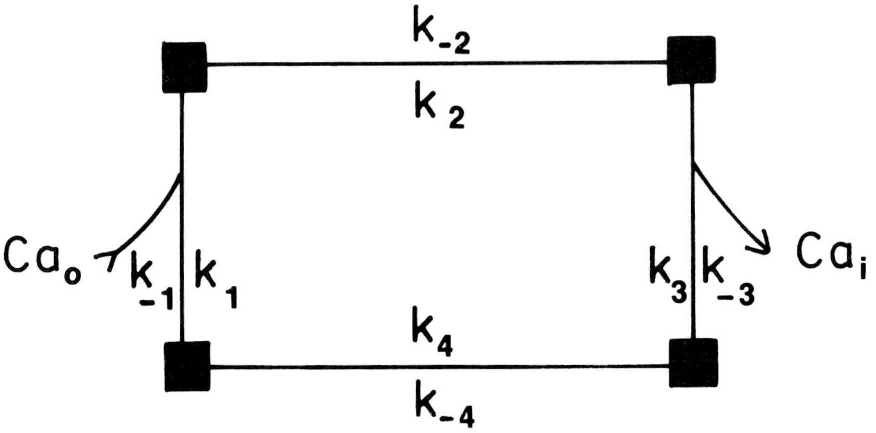

$$\frac{Ca_i}{Ca_o} = \frac{k_1 \cdot k_2 \cdot k_3 \cdot k_4}{k_{-1} \cdot k_{-2} \cdot k_{-3} \cdot k_{-4}}$$

FIGURE 3. Basic calcium transport cycle.

into their low affinity state. The affinity change was proposed to be accomplished by a phosphorylation dephosphorylation cycle (19). This concept had to be abandoned when we found that the membranes possess a reasonable number of high affinity calcium binding sites which saturate at calcium concentrations in the external medium as they are required for pump activity (52 - 55). Hence, phosphorylation has to be assumed to transiently interchange high into low calcium affinity sites. In addition to these interchangeable sites, the transport protein possesses permanent low affinity sites (42, 56). They are not only located at the internal but also at the external section of the transport molecule (56). Their functional role is uncertain. The internal sites might increase the calcium storing capacity of the membranes under physiological condition or they might function as activity modulating sites which, when occupied, suppress activity.

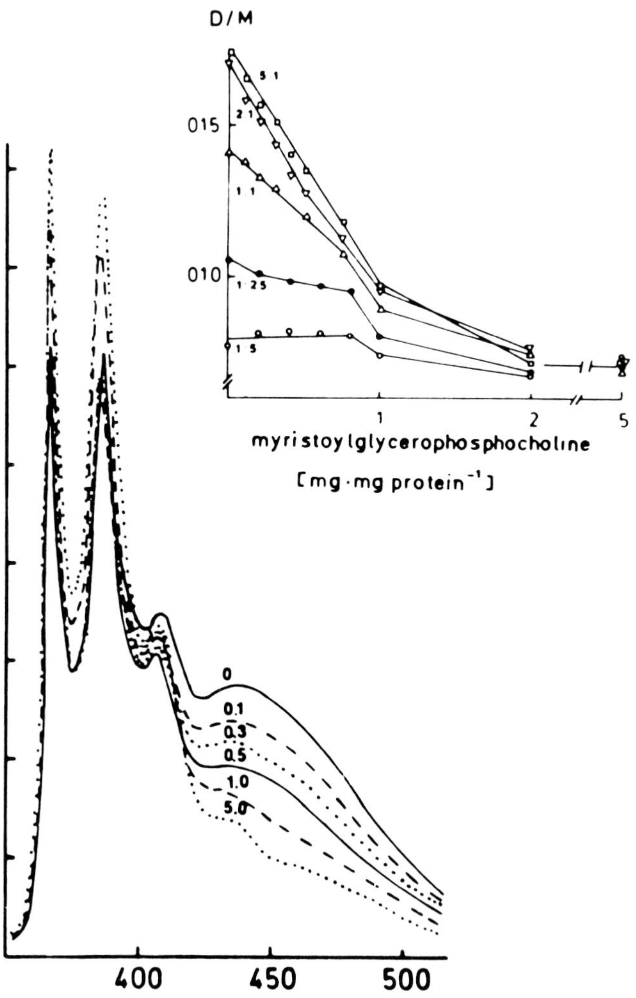

FIGURE 4. Eximer formation in SR membranes by pyrenemaleinimide labeling and its abolition by myristoylglycerophosphoryl choline (60). Increasing amounts of myristoylglycerophosphoryl choline 0 - 5 mg/mg protein were added. The inset shows the relative peak intensities of the pyrenemaleinimide-eximer (450 nm) to the peak intensities of pyrenemaleinimide monomer (sum of absorbancies 376 nm + 397 nm + 416 nm) (D/M) as a function of the myristoylglycerophosphoryl choline concentration at various degrees of labeling with pyrenemaleinimide.

The described concept which emerged from the combination of calcium binding with calcium transport study has further been refined. Recent binding studies (57) furnished results which, however, are at variance with those of the previous studies (52 - 55). The discrepancies concern the number of high affinity calcium binding sites in the transport molecule 1 versus 2 and the mode of its binding independent or cooperative. If there should be only one high affinity calcium binding site in the molecule, the coupling ratio as well as the steep activation profile requires cooperation of at least two neighboring molecules (56). Yet this is in conflict with the observation that the activity of monomeric ATPase preparations retain the steep calcium concentration dependence of native preparations if detergent induced degeneration is avoided (59. 60). In native membranes spacial contacts appear to exist between neighboring ATPase molecules enabling cooperation. When surfacial thiol groups of intact membranes are labeled with pyrenemaleinimide, eximers are formed at a considerable yield (Fig. 4) (61).

Since double-labeling of single molecules and eximer formation by collision could be ruled out, we have to assume that at least molecule pairs exist in the membrane with an intermolecular distance of 0.3 nm. It might be speculated that these dimers were constituents of the molecule clusters visualized in electronmicroscopic images of freeze fractured membranes and of the isolated lipid free transport protein (62, 63). Yet it is difficult to exclude that clustering is a preparation artifact. Therefore the functional importance of the clusters has only hesitatingly been discussed. The suspicion that the observed 9 nm particles might not exist in the native membrane is nourished by the fact that small angle diffractions studies could be interpreted as arising from single molecules embedded in the lipid bilayer (64).

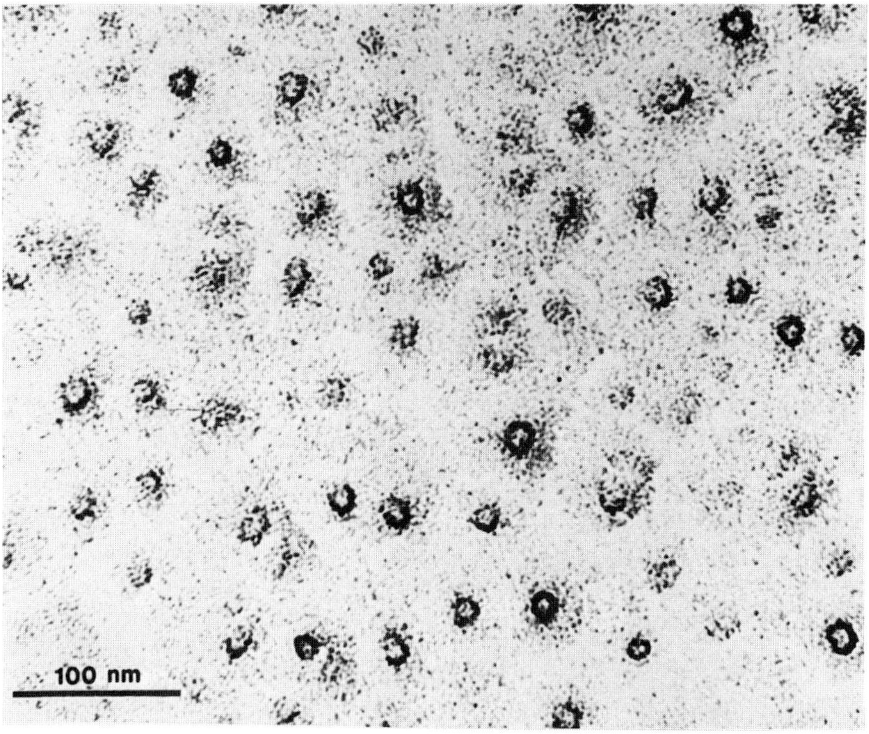

FIGURE 5. Electronmicrograph of delipidated calcium transport ATPase preparation. - The preparation was freeze fractured and rotary shadowed at 25° with platinum carbon. Bar = 100 nm. (Agostini, unpublished results).

IV. PHOSPHATE TRANSFER AND CALCIUM TRANSPORT COUPLING

A. Coupling Near Equilibrium

In recent years many attempts were made to resolve the reaction sequence leading to phosphate liberation and calcium translocation into a sequence of single reaction steps and to assign specific steps of the phosphate transfer sequence to calcium translocation (4 - 7, 65, 66). Most experiments were performed under conditions not requiring energy provision for calcium translocation such as they prevail in initial rate experiments. Under these conditions localisation of the coupling step is inherently uncertain because in an unidirectionally proceeding reaction any step is - at least in theory - endowed with an infinite energy increment. Hence, if in the unidirectionally running system calcium translocation is found to be correlated with any of the phosphate transfer reactions, it cannot be concluded with certainty that this correlation persists if osmotic work is imposed and the total free energy change becomes small.

Therefore we approached the problem by simultaneously measuring calcium movement and phosphate turnover after partial reaction steps or the complete reaction chain have approached equilibrium or steady state near equilibrium (67, 68). In both cases net flux of calcium vanishes. The driving force of the overall reaction as well as of the partial sequences approaches zero when the reaction comes near to equilibrium. This again makes any assignment difficult. Yet, if one can monitor simultaneously calcium movement and phosphate transfer after having dissected the chain in different segments, it should be possible, to localize the translocation segment. At first we consider the complete reaction chain running unidirectionally forward. To this purpose we energize the transport system with ATP or with GTP in the presence of phosphoenolpyruvate and pyruvatekinase. It is the lack of the reactants of the reverse reactions, ADP or GDP and phosphate, which prevents the reversal of any partial reaction step. Net calcium uptake ceases at a persisting calcium turnover when a constant maximal load of approximately 100 nmol/mg is reached.

The rate of calcium turnover depends on the calcium concentration in the medium and on the kind of NTP (Table 3) (68, 69). The same is true for the likewise persisting calcium-dependent NTP hydrolysis. The fact that the vesicles maintain the same load at very different rates of calcium turnover, cannot be explained by passive calcium outflow through membrane leaks. Therefore we suggested a coupling between calcium influx and calcium efflux. Calcium might be carried outward by the pump itself in a step occurring after phosphate

Table 3. NTP driven calcium fluxes at zero NDP and P_i, 22°C (68, 74)

	Calcium influx	Calcium-calcium exchange	Calcium-dependent ATPase	Phosphoprotein
	nmol/mg min			nmol/mg
ATP	4000	180	200	3,4
GTP	2000	70	120	4,5

Table 4. Activation by NDP of NTP driven calcium fluxes at zero P_i (68,74)

	Calcium-calcium exchange NDP		Calcium-dependent ATPase NDP		NTP-NDP exchange	Phosphoprotein
	0	1mM	0	1mM		
	nmol/mg min					nmol/mg
ATP (1mM)	180	310	200	50	2500	1,0
GTP (1mM)	70	190	120	100	220	4,0

Table 5. Activation by NDP and P_i of NTP driven calcium fluxes (68, 74)

	Calcium-calcium exchange	NTP-NDP exchange	$NTP-P_i$ exchange	$NDP+P_i$ calcium efflux	Calcium-dependent ATPase
			nmol/mg	min	
ATP	600	2300	50	200	100
GTP	375	200	60	200	140

release. When the pump is inactive, this pathway has a very low permeability for calcium ions. The results of these experiments show that when the phosphoryl transfer reactions are poised to run unidirectionally forward, calcium is moved outward in an irreversible step which otherwise might not occur. As to be expected the step which carries calcium inward cannot be identified.

Next, we supplemented the system with nucleoside tri-and diphosphates (Table 4). Thereby, the initial phosphoryl transfer reaction becomes reversible giving rise to a transfer of the terminal phosphate residue of nucleoside triphosphates to nucleoside diphosphates. Since the concentration of inorganic phosphate is kept low, reverse reaction steps involving inorganic phosphate incorporation are excluded. The exchange reaction is optimally activated at 0.1 mM free magnesium and at a calcium concentration of approximately 1 μM. The same concentrations are required for optimal transport activity. It has been shown that it is the forward reaction, the formation of the phosphoprotein, which requires the presence of both magnesium and calcium ions in the medium (54, 66, 70). In contrast the reverse reaction the transfer of phosphate to ADP can even occur if both ions were removed from the medium by chelation (31). Microreversibility demands that calcium and magnesium ions must be occluded as constituents in the first phosphoprotein intermediate. Hence, the exchange reaction is indicative for an occluded state of both ions in the phosphoprotein complex. Calcium occlusion has now directly been demonstrated as the inaccessibility of bound calcium to external EGTA (71, 73). Occluded calcium is liberated from the protein when the enzyme is dephosphorylated by ADP, i.e., in the reverse step of the reaction chain. The data in the Table 4 show that, together with the onset of the phosphate exchange reaction, calcium turnover is considerably increased.

This finding indicates that a new pathway for calcium is opened when the first reaction sequence is reversed by NDP supplementation. Concurrently NTP hydrolysis is strongly reduced. Hence, the translocation step can be localized in relation to phosphate release. It has to preceed phosphate liberation. One might suppose that calcium movement is directly coupled to the phosphotransfer reaction. Such a coupling is suggested by the observation showing that phosphoprotein formation and calcium translocation appear to occur simultaneously in initial rate measurements (6). Yet the nonexisting stoichiometry between calcium turnover and the exchange of phosphate between NTP and NDP rather contradict such a coupling. Furthermore, if phosphoprotein formation would directly lead to calcium translocation an affinity reduction of the phosphoprotein for calcium by 3 or 4 orders of magnitude must occur. As a consequence the transfer of the phosphate residue to ADP could only proceed as long as the calcium concentration is high at the intravesicular bindig sites. This, however, is evidently not the case. The rate of ATP - ADP exchange remains fast, when the internal calcium concentration is kept low either by the presence of oxalate or when the vesicles are made leaky (19, 14). We have therefore to assume that calcium translocation must occur in a subsequent reaction step. Hence, following the first phosphorylated intermediate, a second phosphorylated intermediate must arise in the reaction sequence from which calcium is released in the forward running mode of the pump and which is the acceptor of calcium for outward translocation. The transition between the two intermediates must be considered to be the translocation step itself.

This assignment gets further support from the observed flux relation when the reaction sequence is made fully reversible by completing the media with inorganic phosphate (Table 5). The addition of phosphate induces a relatively slow net calcium uptake and considerably stimulates Ca-Ca exchange. The rate of the latter largely surpasses the calcium uptake. In agreement with the proposed assignment Ca-Ca exchange exceeds the rates with which inorganic phosphate is transferred to NDP and the rate of unidirectional calcium release when the pump runs backward synthesizing ATP. One might assume that inorganic phosphate enhances the formation of a phosphopro-

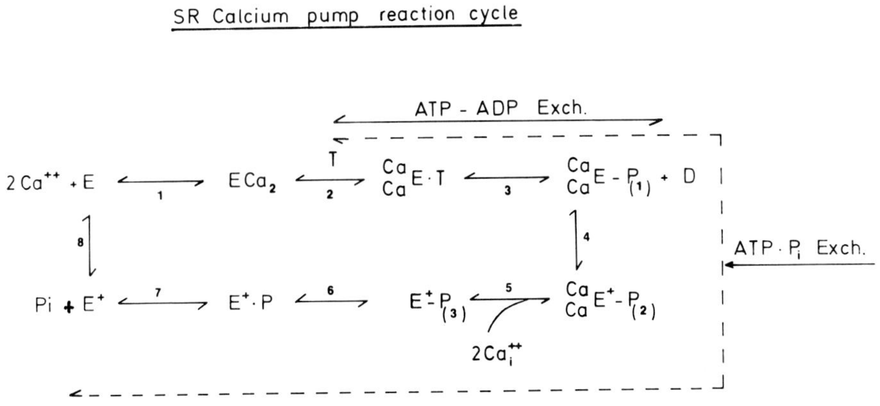

FIGURE 6. Calcium transport reaction scheme

tein which accept internal calcium for translocation. We were, however, unable to detect an increased level of a stable phosphoprotein after addition of inorganic phosphate. The following scheme first suggested by Makinose (67) and refined by de Meis (7) is consistent with the derived reaction sequence.

B. Coupling Far from Equilibrium

1. Coupling in the Forward Mode. The transition between the two phosphorylated intermediates has extensively been analyzed by many groups (73-75). In nonaccumulating membranes the transition leads to a net release of calcium from the protein (55,78-80). However, its demonstration in native membranes needs special provisions such as high magnesium concentrations,low pH values and reduced temperature.(Fig.7) Under these conditions the phosphoprotein fraction predominantly consists of the intermediate whose phosphoryl group cannot be backtransferred to ADP (Fig.7). Hence,the isomerisation step as the transition from the first to the second phosphorylated intermediate is fast compared to preceding and following steps.When the second intermediate is hydrolytically cleaved,calcium is rebound,but a definite time lack is observed (Fig.7). This indicates that a slow step must be located between dephosphorylation and rebinding of calcium.

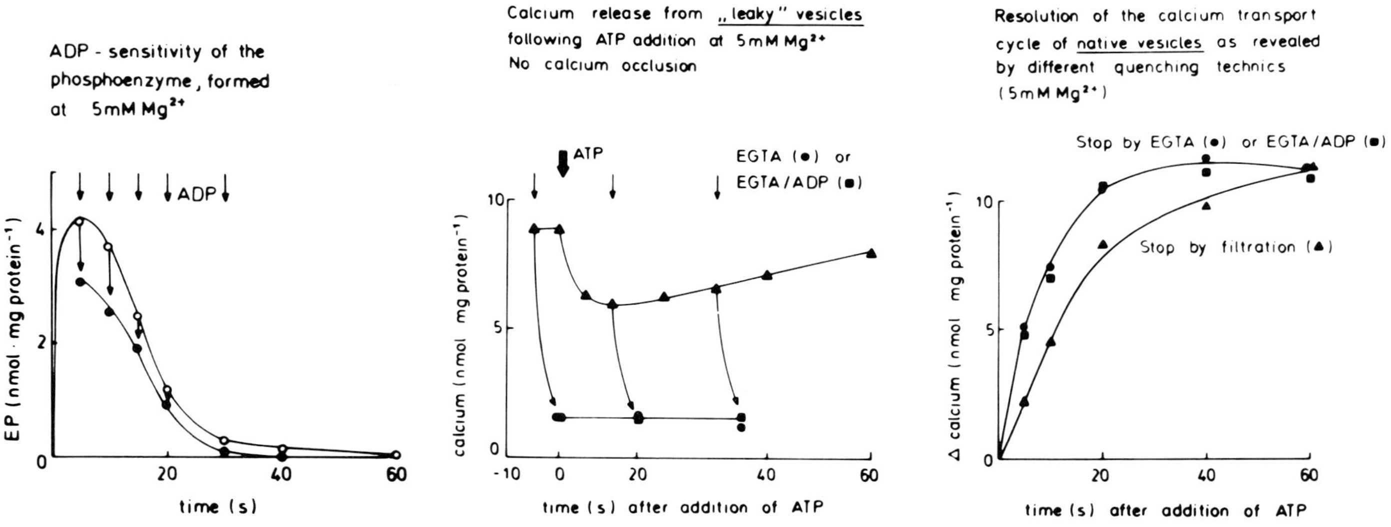

FIGURE 7. Phosphorylation of sarcoplasmic reticulum vesicles with stoichiometrical amounts of ATP at pH 6.1 and 0°C in the presence of 5 mM magnesium and the accompanying calcium movement, Ca^{++}= 0.1 mM. (Waas and Hasselbach, unpublished results).

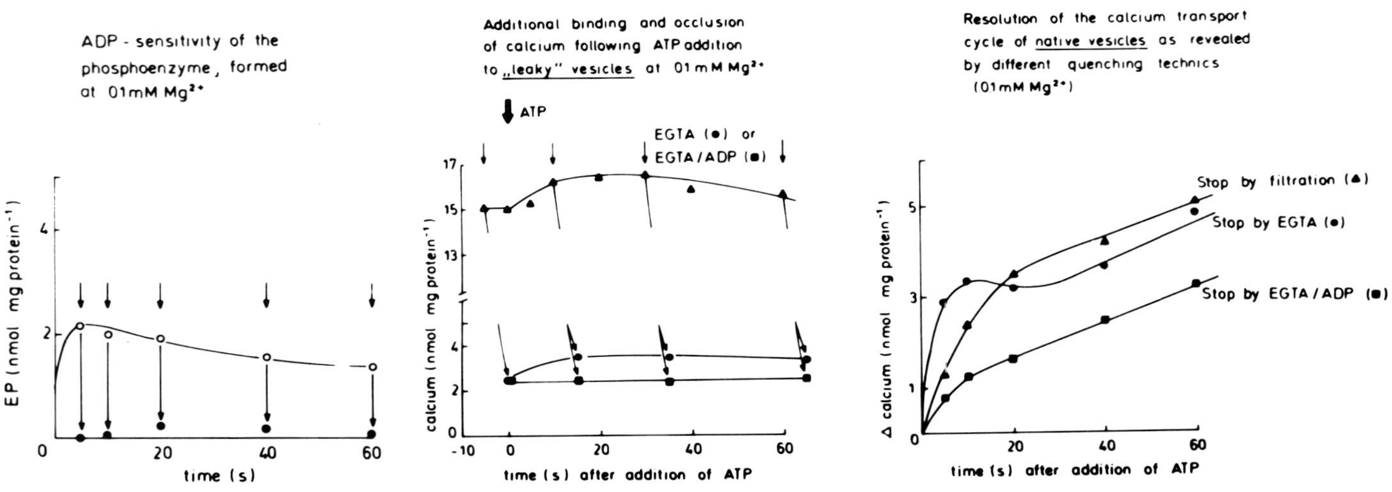

FIGURE 8. Phosphorylation of sarcoplasmic reticulum vesicles with stoichiometrical amounts of ATP at pH 6.1 and 0°C in the presence of 0.1 mM magnesium and the accompanying calcium movement, Ca^{++}= 0.1 mM. (Waas and Hasselbach, unpublished results).

When the same experiment is performed with closed vesicles,calcium is released into the vesicular space. The moment at which calcium becomes discernable inside the vesicles during the cycle depends on the method applied for interrupting the reaction. The calcium uptake increment becomes manifest early in the cycle when the reaction is terminated by removing bound calcium from its external binding sites. Yet,when the reaction is stopped by filtration,calcium uptake appears to be retarded. This retardation indicates a lag between the release of calcium and its rebinding to external sites. When in the same kind of experiments the concentration of magnesium is lowered to 0.1mM, the cycle appears to be locked after the first intermediate is formed. (Fig.8).

The phosphoprotein decays much more slowly and remains ADP-sensitive. In nonaccumulating membranes phosphorylation does not lead to a release of calcium but causes additional calcium binding (80). Most of this calcium resides in the occluded state. When calcium transport performed by closed vesicles is terminated by filtration, the initial rates are significantly higher than when the reaction is terminated by EGTA + ADP. This finding shows in agreement with the results obtained with nonaccumulating membranes that in the initial step of the reaction sequence additional calcium is bound to the membranes.

In summary, the assignment of phosphoryl transfer reaction and calcium translocation reduced from experiments near equilibrium are fully supported by experiments in which calcium translocation transport and phosphoprotein formation are monitored during the reaction cycle unidirectionally driven by ATP. Thus the coupling between calcium translocation and phosphoryl transfer remains invariant when the reaction conditions change from unidirectionality to equilibrium.

2. Coupling in the Reverse Mode. In the forward reaction the pump uses the energy supplied by ATP to drive the transition from the high affinity to the low affinity calcium transport intermediate. In the reverse mode of the pump, thermodynamics re-

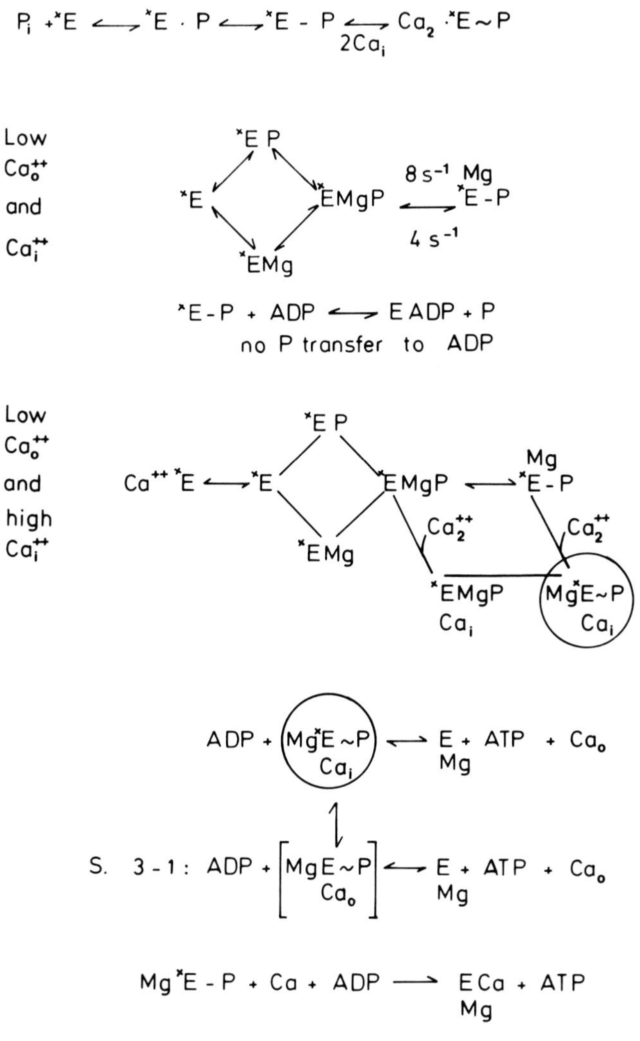

FIGURE 9. Phosphate transfer with and without internal calcium (84 - 87). The intermediate in the circle is evidently identical with the intermediate in brackets as it arises in reaction sequence 1 - 3 of Fig. 6.

quires that ATP formation must be driven by the calcium concentration gradient. Yet, this energy conversion can be subdivided into at least two consecutive reaction steps as first shown by Knowles and Racker (81) and convincingly demonstrated by de Meis and Tume (82). Clear-cut results are only obtained when purified ATPase preparations are used, the low affinity calcium binding sites of which are easily accessible (c. f. 83). In the first step calcium is removed from its externally located high affinity sites whereby inorganic phosphate becomes incorporated into the protein (Fig. 9).
The phosphoprotein which is formed, contains magnesium but no calcium and its phosphate residue cannot be transferred to ADP (84 - 86). Yet, on a sudden increase of the calcium concentration, the complex is converted to a complex which can readily transfer its phosphoryl to ADP. This calcium induced transformation represents the second reaction step. The fact that closed vesicles like ATPase preparation can incorporate inorganic phosphate while only the phosphoprotein in ATPase preparation can be made ADP accessible by calcium addition indicates that phosphorylation by inorganic phosphate converts external high into internal low affinity sites. The subsequent occupation of these sites by calcium must be followed by the translocation of calcium connected with the appearance of external high affinity site and a transferable phosphoryl group.

This concept gets support from experiments with the phosphate analog vanadate. Sarcoplasmic reticulum vesicles or ATPase preparations incorporate vanadate like phosphate when magnesium ions are present and the external high affinity sites of the vesicles are depleted of calcium. (88, 89). When calcium binding is measured, in the presence and in the absence of vanadate, it becomes evident that vanadate shifts the binding profile to the right. ATPase preparations loose their high affinity calcium binding sites nearly completely (Fig. 10).
The extent of vanadate binding can be monitored by either measuring vanadate analytically or by phosphorylation during a short time incubation with ^{32}P-ATP + Ca^{++}. Vanadate binding turns the calcium binding sites into an inward position: On addition of mM concentrations of calcium to ATPase prepara-

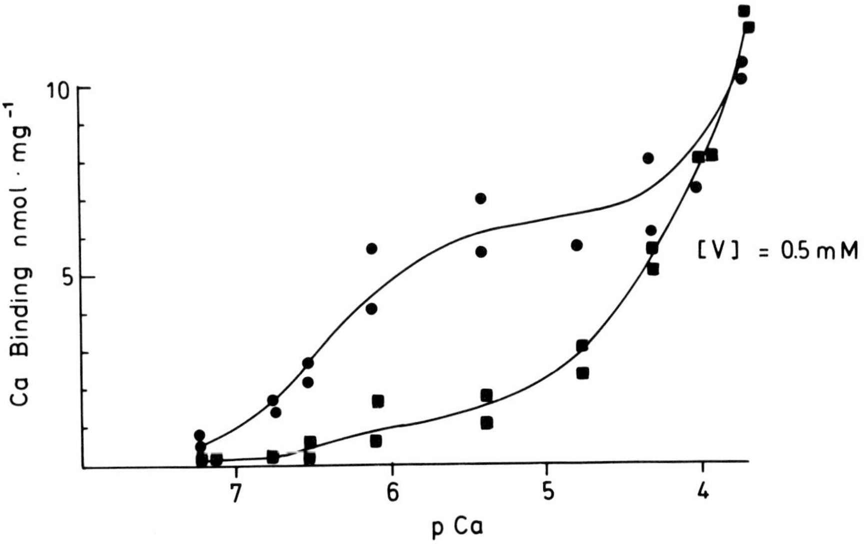

FIGURE 10. Vanadate inhibition of calcium binding to purified ATPase (Medda, Migala and Hasselbach, unpublished results).

tions or vesicles treated with the calcium ionophore A 27183 vanadate is rapidly expelled from the protein. Phosphorylation by ATP reaches maximal values in a fraction of a second. If, however, calcium is added to vanadate inhibited closed vesicles, it takes minutes till optimal phosphoprotein levels are obtained.

That this retardation is due to the slow diffusion of calcium to internal site becomes immediately evident when calcium is added together with the ionophore. The preparation is rapidly phosphorylated to full extent (Fig. 11).

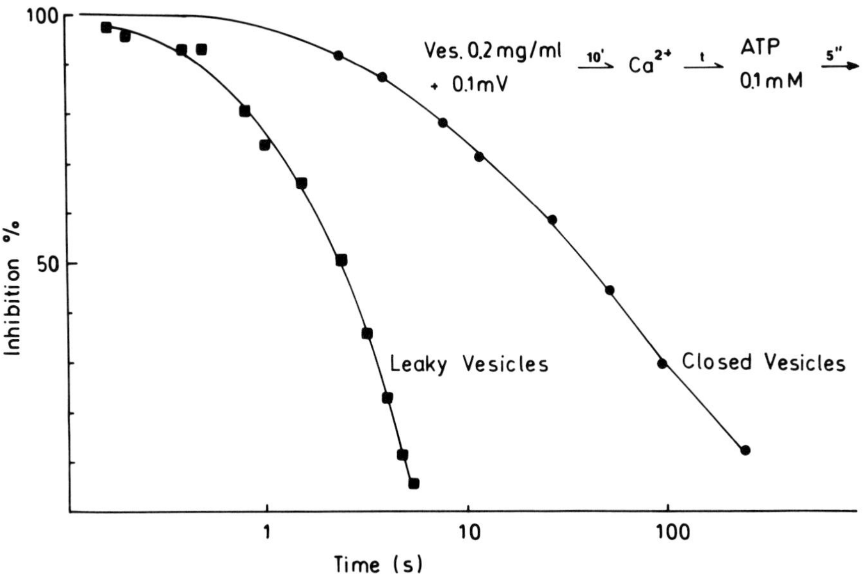

FIGURE 11. Recovery of phosphoprotein formation of vanadate treated closed and leaky vesicle (Medda and Hasselbach, unpublished results).

That low affinity calcium binding must have been occupied is illustrated by Fig. 12. It shows the calcium dependence of the recovery of phosphoprotein formation. The rate with which vanadate is released from closed vesicles remains quite slow even if calcium at mM concentrations is added. Yet, when the same experiments are performed with ionophore treated membranes, the rate of vanadate dissociation increases much more steeply. The rates reach their half maximal value at about 0.5 mM which indicates that vanadate dissociation is initiated by the occupation of the low affinity calcium binding sites. The fact, that the rate of vanadate dissociation is very much accelerated by the ionophore shows that calcium has to reach the interior of the vesicles in order to expel vanadate from the enzyme.

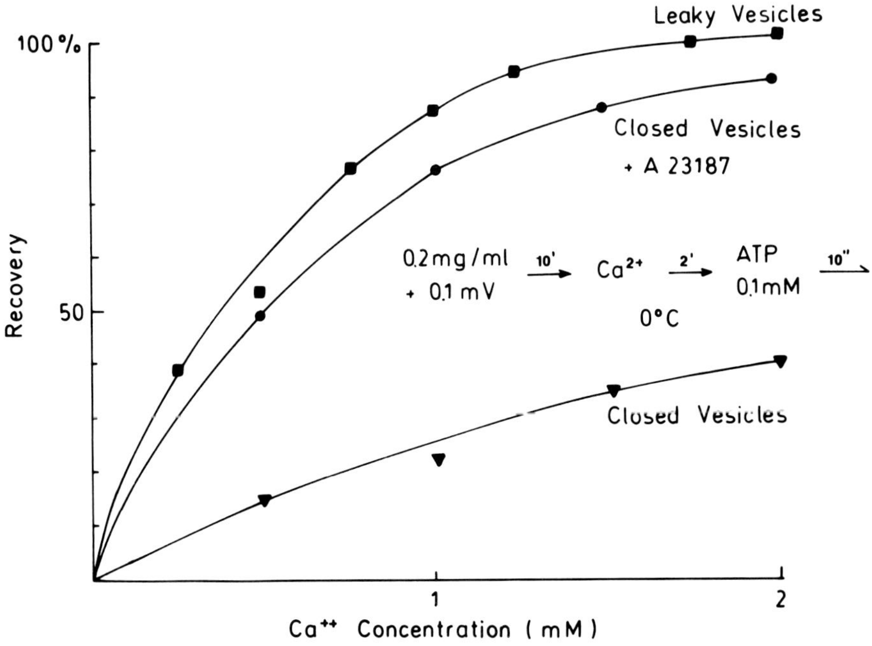

FIGURE 12. Relief of vanadate inhibition of closed and leaky vesicles by calcium. (Medda and Hasselbach, unpublished results).

Summarizing Remarks

The sarcoplasmic reticulum membrane system as a basic element of muscle function has gained general interest in biophysical chemistry due to its high degree of specialization on one hand and the simplicity of its highly efficient calcium transport system on the other hand. In these membranes as well as in their isolated transport protein the chemical and physical events which are connected with calcium translocation can be resolved in time and space most directly and the transport energetics can be evaluated. Yet, as in all other energy transducing systems, the elucidation of the molecular events remains a further challenge.

The most recent developments which allow to resolve calcium binding and release in time during the phosphorylation and dephosphorylation cycle of the transport enzyme has narrowed the historical gap between the apparent contradiction of ATP dependent calcium binding and ATP dependent calcium transport (16, 60).

REFERENCES

1. Endo, M., Physiol. Rev. 57, 71-108 (1977).
2. Hasselbach, W., Oetliker, H., Ann. Rev. Physiol. 45 (1983).
3. MacLennan, D. H., Holland, P. C., Ann. Rev. Biophys. Bioeng. 4, 377-404 (1975).
4. Tada, M., Yamamoto, T., Tonomura, Y., Physiol. Rev. 58, 1-79 (1978).
5. Hasselbach, W., Topics in Current Chemistry 78 1-56, Springer Verlag Berlin Heidelberg New York (1979).
6. Inesi, G., Cell and Muscle Motility, ed. R. M. Dowben and E. W. Shay, 1, 63-97, Plenum Press New York (1981).
7. de Meis, L., Life Science, ed. E.E. Bittar, 2, 1-163 (1981).
8. Marsh, B. B., Biochim. Biophys. Acta, 9, 247-260 (1952).
9. Bendall, J. R., J. Physiol. 121, 232-254 (1953).
10. Hasselbach, W., Weber, H. H., Biochim. Biophys. Acta 11, 160-161 (1953).
11. Portzehl, H., Biochem. Biophys. Acta 26, 373-377 (1957).
12. Ebashi, S., Arch. Biochem. Biophys. 76, 410-423 (1958).
13. Hasselbach, W., Progr. Biophys. Mol. Biol. 14, 167-222 (1964).
14. Lorand, L., Nature 172, 1181-1183 (1953).
15. Portzehl, H., Biochim. Biophys. Acta 24, 477-482 (1957).
16. Nagai, T., Makinose, M., Hasselbach, W., Biochim. Biophys. Acta 43, 223-238 (1960).
17. Hasselbach, W., Makinose, M., Biochem. Z. 333, 518-528 (1961).
18. Hasselbach, W., Makinose, M., Biochem. Biophys. Res. Commun. 7, 132-136 (1962).

19. Hasselbach, W., Makinose, M., Biochem. Z. 339, 94-111 (1963).
20. Hasselbach, W., Makinose, M., Biochem. Z. 343, 360-382 (1965).
21. Weber, A., Winicur, S., J. Biol. Chem. 236, 3198-3202 (1961).
22. Hasselbach, W., Fed. Proc. 23, 909-912 (1964).
23. Costantin, L. L., Franzini-Armstrong, C., Podolski R. J., Science 147, 158-160 (1964).
24. Martonosi, A., Feretos, R., J. Biol. Chem. 239, 648-658 (1964).
25. Kielley W. W., Meyerhof, O., J. Biol. Chem. 176, 591-601 (1948).
26. Hasselbach, W., Ann. New York Acad. Sci. 137, 1041-1048 (1966).
27. Ulbrecht, M., Biochim. Biophys. Acta 57, 455-474 (1962).
28. Makinose, M., Abst. 276, 2. Int. Congress Biophysics, Wien 1966, Verlag Med. Acad., Vienna.
29. Makinose, M., Eur. J. Biochem. 10, 74-82 (1969)
30. Yamamoto, T., Tonomura, Y., J. Biochem. 62, 558-575 (1967).
31. Kanazawa, T., Yamada, S., Yamamoto, T., Tonomura, Y., J. Biochem. 70, 95-123 (1971).
32. Makinose, M., The, R., Biochem. Z. 343, 383-393 (1965).
33. de Meis, L., J. Biol. Chem. 244, 3733-3739 (1969).
34. Inesi, G., Science 171, 901-905 (1971).
35. Weber, A., Herz, R., Reiss, J., Biochem. Z. 345, 329-369 (1966).
36. Barlogie, B., Hasselbach, W., Makinose, M., FEBS Letters 12, 267-268 (1971).
37. Makinose, M., Hasselbach, W., FEBS Letters 12, 271-272 (1971).
38. Somlyo, A. V., Gonzalez-Serratos, H., Shuman, H., McLellan, G., Somlyo, A. P., J. Cell. Biol. 90, 577-594 (1981).
39. Wendt-Gallitelli, M. F., Wolburg, H., Schlote, W., Schwegler, M., Holubarsch, C., Jacob, R., Basic Res. Card. 75, 66-72 (1980).
40. Kometani, T., Kasai, M., J. Membrane Biol. 56, 159-168 (1980).
41. Coray, A., Fry, C. H., Hess, P., McGuigan, J. A. S., Weingart, R., J. Physiol. 305, 60-61 P (1980).

42. Hasselbach, W., Koenig, V., Z. Naturforsch. 35c, 1012-1018 (1980).
43. Guynn, R. W., Veech, R. L., J. Biol. Chem. 248, 6966-6972 (1973).
44. Jencks, W. P., Ad. Enzymol. 51, 75-106, (1980).
45. Dancker, P., Z. Naturforsch. 35c, 431-438 (1980).
46. Kometani, T., Kasai, M., J. Membrane Biol. 41, 295-308 (1978).
47. Meissner, G., J. Biol. Chem. 256, 636-643 (1981).
48. Labarca, P. P., Miller, C., J. Membrane Biol. 61, 31-38, (1981).
49. Feher, J. J., Briggs, F. N., J. Biol. Chem. 257, 10191-10199 (1982).
50. Beil, F. U., von Chak, D., Hasselbach, W., Weber, H. H., Z. Naturforsch. 32c, 281-287 (1977).
51. Tanford, Ch., Proc. Natl. Acad. Sci. (USA) 79, 6161-6165 (1982).
52. Fiehn, W., Migala, A., Eur. J. Biochem. 20, 245-248 (1971).
53. Chevallier, J., Butow, R. A., Biochem. 10, 2733-2737, (1971).
54. Meissner, G., Biochim. Biophys. Acta 298, 906-926 (1973).
55. Ikemoto, N., J. Biol. Chem. 250, 7219-7224 (1975).
56. Miyamoto, H., Kasai, M., J. Biochem. 85, 765-773 (1979).
57. Inesi, G., Kurzmack, M., Coan, C., Levis, D. E., J. Biol. Chem. 255, 3025-3031 (1980).
58. Hill, T., Inesi, G., Proc. Natl. Acad. Sci. (USA) 79, 3978-3982 (1982).
59. Møller, J. V., Lind, K.E., Anderssen, J. P., J. Biol. Chem. 255, 1912-1920 (1980).
60. Lüdi, H., Rauch, B., Hasselbach, W., Z. Naturforsch. 37c, 299-307 (1982).
61. Lüdi, H., Hasselbach, W., Eur. J. Biochem. in press (1983).
62. Deamer, D. W. Baskin, R. J., J. Cell Biol. 42, 296-307 (1969).
63. Nestruck-Goyke, A., Hasselbach, W., Eur. J. Biochem. 114, 339-347 (1981).
64. Brady, G. W., Fein, D. B., Harder, M. E., Spehr, R., Meissner, G., Biophys. J. 34, 13-34 (1981).

65. Froehlich, J. P., Taylor, E. W., J. Biol. Chem. 250, 2013-2021 (1975).
66. Froehlich, J. P., Taylor, E. W., J. Biol. Chem. 251, 2307-2315 (1976).
67. Makinose, M., FEBS Letters 37, 140-143 (1973).
68. Waas, W., Hasselbach, W., Eur. J. Biochem. 116, 601-608 (1981).
69. Katz, A. M., Repke, D. I., Dunnett, J., Hasselbach, W., J. Biol. Chem., 252, 1950-1956 (1979).
70. Makinose, M., Boll, W., Cation Flux Across Biomembranes, ed. Y. Mukohata and L. Packer, 89-100, Acad. Press New York, (1979).
71. Dupont, Y., Eur. J. Biochem. 109, 231-238 (1980).
72. Takisawa, H., Makinose, M., Nature 290, 271-273 (1981).
73. Takakuwa, Y., Kanazawa, T., J. Biol. Chem. 256, 2691-2695 (1981).
74. Ronzani, N., Migala, A., Hasselbach, W., Eur. J. Biochem. 101, 593-606 (1979).
75. Shigekawa, M., Dougherty, J. P., J. Biol. Chem. 253, 1458-1464 (1978).
76. Takakuwa, Y., Kanazawa, T., Biochem. Biophys. Res. Commun. 88, 1209-1216 (1979).
77. Nakamura, Y., Tonomura, Y., J. Biochem. 92, 449-461 (1982).
78. Ikemoto, N., J. Biol. Chem. 251, 7275-7277 (1976).
79. Watanabe, T., Lewis, D., Nakamoto, R., Kurzmack, M., Fronticelli, C., Inesi, G., Biochemistry 20, 6617-6625 (1981).
80. Takisawa, H., Makinose, M., J. Biol. Chem. in press (1983).
81. Knowles, A. F., Racker, E., J. Biol. Chem. 250, 1949-1951 (1975).
82. de Meis, L., Tume, R. K., Biochemistry 16, 4455-4463 (1977).
83. Beil, F. U., von Chak, D., Hasselbach, W., Eur. J. Biochem. 81, 151-164 (1977).
84. Masuda, H., de Meis, L., Biochemistry 12, 4581-4585 (1973).
85. Rauch, B., von Chak, D., Hasselbach, W., Z. Naturforsch. 32c, 828-834 (1977).
86. Punzengruber, C., Prager, R., Kolassa, N., Winkler, F., Suko, J., Eur. J. Biochem. 92, 349-359 (1978).

87. Suko, J., Planck, B., Preis, P. Kolossa, N., Hellmann, G., Conca, W., Eur. J. Biochem. 119, 225-236 (1981).
88. Dupont, Y., Bennett, N., FEBS Letters 139, 237-240 (1982).
89. Pick, U., J. Biol. Chem. 257, No. 2, 6111-6119 (1982).
90. Ebashi, S., Lipmann, F., J. Cell Biol. 14, 389-400 (1962).

Section II

Chemistry and Ultrastructure of the Calcium Pump Protein in Sarcoplasmic Reticulum

STATIC AND TIME-RESOLVED STRUCTURAL STUDIES OF THE Ca^{2+}-ATPase OF ISOLATED SARCOPLASMIC RETICULUM

J.K. Blasie[1,2], L. Herbette[1-3], D. H. Pierce[1], D. Pascolini[1]
V. Skita[1], A. Scarpa[1] and S. Fleischer[4]

Departments of Chemistry and Biochemistry/Biophysics, University of Pennsylvania,[1] Philadelphia, PA

Department of Biology, Brookhaven National Laboratory[2], Upton, NY

Division of Cardiology, University of Connecticut Health Center[3], Farmington, CT

Department of Molecular Biology, Vanderbilt University[4], Nashville, TN

X-ray and neutron diffraction studies of oriented multilayers of isolated light sarcoplasmic reticulum (SR) have provided the separate profile structures of the lipid bilayer and the Ca^{2+}-ATPase molecule within the membrane profile to ~ 10Å resolution. These studies utilized biosynthetically deuterated SR phospholipids incorporated isomorphously into the isolated SR membranes via exchange proteins.

Time-resolved X-ray diffraction studies of these oriented SR membrane multilayers have indicated that significant changes occur in the membrane profile structure within a single-turnover of the Ca^{2+}-transport cycle. These studies utilized the flash-photolysis of caged-ATP to effectively synchronize the ensemble of Ca^{2+}ATPase molecules in the multilayer, synchrotron X-radiation to provide 100-500 millisecond data collection times, and double-beam spectrophotometry to monitor the Ca^{2+}-transport process in the oriented SR membrane multilayer.

ISBN 0-12-260380-X

INTRODUCTION

The ATP-driven active transport of calcium across the sarcoplasmic reticulum (SR) membrane in skeletal muscle causes the physiological relaxation of the myofibrils (1-3). Extremely pure fractions of rabbit skeletal SR can be isolated in the form of unilamellar vesicles; these homogenous vesicular dispersions contain essentially only phospholipid and the Ca^{2+}-ATPase protein (~ 10^5 MW) and are highly functional in terms of their ATP-induced Ca^{2+} uptake and passive Ca^{2+} release properties (4-8). We have shown that these vesicular dispersions of isolated light SR can be formed into highly oriented, hydrated multilayers composed of oriented stacks of the flattened unilamellar vesicles (9).

RESULTS

These oriented multilayers of isolated light SR give rise to extensive lamellar X-ray and neutron diffraction; the separate profile structures (ie. the structures projected onto the normal to the membrane plane) of the phospholipid bilayer and the Ca^{2+}-ATPase molecule within the SR membrane profile have recently been derived to ~ 10Å resolution (see references 10-13 for details) and are shown in figure 1. The derivation of these separate profile structures required the utilization of biosynthetically deuterated SR phospholipids incorporated isomorphously[1] into the isolated SR membranes via phospholipid exchange proteins. The lamellar neutron diffraction data from the oriented multilayers of light SR membranes containing the deuterated SR lipids (phosphatidyl choline only or phosphatidyl choline and phosphatidyl ethanolamine) were analyzed to provide directly by appropriate difference methods the PC and PC + PE bilayer profiles within the SR membrane profile to ~ 30Å resolution. The electron density profile of the light SR membrane derived directly from lamellar X-ray diffraction data from oriented multilayers was subjected to a highly constrained model refinement analysis to provide the separate electron density

---

[1]The isomorphism was checked by X-ray diffraction (~ 10Å resolution), freeze-fracture electron microscopy and ATP-induced Ca^{2+}-transport.

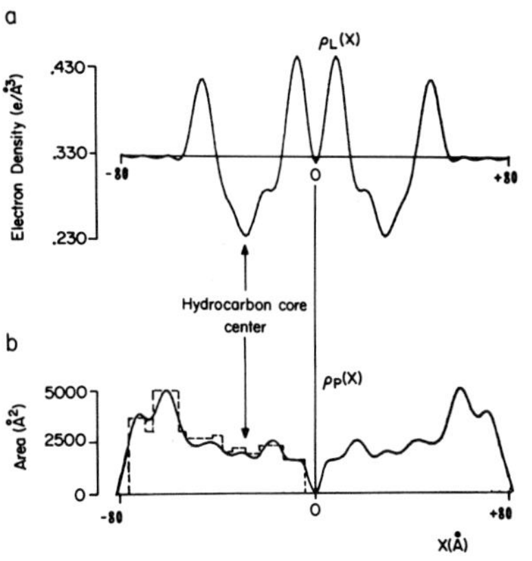

FIGURE 1 (a). The electron density profile of the phospholipid bilayer $\rho_L(x)$ within the SR membrane profile. The bilayer within one SR membrane profile occurs within 0Å ⩽ $|x|$ ⩽ 80Å; the intravesicular water space occurs in the neighborhood of $|x|$ ~ 0Å while the extravesicular water space occurs in the neighborhood of $|x|$ ~ 80Å within the multilayer unit cell profile. (b). The area profile of the Ca^{2+}-ATPase molecule $\rho_p(x)$ within the SR membrane profile. (This profile represents the area occupied by the Ca^{2+}-ATPase molecule as a function of the position x within the SR membrane profile assuming a constant density for the Ca^{2+}-ATPase as a function of (x). The Ca^{2+}-ATPase molecular profile within one SR membrane profile occurs within 0Å ⩽ $|x|$ ⩽ 80Å.

profiles for the phospholipid bilayer and the Ca^{2+}-ATPase molecule within the SR membrane profile to ~ 10Å resolution; the essential constraints were provided by the lower resolution phospholipid bilayer and water distribution profiles within the SR membrane profile provided by the neutron diffraction analysis. The lipid bilayer within the SR membrane is seen to be substantially asymmetric with regard to the distribution of phospholipid molecules in the two monolayers of the bilayer with a polar headgroup separation of ~ 40Å across the bilayer profile; the ratio of the relative number of phospholipid molecules in the inner monolayer (facing the intravesicular water space) versus the outer monolayer (facing the extravesicular water space) is between 54:46 and 55:45, i.e. a distributional asymmetry of 8-10%. The Ca^{2+}-ATPase molecular profile within the SR membrane is seen to span the lipid bilayer extending from $|x| \sim 5Å$ to $|x| \sim 60Å$ in the SR membrane profile with a substantial portion of its mass occurring outside of the lipid polar headgroups on the extravesicular side of the SR membrane profile, ie. from $|x| \sim 60Å$ to $|x| \sim 80Å$. About 20% of the Ca^{2+} ATPase mass occurs within the inner monolayer of the lipid bilayer, about 30% occurs within the outer monolayer and the remaining 50% occurs outside the lipid bilayer on the extravesicular surface of the SR membrane profile. While about 50% of the Ca^{2+}-ATPase mass occurs within the lipid bilayer in the SR membrane profile, only ~ 40% of the Ca^{2+}-ATPase mass occurs within the hydrocarbon core of the lipid bilayer.

An extremely important aspect of this research concerns the Ca^{2+} transport properties of the isolated light SR membranes within these oriented multilayers utilized in the structural (static and time-resolved) studies. We have directly demonstrated that the SR membranes (whose protein content is ~ 65% Ca^{2+}-ATPase) within such hydrated oriented multilayers are fully capable of ATP-induced Ca^{2+}-transport and that the Ca^{2+} transport properties of the less pure SR membranes are identical to those of light SR membranes (whose protein content is >90% Ca^{2+}-ATPase) when the transport properties are expressed as a function of SR protein phosphorylated. These studies of Ca^{2+}-transport kinetics utilized double beam spectrophotometric techniques and Arsenazo III to detect the Ca^{2+}-transport process and the laser flash-photolysis of caged-ATP (14) to effectively synchronize the Ca^{2+}transport cycles among the ensemble of Ca^{2+}-ATPase molecules in the oriented multilayer (15-17). The kinetics of ATP-induced Ca^{2+} transport in the oriented multilayers over the temperature range $-1°C \leq T \leq 21°C$ are essentially identical to those of the vesicular dispersions. In figure 2, we compare the Ca^{2+} transport kinetics for the multilayers and dispersions over the respective temperature ranges $-1°C \leq T \leq 21°C$ and $-2°C \leq T \leq 26°C$. In both cases, the transport kinetics exhibit a Ca^{2+}-ionophore insensitive fast phase (first-order with initial rates of 10^2-10^3 n moles Ca^{2+}/mg ATPase • sec. over the temperature range with an activation energy of $E_a \sim 16$ kcal/mole) and a Ca^{2+}-ionophore sensitive slow phase (first-order with initial rates of 10^0-10^2 n moles Ca^{2+}/mg ATPase • sec. over the temperature range with an activation energy of $E_a \sim 22$ kcal/mole). The fast phase has a stoichiometry of ~ 2 Ca^{2+}/mole phosphorylated enzyme while the slow phase has a stoichiometry of ~ 2Ca^{2+}/mole ATP hydrolyzed. The Ca^{2+}-ionophore insensitive fast phase has kinetics

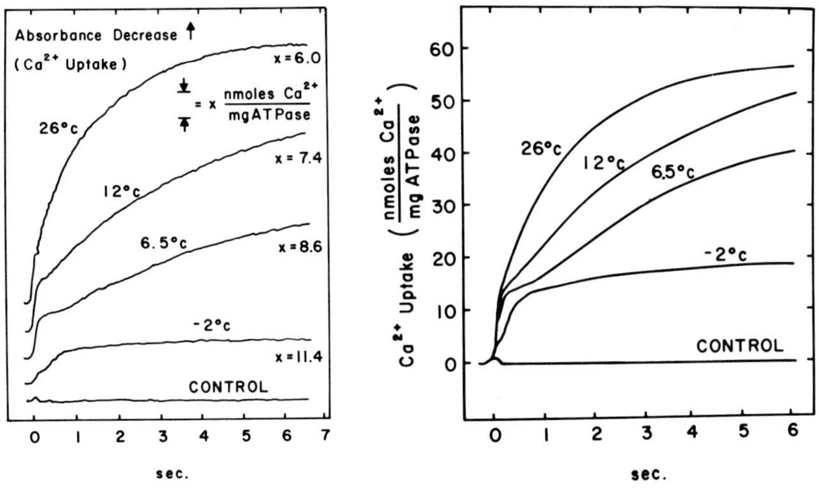

FIGURE 2 (a). The ATP-induced Ca^{2+} transport kinetics (detected with Arsenazo III at 660 nm) for the vesicular dispersions of SR membranes over the temperature range $-2°C \leqslant T \leqslant 26°C$ using a ~ 30 nanosecond UV (347 nm) flash (25-30 mJ) from a frequency-doubled ruby laser. On a time-scale of seconds, the Ca^{2+}-ionophore sensitive slow-phase of Ca^{2+}-uptake is readily apparent while the fast-phase is evident at the lower temperatures. The $[Ca^{2+}]$-calibrations are indicated with the raw data on the left and have been applied to the smoothed data on the right. The controls indicated utilized caged-ADP in place of caged-ATP all other conditions being equivalent.

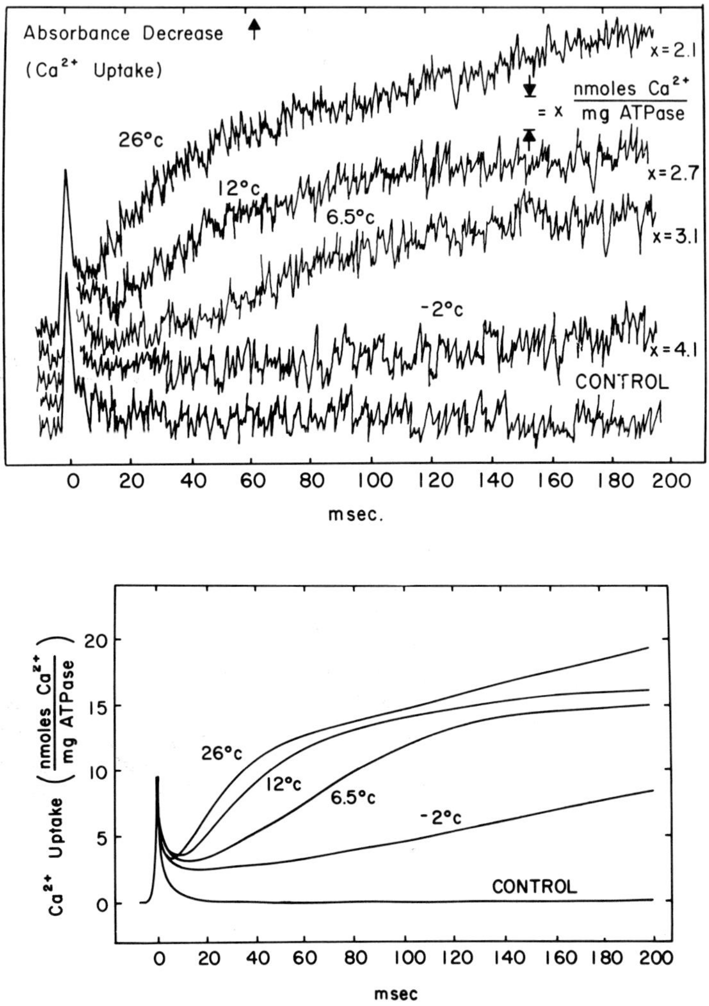

Figure 2 (b). Same as 2(a), but on a millisecond time-scale. The Ca^{2+}ionophore insensitive fast-phase of Ca^{2+}-uptake is readily apparent followed by the slow-phase evident at the higher temperatures. A laser artifact occurs at t=0. The $[Ca^{2+}]$-calibrations are indicated with the raw data above and have been applied to the smoothed data below. The controls indicated utilized caged-ADP in place of caged-ATP, all other conditions being equivalent.

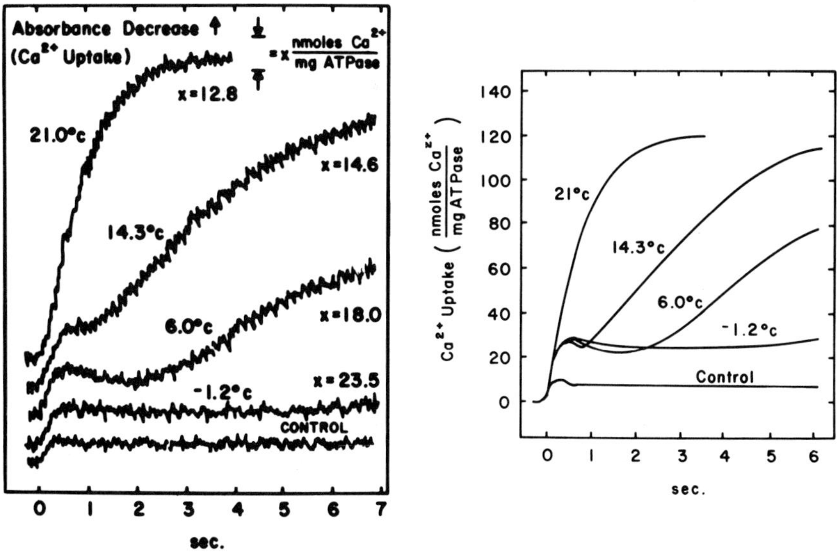

FIGURE 2 (c). The ATP-induced Ca^{2+} transport kinetics (detected with Arsenazo III at 660 vs. 690 nm) for the partially dehydrated oriented multilayers of SR membranes over the temperature range $-1°C \leqslant T \leqslant 21°C$ using a ~ 30 nanosecond UV (347 nm) flash (25-30 mJ) from a frequency-doubled ruby laser. On a time-scale of seconds, the Ca^{2+}-ionophore sensitive slow-phase of Ca^{2+}-uptake is readily apparent while the fast-phase is evident at the lower temperatures. The $[Ca^{2+}]$-calibrations are indicated with the raw data on the left and have been applied to the smoothed data on the right. The controls indicated utilized caged-ADP in place of caged-ATP, all other conditions being equivalent.

essentially identical to those for the formation of the ADP-sensitive, acid-stable phosphorylated enzyme intermediate, as measured independently with a quenched-flow method, and may therefore be consistent with either (a) an ATP-induced increase in the affinity of two Ca^{2+} binding sites on the enzyme or (b) an ATP-induced occlusion of two Ca^{2+} binding sites on the enzyme resulting in their inaccessibility from either surface of the SR membrane. The slow phase may be identified with the translocation of Ca^{2+} across the membrane profile due to its Ca^{2+}-ionophore sensitivity and its relatively high activation energy. However, other kinetic phases of the Ca^{2+}-transport process may be indicated by the somewhat greater lag between the fast and slow phase more readily apparent in the oriented multilayers.

The nature of the time-resolved X-ray diffraction studies of the SR membrane can be summarized briefly as follows: the Ca^{2+}-transport process is initiated essentially synchronously throughout the ensemble of Ca^{2+}-ATPase molecules within an oriented multilayer of isolated SR membranes via the flash-photolysis of caged-ATP. The time-scale of the effective synchronization of the ensemble depends initially upon the duration of the UV light flash required to produce a sufficient quantity of ATP in the multilayer to sustain several turnovers of the Ca^{2+}-ATPase molecules and is ultimately limited to the millisecond time-scale due to the kinetics of the dark-reactions of the photolysis process (14). Lamellar X-ray diffraction is recorded from the multilayer prior to the flash-photolysis and after a variable delay τ following the flash-photolysis using X-ray exposure times which are short compared with the turnover time of the Ca^{2+}-transport cycle which is on the order of a second at 6-8°C. Such diffraction patterns can then be used to investigate changes within the profile structure of the SR membrane occurring at different times (different τ's) within the transport cycle and these changes can ultimately be correlated with the various chemical intermediates in the cycle (15-19). The flash-photolysis of caged-ADP under otherwise identical experimental circumstances provides an essential control to these studies since the same amount of light energy is absorbed and similarly dissipated without producing a useful substrate for the Ca^{2+}-ATPase. The ultimate time-resolution required of such studies depends upon the lifetimes of the various intermediates in the transport cycle. The ultimate time-resolution achievable depends upon the rate of ATP production by the flash-photolysis mechanism (currently limited to the millisecond time-scale utilizing a pulsed laser UV source), and the incident X-ray flux.

We have recently (20) utilized focused synchrotron X-radiation from bending magnets at the Stanford Synchrotron Radiation Laboratory and a SIT image-intensifier area X-ray detector to successfully record changes in the lower order ($h=1\rightarrow8$) lamellar diffraction from hydrated oriented SR membrane multilayers for $\tau=0$ seconds which relaxed after τ >1 minute utilizing 200-500 millisecond X-ray exposure times (see figure 3). The caged-ADP controls did not exhibit such changes in the lamellar diffraction under otherwise identical circumstances. These data were analyzed, utilizing a procedure which takes into proper account the various forms of disorder which can occur in such membrane multi-

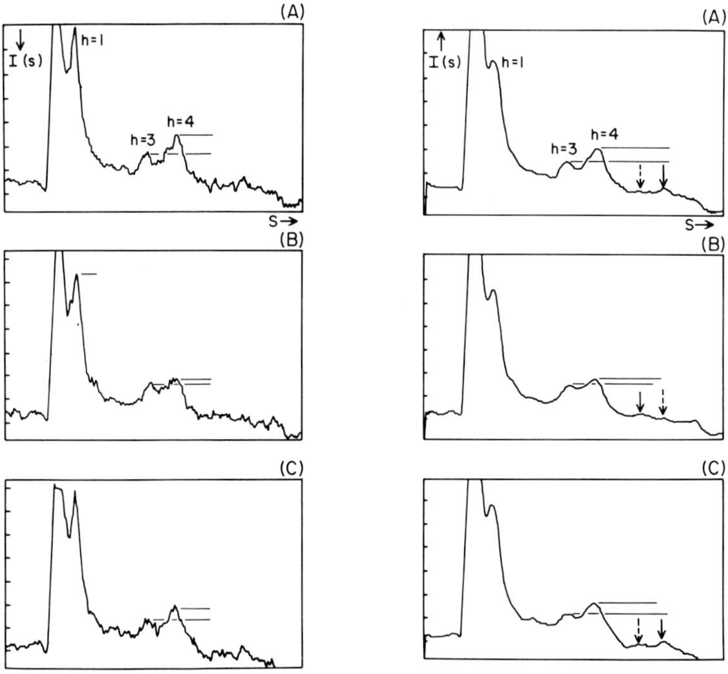

FIGURE 3. The lamellar X-ray diffraction on the left (numerically smoothed on the right) obtained from partially-dehydrated oriented multilayers of isolated light SR membranes using 200 millisecond X-ray exposures: (A) immediately prior to the flash-photolysis (200 millisecond UV flash (300-370 nm) from a Hg-arc optically coupled to the multilayer via a quartz light pipe) of caged-ATP. (B) immediately following the flash-photolysis of caged-ATP (C) 1 minute after the flash-photolysis of caged-ATP. A reversible change in the amplitude of the first order lamellar reflection (h=1) and the ratio of the amplitudes of the third and fourth order lamellar reflections (h=3,4) is readily apparent upon comparison of figure 3B with figures 3A and 3C. Other reversible changes in the higher angle lamellar X-ray diffraction are apparent in the regions indicated by the arrows.

No changes in the lamellar diffraction from such multilayers were observed using caged-ADP in place of caged-ATP (all other conditions being equivalent) within the experimental errors over this range of $S=(2 \sin \Theta)/\lambda$.

layers, with the following results. During the first 200-500 milliseconds following the flash-photolysis of caged-ATP, small but significant changes (at the 10-20% level) occur throughout the membrane profile itself. These changes in the membrane profile are accompanied by a small increase (~ 1-2Å) in the average separation of the two opposed membranes within a flattened vesicle across the intravesicular water space and a small increase (~ 2-3%) in the multilayer lattice disorder without any detectable change in the multilayer lattice periodicity. Consideration of the temperature (7-8°C), τ (0 sec.) and the X-ray exposure times (200-500 milliseconds) utilized and the Ca^{2+}-transport kinetic studies described above would indicate that these changes detected in the membrane profile structure correspond to the Ca^{2+}-ionophore insensitive fast kinetic phase associated with the formation of the ADP-sensitive acid-stable phosphorylated enzyme intermediate.

DISCUSSION

Our results have demonstrated that the SR membrane structure can be resolved into its separate components at moderate resolution. In addition we have developed the necessary technologies to investigate the time evolution of the structure of the SR membrane during the ATP-induced Ca^{2+}transport cycle. These time resolved studies are currently being extended to a time-resolution approaching ~ 1 millisecond using a laser-plasma X-ray source.

REFERENCES

1. Ebashi, S., Endo, M. and Ohtsuki, I. Q. Rev. Biophysics 2:351 (1969).
2. Weber, A., Herz, R. and Reiss, J., J. Gen. Physiol. 46:679(1963).
3. Duggan, P.F. and Martonosi, A., J. Gen. Physiol. 56:147 (1970).
4. Meissner, G., Conner, G.E. and Fleischer, J. Biochim. Biophys. Acta 298:246 (1973).
5. Herbette, L., Marquardt, J., Scarpa, A. and Blasie, J.K., Biophys. J. 20:245 (1977).
6. Martonosi, A. and Feretos, R., J. Biol. Chem. 239:648 (1964).

7. Inesi, G., Ann. Rev. Biophys. Bioeng. 1:191 (1972).
8. deBoland, A.R., Julka, R.J. and Martonosi, A.N. J. Biol. Chem. 250:7501 (1975).
9. Herbette, L., Scarpa, A., Blasie, J.K., Wang, C.T., Saito, A., and Fleischer, S., Biophys. J. 36: 47 (1981).
10. Blasie, J.K., Herbette, L., DeFoor, P.H., and Fleischer, S., Biophys. J. 33:228a (1981).
11. Herbette, L. and Blasie, J.K. in Calcium-Binding Proteins: Structure and Function, (F.L. Siegel, E. Carafoli, R.H. Kretsinger, D.H. MacLennan and R.H. Wasserman, eds.) p. 115. Elsevier North-Holland, Inc. 1980.
12. Blasie, J.K., Herbette, L., Pierce, D., Pascolini, D., Scarpa, A., and Fleischer, S., Ann. N.Y. Acad. Sci. 402:478 (1982).
13. Herbette, L., DeFoor, P. Fleischer, S., Pascolini, D. and Blasie, J.K., manuscript in preparation.
14. McCray, J.A., Herbette, L., Kihara, T. and Trentham, D.R., Proc. Natl. Acad. Sci. USA 77:7237 (1980).
15. Pierce, D.H., Scarpa, A., Topp, M.R. and Blasie, J.K., submitted to Biochemistry.
16. Pierce, D.H., Scarpa, A., Trentham, D.R., Topp, M.R. and Blasie, J.K., submitted to Biophysical Journal.
17. Pierce, D.H., Ph.D. Dissertation, University of Pennsylvania (1982).
18. MacLennan, D.H. and Holland, P.C., Ann. Rev. Biophys. Bioeng. 4:377 (1975).
19. deMeis, L. and Vianna, A.L., Ann. Rev. Biochem., 48: 275 (1979).
20. Herbette, L. and Blasie, J.K., Biophys. J. 33:228a (1981).

CONFORMATIONAL CHANGES OF Ca^{2+},Mg^{2+}-ADENOSINETRIPHOSPHATASE AS DETECTED BY SITE-SPECIFICALLY INTRODUCED FLUORESCENT AND PARAMAGNETIC PROBES AND BY SUSCEPTIBILITY TO TRYPTIC DIGESTION[1]

Masao Kawakita
Kimiko Yasuoka-Yabe[2]
Kimiko Saito
Yasutada Imamura

Department of Pure and Applied Sciences
College of General Education
The University of Tokyo
Meguro-ku, Tokyo, Japan

I. INTRODUCTION

The mechanism of active Ca^{2+} transport across the sarcoplasmic reticulum (SR) membrane, which is carried out by Ca^{2+},Mg^{2+}-ATPase, has been intensively studied in recent years (for review see ref. 1-3). During the transport process, Ca^{2+} is bound to the ATPase from the cytoplasmic side of the membrane and is eventually released from the enzyme into the lumen of the SR. As a result more than 100-fold concentration of Ca^{2+} can be attained in the lumen as compared with the cytoplasm. The driving force for this uphill movement of the ion is derived from ATP and is somehow transmitted by ATPase. Although the physical or molecular basis for the unidirectional movement of the ion which is tightly coupled to ATP hydrolysis

---

[1]Supported in part by grants from the Ministry of Education, Science and Culture of Japan, Itoh Science Foundation and Yamada Science Foundation.
[2]Present address: The Institute of Medical Science, The University of Tokyo

ISBN 0-12-260380-X

is still largely unknown, there is little doubt that conformational transitions of the ATPase molecule underlie the processes of gating and affinity change for Ca^{2+} ions and constitute the essence of the energy coupling mechanism in active transport.

Recently Blasie and his collaborators applied time-resolved X-ray diffraction technique to SR membranes. It may be helpful, after further improvement, in providing information concerning the gross conformational changes of ATPase within a single turnover of the enzyme (4, also see Blasie in this volume). However, an extensive conformational alteration involving a change in helical content of the ATPase polypeptide seems so far rather unlikely, since an attempt to observe circular dichroism changes in SR in various states of the enzyme was unsuccessful (5). Therefore it looks more likely that a local change of the ATPase conformation may play a major role in its catalytic mechanism, although we have to keep in mind that fluctuation of the folded structure might be changing over a significant portion of the molecule, in view of the results showing a decrease in the hydrogen exchange rate on nucleotide binding or under Ca^{2+}-transporting conditions (6,7).

Conformational change within a small portion of a large polypeptide chain, which is the main subject of this article, can be detected by various means. Reactivity of side chain groups of a peptide is often altered when protein conformational change occurs. Thus changes in the rate of chemical modification of sulfhydryl or amino groups were observed on addition of Ca^{2+} and nucleotides to the enzyme and were ascribed to conformational changes of ATPase (8-10). More direct evidence can be obtained through spectroscopic signals of various probes which may be either colored, fluorescent or paramagnetic. Spectral characteristics of such probes are altered when the local environment of their attachment sites is changed as a consequence of protein conformational change, so that they can serve as a sensitive monitor for conformational transitions. The interaction of probes with ATPase can be either covalent or noncovalent. Furthermore a tryptophan residue can serve as an intrinsic fluorescent probe.

Changes in the absorption spectrum of bromcresol purple in response to the addition of Ca^{2+} and ATP were reported (11, 12). A widely used noncovalent fluorescent probe, 8-anilino-1-naphthalene sulfonate, was also tried but the observed enhancement of fluorescence intensity was attributed to the binding of accumulated Ca^{2+} to the SR membrane rather than to a conformational change of ATPase (13). Although noncovalently interacting probes are readily applicable and are also sensitive enough to give us indication that a conformational change is occurring somewhere in the ATPase, a further in depth analysis is often not quite straightforward. This is because the binding sites for such noncovalent probes are usually very difficult to locate precisely on the ATPase peptide. Moreover, when noncovalently interacting probes were employed we have to be aware that the number of bound probe may vary depending on the conditions of measurement. This introduces further complications in the interpretation of experimental results.

The latter complication could be obviated if we use an intrinsic probe or a covalently attached extrinsic probe. Recent studies on the intrinsic fluorescence of ATPase revealed that different intermediate states may differ in their fluorescence intensity (14-21, also see Dupont in this volume). The intrinsic probe has a definite merit in that we are always dealing with totally intact ATPase. However, this is rather inconvenient for the purpose of identifying the portion of the ATPase molecule involved in the conformational change, since there are as many as 19 tryptophanyl residues per mole of ATPase peptide and their environments may not be equivalent to each other's (20). Using extrinsic probes, on the other hand, we would be able to introduce a reporter group to a limited number of sites whose location could be identified precisely on the ATPase molecule. Among those, several spin labels were the most widely used (22-28), although communications using fluorescent probes have also come out recently (29-32).

In order to obtain precise information by using microenvironmental probes efficiently, it is important that at least two requirements be fulfilled. In the first place the labeling reaction must be highly site-specific, because the specificity of the labeling reaction profoundly affects the quality of the information obtained through a particular label. If the labeling occurred rather nonspecifically, then environmental changes felt by individual reporters at various parts of the enzyme molecule could not be detected separately, and a spectral change would only reflect an average of gross conformational changes. In this situation the sensitivity as well as the resolving power of the probe would be necessarily low. On the other hand, if a probe was introduced to a highly specific site, it would detect even the smallest change in its immediate neighborhood. Moreover, if another label was attached to another specific location it might experience a quite different change at the same moment. As a consequence of this the motion of an individual domain of the enzyme molecule could be monitored separately at high resolution. Covalently attached probes would work at their best if used in this way. In this respect it is quite important to have a sample with a label attached to a well-defined site of ATPase. It is also important to stabilize each reaction intermediate of ATPase so that it can be accumulated for a sufficiently long period to allow a detailed conformational analysis.

During the course of recent studies we were able to demonstrate that a highly site-specific labeling of ATPase could be attained by a controlled reaction of sulfhydryl groups with various microenvironmental probes having a maleimide functional group (33). We also devised several procedures for stabilizing individual ATPase intermediates (34, 35). Finally the selectively labeled ATPase was utilized for comparing conformational characteristics of various intermediates which were accumulated by those stabilization procedures described above (33,35). The results of these studies are presented in the following sections.

II. SELECTIVE LABELING OF SULFHYDRYL GROUPS

A. Reactivity of Sulfhydryl Groups Toward N-Ethylmaleimide

ATPase of the SR contains sulfhydryl (SH) groups, some of which are essential for the functioning of the transport system (36-40). Although a number of investigators have utilized them as the target of spin labeling, interpretation of the results is not quite straightforward, since the attachment site of the labels has not been well characterized.

Recently we have carefully examined the reactivity of SH groups toward N-ethylmaleimide (NEM) and the results are summarized in Table I (41). We could distinguish 4 reactive SH groups at pH 7.0 and 30 °C. They are listed in the table in the order of reactivity. Lower pH than that employed

TABLE I. Reactive SH groups of Ca^{2+},Mg^{2+}-ATPase

Designation	Number[a]	Function	Remarks
SH_N	1	———	
SH_D	1	E-P decomposition	less reactive at low Ca^{2+}
$SH_{N'}$	1	———	
SH_F	1	E-P formation	protected by AMP-P(NH)P

[a]Residue / mol ATPase

in previous works was preferred because differences in reactivity among SH groups were more readily discernible at that pH. At NEM concentration of around 50 μM only one SH group was blocked, while enzyme activity was not affected at all by this reaction. Therefore the most reactive one was nonessential, and was named SH_N. At a higher NEM concentration (ca. 1 mM) 4 SH groups reacted in sequence. The second reactive group was essential for ATPase activity, particularly for E-P decomposition, and was named SH_D. Both SH_N and SH_D proved to be quite useful as specific targets of extrinsic probes as shown below. The third reactive group was blocked without further change in enzyme activity, while blocking of the fourth reactive group (SH_F) led to the loss of E-P formation. AMP-P(NH)P specifically protected SH_F against NEM-modification. The remarkable selectivity of NEM-modification as shown above prompted us to use a selective and site-specific labeling of the ATPase molecule by fluorescent and paramagnetic maleimides.

B. Specific Labeling by Extrinsic Probes

Four microenvironmental probes containing a maleimide functional group were chosen. Three of these, N-(1-anilino-naphthyl-4)maleimide (ANM), N-(p-(2-benzimidazolyl)phenyl) maleimide (BIPM), and N-(7-dimethylamino-4-methylcoumarinyl) maleimide (DACM) were fluorescent labels and another, 4-maleimido-2,2,6,6 - tetramethylpiperidinooxyl (MSL) was a spin label. The fluorescent labels were originally developed by Kanaoka and Sekine for studies of the function of SH groups of myosin, and have since been applied to several systems (42-50). The reagents themselves are not fluorescent, but once they are linked to an SH group they become highly fluorescent and serve as excellent reporters of their environment.

Although these reagents were definitely more hydrophobic and were in fact more reactive than NEM, they still retained a selectivity towards the SH groups of ATPase which was essentially similar to that of NEM. Figures 1 and 2 illustrate how a specific labeling of SH_N and SH_D, respectively, with DACM was attained. We were also able to carry out labeling with other fluorescent probes in the same way. MSL was also used for labeling under similar conditions with slight modifications (33). Molar ratio rather than concentration was found to be critical in carrying out a specific labeling of SH_N by the probes. Thus DACM to ATPase ratio of roughly 1.3 was chosen in Figure 1, assuming the ATPase content of the SR to be 70 %. Under this condition DACM reacted rapidly with the SR and approximately 0.8 mol

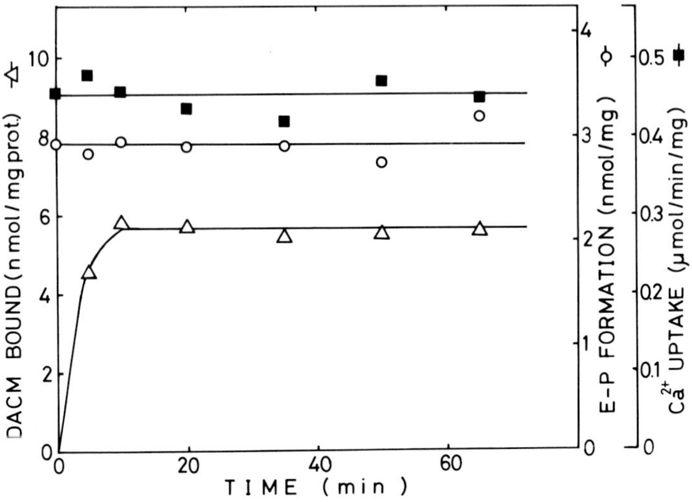

FIGURE 1. Labeling of SH_N with DACM. Rabbit skeletal muscle SR membranes (3.0 mg/ml) prepared as previously described (41) were reacted with 28 μM DACM at 30 °C in a solution containing 40 mM N-tris(hydroxymethyl)methyl-2-aminoethanesulfonic acid (TES)-NaOH buffer (pH 7.0), 0.1 M KCl and 5 mM $MgCl_2$. Aliquots were withdrawn from the reaction mixture at specified time intervals and the amount of bound DACM (△), Ca^{2+} uptake (■) and E-P forming (○) activities were determined as described elsewhere (33).

DACM/mol ATPase was incorporated. Neither Ca^{2+} uptake nor E-P formation was impaired during the course of this reaction, indicating that a selective labeling of SH_N was attained.

SH_D was labeled with DACM in a two-step modification, where SR was first incubated with NEM and then with DACM, as shown in Figure 2. At the end of the first incubation, 1 mol NEM/mol ATPase was bound, while approximately 80 % of the Ca^{2+}-transport activity was retained, showing that most of SH_D still remained intact at this stage. During the second incubation Ca^{2+}-transport activity was rapidly lost upon reaction with DACM, indicating that labeling of SH_D had occurred. Intactness of SH_F could be verified by constant E-P forming activity. The presence of AMP-P(NH)P during the second incubation was necessary to protect SH_F from modification. The amount of protein-bound DACM was 1.5-2.0 mol/mol ATPase. Electrophoretic analysis of SH_N- and

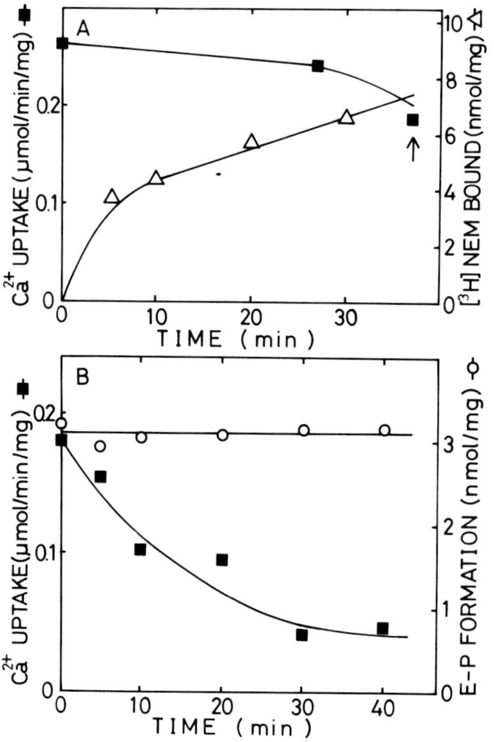

FIGURE 2. Labeling of SH_D with DACM in a two-step reaction. A. The first step: SR membranes (3.0 mg/ml) were treated as in Figure 1 with 70 μM NEM in order to block SH_N completely. [^{3}H]NEM bound (△) and Ca^{2+} uptake activity (■) are shown. At 37 min the reaction was stopped by 6-fold dilution with 40 mM TES-NaOH buffer (pH 7.0), 0.1 M KCl and 5 mM $MgCl_2$. B. The second step: to the diluted reaction mixture 1 mM AMP-P(NH)P and 27 μM DACM were added to label SH_D. Ca^{2+} uptake (■) and E-P forming (O) activities were monitored (33).

SH_D-labeled SR protein on SDS-polyacrylamide gel indicated that the fluorescent label thus incorporated into the SR membrane was almost exclusively located on the ATPase polypeptide (33).

III. STABILIZATION OF REACTION INTERMEDIATES

A. Stabilization of E_1-P by SH_D-Modification

Rather simple procedures such as the adjustment of free Ca^{2+} concentration with ethyleneglycol bis(β-aminoethyl ether)-N,N,N',N'-tetraacetic acid (EGTA) and/or addition of ATP, ADP and their analogs were used conveniently to cause accumulation of certain intermediates. Thus intermediates (or their analogs) such as E, E^{Ca}, $E_{AMP-P(NH)P}$, $E^{Ca}_{AMP-P(NH)P}$ were easily obtained. Although AMP-P(NH)P is not absolutely nonhydrolyzable by ATPase (51), the rate of E-P formation by this substrate is extremely slow and we may safely assume that most of the ATPase would exist as $E^{Ca}_{AMP-P(NH)P}$ complex in the presence of this ATP analog (M. Kawakita, unpublished observation). Maintenance of E_1^{Ca}-P for a sufficient period, however, was a little more difficult. Although it can be accumulated during the steady state of ATP hydrolysis, it will soon disappear when ATP is exhausted and will not usually last long enough to allow a detailed conformational study. This can be a serious problem particularly in studies using the spin-labeled ATPase, because ESR measurements take longer time and need higher enzyme concentration. Recently we reported that by using adenosine-5'-O-(3-thiotriphosphate) (ATPγS), a fairly stable E-P analog $E\overset{P}{\underset{S-S}{=}}$ could be obtained (34).

The E-P intermediate could also be stabilized in another way as shown below. Modification of SH_D results in the impairment of E-P decomposition as mentioned above. Further analysis revealed that the blocking of SH_D causes a severe inhibition of the conversion of E_1^{Ca}-P to E_2-P as shown in Figure 3, thus giving rise to a long-lived E_1^{Ca}-P intermediate. With native ATPase, E-^{32}P accumulated during the steady state was ADP-sensitive (E_1^{Ca}-P) and decomposed rapidly when ADP was added together with EGTA. E-P decomposition was accompanied by formation of [γ-^{32}P]ATP through the backward reaction (not shown). This ADP-sensitivity was rapidly lost as a result of the transition from E_1^{Ca}-P to E_2-P. On the other hand when SH_D-modified SR was used, E_1^{Ca}-P once formed still retained its ADP-sensitivity even 20 sec after addition of EGTA. The moderate decrease in E_1^{Ca}-P seen in Figure 3B is likely to be due to the reverse reaction with ADP produced during the preincubation. This could be prevented and E_1^{Ca}-P could stably accumulate in the reaction mixture, if a sufficient amount of ATP and Ca^{2+} were present.

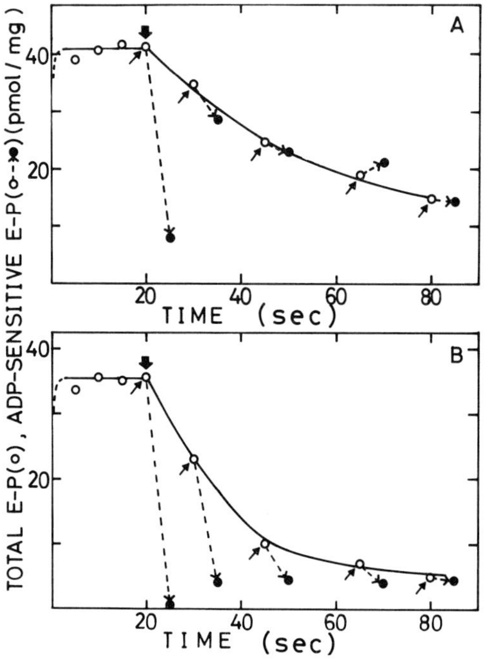

FIGURE 3. Inhibition of E_1^{Ca}-P to E_2-P conversion by SH_D-modification. Intact SR membrane (A) or SH_D-modified SR membrane (B) (1 mg/ml respectively) was phosphorylated at 0 °C with 0.1 μM [γ-^{32}P]ATP in a mixture containing 20 mM imidazole-HCl buffer (pH 7.0), 50 mM KCl, 0.1 mM $CaCl_2$ and 0.2 % Triton X-100. After stopping E-P formation with 2 mM EGTA (⬇), the decrease in the amount of E-^{32}P was followed. ADP-sensitivity of E-P was tested by adding 1 mM ADP at specific times (↗) and measuring the amount of E-^{32}P remaining 5 sec later (35).

B. Effect of Adding ADPβS to E-P

Another interesting intermediate was obtained by using SH_D-modified SR and adenosine-5'-O-(2-thiodiphosphate) (ADPβS). As explained above and also shown in Figure 4, when the forward reaction was stopped by the removal of free Ca^{2+}, the amount of phosphoenzyme decreased rather rapidly through the backward reaction. Addition of ADPβS together with EGTA caused a marked stabilization of E-^{32}P so that the level of E-^{32}P remained high for several minutes. In the presence of excess ADPβS a complex between ADPβS

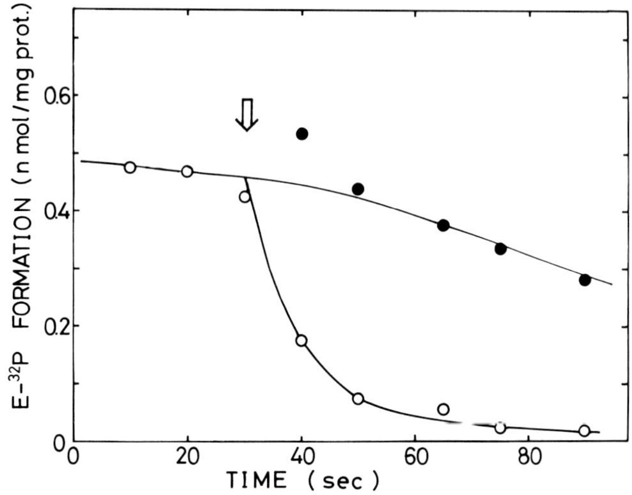

FIGURE 4. Effect of ADPβS on the stability of the E-P intermediate. SH_D-modified SR was phosphorylated as in Figure 3 using 6.7 μM [γ-^{32}P]ATP. After E-P formation was stopped (⇩) by adding 2.7 mM EGTA with (●) or without (○) 1.0 mM ADPβS, the decrease in the amount of E-^{32}P was followed (35).

and E_1^{Ca}-P, E_1^{Ca}-P·ADPβS, was probably formed. Although it is not completely inert, as reported recently (52), the reverse reaction to yield ATPβS occurs only very slowly. Since the decomposition of E_1^{Ca}-P in the forward direction had already been blocked by SH_D-modification, addition of ADPβS to E_1^{Ca}-P led to an accumulation of E_1^{Ca}-P·ADPβS.

IV. CONFORMATIONAL CHANGES OF THE ATPase

A. Studies with Fluorescent Probes

1. Ca^{2+} Binding-Induced Change. The fluorescence spectrum of the SR labeled at SH_D with ANM (ANM(SH_D)-labeled SR, similar notation will be observed in the following and also for other labels) is shown in Figure 5A. On removal of free Ca^{2+} the fluorescence intensity was increased by about 10 % without detectable shift of emission maximum at 430 nm. Figure 5B shows that the Ca^{2+}-dependent change is rapid and readily reversible on repeated addition and

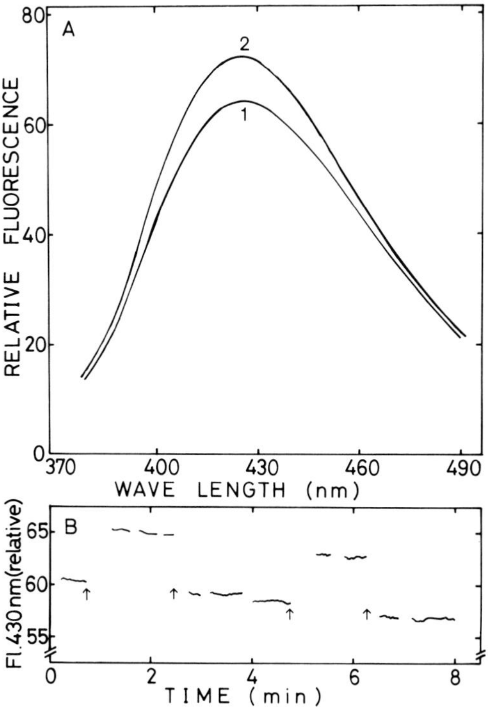

FIGURE 5. Ca^{2+}-dependent change in the fluorescence of ANM(SH_D)-labeled SR. Labeled SR was prepared according to a method similar to that described in Figure 2. A. Fluorescence emission spectra of labeled SR (64 µg/ml) at 24 °C excited at 350 nm in a solution containing 40 mM TES-NaOH buffer (pH 7.0), 0.1 M KCl and 5 mM $MgCl_2$ were recorded before (1) and after (2) the addition of 1 mM EGTA. B. Fluorescence intensity at 430 nm of the sample as in A was followed. At the times indicated by the arrows, 0.5 mM EGTA, 0.5 mM $CaCl_2$, 2.9 mM EGTA and then 2.9 mM $CaCl_2$ were added one at a time in this order (33).

removal of Ca^{2+} ions. Figure 6 further shows that the change was caused by Ca^{2+} at a concentration range between 10^{-7} - 10^{-6} M. The Ca^{2+} concentration range is similar to that required for the activation of ATPase and is the one where the binding of Ca^{2+} to the transport site takes place. Since the label was covalently attached to the enzyme, a change in fluorescence intensity could only be caused by a change in the fluorescence quantum yield. It seems likely that a conformational change has occurred in the neighborhood

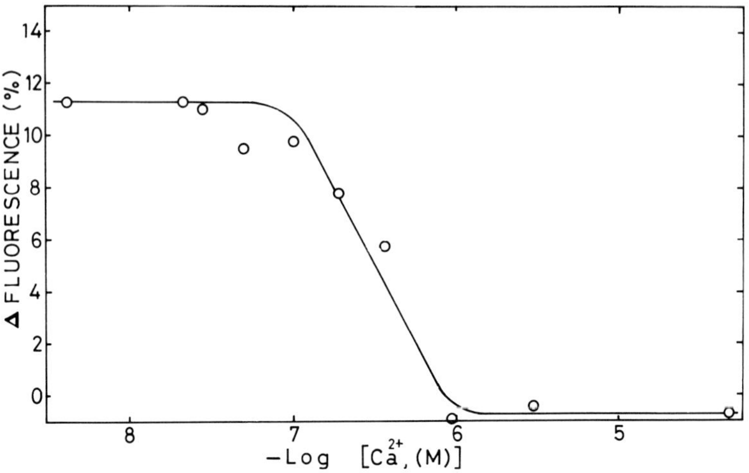

FIGURE 6. The Ca^{2+} concentration dependence of the fluorescence intensity of ANM(SH_D)-labeled SR. Fluorescence intensity of the labeled SR (41 μg/ml) at 430 nm was measured as in Figure 5 at various Ca^{2+} concentrations as indicated. The increment is expressed as the percentage of that in the presence of 10 μM Ca^{2+} (33).

of SH_D as a consequence of the binding of Ca^{2+} to the transport site. This is consistent with the recent observations in which other probes were utilized (14-21,25,29,30).

A similar Ca^{2+}-dependent change in fluorescence intensity was also observable with ANM(SH_N)-labeled SR. As shown in Figure 7, however, the magnitude of the change was smaller than SH_D-labeled SR, being about half as much. Probably the local environment of the label and its change on Ca^{2+} binding were different and these differences were recognized by the fluorophore introduced to each site. If this is the case, the present result implies that changes occurring simultaneously at distinct sites on ATPase can be monitored separately in high resolution by means of site-specific labeling. Studies in progress in our laboratory indicate that both SH_N and SH_D are located on the 30K dalton tryptic fragment as obtained by the procedure of Stewart and MacLennan (59) (K. Saito, unpublished observation). The qualitatively similar responses to Ca^{2+} binding observed with both SH_D- and SH_N-labeled SR preparations may be relevant to the proximity of these groups in the ATPase molecule.

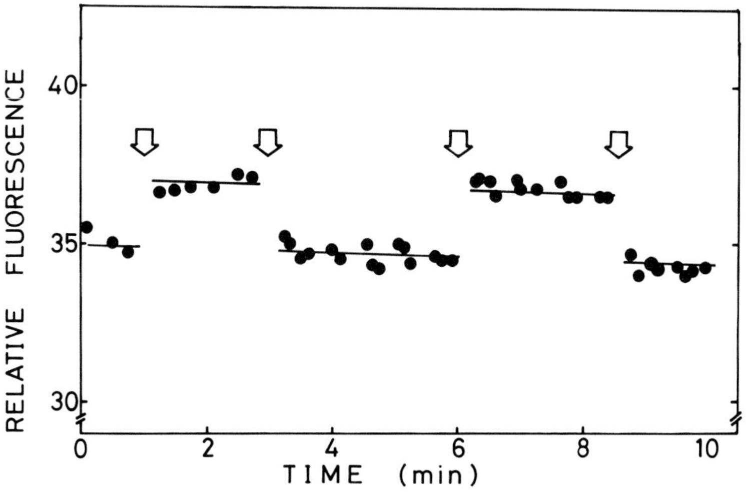

FIGURE 7. Ca^{2+}-dependent change in the fluorescence of ANM(SH_N)-labeled SR. The labeled SR was prepared as outlined in the legend to Figure 1, and the fluorescence intensity at 430 nm was measured as in Figure 5. The reversible change of the fluorescence intensity in response to Ca^{2+} concentration was demonstrated by a sequential addition of 0.5 mM EGTA, 0.5 mM $CaCl_2$, 2.9 mM EGTA, and 2.9 mM $CaCl_2$, one by one at each arrow, as indicated(33).

It should be noted here that the Ca^{2+}-dependent conformational change was detected only when ANM was used as the probe. Neither DACM nor BIPM responded to the removal of Ca^{2+} with a detectable fluorescence change. It is in sharp contrast to the nucleotide-dependent change as described below, where essentially similar results were obtained with any one of the three probes.

2. Conformational Change Induced by Nucleotide Binding. The fluorescence spectrum of DACM(SH_D)-labeled SR had a maximum at around 470 nm when excited at 390 nm. ATP, ADP and their analogs did not shift the maximum but affected the fluorescence intensity. Thus addition of AMP-P(NH)P led to a decrease in the fluorescence intensity by 2.5 % (Figure 8A). Simple dilution by adding an equivalent volume of H_2O affected the fluorescence only negligibly. Moreover, dependence on AMP-P(NH)P concentration (Figure 8B) clearly indicates that the change reflects a specific binding of the nucleotide. Approximately 10 μM AMP-P(NH)P was found to be sufficient to induce the maximal change. This is

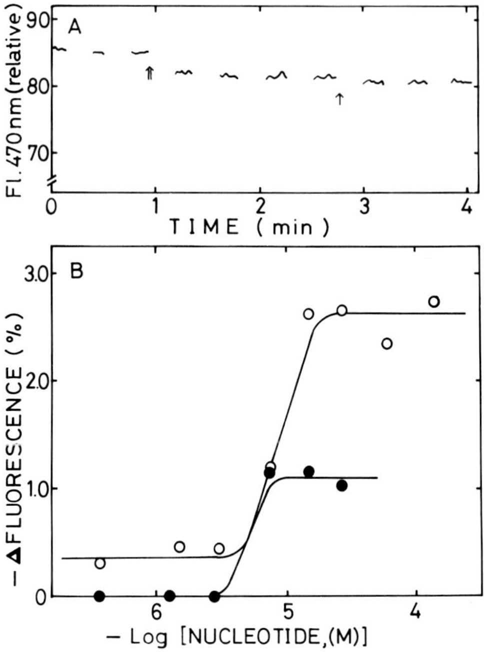

FIGURE 8. Effect of AMP-P(NH)P and ADP binding on the fluorescence emission of DACM(SH_D)-labeled SR. A. Fluorescence intensity of the labeled SR (41 μg/ml) at 470 nm was measured as in Figure 5 in the presence of 0.6 mM $CaCl_2$ and 0.6 mM EGTA. To 1 ml of the sample, 5 μl of AMP-P(NH)P solution (to a final concentration of 0.26 mM) (⇓) or 5 μl of H_2O (↓) was added as indicated. The sample was irradiated by an excitation beam at intervals to minimize possible bleaching. B. Fluorescence quenching caused by various concentrations of AMP-P(NH)P (○) or ADP (●) was measured at 470 nm as in A. The magnitude of the quenching in terms of percentage of initial intensity is shown as a function of nucleotide concentration (35).

roughly equivalent with the concentration necessary for the saturation of the substrate binding site by AMP-P(NH)P (M. Kawakita, unpublished observation). It is likely that the fluorescence change is a consequence of the conformational change induced by AMP-P(NH)P binding to the catalytic site.

Figure 8B also shows that ADP caused a similar but less pronounced change. ATPγS and ADPβS also reduced the fluorescence intensity of DACM(SH_D)-labeled SR. It is interesting to note that in this case again ATPγS had a larger, while ADPβS had a smaller effect (K. Yasuoka-Yabe, unpublished

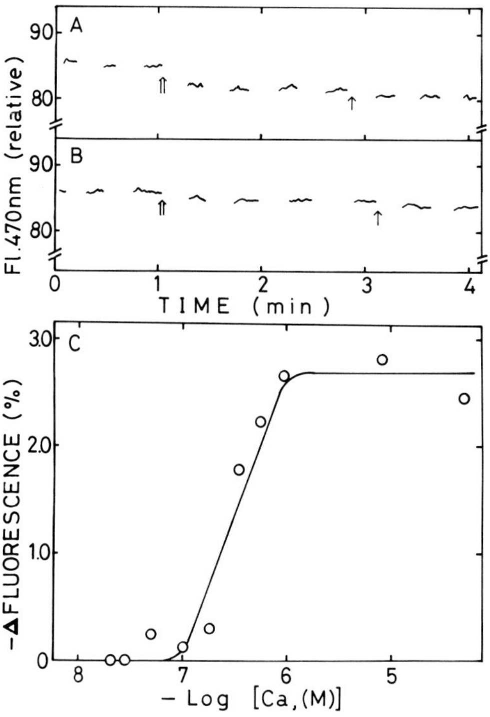

FIGURE 9. Ca^{2+} requirement for fluorescence quenching of DACM(SH_D)-labeled SR induced by AMP-P(NH)P. Experiments as in Figure 8A at 20 μM Ca^{2+} (A) and in the presence of 0.6 mM EGTA without $CaCl_2$ (B) were carried out in parallel to show the Ca^{2+}-dependence of the effect of AMP-P(NH)P. In C. similar experiments were performed at various concentrations of Ca^{2+} in the presence of 0.18 mM AMP-P(NH)P and the change in fluorescence was shown as a function of Ca^{2+} concentration as in Figure 8 (35).

observation). These differences might be related to a feature which distinguishes between ATP-bound (or energized) and ADP-bound (or non-energized) states of the transport system.

It is important that in the absence of Ca^{2+} ion, AMP-P(NH)P did not affect the fluorescence of bound DACM as can be seen in Figure 9B. Although Ca^{2+} alone did not affect the DACM fluorescence, as mentioned in the preceding section, its presence is absolutely required for the nucleotide-induced change. In fact it is better described as a nucleotide- and Ca^{2+}-dependent fluorescence quenching. When examined at a fixed AMP-P(NH)P concentration, the magnitude of the quenching was determined by Ca^{2+} concentration as shown in Figure 9C. It is evident that 10^{-6} M or more Ca^{2+} was required and this indicates that the occupancy of the transport site by Ca^{2+} is the prerequisite to the AMP-P(NH)P-induced quenching of DACM fluorescence. Essentially similar changes were also induced in either BIPM- or ANM-labeled SR. Moreover, the fluorescence was affected similarly when the label was attached to either SH_N or SH_D. These results suggest that there may be some qualitative difference between the changes caused by Ca^{2+} alone and by Ca^{2+} and AMP-P(NH)P. The physical basis of the difference, however, remains open for future studies.

Up to this point using mainly AMP-P(NH)P, which mimics ATP but virtually unhydrolyzable, we could detect differences between E and E^{Ca}, E^{Ca} and $E^{Ca}_{AMP-P(NH)P}$ (which might be assumed to be analogous to E^{Ca}_{ATP}). Furthermore the difference between $E_{AMP-P(NH)P}$ and $E^{Ca}_{AMP-P(NH)P}$ was also demonstrated. This is in line with the well established fact that the activation of the catalytic site of ATPase never occurs in the absence of Ca^{2+}, even though ATP is able of interacting with the enzyme (53).

3. Fluorescence Change Accompanying the E_1^{Ca}-P Formation. When AMP-P(NH)P was replaced by ATP, the true substrate, ATP in E^{Ca}_{ATP} complex reacted rapidly with the ATPase to yield E_1^{Ca}-P, the major intermediate accumulated in the steady state. The transition from E^{Ca} to E_1^{Ca}-P was also accompanied by a fluorescence quenching indistinguishable from the one accompanying the E^{Ca} to $E^{Ca}_{AMP-P(NH)P}$ transition (Figure 10). BIPM(SH_N)-labeled SR was employed in this particular experiment, although other labels could work as well. It gave a fluorescence spectrum with a peak at around 360 nm when excited at 310 nm. When ATP was added to this preparation it was rapidly and exhaustively hydrolyzed, since the ATPase was still fully competent even after being labeled at SH_N. As shown in the figure, addition of a limited amount of ATP in the presence of 27 μM Ca^{2+} caused a rapid quenching by about 2.5 % followed by a slow recovery to the original level. The change could be reproduced on repeated addition of ATP. The base line drift is due to photobleaching of BIPM. The amount of E-P intermediate as determined in a parallel experiment followed exactly the same time course as the fluorescence change, clearly indicating the quenching to be due to E-P formation. In addition, in the absence of Ca^{2+} neither E-P formation nor fluorescence quenching could be observed.

Almost identical fluorescence quenching was observed when ATP was added to SH_D-labeled SR. In this case, however,

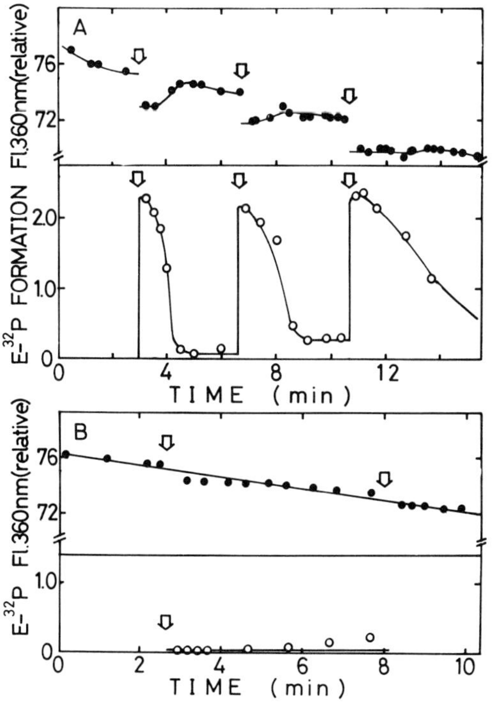

FIGURE 10. Fluorescence quenching of BIPM(SH_N)-labeled SR accompanying the conversion of E^{Ca} to E_1^{Ca}-P. A. Fluorescence emission at 360 nm of the labeled SR (32 μg/ml) excited at 310 nm was measured in the presence of 1 mM each of $CaCl_2$ and EGTA (upper panel). At times indicated by arrows, 4 μM, 4 μM, and 14 μM ATP was added successively to the reaction mixture. In a parallel sample (lower panel) E-^{32}P formation was measured by using [γ-^{32}P]ATP (35). B. Similar experiment as in A except that Ca^{2+} was omitted. Each arrow indicates the addition of 4 μM ATP.

the fluorescence did not recover to the original level quickly, since the labeled SR had lost ATPase activity and E_1^{Ca}-P could be maintained for a long time. As mentioned in a preceding section accumulated E_1^{Ca}-P complex was further transformed into E_1^{Ca}-P·ADPβS by adding excess ADPβS. However, addition of ADPβS did not affect the fluorescence of E_1^{Ca}-P significantly. It seems that the three intermediates, E^{Ca}_{ATP} ($E^{Ca}_{AMP-P(NH)P}$), E_1^{Ca}-P·ADP (E_1^{Ca}-P·ADPβS) and E_1^{Ca}-P shared some common characteristics and the differences between them have so far not been recognized by our fluorescence labeling experiments.

Recently Miki *et al.*(32) reported that they could distinguish between $E^{Ca}_{AMP-P(NH)P}$ and E^{Ca}_{1}-P based on the fluorescence intensity of ANM-labeled SR. The apparent discrepancy with our present results could be explained by the fact that their ANM label was located on the 20K dalton tryptic fragment (32) whereas ours was exclusively on the 30K dalton tryptic fragment. The different behavior of the same label depending on its attachment site is quite interesting and may be very important for full understanding of the transport process.

B. Spin Labeling Studies

1. Conventional ESR Spectra of SR Labeled with MSL. Conformational characteristics of various intermediate states of the ATPase was studied further by means of spin-labeling. The SR was labeled with MSL at SH_N, SH_D and SH_F, respectively, and their conventional ESR spectra were measured at 24 °C. The probe was found to be strongly immobilized irrespective of the attachment site (not shown) and neither Ca^{2+} nor ATP caused any significant spectral change (35). Previous reports from other laboratories using MSL as a probe claimed that the ESR signal was composed of strongly and weakly immobilized components and that the ratio of two components varied on addition of ATP (23,24). Site-specific labeling revealed that MSL(SH_D)-labeled SR gave only a strongly immobilized signal. Although the ESR spectrum of MSL(SH_N)-labeled SR was in fact composed of two components, the weakly immobilized component seems to be ascribable to the probe bound to protein(s) other than ATPase, since MSL(SH_N)-labeled purified ATPase vesicles showed only the strongly immobilized component (see Figure 11 and also ref. 54). It may be that we have to be rather careful in interpreting the ratio of weakly and strongly immobilized signals.

Several laboratories preferred iodoacetamide spin labels rather than MSL and evidence for Ca^{2+}-dependent or Ca^{2+} and nucleotide-dependent change in the ESR spectra has been accumulating in recent years (22,25,26,28). Although the site specificity of the derivatization is not very well documented in the studies cited here, the attachment sites of iodoacetamide derivatives may be different from those of the maleimide derivatives. Our preliminary results have in fact suggested that the SH group with the highest reactivity toward iodoacetamide is distinct from SH_N (T. Nakamura and M. Kawakita, unpublished observation).

2. Saturation Transfer ESR Measurements. Strongly immobilized spin labels as obtained in the preceding section seemed to be suitable for saturation transfer ESR studies (55). Whereas conventional ESR spectra of a nitroxide spin-label are sensitive to rather rapid motion with a rotational correlation time of the order of 10^{-11} to 10^{-7} sec, saturation transfer ESR spectra are sensitive to much slower motion so that the limit of the spin-label method can be extended up to the order of 10^{-3} sec. Since the rotational correlation time of the whole protein molecule or its sub-molecular domains usually falls into the range covered by this method, it has been applied to a number of systems including the ATPase of SR membranes (55-57).

In the present study purified ATPase vesicles were used mainly to eliminate complications arising from the contamination showing weakly immobilized components. Typical spectra, both conventional and saturation transfer (measured at 4 °C), of SH_N- and SH_D-labeled ATPase vesicles in their

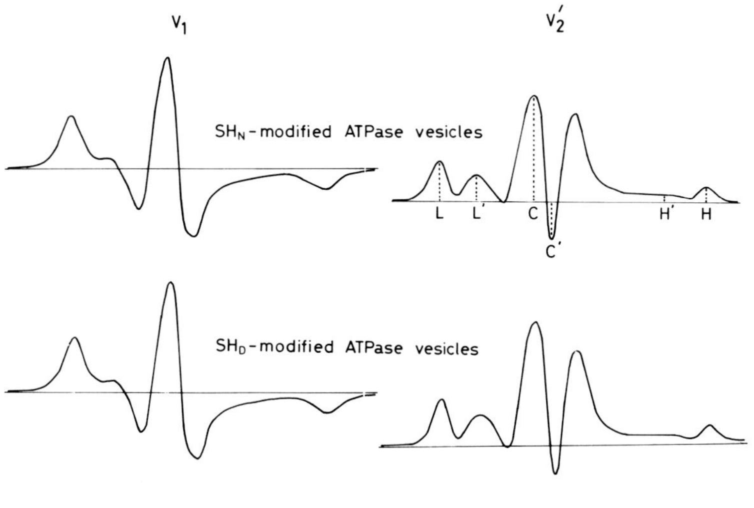

FIGURE 11. Conventional (V_1) and saturation transfer (V_2') ESR spectra of MSL-labeled ATPase vesicles. Purified ATPase vesicles were prepared by solubilization of SR with Triton X-100 followed by the removal of Triton X-100 by passing the sample through a Bio-Beads SM-2 (Bio-Rad) column. ESR spectra of MSL(SH_N)-labeled ATPase (66 mg/ml, upper traces) and MSL(SH_D)-labeled ATPase (77mg/ml, lower traces), both in 40 mM Tris-maleate buffer (pH 6.5), 0.1 M KCl, 5 mM $MgCl_2$ and 0.3 M sucrose, were recorded at 4 °C. The length of each baseline corresponds to 100 gauss (35).

E^{Ca} form are presented in Figure 11. The low field peak height ratio L'/L is somehow related to the conformation or molecular motion of ATPase.

Table II summarizes the results of the measurements carried out under appropriate conditions where individual intermediates of ATPase were allowed to accumulate. MSL(SH_D)-labeled ATPase was used in these experiments so that the enzyme could be stably maintained in the E_1^{Ca}-P state when ATP was added. An L'/L value of 0.61 was obtained in Experiment 1 for the E^{Ca} state. Removal of Ca^{2+} to give rise to E, or addition of ATP to yield E_1^{Ca}-P did not affect the L'/L value. The amount of ATP added was sufficient to maintain the ATPase in E_1^{Ca}-P state throughout the time during the ESR measurement. It is interesting that the addition of ADPβS to the accumulated E-P intermediate, which led to the formation of the E_1^{Ca}-P·ADPβS complex, caused a significant increase in the peak height ratio. Addition of ADPβS alone did not result in a high L'/L value. Moreover ATP and ADPβS added together in the presence of excess EGTA was totally ineffective. This clearly indicates that the high L'/L ratio is characteristic of E_1^{Ca}-P·ADPβS complex. Binding of AMP-P(NH)P also resulted in a high peak height ratio state

TABLE II. Effect of various nucleotides on MSL(SH_D)-labeled ATPase vesicles. The labeled ATPase vesicles (71 mg/ml in Experiment 1 and 39 mg/ml in Experiment 2) were prepared and their saturation transfer ESR spectra were measured as in Figure 11. Values for τ_2 were estimated from corresponding L'/L values according to Kusumi et al.(58). Where ATP alone was added, its concentration was 18 mM. Concentration of other nucleotides as well as ATP when added together with ADPβS, was 12 mM, and that of EGTA was 5 mM (35).

Additions	L'/L	τ_2
Experiment 1		($\times 10^{-6}$ sec)
None	0.61 ± 0.02	30
ATP	0.59 ± 0.01	28
ADPβS + ATP	0.70 ± 0.02	48
ADPβS	0.59 ± 0.03	28
EGTA + ADPβS + ATP	0.59 ± 0.01	28
EGTA	0.60 ± 0.01	29
Experiment 2		
None	0.59 ± 0.02	28
ADPβS + ATP	0.64 ± 0.03	36
AMP-P(NH)P	0.67 ± 0.02	42

as shown in experiment 2. Higher L'/L values characteristic of these intermediates as distinct from the others may further suggest a similarity in the conformations of $E^{Ca}_{AMP-P(NH)P}$ and E_1^{Ca}-P·ADPβS. Similar results as shown in Table II could also be obtained using SH_D-labeled SR, though the change was quantitatively smaller as compared to the results with the purified ATPase vesicles.

C. Effects of Substrates on Susceptibility to Trypsin

Conformational differences of ATPase can also be discriminated by its susceptibility to tryptic digestion. It has been well established that tryptic digestion of SR membranes in the presence of 1 M sucrose results in a limited hydrolysis of ATPase yielding 55K, 45K, 30K and 20K dalton fragments, of which the latter two are derived from 55K dalton component by a second cleavage (59-61). In addition, another fragment was derived from the 30K component on further incubation (K. Saito, unpublished observation), and its accumulation could be noted on a SDS-polyacrylamide gel slightly below the 30K band (see Figure 12D).

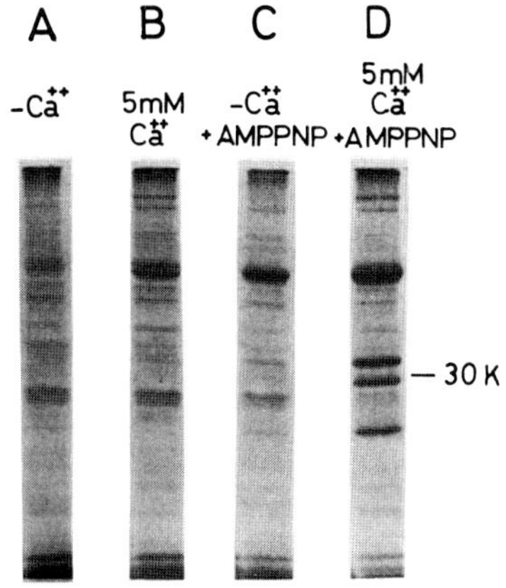

FIGURE 12. The effect of Ca^{2+} and AMP-P(NH)P on tryptic digestion of SR membranes. SR membranes (1 mg/ml) were digested with trypsin (10 μg/ml) for 15 min at 30 °C in 20 mM Tris-HCl buffer (pH 7.0), 50 mM KCl. Further additions as follows were made to each sample: A. 1 mM EGTA, B. 5 mM $CaCl_2$, C. 1 mM EGTA and 1 mM AMP-P(NH)P, or D. 5 mM $CaCl_2$ and 1 mM AMP-P(NH)P. After digestion, the samples were analyzed by SDS-polyacrylamide electrophoresis according to Laemmli (62).

In the absence of sucrose Ca^{2+}-free ATPase is highly susceptible to trypsin, so that well defined fragments did not accumulate under the conditions shown in Figure 12A. Addition of Ca^{2+} or AMP-P(NH)P alone (Figure 12B and C) retarded the hydrolysis only slightly, though a trace amount of 30K and 20K dalton fragments were detectable when AMP-P(NH)P was added. In contrast, when both Ca^{2+} and AMP-P(NH)P were present, the enzyme became markedly resistant to trypsin, and the same fragments which can be stabilized in sucrose were accumulated. This could be taken as good evidence for a conformational change of ATPase accompanying the formation of the $E^{Ca}_{AMP-P(NH)P}$ complex, which is consistent with the results of fluorescence and spin labeling studies described in the preceding sections.

Even though the effect of Ca^{2+} binding was not readily recognizable in Figure 12, another experiment carried out in the presence of 1 M glycerol demonstrated that Ca^{2+} affected the appearance of sub-30K fragments. The sub-30K region is actually composed of three peptides of distinct size, all of which have been shown to be SH_N-containing

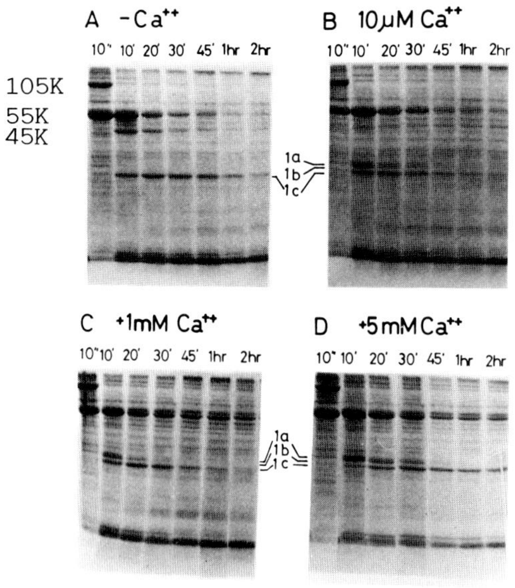

FIGURE 13. Effect of Ca^{2+} on the appearance of sub-30K fragments. SR membranes were treated with trypsin as in Figure 12 at various Ca^{2+} concentrations for various time periods. All the samples contained 1 M glycerol. The samples after digestion were analyzed as in Figure 12.

peptides (K. Saito, unpublished observation). They are designated in Figure 13 as 1a, 1b, and 1c. Appearance of fragments 1a and 1b were significantly affected by whether the transport site is occupied by Ca^{2+} or not. Thus in the presence of 10 μM Ca^{2+} these fragments are definitely more stable than in its absence (Figure 13A and B). It is interesting that higher Ca^{2+} concentrations did not further affect the results (Figure 13 C and D).

V. CONCLUSION

By using fluorescent and paramagnetic probes which are introduced to specific locations on the ATPase molecule, we have been able to gather systematic information concerning the conformational similarity and dissimilarity among various intermediates of ATPase. There was some indication that the same probe when attached to distinct sites on ATPase behaves differently depending on the local environment of its attachment site. By virtue of its high sensitivity and resolving power, this approach might enable us to analyze ATPase conformation in terms of the interaction between sub-molecular domains.

Accumulating evidence suggests that the ATPase molecule is composed of a few relatively hydrophilic stretches connected by intramembranous hydrophobic core regions, and that each of those might be regarded as a distinct domain (63-66). During the catalytic cycle each domain might behave as an individual unit working in concert with the others to carry out the overall transport reaction. If this were to

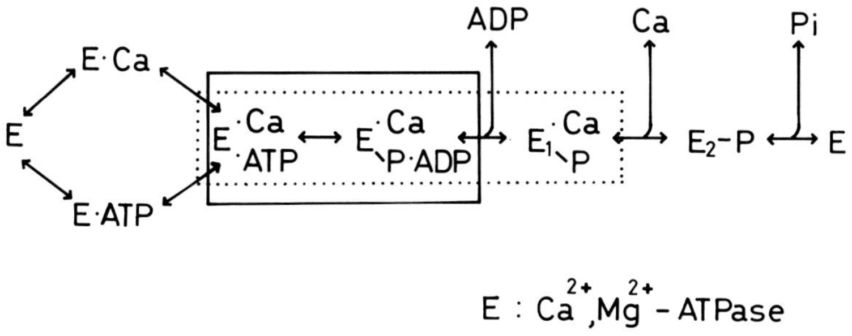

FIGURE 14. A schematic representation of the reaction intermediates of Ca^{2+},Mg^{2+}-ATPase and possible conformational relationships among them. See text for details.

be the case, dissection of the enzyme in terms of segmental motion followed by the reconstruction of the overall picture would be a promising approach to a better understanding of the mechanism of the energy coupling in this active transport system.

Conformational relationships among various intermediate complexes as suggested from the present study are summarized in Figure 14. Fluorescence labeling experiments revealed that the three intermediates enclosed by the dotted line shared common characteristics by which they could be distinguished from the others. Their apparent relatedness is interesting because Ca^{2+} is believed to move across the membrane somewhere in transit from E^{Ca}_{ATP} to E^{Ca}_{I}-P. At the same time possible difference in E^{Ca}_{I}-P·ADP and E^{Ca}_{I}-P as suggested from the saturation transfer ESR study is also interesting. Future studies on the significance of the dissociation of ADP may lead to deeper insight on the Ca^{2+} transport process.

Finally, distinct conformational characteristics of various intermediates were shown to be reflected also in their susceptibility to proteolytic cleavage. If we could identify the site which showed increased or decreased susceptibility in a particular intermediate complex, we would have useful information concerning the change in the three-dimensional structure of the enzyme. This kind of approach to protein conformational change, which has been attempted so far only poorly in the ATPase system, would be complementary to the one utilizing extrinsic probes and may also turn out to be useful in future study.

ACKNOWLEDGMENTS

We are grateful to Dr. S. Ohnishi of Kyoto University for suggesting the saturation transfer ESR study and also for kindly making ESR facilities available. Kind instructions of Drs. T. Sakaki and A. Tsuji of Kyoto University during the ESR experiments are deeply appreciated. We also thank Dr. Y. Kaziro of The University of Tokyo for valuable discussions.

REFERENCES

1. Tada, M., Yamamoto, T., and Tonomura, Y., Physiol. Rev. 58: 1 (1978).
2. Hasselbach, W., Biochim. Biophys. Acta 515: 23 (1978).
3. deMeis, L., and Vianna, A. L., Ann. Rev. Biochem. 48: 275 (1979).

4. Blasie, J. K., Herbett, L., Pierce, D., Pascolini, D., Scarpa, A., and Fleischer, S., Ann. New York Acad. Sci. in press.
5. Mommaerts, W. F. H. M., Proc. Natl. Acad. Sci., U.S.A. 58: 2476 (1967).
6. François, C., Biochim. Biophys. Acta 173: 86 (1969).
7. Anzai, K., Kirino, Y., and Shimizu, H., J. Biochem. 90: 349 (1981)
8. Murphy, A. J., J. Biol Chem. 253: 385 (1978).
9. Ikemoto, N., Morgan, J. F., and Yamada, S., J. Biol. Chem. 253: 8027 (1978)
10. Kawakita, M., Yasuoka, K., and Kaziro, Y., in "Cation Flux across Biomembranes" (Y. Mukohata and L. Packer, eds.), p. 119. Academic Press, New York, 1979.
11. Nakamaru, Y., and Schwartz, A., Eur. J. Biochem. 34: 159 (1973).
12. Nakamaru, Y., and Sato, C., J. Biochem. 91: 537 (1982).
13. Vanderkooi, J. M., and Martonosi, A., Arch. Biochem. Biophys. 144: 99 (1971).
14. Dupont, Y., Biochem. Biophys. Res. Commun. 71: 544 (1976).
15. Dupont, Y., Biochem. Biophys. Res. Commun. 82: 893 (1978).
16. Dupont, Y., and Leigh, J. B., Nature 273: 396 (1978).
17. Dupont, Y., and leMaire, M., FEBS Lett. 115: 247 (1980).
18. Dean, W. L., and Gray, R. O., J. Biol. Chem. 255: 7514 (1980).
19. Guillain, F., Gingold, M. P., Buschlen, S., and Champeil, P., J. Biol. Chem. 255: 2072 (1980).
20. Verjovski-Almeide, S., J. Biol. Chem. 256: 2662 (1981).
21. Verjovski-Almeida, S., and Silva, J. L., J. Biol. Chem. 256: 2940 (1981).
22. Landgraf, W. C., and Inesi, G., Arch. Biochem. Biophys. 130: 111 (1969).
23. Nakamura, H., Hori, H., and Mitsui, T., J. Biochem. 72: 635 (1972).
24. Pang, D. C., Briggs, F. N., and Rogowski, R. S., Arch. Biochem. Biophys. 164: 332 (1974).
25. Champeil, P., Bastide, F., Taupin, C., and Gary-Bobo, C. M., FEBS Lett. 63: 270 (1976).
26. Coan, C. R., and Inesi, G., J. Biol. Chem. 252: 3044 (1977).
27. Champeil, P., Büschlen-Boucly, S., Bastide, F., and Gary-Bobo, C., J. Biol. Chem. 253: 1179 (1978).
28. Coan, C., Verjovski-Almeida, S., and Inesi, G., J. Biol. Chem. 254: 2968 (1979).
29. Pick, U., and Karlish, S. J. D., Biochim. Biophys. Acta 626: 255 (1980).
30. Pick, U., FEBS Lett. 123: 131 (1981).
31. Andersen, J. P., Møller, J. V., and Jørgensen, P. L., J. Biol. Chem. 257: 8300 (1982).

32. Miki, K., Scott, T. L., and Ikemoto, N., J. Biol. Chem. 256: 9382 (1981).
33. Yasuoka-Yabe, K., and Kawakita, M., submitted to J. Biochem.
34. Yasuoka, K., Kawakita, M., and Kaziro, Y., J. Biochem. 91: 1629 (1982).
35. Yasuoka-Yabe, K., Tsuji, A., and Kawakita, M., submitted to J. Biochem.
36. Hasselbach, W., and Seraydarian, K., Biochem. Z. 345: 159 (1966).
37. Yoshida, H., and Tonomura, Y., J. Biochem. 79: 649 (1976).
38. Murphy, A. J., Biochemistry 15: 4492 (1976).
39. Thorley-Lawson, D. A., and Green, N. M., Biochem. J. 167: 739 (1977).
40. Yamada, S., and Ikemoto, N., J. Biol. Chem. 253: 6801 (1978).
41. Kawakita, M., Yasuoka, K., and Kaziro, Y., J. Biochem. 87: 609 (1980).
42. Kanaoka, Y., Machida, M., Ando, K., and Sekine, T., Biochim. Biophys. Acta 207: 269 (1970).
43. Kanaoka, Y., Machida, M., Machida, M. I., and Sekine, T., Biochim. Biophys. Acta 317: 563 (1973).
44. Machida, M., Ushijima, N., Takahashi, T., and Kanaoka, Y., Chem. Pharm. Bull. 25: 1289 (1977).
45. Machida, M., Machida, M. I., and Kanaoka, Y., Chem. pharm. Bull. 25: 2739 (1977).
46. Kameyama, T., Katori, T., and Sekine, T., J. Biochem. 77: 361 (1975).
47. Arai, K., Arai, T., Kawakita, M., and Kaziro, Y., J. Biochem. 77: 1095 (1975).
48. Ohyashiki, T., Sekine, T., and Kanaoka, Y., Biochim. Biophys. Acta 351: 214 (1974).
49. Ohyashiki, T., Kanaoka, Y., and Sekine, T., Biochim. Biophys. Acta 420: 27 (1976).
50. Ohyashiki, T., and Sekine, T., Biochim. Biophys. Acta 576: 51 (1979).
51. Taylor J. S., J. Biol. Chem. 256: 9793 (1981).
52. Pintado, E., Scarpa, A., and Cohn, M. J., J. Biol. Chem. 257: 11346 (1982).
53. Meissner, G., Biochim. Biophys. Acta 298: 906 (1973).
54. Hidalgo, C., and Thomas, D. D., Biochem. Biophys. Res. Commum. 78: 1175 (1977).
55. Hyde, J. S., in "Methods in Enzymol." vol.49 (C. H. W. Hirs and S. N. Timasheff, eds.), p.480 Academic Press New York, 1978.
56. Thomas, D. D., and Hidalgo, C., Proc. Natl. Acad. Sci., U.S.A. 75: 5488 (1978)

57. Kaizu, T., Kirino, Y., and Shimizu, H., J. Biochem. 88: 1837 (1980).
58. Kusumi, A., Sakaki, T., Yoshizawa, T., and Ohnishi, S., J. Biochem. 88: 1103 (1980).
59. Stewart, P. S., and MacLennan, D. H., J. Biol. Chem. 249: 985 (1974).
60. Thorley-Lawson, D. A., and Green, N. M., Eur. J. Biochem. 59: 193 (1975).
61. Scott, T. L., and Shamoo, A. E., J. Memb. Biol. 64: 137 (1982).
62. Laemmli, U. K., Nature 227: 680 (1970).
63. Allen, G., Proc. FEBS 11th. Meeting, Copenhagen, Symposium A4, vol. 45, p. 159 (1977).
64. Allen, G., Biochem. J. 187: 591 (1980).
65. Klip, A., Reithmeier, R. A. F., and MacLennan, D. H., J. Biol. Chem. 255: 6562 (1980).
66. MacLennan, D. H., Reithmeier, R. A. F., Shoshan, V., Campbell, K. P., LeBel, D., Herrman, T. R., and Shamoo, A. E., Ann. New York Acad. Sci. 358: 138 (1980).

STRUCTURAL ANALYSIS OF THE Ca^{2+} + Mg^{2+}-ATPase OF SARCOPLASMIC RETICULUM

David H. MacLennan
Banting and Best Department of Medical Research
University of Toronto, C.H. Best Institute
112 College Street, Toronto, Ontario

Reinhart A.F. Reithmeier
Department of Biochemistry
University of Alberta
Edmonton, Alberta

Studies of the structure of the ATPase of sarcoplasmic reticulum are of considerable interest since the molecule catalyzes the movement of Ca^{2+} ions across a membrane that is several nm thick. This enzymatic function must ultimately be rationalized in terms of a mechanism based on protein structure. Complete primary structural analysis (Allen et al., 1980a) and time resolved changes in low angle X-ray diffraction patterns (Blasie et al. this volume) are being pursued vigorously and these experiments will greatly advance our knowledge of the ATPase structure. This paper reviews some of the approaches that have led us to our present state of knowledge of the structure of the protein.

Deamer and Baskin (1969) observed that freeze fracture of sarcoplasmic reticulum membranes exposed 9 nm intramembrane particles associated with the cytoplasmic leaflet. They suggested that these particles might represent the ATPase. When membranes were reformed from the purified ATPase and its associated phospholipid (MacLennan et al., 1971) 9 nm particles were observed in freeze fractured vesicles proving that they were constituted of buried sectors of the ATPase molecule. The fact that these particles were asymmetrically located in the original vesicle (Deamer and Baskin, 1969) indicates that very little of the ATPase protein was exposed on the luminal surface of

ISBN 0-12-260380-X

the membrane. In fact, when the ATPase was scrambled by reconstitution, the fracture patterns were symmetrical (MacLennan et al., 1971; Packer et al., 1974).

Ikemoto et al. (1968) and Inesi and Asai (1968) discovered 4 x 6 nm particles on the surface of sarcoplasmic reticulum membranes. These particles were also observed in purified, vesicular preparations of the ATPase (Migala et al., 1973, Thorley-Lawson and Green, 1973; Stewart and MacLennan, 1974) and apparently represent cytoplasmic extensions of the ATPase. There are about three to four times as many surface particles as intramembrane globules (Jilka et al., 1975). This may reflect the location of more than one ATPase molecule in each intramembrane particle or it may reflect the fact that several, separate cytoplasmic extensions can be visualized for each ATPase molecule. Negative staining does not detect asymmetry of the membrane but positive staining with tannic acid, Saito et al (1978) shows that the surface particle are asymmetrically located on the cytoplasmic surface of the membrane.

Studies with X-ray diffraction have also shown the ATPase to be asymmetrically localized in the sarcoplasmic reticulum membrane (Worthington and Liu, 1973; Dupont et al., 1973; Herbette and Blasie, 1980). Using a combination of X-ray and neutron diffraction, Herbette and Blasie (1980) observed that the shape of the ATPase was a tapered rectangle, wider at the cytoplasmic surface. Some 38% of the mass was found inside the bilayer with essentially no extension to the luminal surface, but with 62% of the mass extending into the cytoplasm. Le Maire et al (1981) proposed a model for the shape of the ATPase based on solution X-ray and sedimentation equilibrium studies. Their model consists of two joined, coaxial cylinders. An elongated region with a height of 7 nm and a diameter of 4 nm represents the portion of the ATPase that is embedded in the membrane. The headpiece has a height of 3 nm and a diameter of 7.5 nm, and represents the portion of the ATPase that is exposed to the cytoplasm.

Morphological studies, therefore, are consistent with a model in which the ATPase is asymetrically located in the sarcoplasmic reticulum membrane with part of the molecule buried in hydrophobic regions and part extending into the cytoplasm. This model must be rationalized with the primary structure of the ATPase.

Studies of the primary structure of the ATPase began with the observation that exposure of intact sarcoplasmic reticulum to trypsin for a few seconds results in cleavage of the ATPase into two fragments, an NH_2 terminal fragment of about 55,000 and a COOH-terminal fragment of about 45,000 daltons, without any significant change in the ATPase or Ca^{2+} transport function of the molecule (Migala et al., 1973; Thorley-Lawson and Green, 1973; MacLennan and Stewart, 1974) (See Fig. I). Further incubation results in the slower cleavage of the 55,000 dalton fragment into a 25,000 dalton fragment and a 30,000 dalton fragment. The three fragments have been isolated and partially characterized. They are all strongly membrane-associated, suggesting that they contain transmembrane, hydrophobic sequences (Rizzolo and Tanford, 1978). In addition, these tryptic fragments remain tightly associated with one another (MacLennan et al., 1976). Therefore, there must be extensive regions of protein-protein interactions between the fragments.

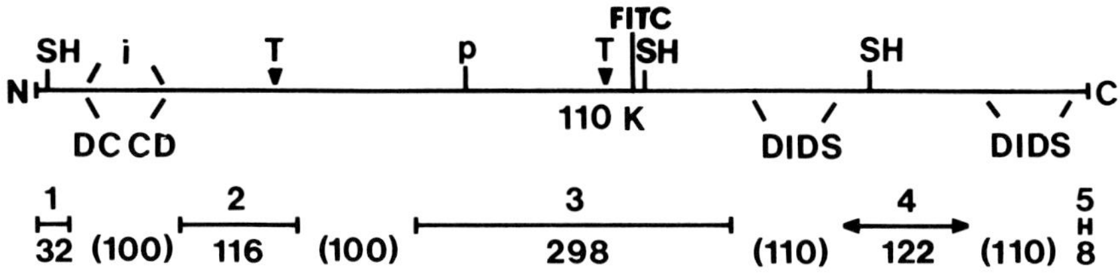

FIGURE 1. Location of various sites and known hydrophilic sequences (1-5) within the ATPase (molecular weight 110,000). The top line indicates the position of the ionophoric site (i), tryptic cleavage sites (T), phosphorylation site (p), cytoplasmic cysteine residues (SH), fluorescein isothiocyanate binding site (FITC), and probable locations of the dicyclohexylcarbodiimide (DCCD) and 4,4'-diisothiocyano-2,2'-stilbene disulfonate (DIDS) binding sites, relative to the amino (N) and carboxyl (C) terminal amino acid residues. The lower line indicates the position of five known hydrophilic sequences within the ATPase molecule. The exact position of sequence 4 is uncertain. The numbers below the lower lines indicate the number of amino acids present in each sequenced region and the number of amino acids (in parentheses) assigned to each unsequenced hydrophobic region. We have arbitrarily assigned 110 amino acids to unsequenced regions on either side of sequence 4.

Different functional sites have been associated with the various fragments. Shamoo and MacLennan (1974) found that the ATPase, when incorporated into a phospholipid bilayer, promoted the transmembrane passage of cations over anions with a selectivity of about 5-fold. The protein also showed selectivity among cations translocated in the order $Ba^{2+} > Ca^{2+} > Sr^{2+} > Mg^{2+} > Mn^{2+} > Zn^{2+}$. This activity was found to be associated with the 25,000 dalton NH_2-terminal fragment of the enzyme (Shamoo et al., 1976). This fragment also binds dicyclohexylcarbodiimide (DCCD) an inhibitor of ATP hydrolysis and of Ca^{2+} uptake (Pick and Racker, 1979). Since DCCD binding is competitive with Ca^{2+}, this site may be related to a Ca^{2+} binding or translocation site in the molecule.

An anion transport inhibitor, 4,4'-diisothiocyano-2,2'-stilbene disulfonic acid (DIDS) strongly inhibits anion transport in sarcoplasmic reticulum as well as oxalate or phosphate supported Ca^{2+} uptake (Campbell and MacLennan, 1980). When sarcoplasmic reticulum was reacted with 10 μM $[^3H]H_2$DIDS at pH 8.0 for 20 min the reagent bound selectively to the ATPase. Digestion of the sample showed that DIDS was bound to hydrophobic peptides in the 45,000 dalton fragment of the enzyme (Campbell, K.P., Reithmeier, R.A.F. and MacLennan, D.H., unpublished observations). Thus it appears that elements of the anion channel in the ATPase are located in transmembrane segments of the 45,000 dalton fragment. Interestingly, DIDS inhibits anion transport in human red blood cells by binding to a hydrophobic peptide region located in the carboxyl-terminal half of Band 3 (Knauf, 1979).

Fluorescein isothiocyanate also binds to the 45,000 dalton fragment of the ATPase, inhibiting the enzyme (Pick, 1981; Pick and Bassilian, 1981). Mitchinson et al., 1982 have isolated the FITC-binding peptide and shown it to be located in a hydrophobic region near the NH_2-terminal end of the 45,000 dalton fragment. FITC appears to compete with ATP for a binding site and the sequence surrounding the FITC binding site suggests a nucleotide binding sequence. The site of phosphorylation of the enzyme is located in the middle of the 30,000 dalton fragment showing that the phosphoryl group of ATP binds in this region (Thorley-Lawson and Green, 1973) and that the ATP binding site is composed of different peptide sequences located in widely separated regions of the chain that must fold in close proximity.

The NH_2-terminal methionine of the ATPase is acetylated (Tong 1977). We have studied the NH_2 and COOH terminal sequences of the tryptic fragments and found that the 25,000 dalton fragment is acetylated and that the fragments are created by cleavage of single Arg-Ala bonds with no deletion of peptides from the overall sequence (Klip et al., 1980; Tong, 1980).

Allen and Green have carried out extensive sequence analysis of the protein (Allen, 1978, 1980a,b; Allen et al., 1980a,b; Green et al., 1980). They found that cleavage of the ATPase with proteolytic enzymes or with cyanogen bromide released soluble and insoluble peptides. The soluble peptides were sequenced and aligned in five stretches, a 32 residue NH_2-terminal sequence, an 8 residue COOH terminal sequence and 3 internal sequences of 116, 298 and 122 residues. The 116 residue sequence could be localized from the sequence around the 25,000-30,000 cleavage point and the 298 residue fragment could be localized by its correspondence with sequences around the 30,000-45,000 cleavage point (Klip et al., 1980). The 122 residue sequence could only be localized in the 45,000 fragment because of size constraints but its location in this fragment is not precisely known (MacLennan et al., 1980; Green et al., 1980).

When 4 of the 5 sequences were positioned it was apparent that about 100 unknown residues separated sequences 1 and 2 and sequences 2 and 3. Moreover, about 220 residues were also missing in the 45,000 dalton fragment and these could represent two unknown regions of about 110 amino acids each. The five known sequences contained about 52% non polar amino acids in agreement with their water solubility. The four unknown sequences were composed of about 67% non polar amino acids (Green et al., 1980). This suggested that the sequenced regions of the molecule consisted of hydrophilic, cytoplasmic extensions of the ATPase whereas the unsequenced material consisted of hydrophobic, membrane associated regions of the protein. When the ATPase was extensively digested while still in membranous form, surface particles were digested away but intramembrane particles remained behind (Stewart and MacLennan, 1974). Peptides released from membranous ATPase after extensive proteolytic digestion were isolated and all were found to be located in the five hydrophilic sequences (Green et al., 1980). The five sequenced regions therefore represent portions of the ATPase that are exposed to the cytoplasm.

The view that the five hydrophilic regions are exposed on the cytoplasmic surface of the membrane suggests experiments for determining the localization of these sites to the cytoplasmic surface. We have been interested in the biosynthesis of the sarcoplasmic reticulum membrane and we have investigated the way in which the ATPase is formed. Our initial studies (Greenway and MacLennan, 1978) showed that the ATPase is formed on membrane bound polyribosomes. We then became interested in whether the protein was made with or without an NH_2-terminal signal sequence. When the protein was translated *in vitro* from polyadenylated mRNA it had the same molecular weight and methionine-labelled peptide map as the mature ATPase (Reithmeier _et al_., 1980). The methionine residue at the NH_2-terminus of the newly synthesized enzyme was donated by initiator methionyl-tRNA $_f^{Met}$ and it became acetylated during translation. These results suggested that the ATPase was made without an NH_2-terminal signal sequence. Further confirmation of the lack of a signal sequence in the ATPase was provided by Mostov _et al_., (1981) when they showed that Cys was incorporated into position 12 and that Ala was incorporated into positions 3,4 and 14 of the newly synthesized enzyme, the same positions as in the mature enzyme.

If the ATPase were made without a signal sequence, then the NH_2-terminus would not be expected to penetrate the membrane but to remain on the cytoplasmic surface. We have investigated the location of the NH_2-terminal sequence by the use of non-penetrating sulfhydryl reagents (Reithmeier and MacLennan, 1981). We found that Cys 12 was one of a total of 10-11 Cys groups that could be labelled with N[^{3}H]-ethylmaleimide (NEM) under saturating conditions. This was shown by the isolation of a 31 residue, NH_2-terminal peptide from a tryptic digest of the succinylated ATPase prepared from N[^{3}H]ethylmaleimide labelled vesicles. Reaction of the vesicles with glutathione maleimide, parachloromercuribenzoic acid or parachloromercuriphenylsulfonic acid, membrane impermeant reagents, prevented further reaction of sulfhydryl groups with NEM. This result indicated that all Cys groups that react with NEM are cytoplasmically located. These include the NH_2-terminal portion and residues located in peptides originating in sequences 3 and 4.

These observations, together with the observation that the ATPase is synthesized without a signal sequence, strongly suggest that the NH_2-terminus of the ATPase (sequence 1) is cytoplasmically located. Sequence 2 is clearly cytoplasmic since it contains a very sensitive trypsin cleavage site. Sequence 3 is also cytoplasmic since it contains another exposed trypsin site, a site of binding for the phosphoryl group of ATP, and an NEM reactive Cys group. Sequence 4 is clearly cytoplasmic since it contains a site of FITC binding and an apparent nucleotide binding site as well as an NEM reactive Cys group. The COOH terminal sequence (sequence 5) is almost certainly cytoplasmic since it is hydrophilic and, since it would be the last sequence to leave the ribosome during synthesis, it would be expected not to penetrate the membrane but to remain cytoplasmically located.

These observations suggest a probable folding pattern for the ATPase. The NH_2-terminal 32 amino acids are in the cytoplasm. The last few of these amino acids are basic (Allen, 1978) and may provide a site of interaction with negatively charged phospholipid head groups at the membrane surface. The peptide chain would then enter the hydrophobic region of the membrane and pass through as a regular structure such as an α-helix. This view is based in the arguments of Kennedy (1978) that the inability of the peptide chain to react with H_2O in a hydrophobic region would force intrachain hydrogen bonding which would be optimal in regular structures such as an α-helix or a β-helix. Dunker _et al_., (1980) have, in fact, shown that membrane associated regions of the sarcoplasmic reticulum are rich in α-helices. Kennedy (1978), has also pointed out that it is unlikely that a peptide chain could turn within a hydrophobic environment. Therefore, we assume that the chain passes directly through the phospholipid bilayer and turns within the aqueous or polar head group region in the lumen of the membrane.

After turning in this luminal space, the peptide chain reenters the hydrophobic sector, passing through in a regular structure and extending into the cytoplasmic region of sequence 2. This pattern would be repeated between sequences 2 and 3, 3 and 4 and 4 and 5 so that the chain probably stitches back and forth eight times in four transmembrane loops. These four loops would comprise some 400 amino acids. Since about 23 amino acids in an α-helix would be required to pass through the thickness of the membrane, each loop must contain _at least_ 50 amino acids (allowing extra residues for luminal turns). Although, it is possible that 4 transmembrane passages could occur in each loop, there is probably only enough protein for two passages, for a total of 8 transmembrane segments.

There is no evidence from either structural or labelling studies that the ATPase chain does, in fact, enter into the luminal surface of the sarcoplasmic reticulum and, therefore, at best, only a small portion of the ATPase could be exposed on the luminal side of the membrane. However, Ca^{2+} and anions do cross the membrane (Hasselbach, 1964) and, therefore, we can deduce that channels must be created across the thickness of the membrane.

It is not clear whether all channels are present in a single ATPase molecule or whether these are formed by interacting chains in an oligomer. The monomeric ATPase is capable of ATP hydrolysis with its attendent conformational changes leading to Ca^{2+} translocation (Dean and Tanford, 1978, Moller _et al_., 1980, Tanford this volume). The monomeric ATPase undergoes the same Ca^{2+} induced fluorescence changes as the membrane-associated form of the enzyme (Dean and Gray 1980). Fleischer and colleagues (this volume) have shown, however, that the Ca^{2+} transporting unit is a dimer. Therefore, it is possible that the transmembrane passages of two ATPase molecules cooperate to produce the channels for ion movement.

Recombinant DNA Studies

The fact that the primary sequence of the hydrophobic regions of the ATPase is so difficult to obtain has led us to the view that the primary sequence will be more easily obtained from DNA sequences coding for the ATPase. Accordingly, we have begun a program to clone cDNA coding for the ATPase. Our first experiments were to investigate whether neonatal rabbit mRNA could be used to synthesize the ATPase in a cell-free translation system. This proved to be an excellent source and so we have cloned cDNA from total and size selected rabbit mRNA into the PstI site of the plasmid PBR322.

Our next task was to identify the clone containing the ATPase sequence. We have used several approaches. Our most promising approach has been to use synthetic oligomeric probes (cf. Mevarech *et al*., 1979) corresponding to DNA sequences in the ATPase, deduced from known amino acid sequences, (studies in collaboration with N.M. Green). Tetradodecamer probes (n=14) were heavily labelled with ^{32}P in a kinase reaction. These probes were then hybridized directly to colonies at room temperature in high salt and subsequently washed at 30°C in high salt. A single colony has been detected by this method in some 5,000 cDNA clones (C. Brandl and D.H. MacLennan, unpublished observations). This colony contains a plasmid with a 120 bp insert that, upon sequence analysis (Maxam and Gilbert, 1977), corresponds to a part of the sequence of the ATPase. We anticipate that developments in sequence analysis of DNA coding for the ATPase will now be very rapid and that the entire primary structure of the ATPase will soon be known.

ACKNOWLEDGMENTS

Research originating in the Authors' laboratory was supported by grants from the Medical Research Council of Canada, the Muscular Dystrophy Association of Canada and the Ontario Heart Foundation.

REFERENCES

Allen, G. (1978). FEBS Symp., 45, 159-168.

Allen, G. (1980a). Biochem. J. 187, 545-563.

Allen, G. (1980b). Biochem. J. 187, 565-575.

Allen, G., Bottomley, R.C., and Trinnaman, B.J. (1980a). Biochem. J. 187, 577-589.

Allen, G., Trinnaman, B.J., and Green, N.M. (1980b). Biochem. J. 187, 591-616.

Campbell, K.P. and MacLennan, D.H. (1980) Ann. N.Y. Acad. Sci. 358.

Deamer, D.W., and Baskin, R.J. (1969). J. Cell Biol. 42, 296-307.

Dean, W.L., and Gray, R.D. (1980). J. Biol. Chem. 255, 7514-7516.

Dean, W.L., and Tanford, C. (1978). Biochemistry 17, 1683-1690.

Dunker, A.K., Toogood, K.C., Stuart, G.W., and Williams, R.W. (1980). Federation Proc. 39, 1663.

Dupont, Y., Harrison, S.C., and Hasselbach, W. (1973). Nature (London) 244, 555-558.

Green, N.M., Allen, G., and Hebdon, G.M. (1980). Ann. N.Y. Acad. Sci. 358, 149-158.

Greenway, D.C., and MacLennan, D.H. (1978). Can. J. Biochem. 56, 452-456.

Hasselbach, W. (1964). Prog. Biophys. Mol. Biol. 14, 167-222.

Herbette, L., and Blasie, J.K. (1980). In Calcium Binding Proteins: Structure and Function (F.L. Siegel, E. Carafoli, R.H. Kretsinger, D.H. MacLennan, and R.H. Wasserman, eds.), pp. 115-120, Elsevier/North-Holland, Amsterdam.

Ikemoto, N., Streter, F.A., Nakamura, A., and Gergely, J. (1968). J. Ultrastruct. Res. 23, 216-232.

Inesi, G., and Asai, H. (1968). Arch. Biochem. Biophys. 126, 469-477.

Jilka, R.L., Martonosi, A.N., and Tillack, T.W. (1975). J. Biol. Chem. 250, 7511-7524.

Kennedy, S.J. (1978). J. Membr. Biol. 42, 265-279.

Klip, A., Reithmeier, R.A.F., and MacLennan, D.H. (1980). J. Biol. Chem. 255, 6562-6568.

Knauf, P.A. (1979). Current Topics in Membranes and Transport. 12, 249-363.

Le Maire, M., Moller, J.V. and Tardieu, A. (1981). J. Mol. Biol. 150, 273-296.

MacLennan, D.H., Seeman, P., Iles, G.H. and Yip, C.C. (1971). J. Biol. Chem. 246, 2702-2710.

MacLennan, D.H., Khanna, V.K. and Stewart, P.S. (1976). J. Biol. Chem. 251, 7271-7274.
MacLennan, D.H., Reithmeier, R.A.F., Shoshan, V., Campbell, K.P., and LeBel, D. (1980). Ann. N.Y. Acad. Sci. 358, 138-148.
Maxam, A.M., and Gilbert, W. (1977). Proc. Natl. Acad. Sci. USA. 74, 560-564.
Mevarech, M., Noyes, B.E. and Agarwal, K.L. (1979). J. Biol. Chem. 254, 7472-7475.
Migala, A., Agostini, B., and Hasselbach, W. (1973). Naturforsch. 28, 178-182.
Mitchinson, C., Wilderspin, A.F., Trinnaman, B.J. and Green, N.M. (1982). FEBS Lett. 146, 87-92.
Moller, J.V., Lind, K.E., and Andersen, J.P. (1980). J. Biol. Chem. 255, 1912-1920.
Mostov, K.E., De Foor, P., Fleischer, S. and Blobel, G. (1981). Nature 292, 87-89.
Packer, L., Mehard, C.W., Meissner, G., Zahler, W.L., and Fleischer, S. (1974). Biochim. Biophys. Acta. 363, 159-181.
Pick, U. (1981). Eur. J. Biochem. 121, 187-195.
Pick, U., and Racker, E. (1979). Biochemistry 18, 109-113.
Pick, U. and Bassilian, S. (1981) FEBS. Lett. 123, 127-130.
Reithmeier, R.A.F., and MacLennan, D.H. (1981). J. Biol. Chem., 256, 5957-5960.
Reithmeier, R.A.F., de Leon, S., and MacLennan, D.H. (1980). J. Biol. Chem. 255, 11839-11846.
Rizzolo, L.J., and Tanford, C. (1978). Biochemistry 17, 4044-4048.
Saito, A., Wang, C.T., and Fleischer, S. (1978). J. Cell Biol. 79, 601-616.
Shamoo, A.E., Ryan, T.E., Stewart, P.S., and MacLennan, D.H. (1976). J. Biol. Chem. 251, 4147-4154.
Shamoo, A.E. and MacLennan, D.H. (1974). Proc. Nat. Acad. Sci. U.S.A. 71, 3522-3526.
Stewart, P.S., and MacLennan, D.H. (1974). J. Biol. Chem. 249, 985-993.
Thorley-Lawson, D.A., and Green, N.M. (1973). Eur. J. Biochem. 249, 985-993.
Tong, S.W. (1977). Biochem. Biophys. Res. Commun. 74, 1242-1248.
Tong, S.W. (1980). Arch. Biochem. Biophys. 203, 780-791.
Worthington, C.R., and Liu, S.C. (1973). Arch. Biochem. Biophys. 157, 573-579.

PROTEIN-LIPID INTERACTION IN SARCOPLASMIC RETICULUM AND PHOSPHOLIPID-REPLACED MEMBRANES[1]

Yutaka Kirino
Ken-ichiro Higashi

Faculty of Pharmaceutical Sciences
University of Tokyo
Bunkyo-ku, Tokyo

I. INTRODUCTION

Elucidation of protein-lipid interactions is of great importance to understanding the organization and function of biomembranes including sarcoplasmic reticulum (SR). The membrane lipid can influence the physical state and function of intrinsic membrane proteins. On the other hand, the presence of intrinsic proteins can modify the properties of membrane lipids (See ref. 1 - 6 for recent review). One of the best systems for studying the protein-lipid interactions is provided by the membrane reconstituted from purified SR Ca-ATPase and a chemically well-defined phospholipid. The method of Ca-ATPase reconstitution using exogenous phospholipids has been well established and biochemical and enzymatic investigations have been extensively carried out (1, 7 - 12). Since the thermotropic phase transition of pure hydrated phosphatidylcholine is well characterized (13), it is a very convenient way, in studies of protein-lipid interactions, to investigate the influence of temperature variation on the physical state of Ca-ATPase protein and the lipid and on the enzyme activity in reconstituted membrane systems. Among the synthetic phosphatidylcholines, dimyristoyllecithin

[1] This study was supported in part by a Grant-in-Aid for Scientific Research (No. 57222005) from the Ministry of Education, Science and Culture of Japan.

ISBN 0-12-260380-X

(DML) and dipalmitoyllecithin (DPL) are most frequently used, for the gel-to-liquid crystalline phase transition temperature of the two is in a temperature range compatible with the Ca-ATPase activity measurement (23°C for DML and 42°C for DPL).

There seems wide agreement that the "fluid" state of membrane lipid is essential for the activity of membrane-bound enzymes. This seems true for Ca-ATPase recombinant with DML since its Ca-ATPase activity is reported to become appreciable above about 20°C (12). On the other hand, Ca-ATPase activity of membrane reconstituted with DPL is appreciable at 30°C (10,12), well below the phase transition temperature of pure DPL. This "unusual" property of DPL-ATPase complex together with the result of a fatty acid spin label study led Hesketh et al. (10) to the interpretation that annular lipids have a thermodynamic property distinct from bulk lipids and that the former determine Ca-ATPase activity. In spite of the fact that this was based on the "anomalous" property of the particular complex of Ca-ATPase with DPL, it contributed to promotion of the concept of the boundary lipid in biological membranes proposed by Jost et al. (14).

The interpretation of Hesketh and his colleagues (10) was criticized by Hoffman et al. (15). They proposed that the melting of protein-rich patches in DPL-ATPase complex at 30°C brings about an abrupt increase in the rotational mobility of the Ca-ATPase protein and in the Ca-ATPase activity at the same temperature. However, their result of rotational mobility of the Ca-ATPase in SR membrane measured by the transient dichroism technique (16) does not seem to be consistent with that obtained with the same technique by Bürkli and Cherry (17) nor with the result by saturation transfer electron spin resonance (ST-ESR) (18,19). Therefore, it is highly desirable to measure by various methods temperature-dependent changes in dynamic properties of bulk and boundary lipids and of Ca-ATPase proteins and to elucidate a structure-function relationship in the "unusual" DPL-ATPase as well as the "normal" DML-ATPase complex.

We have studied the temperature-dependence of rotational mobility of the Ca-ATPase protein molecule and alkyl chain flexibility of bulk and boundary lipids using ESR-spin labeling techniques. Also studied was the structural fluctuation of the Ca-ATPase molecule measured by hydrogen-deuterium exchange kinetics (20).

The chemical structures of the three types of spin labels used in this study are shown in Fig. 1. Locations of these labels in the membrane are also shown diagrammatically. MSL(d) is a deuterated derivative of MSL, which binds rigidly

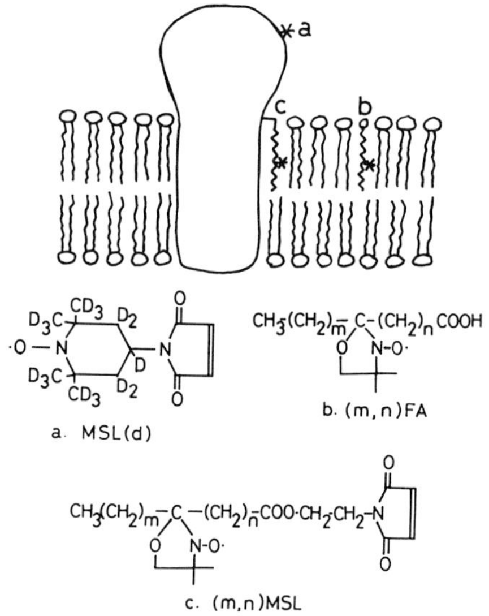

FIGURE 1. Diagrammatic representaion of the locations of three kinds of spin labels in the membrane and their chemical structures.

to a hydrophilic part of Ca-ATPase (18,19,21) and reports its rotational mobility. A fatty acid spin label, (m,n)FA, was used to probe fluidity of the bulk lipid of the membrane. A long chain derivative of MSL, (m,n)MSL, is a probe of boundary lipid (22) which is defined here as lipid proximate to intrinsic proteins. This label binds at its terminal MSL moiety to an SH group in the hydrophilic/hydrophobic interface of the Ca-ATPase protein molecule (see below).

II. MATERIALS AND METHODS

Fragmented sarcoplasmic reticulum (FSR) was prepared from rabbit white muscles of back and hind legs as described previously (18). The lipid replacement of the FSR membranes was at first performed by the procedure of Warren et al. (7) with a slight modification: the incubated mixture of SR membranes, deoxycholate, and exogenous lipid was extensively dialyzed (24 hrs at $4^{\circ}C$) before sucrose density gradient centrifugation. According to this method, it was difficult to replace with a single step more than 85 % of the endogenous lipid with exogenous lipids. The lipid to protein ratio was usually about 80 mol/mol of ATPase, lower than the 100 - 120 mol/mol for native SR. Later we used the procedure of Moore et al. (12) which involves delipidation of Ca-ATPase with deoxycholate and poly(ethylene glycol)-6000. More than 95 % of the endogenous lipid was replaced with a single step and the lipid to protein molar ratio of resultant membrane was usually 150. The work described in this report was, therefore, carried out using the membranes reconstituted by the method of Moore et al. (12) unless otherwise stated. The formation of vesicles after membrane reconstitution was confirmed by negative-stain electron micrography.

Ca-ATPase activity was determined as described previously (20). The composition of assay medium is 0.1 M KCl, 5 mM $MgCl_2$, 20 mM MOPS, 0.1 mM [γ-^{32}P-]ATP and 10 μM free Ca buffered with 1 mM EGTA-Ca.

Spin labeling of membrane with (m,n)FA was carried out by incorporating the spin label in the usual way at the ratio of one molar % of the membrane phospholipid. The procedure of spin labeling with (m,n)MSL was as described previously (23). DML-ATPase labeled with (m,n)MSL was prepared either by lipid exchange after spin labeling of SR membrane or by spin labeling after lipid exchange. No difference was observed in ESR spectra between the samples prepared by the two methods. The amount of label (m,n)MSL bound to the Ca-ATPase was a little less than one mol/mol of ATPase. Labeling of the Ca-ATPase with MSL(d) was carried out either by the method reported previously (18) or after the procedure of Kawakita et al. which was used to label a thiol group named SH_D (24). MSL(d) was synthesized according to Venkataramu et al. (25). ESR spectra were recorded by a JEOL FE-3XG spectrometer equipped with a second harmonic unit for saturation transfer spectroscopy.

Hydrogen-deuterium exchange was carried out as reported previously (20). The reaction was initiated by mixing deuterium oxide with a lyophilized membrane sample. The reaction was followed by measuring the amount of undeuterated peptide groups which is proportional to X(t) = A(amide II)/A(amide I), the ratio of the absorbance of the amide II band at 1545 cm^{-1} to that of the amide I band at 1645 cm^{-1} in infrared absorption spectra (20).

III. RESULTS

A. Temperature-Dependence of Ca-ATPase Activity

The Ca-ATPase activity of DML-ATPase and SR membranes was measured as a function of temperature (Fig. 2). Since DML-ATPase vesicles do not accumulate Ca during ATP hydrolysis, the ATPase activity of SR membranes was measured in the presence of 8 μM Ca ionophore A23187 to prevent Ca accumulation. In agreement with the results of previous investigators (12), the ATPase activity of DML-ATPase becomes appreciable when the temperature is above about 20°C. In the inset of Fig. 2 is plotted the ratio of the activity of DML-ATPase to that of SR. The activity of DML-ATPase is seen to increase more rapidly than that of SR as the temperature exceeds 20°C. The phase transition temperature of pure DML is known to be 23 $^{\circ}$C (13), which was confirmed in the present study.

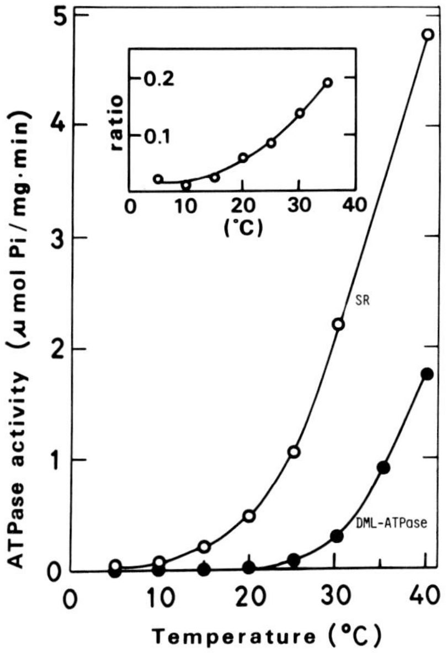

FIGURE 2. Ca-ATPase activity of SR membrane or DML-ATPase recombinant as a function of temperature. The ratio of the activity of DML-ATPase to that of SR is shown in the inset.

Also confirmed was the reports of previous investigators (10,12) that the Ca-ATPase activity of DPL-ATPase membranes is appreciable when the temperature is above 30°C, well below the transition temperature of pure DPL (42°C).

B. Spin Labeling with (m,n)MSL

Mode of the reaction of spin label (m,n)MSL with the Ca-ATPase of SR membranes was studied since this type of label has been introduced only recently (22) and its reactivity has not been fully investigated. An FSR suspension in a solution of 0.1 M KCl and 10 mM MOPS (pH 7.0) was incubated with (5,10)MSL, (10,5)MSL or (12,3)MSL at 0°C. When the suspension was centrifuged after brief incubation and ESR spectrum was measured, no radical was detected in the supernatant, all the radicals being found in the pellet. Washing the pellet with a solution containing 2 % bovine serum albumin (BSA, fatty acid free) removed most of the radicals from the membrane. Incubation for a longer period increased the amount of the radicals which could not be removed with BSA. These findings indicate that (m,n)MSL labels are rapidly incorporated in the lipid bilayer of SR membranes and that during lateral diffusion in the membrane they collide with the Ca-ATPase molecule making a covalent bond.

The pretreatment of SR membranes with a large excess of N-ethylmaleimide (NEM) prior to the labeling with (m,n)MSL had little effect on the amount of bound (m,n)MSL. This result indicates that (m,n)MSL binds to an SH group or SH groups of the Ca-ATPase which are not accessible to NEM. Labeling with (m,n)MSL increased the membrane permeability to Ca ions but had little effect on the Ca-ATPase activity.

Ascorbate is known to penetrate into lecithin or biological membranes to a certain depth depending on temperature (26). Spin-labeled SR was incubated with 5 mM sodium ascorbate at 0°C and the ESR signal intensity was monitored. No reduction of the radical took place within 2 hr for (5,10)FA- and (5,10)MSL-labeled SR while a gradual decrease of the signal was observed for (12,3)FA- and (12,3)MSL-labeled membranes. Namely, little difference was observed between the reduction rates of the corresponding (m,n)MSL and (m,n)FA labels in the membrane, indicating that the nitroxide moiety of (m,n)MSL labels bound to the Ca-ATPase and of the corresponding (m,n)FA labels are at a similar depth in the membrane bilayer. Therefore, it is most likely that the long-chain labels bound to the Ca-ATPase molecules reside in lipid layers next to the protein molecules and can be used to monitor the state of the boundary lipids.

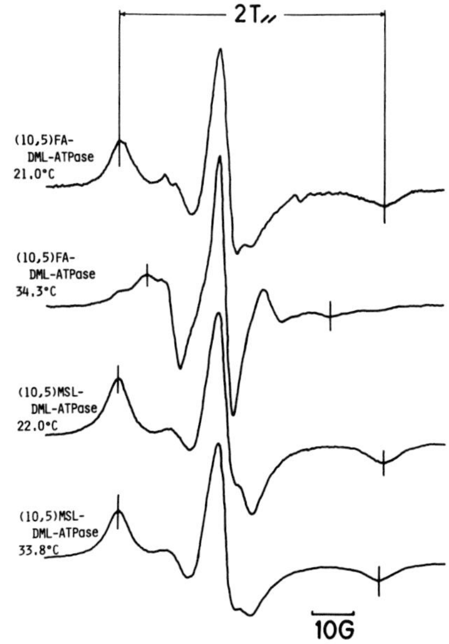

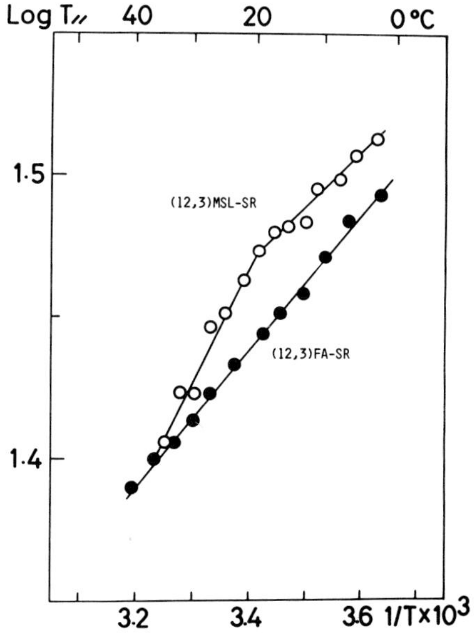

FIGURE 3 (left). ESR spectra of spin labeled DML-ATPase membranes.

FIGURE 4 (right). A logarithmic plot of $T_{//}$ of ESR spectra against the reciprocal of absolute temperature obtained for SR membranes labeled with (12,3)FA (•) or (12,3)-MSL (o) (23).

C. Conventional Electron Spin Resonance Study

ESR spectra were taken in the temperature range of 4 - 45°C for SR and reconstituted membranes labeled with (m,n)FA or (m,n)MSL. Examples of the spectra obtained for DML-ATPase labeled with (10,5)FA or (10,5)MSL at a low and a high temperature are shown in Fig. 3. Essentially the same spectra were obtained when membranes were labeled with (12,3)-FA or (12,3)MSL. The separation between the outermost peaks is known as $2T_{//}$ and is a measure of the flexibility of hydrocarbon chains of lipid molecules. A smaller value of $T_{//}$ means a greater flexibility.

In Fig. 4 is shown a temperature-dependence of the $T_{//}$ value obtained for bulk and boundary lipids of SR membranes. No break was observed in the temperature-dependence profile of bulk lipid of SR membranes while that of boundary lipid of SR membranes exhibited a break at about 20°C. This break can be related to the conformational change of the Ca-ATPase molecule that was first reported by us and later by other investigators (18,20,27,28). It has been concluded from the result shown in Fig. 4 that the change in the conformation of the ATPase protein molecule is accompanied by a change in the physical state of the boundary lipids, but not of the bulk lipids (23). It should be noted that the hydrocarbon chain flexibility in the boundary lipid of SR membranes is as high as in bulk lipid at higher temperatures.

Although (m,n)FA labels in DML-ATPase recombinant give a single component ESR spectrum at low temperatures, they exhibit a composite spectrum of two components at higher temperatures (Fig. 3). The minor component has a larger $T_{\parallel}$ value than the major component. The $T_{\parallel}$ value of the former is quite similar to that of the spectrum of (m,n)MSL-labeled membranes. This fact and the lipid/protein ratio (150 mol/mol) of this reconstituted membrane leads to the assignment that the main component arises from an (m,n)FA label in

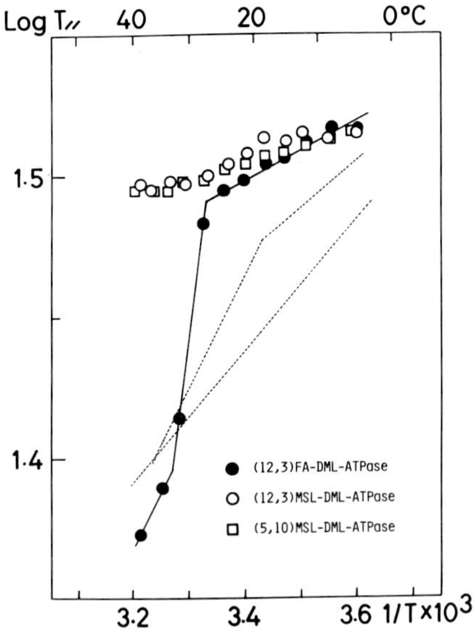

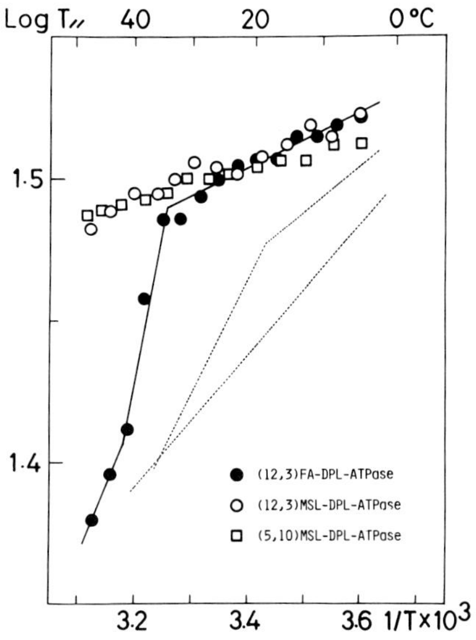

FIGURE 5 (left). The temperature-dependence of $T_{\parallel}$ value of ESR spectra of DML-ATPase labeled with (12,3)FA, (12,3)MSL or (5,10)MSL (29).

FIGURE 6 (right). The temperature-dependence of $T_{\parallel}$ value of ESR spectra of DPL-ATPase labeled with (12,3)FA, (12,3)MSL or (5,10)MSL (29).

bulk DML and the minor component from a label in boundary DML lipid. A temperature-dependence plot for the $T_{/\!/}$ value of the main component exhibits a jump at about 29°C as shown in Fig. 5 (●), which can be considered to result from the gel-to-liquid crystalline phase transition of the bulk DML. This phase transition temperature is about 6°C higher than that of pure DML.

The fluidity of bulk DML in the liquid crystalline state is about the same as or even somewhat greater than endogenous SR lipid. On the other hand, boundary lipid layer probed by (m,n)MSL-type spin labels was found to be strongly immobilized even at a higher temperature range (o, □ in Fig. 5). Since the boundary lipid probes with a nitroxide group at different positions along the hydrocarbon chain report similar immobilization, the entire hydrocarbon chain seems to be strongly immobilized in the boundary layer even when bulk lipid is in a fluid state.

Quite similar results were obtained for membranes of which lipid was replaced with DPL, as shown in Fig. 6. The bulk lipid underwent the gel-to-liquid crystalline phase transition at about 39°C, which is 3°C lower than that of pure DPL. The fluidity of liquid crystalline bulk DPL is similar to that of SR lipid. The hydrocarbon chain of the boundary lipid was strongly immobilized throughout the temperatures studied.

D. Saturation Transfer Electron Spin Resonance Study

Ca-ATPase molecules of SR membranes were labeled with MSL(d) at pH 7.0 (24) after treatment with N-ethylmaleimide. Pretreatment is necessary to block fast-reacting SH groups of Ca-ATPase or other proteins in SR membranes which otherwise result in ESR spectra due to spin labels which are mobile independent of the protein molecule. A typical example of ordinary V_1 spectra of MSL(d)-labeled SR membranes is shown in Fig. 7. As is shown from the spectrum, the deuterated spin label gives narrower linewidth, better resolution and higher sensitivity (about 60 % higher) than the usual non-deuterated MSL. The spectrum consists mainly of a strongly immobilized component although a very small fraction of a mobile component (indicated by arrows in the figure) exists.

The saturation transfer ESR spectrum (V_2') of the same sample taken at 20°C is also reproduced in Fig. 7. The high field portion of the spectrum is completely free from the influence of the mobile component because of its narrow linewidth. Therefore, the peak height ratio of H"/H was taken as the measure of rotational mobility of the Ca-ATPase

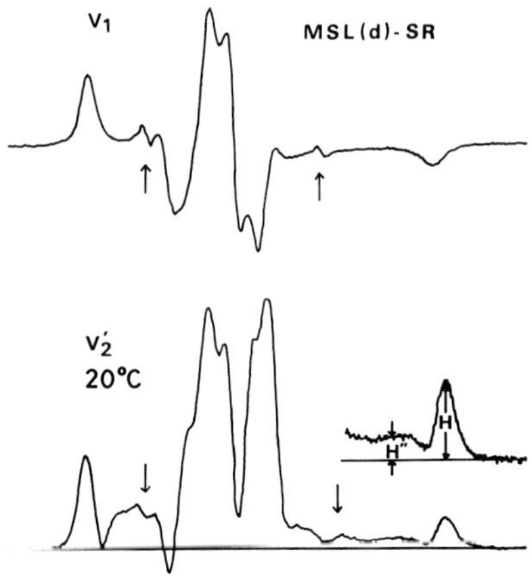

FIGURE 7. The ordinary V_1 spectrum and the ST-ESR V_2' spectrum of SR membranes labeled with MSL(d). Arrows indicate the signal due to weakly immobilized spin labels.

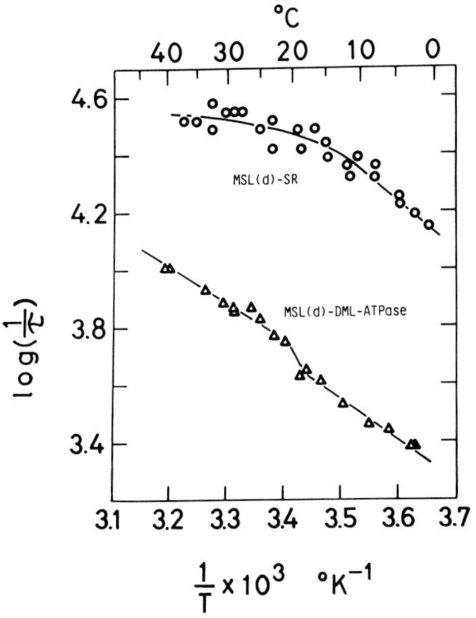

FIGURE 8. An Arrhenius plot for rotational mobility of the Ca-ATPase in SR (o) or in DML-ATPase (Δ) membranes (29).

molecule protein (30) and an effective rotational correlation time (τ) was deduced using the calibration curve made from the spectra of MSL(d)-labeled hemoglobin.

The temperature-dependence of the rotational mobility of Ca-ATPase in SR membrane is shown (o) in the form of an Arrhenius plot in Fig. 8. Effective correlation time of about 30 μsec was obtained at 20°C and this result is in accordance with the result of Thomas et al. (19). The change in activation energy at about 20°C may be due to a conformational change in the Ca-ATPase molecule (18,20,27,28). When the SR membrane was labeled at pH 8.2 (18), lower rotational mobility was detected although the temperature-dependence was not much different. The τ value at 20°C was about 60 - 200 μsec. The scattered values result from the fact that the membrane labeled at pH 8.2 undergoes a gradual irreversible change in τ value when it is kept at high temperatures during the ST-ESR measurements. The difference in τ values of samples labeled at different pH values could be due to the difference in labeled SH groups, or in the conformation or aggregation state of the Ca-ATPase molecules depending on the labeling conditions. All the results reported below were obtained with membranes labeled at pH 7.0.

Figure 8 also shows an Arrhenius plot for rotational mobility of MSL(d)-labeled DML-ATPase recombinant (▲). Although a very slight jump in the mobility was noted at about 20°C, none could be detected corresponding to the phase transition of bulk DML at about 29°C. It was also noted that the Ca-ATPase in DML recombinant has a rotational mobility lower than in native SR membrane even at high temperatures where the former has a little higher bulk lipid fluidity than the latter (see above, III-C). The very slight jump observed at about 20°C may be related to a conformational change in the Ca-ATPase molecule rather than as a result of the gel-to-liquid crystalline pahse transition of bulk DML.

Similar results were obtained with DPL-ATPase recombinants (Fig. 9). The Arrhenius plot for the mobility of the Ca-ATPase was almost linear. No abrupt change in mobility was found throughout the temperature range of 4 - 45 °C: that is, no mobility jump corresponding to the phase transition of bulk DPL at about 39°C could be detected. The rotational mobility of the Ca-ATPase molecule is much lower in DPL-replaced membranes than in SR membranes even at temperatures where bulk DPL is in a fluid state.

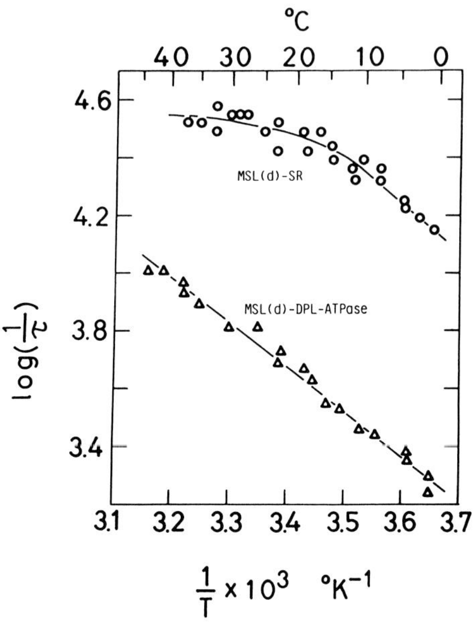

FIGURE 9. An Arrhenius plot for rotational mobility of the Ca-ATPase in DPL-ATPase membrane (Δ) in comparison with that in SR membrane (o) (29).

E. Hydrogen-Deuterium Exchange Reaction Kinetics

The structural fluctuation of the Ca-ATPase protein was examined by the technique of hydrogen-deuterium exchange kinetics. The exchange reaction can be expressed (31), in general, as

$$N \underset{k_2}{\overset{k_1}{\rightleftarrows}} D \xrightarrow{k_e} D^* \rightleftarrows N^* \qquad (1)$$

where N is a "native" conformation of the protein in which peptide groups are hydrogen-bonded and D is a "denatured" conformation in which peptide hydrogens are exchangeable with the rate constant k_e. The superscript * refers to the deuterated products. The rate constant k_e is given empirically as a function of pH and temperature c in Celsius (31),

$$k_e = 50 \times (10^{-\mathrm{pH}} + 10^{\mathrm{pH}-6.0}) \times 10^{0.05(c-20)} \quad \mathrm{min}^{-1} \qquad (2)$$

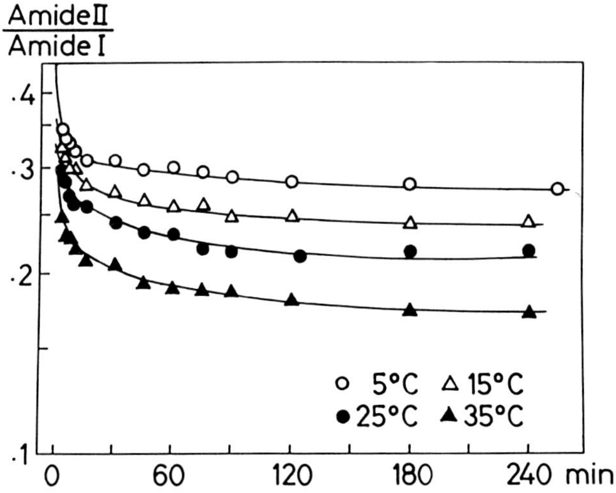

FIGURE 10. Semilogarithmic plots of time courses of the hydrogen-deuterium exchange reaction in DML-ATPase recombinant membranes.

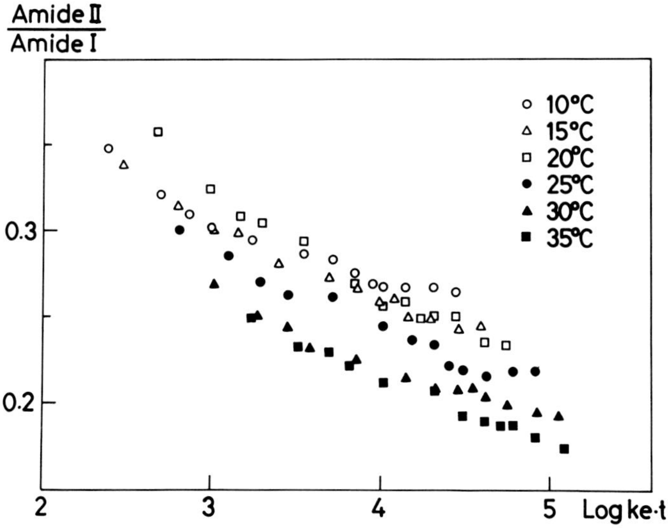

FIGURE 11. Relaxation spectra of the hydrogen-deuterium exchange reaction at various temperatures in DML-ATPase membranes.

Usually the fluctuation is very rapid, i.e.,

$$k_1 + k_2 \gg k_e \tag{3}$$

Then, an observed rate constant of the exchange reaction, k, can be expressed as

$$k = \gamma k_e \tag{4}$$

where γ is the fluctuation amplitude defined as

$$\gamma = (D)/((N) + (D)) \tag{5}$$

Figure 10 is a semilogarithmic presentation of the time course of the exchange reaction for DML-ATPase at four temperatures. The plots are not linear, meaning that there are many classes of peptides whose NH groups exchange hydrogen for deuterium with different rate constants. To analyze such curves, the "relaxation spectrum" method has been introduced by Willumsen (32) and Hvidt and Wallevik (33). The relaxation spectrum is a kind of spectrum of γ and is obtained by a plot of X(t) against $\log(k_e t)$ as shown in Fig. 11. Curves shifted toward the left (i.e., toward a higher value of γ) with an increase in temperature. The shift of the spectrum of γ with temperature is very large between 25°C and 30°C. This abrupt change in temperature-dependence of γ seems to correspond to the phase transition of bulk DML in DML-ATPase membranes.

IV. DISCUSSION

The effect of intrinsic protein on the phase transition of the synthetic phospholipid has been investigated with various systems using various techniques (13). In general, the phase transition temperature of a protein-lipid complex is either higher or lower than that of the pure lipid as observed experimentally (13) and predicted theoretically (34). The result seems to depend not only on the protein-lipid system but also on the method used. This is probably because different methods measure different phenomena accompanying the phase transition. The spin label technique using an (m,n)FA type spin label measures the alkyl chain flexibility of lipid in the protein-lipid system. The phase transition temperature of the bulk lipid was found to be 29°C for DML-ATPase and 39°C for DPL-ATPase while the temperature at which appreciable Ca-ATPase activity was observed was about

20°C for the former and 30°C for the latter. The lowest temperature for the enzyme activity is about 9 degrees lower than the bulk lipid phase transition temperature in both systems; therefore, DPL-ATPase is not unusual. A similar conclusion was reached by Moore et al. based on their fluorescence depolarization data (12).

In contrast to the bulk lipid, the boundary lipid of the reconstituted membranes probed by (m,n)MSL spin labels exhibited no transition in its temperature profile of alkyl chain flexibility (Figs. 5, 6). Similarly, the Arrhenius plot for the rotational mobility of the Ca-ATPase protein was linear in the reconstituted systems (Figs. 8, 9). No abrupt change was detected in the motional state of the Ca-ATPase or the boundary lipid corresponding to the steep increase in ATPase activity.

The Ca-ATPase and the boundary lipid in the reconstituted membranes were strongly immobilized even at a high temperature where the bulk lipid fluidity was as high as in the native SR membrane. This suggests the aggregation or oligomerization of the Ca-ATPase in the reconstituted membranes. A similar temperature profile has been reported for membranes reconstituted from DML and rhodopsin or erythrocyte band 3 protein when the lipid/protein ratio is low (35 - 37).

The linear temperature-dependence of the rotational mobility of the Ca-ATPase in the reconstituted membrane as measured by ST-ESR shown in Figs. 8 and 9 is quite different from the result of Hoffman et al. who observed a jump at 30 °C of more than two orders of magnitude in the rotational mobility of the Ca-ATPase in DPL-ATPase membrane using the transient dichroism technique (15). The membrane samples of the two groups may be different as revealed by each technique since the preparation methods are different. They reconstituted Ca-ATPase with DPL by repeating a procedure essentially the same as that of Hesketh et al. (10) while we used the procedure of Moore et al. (12).

Our interpretation of the results obtained in the present study is summarized as follows. The Ca-ATPase molecules in membrane reconstituted with DML or DPL are aggregated, of which the rotational mobility is very low out of the detection limit of ST-ESR throughout the temperature range studied. What we see in Figs. 8 and 9 is the temperature-dependence of residual mobility probably attributable to restricted conical motion of MSL(d) spin label independent of the protein molecule or of segmental motion of the protein molecule. What is probed by (m,n)MSL spin labels is lipids trapped in the protein aggregates. The minor component of the (m,n)FA signals is also due to spin labels in lipids trapped between the protein molecules. The main component of the (m,n)FA signals represents the fluidity of bulk lipids which is modified by the protein through the rapid exchange of bulk lipids with annular lipids not trapped between but in contact with the protein molecules. The exchange rate has recently been estimated sufficiently large by Devaux and his colleagues (38).

The results of ESR studies, therefore, lead to the conclusion that there is little direct correlation between the Ca-ATPase activity and the rotational mobility of the enzyme or the alkyl chain flexibility of the boundary lipid surrounding the enzyme. This differs from the notion by Hidalgo et al. who stressed the importance of protein rotational mobility in the enzymatic activity of Ca-ATPase (39).

The result of the hydrogen-deuterium exchange reaction technique (Fig. 11) revealed a transition at about 27°C in the structural fluctuation amplitude or the internal flexibility of the Ca-ATPase molecule in DML-ATPase membrane. This differs from the conformational change in the Ca-ATPase molecule at about 18°C reported previously (20), but seems to be related to the phase transition of bulk DML. This result together with the result of ST-ESR may indicate that the internal flexibility of the enzyme protein molecule is more essential than the rotation in the function of the Ca-ATPase of SR.

ACKNOWLEDGMENTS

One of the authors (Y. K.) is sincerely grateful to the late Professor Yuji Tonomura for his kind introduction to the field of sarcoplasmic reticulum 7 years ago and for his encouragement during the course of this work. The authors are very grateful to Professor H. Shimizu for his encouragement and critical discussion and wish to thank Dr. M. Nakanishi and Professor M. Tsuboi for valuable advice on the hydrogen-deuterium exchange experiments, Professor T. Akiyama for advice on the synthesis of MSL(d) and Dr. J. Usukura for electron microscopy.

REFERENCES

1. Bennett, J. P., McGill, K. A. and Warren, G., B., *Curr. Topics Membr. Transport 14*:127 (1980).
2. Edidin, M., in "*Membrane Structure*" (J. B. Finean and R. H. Mitchell, eds.), p. 37. Elsevier, Amsterdam, 1981.

3. Quinn, P. J., *Progr. Biophys. Molec. Biol. 38*:1 (1981).
4. Marsh, D. and Watts, A., in "*Lipid-Protein Interactions, Vol. 2*" (P. C. Jost and O. H. Griffith, eds.), p. 53. John Wiley & Sons, New York, 1982.
5. Seelig, J., Seelig, A. and Tamm, L., in "*Lipid-Protein Interactions, Vol. 2*" (P. C. Jost and O. H. Griffith, eds.), p. 127. John Wiley & Sons, New York, 1982.
6. Chapman, D., in "*Biological Membranes, Vol. 4*" (D. Chapman, ed.), p. 179. Academic Press, London, 1982.
7. Warren, G. B., Toon, P. A., Birdsall, N. J. M., Lee, A. G., and Metcalfe, J. C., *Proc. Natl. Acad. Sci. U.S.A. 71*:622 (1974).
8. Warren, G. B., Toon, P. A., Birdsall, N. J. M., Lee, A. G., and Metcalfe, J. C., *Biochemistry 13*:5501 (1974).
9. Nakamura, H., Jilka, R. L., Boland, R. and Martonosi, A. N., *J. Biol. Chem. 251*:5414 (1976).
10. Hesketh, T. R., Smith, G. A., Houslay, M. D., McGill, K. A., Birdsall, N. J. M., Metcalfe, J. C. and Warren, G. B., *Biochemistry 15*:4145 (1976).
11. Bennett, J. P., Smith, G. A., Houslay, M. D., Hesketh, T. R., Metcalfe, J. C., and Warren, G. B., *Biochim. Biophys. Acta 513*:310 (1978).
12. Moore, B. M., Lentz, B. R., Hoecli, M. and Meissner, G., *Biochemistry 20*:6810 (1981).
13. Silvius, J. R., in "*Lipid-Protein Interactions, Vol. 2*" (P. C. Jost and O. H. Griffith, eds.), p. 239. John Wiley & Sons, New York, 1982
14. Jost, P. C., Griffith. O. H., Capaldi, R. A. and Vanderkooi., G., *Proc. Natl. Acad. Sci. U.S.A. 70*:480 (1973).
15. Hoffman, W., Sarzala, M. G., Gomez-Fernandez, J. C., Goni, F. M., Restall, C. J., Chapman, D., Heppeler, G., and Kreutz, W., *J. Mol. Biol. 141*:119 (1980).
16. Hoffman, W., Sarzala,M.G. and Chapman, D., *Proc. Natl. Acad. Sci. U.S.A. 76*:3860 (1979).
17. Bürkli, A. and Cherry, R. J., *Biochemistry 20*:138 (1981).
18. Kirino, Y., Ohkuma, T. and Shimizu, H., *J. Biochem. 80*: 111 (1978).
19. Thomas, D. D. and Hidalgo, C., *Proc. Natl. Acad. Sci. U.S.A. 75*:5488 (1978).
20. Anzai, K., Kirino, Y., and Shimizu, H., *J. Biochem. 84*:815 (1978).
21. Kaizu, T., Kirino, Y. and Shimizu, H., *J. Biochem. 88*:1837 (1980).
22. Favre, E., Baroin, A., Bienvenue, A. and Devaux, P. F., *Biochemistry 18*:1156 (1979).

23. Kirino, Y., Higashi, K., Matsui, M. and Shimizu, H., *J. Biochem.* *89*:975 (1981).
24. Kawakita, M., Yasuoka, K., and Kaziro, Y., *J. Biochem.* *87*:609 (1980).
25. Venkataramu, S. D., Pearson, D. E., Beth, A. H., Perkins, R. C., Park, C. R. and Park, J. H., *J. Labelled Compd. Radiopharm.* *18*:371 (1981).
26. Rousselet, A., Colbeau, A., Vignais,P., and Devaux, P.F., *Biochim. Biophys. Acta* *426*:372 (1976)
27. Lippert, J. L., Lindsay, R. M. and Schultz, R., *J. Biol. Chem.* *256*:12411 (1981).
28. Quinn, P. J., *J. Physiol.* *326*:13P (1982).
29. Higashi, K. and Kirino, Y., *J. Biochem.* submitted.
30. Thomas, D. D., Dalton, L. R. and Hyde, J. S., *J. Chem. Phys.* *65*:3006 (1976).
31. Hvidt,A. and Nielsen, S. O., *Adv. Protein Chem.* *21*:287 (1966).
32. Willumsen, L., *Biochim. Biophys. Acta* *126*:382 (1966).
33. Hvidt, A. and Wallevik, K., *J. Biol. Chem.* *247*:1530 (1972).
34. Marčelja, S., *Biochim. Biophys. Acta* *455*:1 (1976).
35. Davoust, J., Bienvenue, A., Fellmann, P. and Devaux, F. P., *Biochim. Biophys. Acta* *596*:28 (1980).
36. Kusumi, A., Sakaki, T., Yoshizawa, T. and Ohnishi, S., *J. Biochem.* *88*:1103 (1980).
37. Sakaki, T., Tsuji, A., Chang, C.-H. and Ohnishi, S., *Biochemistry* *21*:2366 (1982).
38. Seigneuret, M., Davoust, J., Herve, P. and Devaux, F.P., *Biochimie* *63*:867 (1981).
39. Hidalgo, C., Thomas, D. D., and Ikemoto, N., *J. Biol. Chem.* *253*:6879 (1978).

SARCOPLASMIC RETICULUM AND OTHER MEMBRANES IN THE REGULATION OF SKELETAL MUSCLE CONTRACTION AND RELAXATION, A MULTILEVEL APPROACH

Sidney Fleischer[1]

Department of Molecular Biology
Vanderbilt University
Nashville, Tennessee

I. INTRODUCTION

The skeletal muscle fiber is a giant multinucleate cell, extending cms in length, and up to 100μ in diameter. The bulk of the mass (>80%) of the fast twitch skeletal muscle fiber is occupied by myofibrils which run the length of fiber. There can be hundreds of myofibrils in a single muscle fiber. The myofibril consists of a linear array of sarcomeres (approximately 2.5 μ long and 1 μ in diameter), the structural units which carry out muscle contraction and relaxation. The number of sarcomeres in a fibril depends on the length of the fiber (1).

Muscle contraction and relaxation are controlled by the intrafiber free Ca^{++} concentration, $[Ca^{++}]_i$, in the myoplasm which bathes the muscle filaments of the sarcomeres. In turn, the $[Ca^{++}]_i$ is controlled by a network of membranes which regulate and coordinate the process in time and space. The anatomy of a portion of a muscle fiber is diagrammatically illustrated in Figure 1. Most of the fiber consists of myofibrils; only a small portion of the mass is referable to membranes

[1]Supported in part by Grants from NIH AM14632, AM21987 and the Muscular Dystrophy Association.

[2]Abbreviations used are: CPM, calcium pump membrane; CPP, calcium pump protein; JFM, junctional face membrane; L/P, molar ratio of phospholipid to CPP; SR, sarcoplasmic reticulum; TT, transverse tubule; TC, terminal cisternae.

ISBN 0-12-260380-X

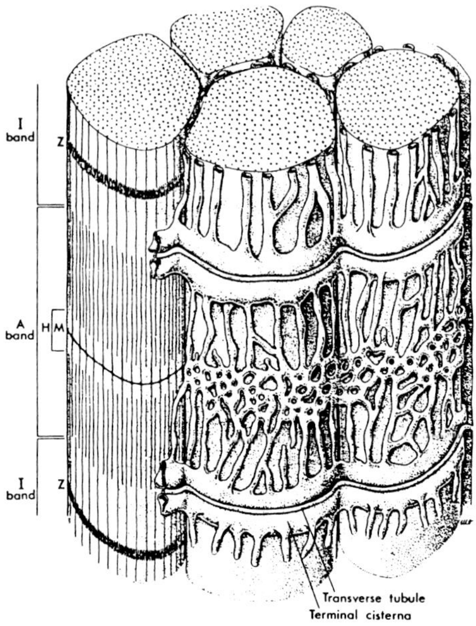

Figure 1. Diagram of a portion of a mammalian striated muscle fiber. Two transverse tubules supply one sarcomere. The transverse tubules invaginate from the plasma membrane, close to the line where the A and I bands meet. Two terminal cisternae of SR are junctionally associated with one transverse tubule and connect with the longitudinal sarcotubules of SR located around the A band. The triple structure, seen in cross-section of two terminal cisternae (one from each adjacent sarcomere) flanking the transverse tubule, constitutes a triad (see Figure 2). The SR-transverse tubule network surrounds each sarcomere. From "Histology" by A.D. Ham and D.H. Comack (1).

which regulate the entire process. The transverse tubules are derived by invagination from the plasmalemma. One sarcomere is supplied by two transverse tubules, which in mammals are on the line between the A and I bands. The action potential travels the length of the fiber via the plasma membrane and transversely to within the fiber by way of the transverse tubule. The signal is then transferred across the triad junction to the terminal cisternae of sarcoplasmic reticulum resulting in calcium release. The membrane network surrounds each myofibril so that calcium uptake or release by the sarcoplasmic reticulum [2] is in close proximity to the site of action, i.e., the sarcomere. The SR consists of two morphologically distinct regions. The terminal cisternae is the portion in junctional association, which together with the transverse tubule forms the triad junction, and the longitudinal cisternae consisting of sarcotubular structures anastomosing to give a fenestrated appearance. The terminal and longitudinal cisternae of SR are contiguous and form a single compartment involved in storing Ca^{++}.

A longitudinal section of the muscle fiber is shown in the electron micrograph in Figure 2. The transverse tubule can be observed in junctional association with the terminal cisternae of sarcoplasmic reticulum via the so called "feet" structures (3,4). The terminal cisternae are morphologically distinct from the longitudinal cisternae, in that they contain junc-

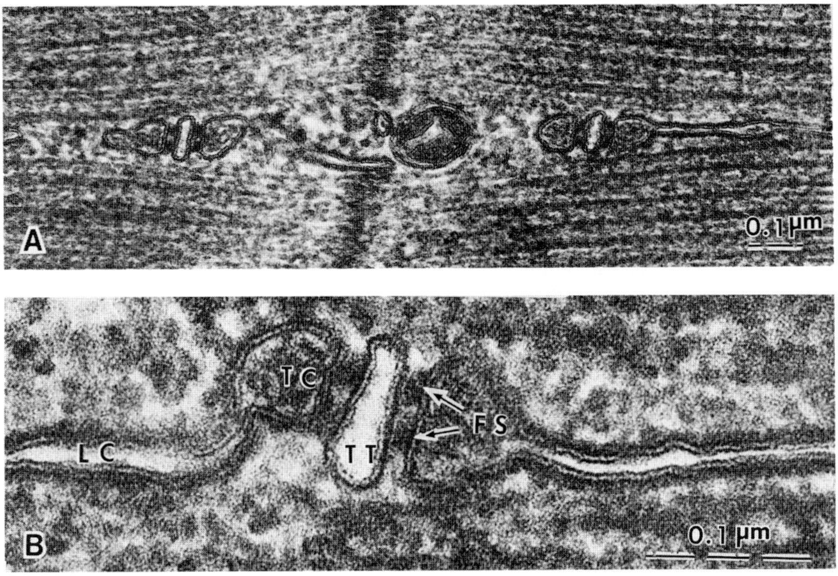

Figure 2. Longitudinal section of rabbit skeletal muscle in thin section with emphasis on the triad region. The sample was fixed with tannic acid which enhances contrast so that the asymmetry of the SR membrane is readily visualized. A) Low power field; B) Higher magnification of cross-section of the triad region. The transverse tubule (TT) is junctionally associated with the terminal cisternae (TC) of SR, the gaps being bridged by feet structures (FS). Note dense contents within the terminal cisternae but not in the transverse tubule or longitudinal cisternae (LC)(5).

tional face membrane and electron opaque contents. The more typical SR membrane can be observed *in situ*, as a highly asymmetric trilayer membrane (70, 20, 20A°). This thin section was prepared using tannic acid enhancement (5).

The recognition that: 1) the myoplasmic free Ca^{++} concentration regulated muscle contraction and relaxation; and 2) fragmented sarcoplasmic reticulum vesicles, capable of energized Ca^{++} uptake, was the "relaxing factor" (6,7) advanced and transformed the problem, in part, from anatomy and muscle physiology to membrane molecular biology. The sarcoplasmic reticulum membrane could be studied as a specialized membrane system capable of energized Ca^{++} transport. Sarcoplasmic reticulum is one of the most studied membrane systems (8-14). In this limited overview, the progress of our laboratory in the study of SR and other muscle membranes will be reviewed as they relate to muscle contraction (15,16) and relaxation.

II. MOLECULAR BIOLOGY APPROACH TO THE STUDY OF SARCOPLASMIC RETICULUM

The focus of the studies in this section will be on the reconstitution approach to correlate membrane composition with structure and function and to characterize the nature of lipid-protein interactions in the SR membrane.

A. Isolation and Characterization of Purified SR

For such a program to become viable it was necessary to isolate purified and functional SR and in sizeable quantity for study (hundreds of mg protein). Such a procedure was developed using zonal rotors (17). In our preparation of SR, three proteins predominated. The calcium pump protein (also referred to as the Ca^{2+} ATPase) (18), the calcium binding protein (calsequestrin, 19) and a component with M_r of 55 kilodaltons, designated M_{55} (17). The protein and lipid content accounted for 60% and 40% of the mass, respectively. The lipid composition was also relatively simple, consisting largely of phospholipid of which phosphatidylcholine is the major component (20).

Purified sarcoplasmic reticulum could be subfractionated into heavy and light fractions of SR based on their isopycnic densities (Figure 3). The heavy fraction contains electron opaque material within its compartment, and appears to have been derived from the terminal cisternae whereas the light fraction is devoid of electron opaque contents and appears to be referable to the longitudinal cisternae (15,21). Polyacrylamide gels of purified SR (center) and the two subfractions, light (left) and heavy SR (right) are shown (Figure 4). For each, the calcium pump protein of about 115 kilodaltons (13,20), is the major protein component of sarcoplasmic reticulum membrane. The heavy fraction contains, additionally, the calcium binding protein, the major protein component within the compartment of the heavy SR fraction (see Section IIB below) (15,21).

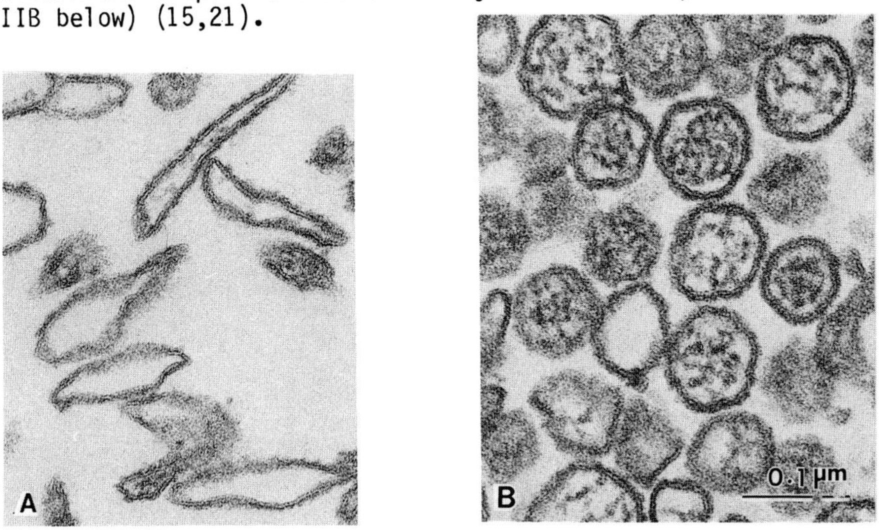

Figure 3. Electron micrographs of isolated (A) light and (B) heavy sarcoplasmic reticulum fractions (15,21).

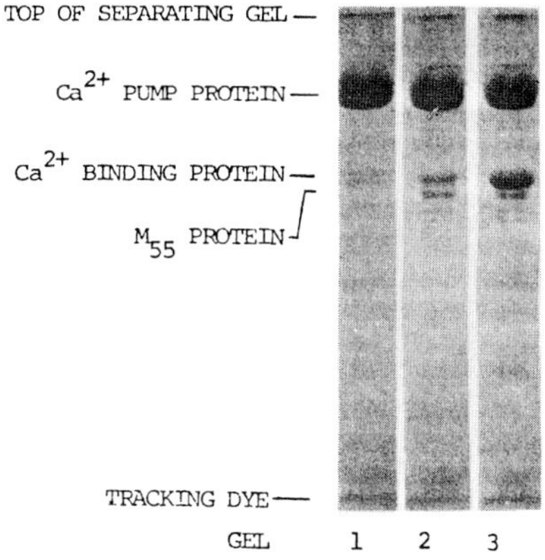

Figure 4. Polyacrylamide gel electrophoresis (PAGE) of SR fractions. The samples are: (1) light SR; (2) normal SR; and (3) heavy SR (15,21).

B. Disassembly of SR and Characterization of Components

The heavy SR can be made leaky so that the compartmental contents are released into the supernatant. The calcium binding protein is then readily purified. It was found to have a high capacity to bind Ca^{2+}. The calcium binding protein is the major protein component within the compartment of the terminal cisternae (17,21) and undoubtedly serves as the binding reservoir for Ca^{2+}. The calcium binding protein appears to be identical with calsequestrin which was purified and characterized by MacLennan and coworkers (19).

Light SR consists of membrane vesicles apparently devoid of polypeptide compartmental contents. Its membrane consists mainly of calcium pump protein (90% or more of the protein) (15,21). The purified calcium pump protein, contains two specific Ca^{2+} binding sites, one ATP binding site, and forms the phosphoenzyme intermediate (17,22). It could be reconstituted to form functional membrane vesicles (23), capable of energized Ca^{2+} accumulation. Thus, the calcium pump protein, possessed the key characteristics for calcium pumping identifying it as the calcium pumping molecule (23). The membrane of the longitudinal cisternae consists of a single type of membrane, designated the "Ca^{2+} pump membrane".

C. Reconstitution of Functional SR Membrane Vesicles

We have relied heavily on the dissociation and reconstitution approach to characterize the SR membrane. The membrane is solubilized to monodispersity using detergent. Removal of the detergent leads to reformation of membrane vesicles. The reconstituted membrane vesicles are capable of energized Ca^{2+} accumulation. The main difference between the normal and reconstituted calcium pump membrane is that the alignment of the pumps have become bidirectional [See below and Section IIG]. The reconstitution procedure developed in our laboratory (23) had three key characteristics which distinguished it from that of others (24): (a) the reconstituted membrane vesicles like normal SR consist mainly of protein (60 percent of the mass) and are not liposomes mildly doped with trace amounts of calcium pump protein; (b) the reconstituted SR does not require trapping agents such as oxalate within its compartment during vesicle formation in order to achieve subsequent energized calcium pumping; and (c) the reconstituted membrane vesicles are capable of energized calcium uptake in the absence of a trapping agent.

The reconstituted preparations display the highest activity reported for reconstituted sarcoplasmic reticulum systems. The rates of phosphoenzyme formation from ATP and hydrolysis as well as steady-state levels match the values obtained with normal SR vesicles. Rapid calcium transport and repeated cycles of ATP synthesis by reversal of the pump are also obtained. However, the efficiency of Ca^{2+} transport and ATP synthesis from a Ca^{2+} gradient are approximately three fold lower than in native vesicles. Time resolution of the phosphorylation reaction by the rapid quench technique can be used to estimate the orientation of the calcium pump in the membrane. Such studies indicate that the calcium pump protein is largely bidirectionally oriented in reconstituted SR vesicles (25).

We have studied the dissociation and reconstitution process for the SR membrane using deoxycholate. Dissociation involves, 1) extraction of bilayer phospholipid until a limit membrane is reached with a L/P of 48, and 2) solubilization of CPP-phospholipid complexes. Reassembly involves the reverse process. Removal of detergent leads to 3) formation of a limit membrane, and 4) vesicle formation with increase in L/P as phospholipid gradually inserts. In this way, the phospholipid content can be controlled to form membranes in the same range, as well as both lower and higher L/P than normal SR (L/P = 115) (26).

The reconstitution technology developed in our laboratory enables the preparation of membranes consisting of calcium pump protein and with varying phospholipid to protein ratios (26, 27). Further, the technology has been developed to prepare functional membranes with one type of phospholipid class and molecular species (28) [See Section II D]. The technology to prepare membranes of defined composition and to vary the L/P especially in the range of high protein content have made possible key correlations to characterize the Ca^{2+} pump membrane and are discussed, in part, in the following sections.

D. Lipid Requirement for the Ca^{2+} Pump Cycle

Lipid is required for Ca^{2+} stimulated ATPase activity (29, 30), although lipid specificity does not appear to be stringent since the detergent $C_{12}E_8$ can serve to activate this enzymic activity (31). We found previously, using the detergent exchange procedure, that Ca^{2+} pumping is equally good with lecithin as with SR phospholipids (28). As already mentioned, the detergent treatment leads to loss of vectorial alignment of the calcium pump protein in the membrane (5).

We are reinvestigating the lipid specificity using the general phospholipid exchange protein (32). This protein (33) was used to exchange phospholipid from the SR membrane vesicles with added phospholipid vesicles. Phospholipid exchange as a function of time is shown in Figure 5A. The rate of exchange is biphasic. The rapid phase is interpreted to represent exchange in the outer half of the bilayer of the SR membrane and extrapolates to 60% non-exchanged. This indicates that there is an uneven distribution of phospholipid in the two halves of the bilayer of the SR membrane, i.e., approximately 40% is in the outer leaflet and 60% in the inner leaflet. The slow phase is interpreted to reflect the exchange into the inner half of the bilayer of the phospholipid which is rate limited by the phospholipid flip-flop rate (transbilayer exchange). The half-time for the flip-flop rate varies from 15-28 hours depending on the type of phosphatidylcholine. In Figure 5B, phosphatidylethanolamine has been exchanged into the SR membrane. The ratio of PE/PC in the membrane has been increased nine-fold with a concomitant marked decrease in calcium pumping. Transport activity could be re-

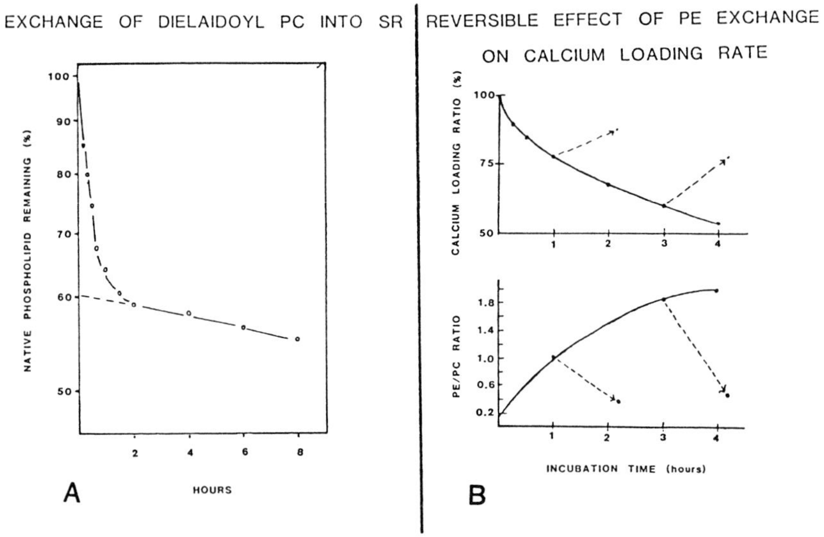

Figure 5. Exchange of phospholipids in SR vesicles with added liposomes using the nonspecific phospholipid exchange protein. A) Sealed SR vesicles were incubated with 100-fold excess of dielaidoyl phosphatidylcholine (PC) liposomes. B) Effect of phosphatidylethanolamine (PE) exchange on the Ca^{2+} loading rate of SR vesicles. SR vesicles were incubated with 100-fold excess of soybean PE liposomes. At indicated times, SR membranes were reisolated and analyzed. Some samples were again exchanged with dioleoyl PC (dashed lines)(32).

stored by exchanging back the normal SR lipids or phosphatidylcholine into the membrane. These studies indicate that there is some specificity when Ca^{2+} pumping, the holoactivity, is measured. The vectorial alignment of the Ca^{2+} pump protein was not altered by the double exchange with phospholipid (32).

E. Oligomeric Size of the Calcium Pump Protein in the Membrane

Target inactivation analysis was used to determine the oligomeric size of the calcium pump protein in the sarcoplasmic reticulum membrane. With this method, loss of enzymic activity is correlated with radiation flux. Since the calcium pump protein is the major component of the membrane, we could also follow the fragmentation of the CPP as a function of radiation dose by polyacrylamide gel electrophoresis. The fragmentation of the CPP and the loss of enzymic activity gave a target size for the CPP of between 210-250 kilodaltons. This is convincing evidence that the Ca^{2+} pump protein exists as a dimer in the membrane (35).

F. Correlation of Composition with Structure

Membrane vesicles were reconstituted with different lipid to protein ratios and are shown in the freeze-fracture electron micrographs in Figure 6. The particle density can be observed to increase as the protein content is increased, (Figure 6F to 6B), whereas the extent of smooth background correlates with the lipid content of the membrane. In the limit, no particles are observed in phospholipid vesicles. The particle density observed by freeze-fracture electron microscopy was found to be directly proportional to the content of calcium pump protein in the membrane. It can be calculated that there are approximately 2 molecules of calcium pump protein per particle, apparently a single pump unit (27).

G. Orientation of the Calcium Pump Protein in the Membrane

1. Electron Microscopy Approach. The orientation of the CPP in the SR membrane has been deduced from electron microscopic studies. The normal sarcoplasmic reticulum membrane, viewed in thin section using tannic acid enhancement, is a highly asymmetric trilayer (70, 20, 20 A°), the outer band of the trilayer being much wider than the middle and inner bands (Figure 7). The reconstituted membrane of high protein con-

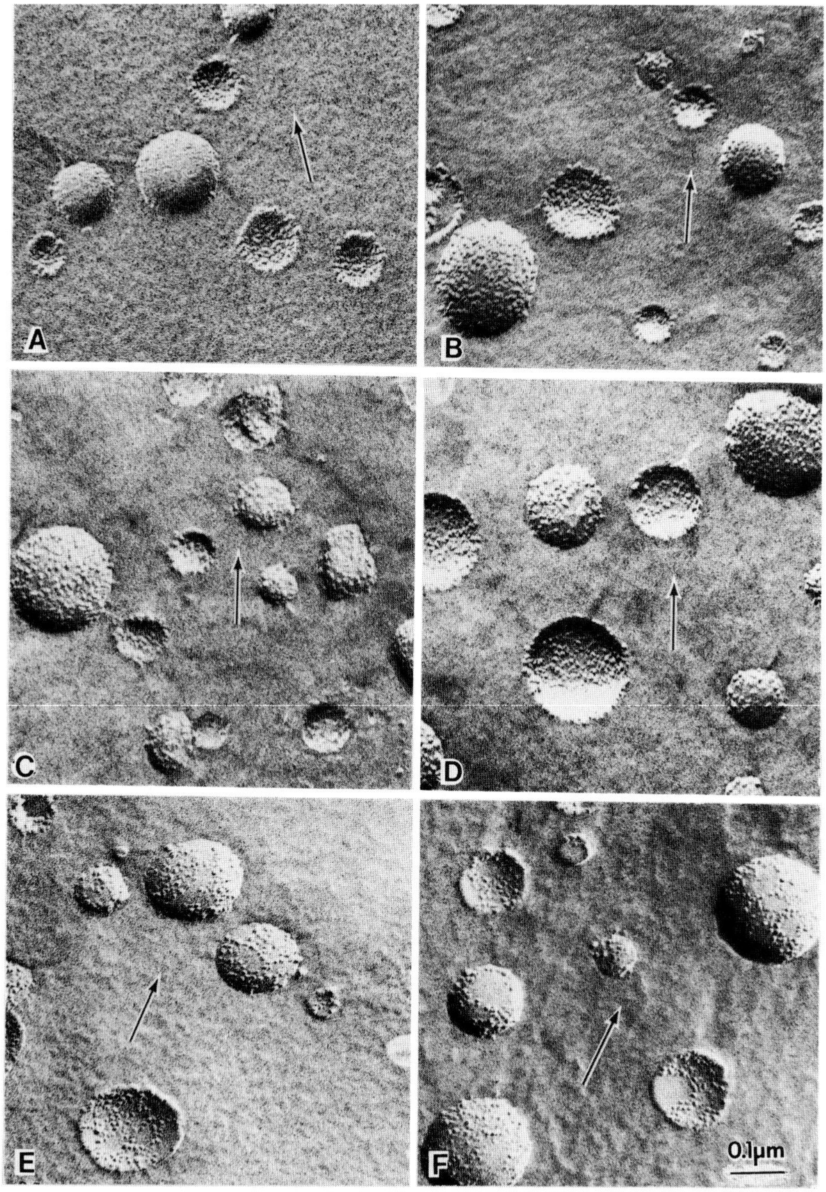

Figure 6. Freeze-fracture electron micrographs of normal SR (A), and reconstituted SR of varying phospholipid content (B) 0.38; (C) 0.57; (D) 0.78; (E) 1.03; and (F) 1.29 μmoles phospholipid/mg protein. The value for normal SR is 0.78 (27).

tent gives a symmetric trilayer, with broad outer and inner bands, (70, 20, 70 A°). As the lipid content of the reconsti-

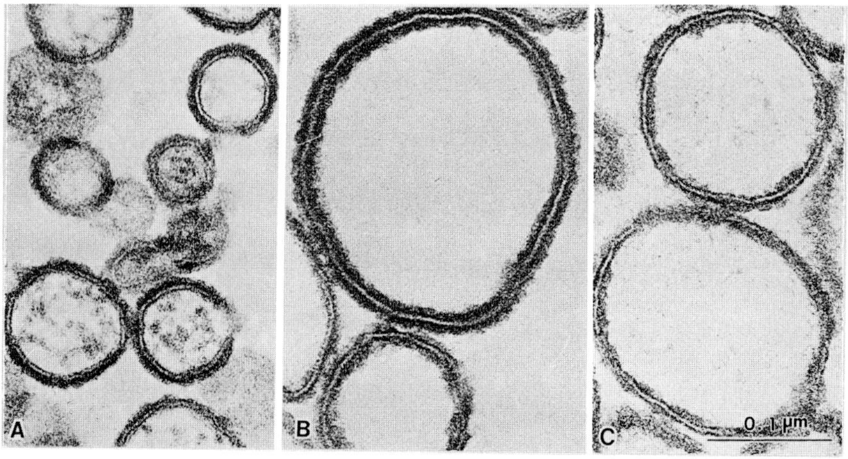

Figure 7. SR preparations as visualized in thin section. Tannic acid was used to achieve enhanced contrast of the SR membranes. A is normal SR; B and C are reconstituted SR with higher and lower protein content, respectively (5).

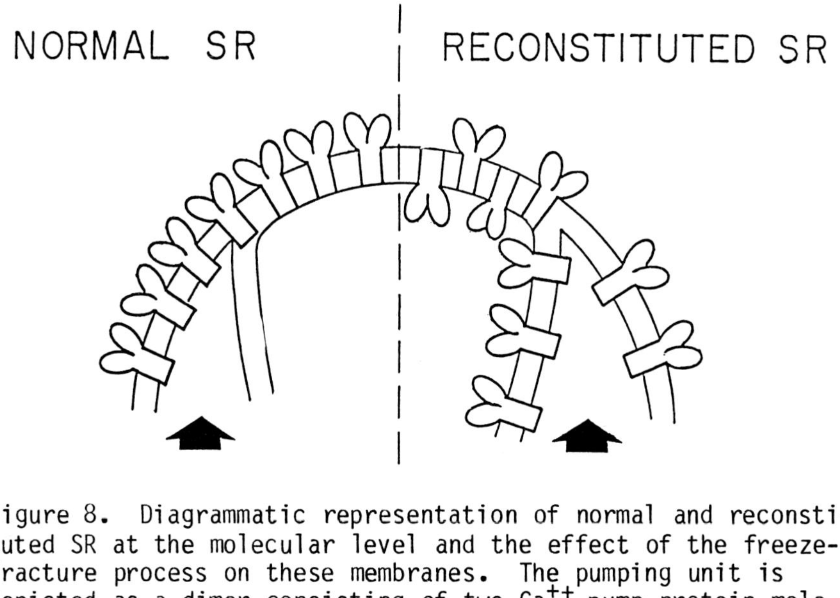

Figure 8. Diagrammatic representation of normal and reconstituted SR at the molecular level and the effect of the freeze-fracture process on these membranes. The pumping unit is depicted as a dimer consisting of two Ca^{++} pump protein molecules (5).

tuted membrane is decreased, the outer and inner bands become irregular and approach a simple symmetric trilayer (20, 20, 20 A°) characteristic of phospholipid (not shown). Since the CPP is the major component of the normal SR membrane, the broad outer band is referable to CPP protruding from the cytoplasmic face. In the reconstituted membrane, the broad outer and inner bands reflect the bidirectional orientation of the calcium pump protein. The pump must be transmembrane since Ca^{2+} is translocated across the membrane, from one compartment to the other (5).

The CPP in normal SR is indicated to be oriented transmembrane and unidirectionally oriented with a major portion extending out from the cytoplasmic face (Figure 8). The pump is shown to be a dimer in the membrane based on the target inactivation analysis [Section IIE].

2. X-Ray and Neutron Diffraction of Oriented Multilayers of SR. In a complementary approach, the orientation of the CPP was obtained using X-ray and neutron diffraction. These are joint studies with Leo Herbette, Toni Scarpa and Kent Blasie [see reference 36]. The basis for this approach is outlined below.

1. The CPP is the main protein constituent (>90%) of the SR membrane.
2. Low angle X-ray diffraction was carried out on oriented multilayers of SR and a total electron density profile [ED (total)] was obtained.
3. ED (total) consists of contributions from the three main constituents of the membrane, i.e., the calcium pump protein, phospholipid and water.

ED (total) = ED (CPP) + ED (phospholipid) + ED (water)

4. Neutron diffraction, which is sensitive to the difference between deuterium and hydrogen atoms, was carried out on oriented multilayers of SR membranes containing protonated vs deuterated water and phospholipid, i.e., the protonated water was exchanged with deuterated water and protonated phospholipid was exchanged for deuterated phospholipid using the phospholipid exchange protein.
5. The electron density referable to the water and phospholipid were scaled out from the ED (total). In this way, the ED profile of the CPP in the membrane and its orientation with respect to the phospholipid bilayer were determined.

The resolved electron density profiles of the phospholipid and calcium pump protein in the SR membrane are shown in Fig-

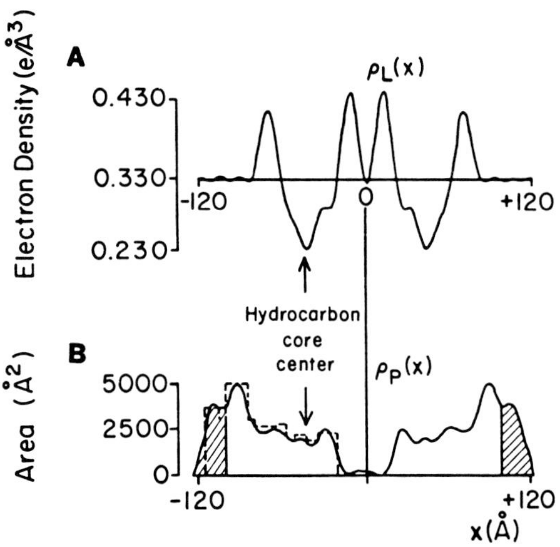

Figure 9. Electron density profile of phospholipid and calcium pump protein (CPP) within the SR membrane profile. A) The electron-density profile of the phospholipid bilayer, $\rho_L(x)$. B) The area profile of the CPP, $\rho_P(x)$. (This profile represents the area occupied by the CPP as a function of the position x within the SR membrane profile, assuming a constant density for the CPP as a function of x.)(36).

ure 9. For each, there are two membranes per unit cell. The phospholipid profile (upper) is characteristic of a phospholipid bilayer, having two outer regions of higher density and a central region of lower density. The higher density regions are referable to the polar head groups of the phospholipid and the central region is referable to hydrocarbon. The electron density profile of the CPP in the membrane does not coincide with the center of the bilayer. The CPP is transmembrane, with a major portion extending out from the cytoplasmic (outer) face of the membrane.

Thus, two independent approaches, diffraction analysis and electron microscopy lead to the same conclusion. The CPP is transmembrane and vectorially oriented.

H. Motional Characteristics of the Phospholipid and the Calcium Pump Protein (CPP) in the Membrane

The motional characteristics of phospholipids have been studied using magnetic resonance approaches. Deuterium NMR is ideal for such studies in that: 1) there exists a theoretical basis for interpretation of such data so that both structural and motional information can be obtained; and 2) the deuterium and hydrogen atoms are structurally equivalent having identical Van der Waals radii. Thus, replacing a hydrogen with a deuterium in the molecule is structurally non-perturbing.

Deuterium NMR spectra of phospholipid vesicles containing selectively deuterated dioleoylphosphatidylcholine ([9,10-^{2}H] DOPC) deuterated at the cis double bond, is shown in Figure 10A. In these studies, the deuterium is a reporter for the motion in the hydrophobic domain. Three distinct quadrupole splittings can be observed. The smallest and middle pairs are referable to the C-10 of the sn-2 and sn-1 acyl chains, respectively. The largest splitting is referable to a superposition of the two C-9 deuterons. Qualitatively, a similar spectrum (Figure 10B) containing three pairs of quadrupolar splittings is observed for reconstituted membranes of sarcoplasmic reticulum exchanged with the same deuterated phospholipid. This important result means that the lipid in the reconstituted SR membrane approximates a homogeneous lipid environment. That is to say, the lipid at the boundary of the CPP is in rapid equilibrium with the bulk bilayer phospholipid in the membrane. The reconstituted membrane, in this experiment, was prepared with a high protein content having a lipid to CPP molar ratio (L/P) of 64, or nearly half that in normal

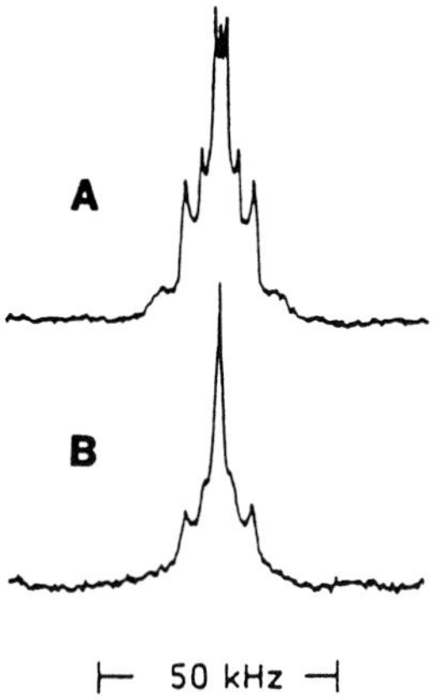

Figure 10. D-NMR spectra of functional reconstituted sarcoplasmic reticulum membrane vesicles (R-SR) exchanged with 1,2-di[9,10-$^2H_2$]oleoyl-sn-glycero-3-phosphocholine ([9,10-$^2H_2$] DOPC). The phospholipid is in the liquid crystalline phase. A) Pure [9,10-$^2H_2$]-DOPC, dispersed in water, B) R-SR in buffer. Lipid-to-protein ratio, L/P, is about 64 mole of phospholipid per mole of Ca^{2+} pump protein (28).

SR (L/P=115). Such high protein content membranes are favorable for detection of the influence of the protein on the motion of the phospholipid. It has been estimated that the amount of lipid needed to form a coherent lipid boundary is 30 moles PL/mole CPP (34). A separate long-lived lipid environment referable to the CPP would account for nearly half of the lipid in such a high protein content membrane, but none was detected (28).

The quadrupolar splitting provides structural information on the arrangement of the lipid. The order parameter can be directly calculated from the quadrupolar splitting. A decrease in the splitting is equivalent to increased disorder. It can be noted that the deuterium quadrupolar splitting is somewhat reduced for the recombinant membrane as compared with the phospholipid vesicles (Figure 10). Hence, the presence of CPP in the bilayer results in some increased disorder, equivalent to about 20%. This is a relatively small effect considering the high protein content of the reconstituted membrane.

A similar comparison of recombinant membrane and phospholipid vesicles is given, but with dielaidoylphosphatidylcholine (DEPC), also labelled with deuterium in the fatty acyl moiety (Figure 11). DEPC has a melting transition at 10-11°C so that spectra obtained at 20° and 4° are in the liquid crystalline and gel states, respectively. Only a single pair of

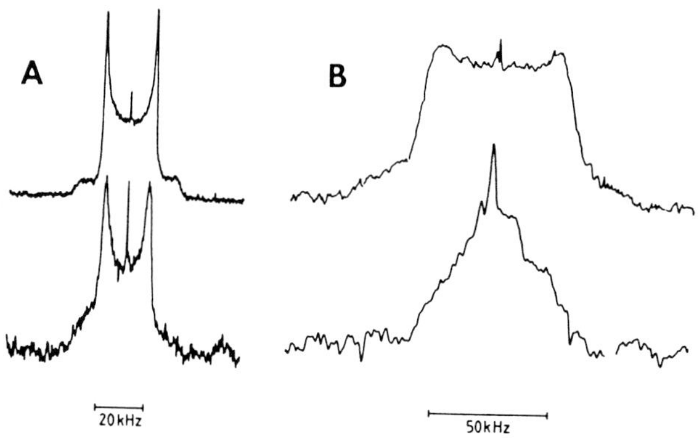

Figure 11. D-NMR spectra of reconstituted functional SR. The natural lipids were exchanged to the extent of 99% with 1,2-dielaidoyl-sn-glycero-3-phosphocholine (DEPC), deuterated in both chains at the 9,10 position. The phase transition of DEPC occurs at about 10°C. The L/P is 51. A) Spectra taken above the phase transition temperature (25°C). B) Spectra taken below the phase transition temperature (4°C). Upper spectra are DEPC; lower spectra are reconstituted SR (28).

quadrupolar splitting is observed in the liquid crystallize state, (Figure 11A). As noted in Figure 10 with DOPC, there is a small decrease in quadrupolar splitting in the recombinant as compared with the phospholipid vesicles alone. The spectra, of lipid in the gel state, (Figure 11B) in the presence or absence of the calcium pump protein, reflect superpositions of multiple quadrupolar splittings, are indicative of a heterogeneous lipid environment. We conclude that there is a single homogenous lipid environment in the liquid crystalline state on the time-scale of NMR ($\sim 10^5$/sec).

The rotational mobility of the phospholipid is proportional to the spin lattice relaxation time (T_1). Spin lattice relaxation times were obtained using the inversion recovery technique. The data as shown in Figure 12. For the recombinant membranes, the T_1 is smaller indicating somewhat slower motion. That is, the rotational mobility of the phospholipid is reduced somewhat by the high protein content in the membrane.

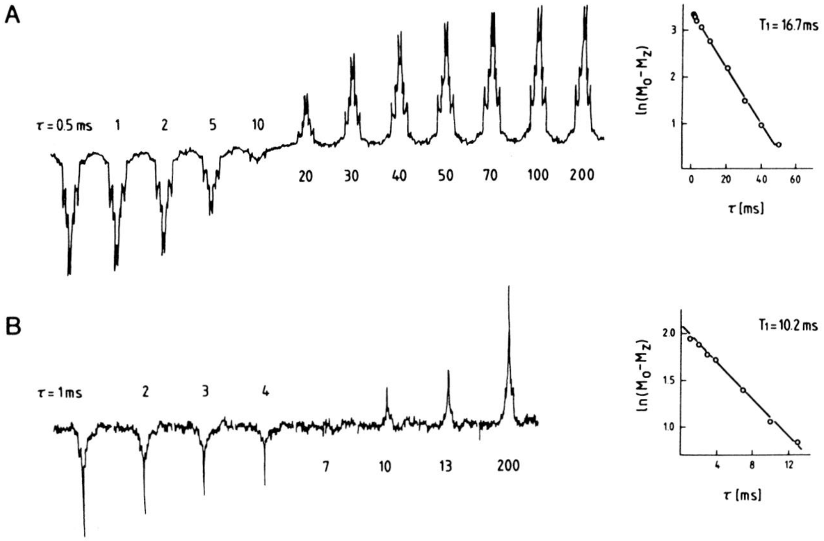

Figure 12. Measurements of the deuterium T_1 relaxation time by the inversion recovery technique. (A) DOPC. 1,2-Di[9,10-$^2H_2$]oleoyl-sn-glycero-3-phosphocholine was dispersed in excess water. Only the relevant part of the spectra is shown in the figure (31°C). (B) SR reconstituted with [9,10-$^2H_2$] DOPC and dispersed in deuterium-depleted buffer (15°C)(28).

D-NMR studies of reconstituted sarcoplasmic reticulum lead to a number of important conclusions (28).

(1) A homogeneous lipid environment is indicated on the time-scale of D-NMR (10^5/sec). That is to say, boundary lipid and bilayer lipid are in rapid exchange on this time-scale.
(2) Structural information is obtained from the quadrupolar splitting and indicate that the phospholipid in the reconstituted membrane is somewhat more disordered (15-20%) than in liposomes.
(3) The T_1 relaxation times yield information about motion. The rotational motion of the phospholipid in reconstituted SR membranes is somewhat slower (15%-20%) than in phospholipid vesicles alone.
(4) To a first approximation, the motion and orientation of the phospholipid in the membrane is that of a bilayer.

In these studies, the deuterium was a reporter for the hydrophobic portion of the phospholipid, since the deuterium probe was in the fatty acyl moiety of the phospholipid. Qualitatively, similar results were obtained in studies of these same samples, using either phosphorus NMR (28,37), or proton NMR, observing resonances in both the polar and hydrophobic regions (38). Therefore, similar conclusions can be made regarding the motional characteristics of the polar moiety of the phospholipid.

I. Summary - Characteristics of the Calcium Pump Containing Membrane of SR

The calcium pump protein is the main constituent (90% of the protein) of the calcium pump membrane of SR and is the calcium pumping molecule. It exists as a dimer in the membrane. The calcium pump protein is transmembrane and vectorially oriented with a major portion extending out from the cytoplasmic face. To a first approximation, the phospholipid is typical of a bilayer with boundary lipid in rapid exchange with bilayer phospholipid. The flip-flop rate of the phospholipid in the membrane is slow ($t_{1/2}$ 1 day). The turnover number of the calcium pump protein (10^2) is several orders of magnitude slower than its rotational motion which in turn is considerably slower than the motion of the phospholipid (39).

III. MEMBRANES IN THE REGULATION OF CALCIUM RELEASE

The mechanism, whereby the action potential of the plasmalemma of the muscle fiber triggers Ca^{2+} release, remains the key unsolved problem in excitation-contraction coupling and is one of the most intriguing research areas in molecular and cell biology. There is now active interest in understanding the Ca^{2+} release process (for example, see 10, 40-42).

Our approach to the study of muscle contraction and relaxation has been to isolate well-defined muscle membranes in order to characterize the membrane and molecular machinery involved in the regulation of myoplasmic calcium levels. The membranes of skeletal muscle are tabulated in Table I. They include the plasmalemma, the transverse tubule, the sarcoplasmic reticulum, both terminal cisternae and longitudinal cisternae and triads (the junctional association of transverse tubule with terminal cisternae). The purification of a membrane is a first and essential step toward its characterization. The characteristics of these isolated membrane fractions are described sequentially. This section will focus on the characterization of skeletal muscle membrane fractions with view toward the calcium release process.

TABLE I. SKELETAL MUSCLE MEMBRANES INVOLVED IN MUSCLE CONTRACTION AND RELAXATION

1. Plasmamembrane
2. Transverse tubule
3. Sarcoplasmic Reticulum
   a. terminal cisternae
   b. longitudinal cisternae
4. Triads - junctional association of transverse tubule and terminal cisternae

An electron micrograph of isolated inside-out plasmalemma vesicles is shown in Figure 13A. Such vesicles are capable of generating a sodium gradient comparable to that in the intact fiber (Figure 13B). The gradient can be dissipated by adding the ionophore monensin. Preincubation of the fraction with batrachotoxin, which opens Na^+ channels in excitable tissues, prevents most of the Na^+ uptake, confirming the presence of toxin activated Na^+ channels in these plasmalemma vesicles (43).

Transverse tubules have been isolated which retain a tubular appearance (Figure 14A). Electron dense material is frequently observed at the nipple-like ends of the tubules (Figure 14B). On occasion, structures are observed at the surface of the tubules (arrows) suggesting a role in junctional association (44,46).

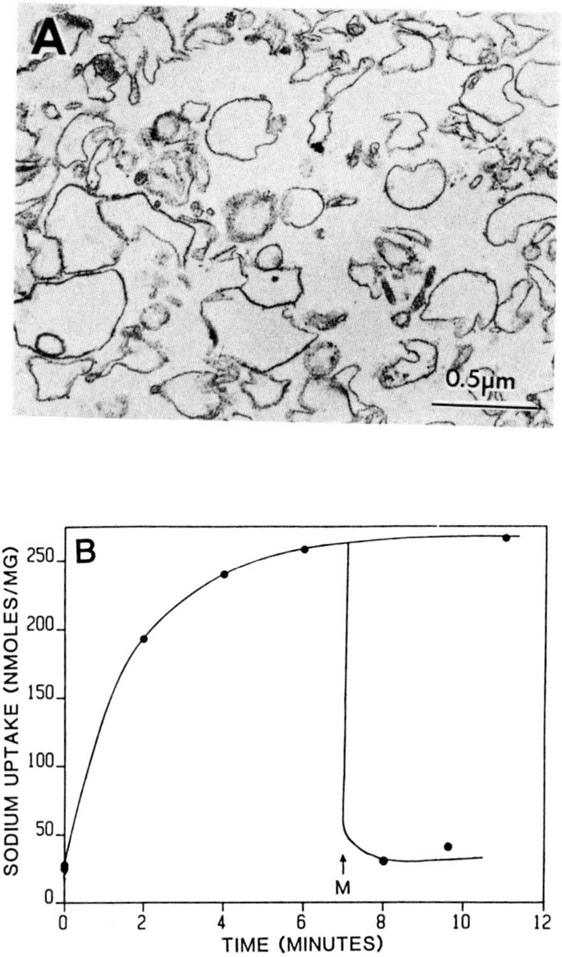

Figure 13. Characterization of isolated plasma membrane vesicles. A) Electron micrograph; B) Active transport of sodium ions into plasma membrane vesicles. The active accumulation of ^{22}Na was measured using the filtration technique. The arrow indicates the addition of monensin (M)(43).

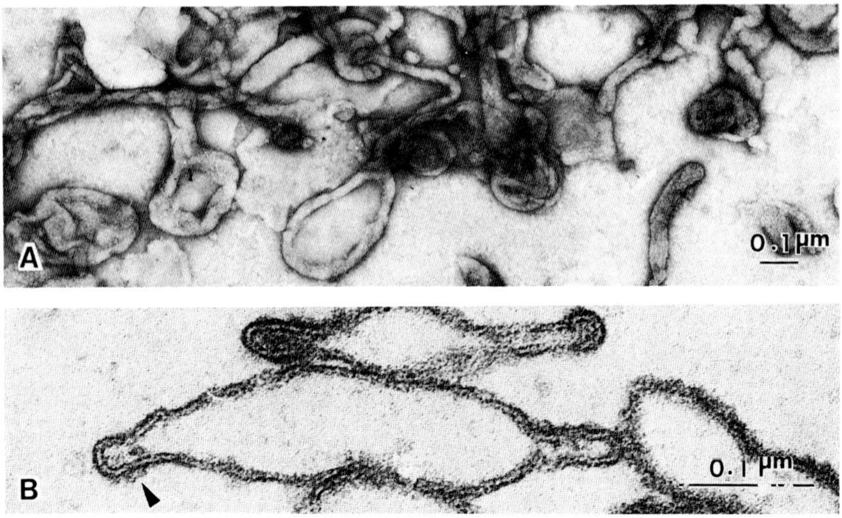

Figure 14. Isolated fraction of transverse tubules. A) Negatively stained preparations at low magnification (44). B) Transverse tubule at higher magnification in thin section. The transverse tubules are elongate with electron dense material at the nipple-like ends which contain structures (arrowhead) at the surface suggesting a role in junctional association (46).

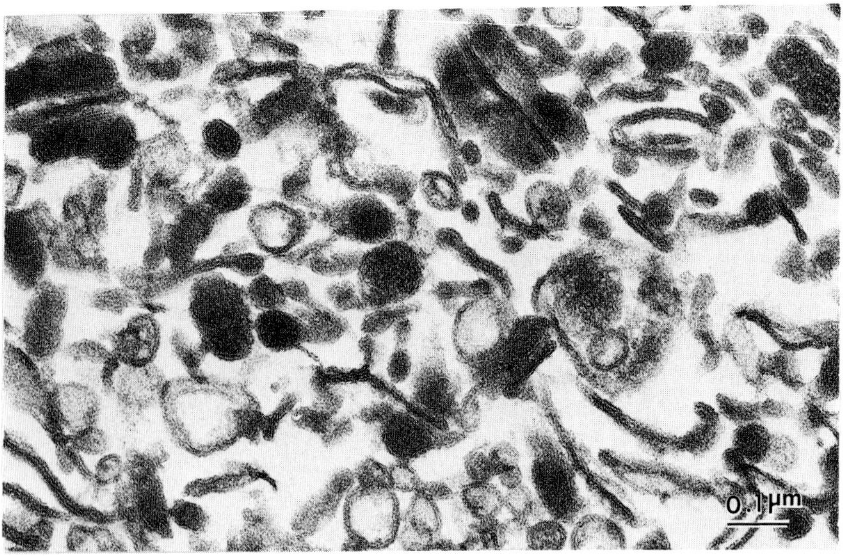

Figure 15. Electron micrograph of isolated triads in thin section (45,46).

We have succeeded in isolating an enriched triad fraction (Figure 15). A single transverse tubule can be junctionally associated with one to several vesicles of terminal cisternae. In the purified triads, most of the vesicles are junctionally associated as analyzed by particle counting of electron micrographs. Thin section electron microscopy reveals that the junctional association and "feet" structures are largely preserved in the isolated triads. The junctional association of the triads can also be observed using freeze-fracture electron microscopy (45,46).

A three dimensional perspective of the isolated triad is obtained by negative staining (Figure 16A). The transverse

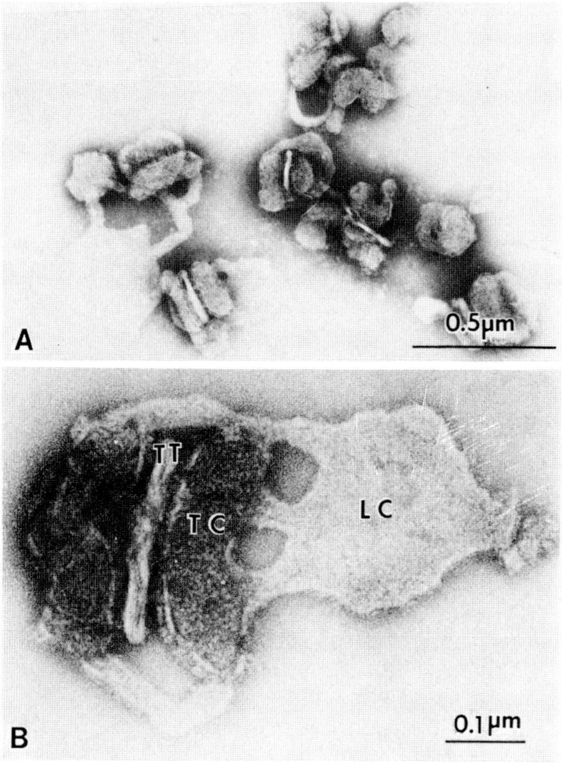

Figure 16. Electron micrograph of isolated triads visualized by negative staining. A) Isolated triads at low magnification. B) Isolated triad structure with associated longitudinal cisterna (LC). Fenestrations can be observed near the attachment of the longitudinal cisterna to the terminal cisterna (TC)(46).

tubule appears electron opaque and is junctionally associated with the terminal cisternae on both sides. In the triad, observed at higher magnification (Figure 16B), the anastomosing of regions of longitudinal and terminal cisternae (46) give rise to the fenestrated appearance between the two types of SR. The retention of morphology by the isolated triads and the ability of the component organelles to carry out transport suggest that the triads may be valuable in the study of the Ca^{2+} release process (47).

Recently, we have isolated a fraction enriched in terminal cisternae with morphologically well defined junctional face membrane and feet structures (48). Thin section and negative staining of terminal cisternae vesicles, (Figure 17), give comparable information. The terminal cisternae contain two types of membranes, which can be observed in the same vesicle, the calcium pump containing membrane and the junctional face membrane with characteristic "feet" structures. The "feet" structures extend some 120 A° from the surface of the membrane and are about 230 A° wide. The space between the "feet" structures is devoid of the surface material characteristic of the calcium pump containing membrane.

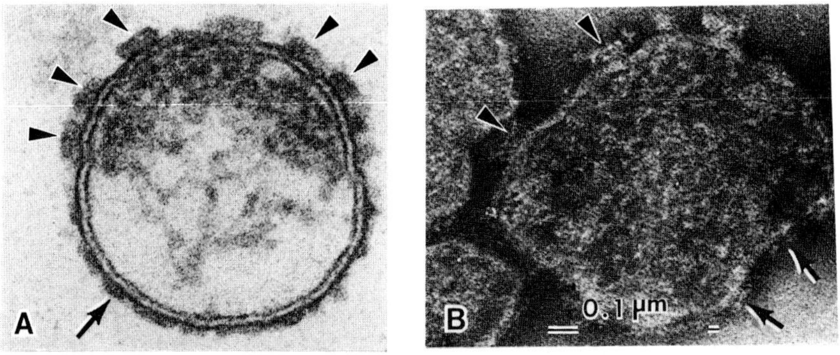

Figure 17. Comparison of isolated terminal cisternae vesicles by thin section and negative staining electron microscopy. Two types of membranes are discernable in the same vesicle, the junctional face membrane containing the junctional feet structures (arrowheads) and the calcium pump membrane (arrows).

Tangential sections of the terminal cisternae pellet reveal the shape and arrangement of the feet structures at the junctional face membrane, (Figure 18A). The feet are square-like structures organized in checkerboard array. They can be

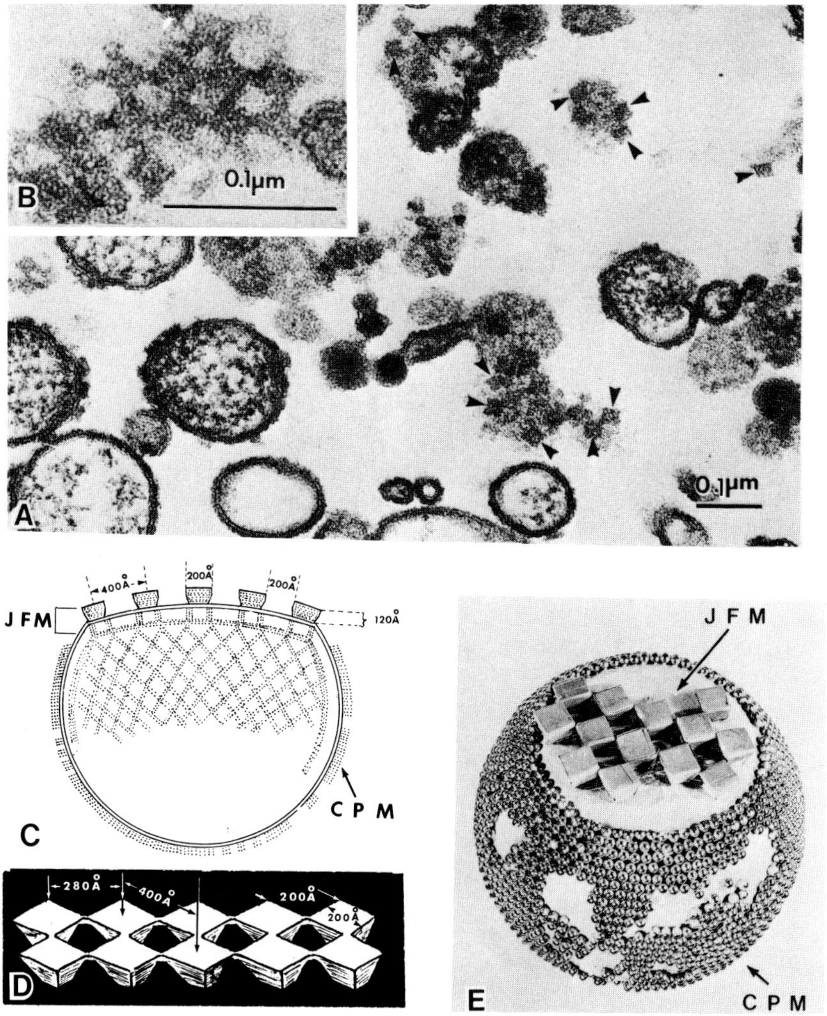

Figure 18. Structure of isolated terminal cisternae. A) Numerous feet structures are observed (arrowheads) in thin section, approximately tangential to the junctional face. B) The arrays of alternating square-shaped feet, 20 nm on each side, and spaces give a checkerboard-like lattice. Diagrammatic representations of terminal cisternae and the junctional face membrane are shown in C-E. The junctional feet and intervening spaces are depicted as square-shaped, with a checkerboard lattice arrangement (D and E). Two types of membranes are shown (C and E). The junctional face membrane (JFM) at the upper surface contains the feet structures. The remainder is Ca^{2+} pump membrane (CPM)(48).

observed throughout the section, as indicated by arrowheads. Figure 18B gives a close-up view of the checkerboard array of the "feet" structures.

The anatomy of the terminal cisternae is diagrammatically represented in Figures 18C,D,E. The terminal cisternae consist of two types of membranes, the calcium pump containing membrane and the junctional face membrane containing "feet" structures in checkerboard array. The calcium pump membrane consists mainly of calcium pump indicated by particles. The compartment of the terminal cisternae contains calcium binding protein, which is anchored to the junctional face membrane. The calcium binding protein serves as a reservoir for calcium. By contrast, the longitudinal cisternae consist essentially of the calcium pump containing membrane and are devoid of such compartmental contents (48).

Terminal cisternae vesicles are quite leaky to calcium ions so that the calcium loading rate energized by ATP is quite low. However, addition of ruthenium red (42,49) enhances the calcium accumulation rate five to ten fold (50). Ruthenium red has no effect on light SR and there is only a small enhancement of Ca^{2+} loading rate in isolated triads (50). These results suggest that terminal cisternae contains calcium release channels which are usually maintained in the closed state by the transverse tubule.

Our working hypothesis regarding calcium release from skeletal muscle is that calcium release occurs in the terminal cisternae via calcium release channels probably located in the junctional face membrane. The transverse tubule normally maintains the calcium channels in the closed state. Calcium release occurs by gating of the calcium release channels in the terminal cisternae via depolarization of the transverse tubule.

ACKNOWLEDGMENTS

The author is grateful to his many coworkers and collaborators who participated in these studies (cf. references). Special thanks go to Akitsugu Saito who carried out the electron microscopy and provided the illustrations and J. Oliver McIntyre for his comments on the manuscript.

REFERENCES

1. Peachey, L.D., Adrian, R.H. and Geiger, S.R. (editors), "Skeletal Muscles, Handbook of Physiology, Am. Physiol. Soc., Bethesda, MD (1983).
2. Ham, A.W. and Cormack, D.H., "Histology" Eighth Edition, J.P. Lippincott Co., Philadelphia and Toronto, pp 554 (1979).
3. Franzini-Armstrong, C., Fed. Proc. 39, 2403-2409 (1980).
4. Franzini-Armstrong, C., J. Muscle Res. Cell Motil. 4, 233-252 (1983).
5. Saito, A., Wang, C-T. and Fleischer, S., J. Cell Biol. 79, 601-616 (1978).
6. Hasselbach, W., Progr. Biophys. Mol. Biol. 14, 167-222 (1964).
7. Ebashi, S., Endo, M. and Ohtsuki, I., Q. Rev. Biophys. 2, 351-384 (1969).
8. Tada, M., Yamamoto, T. and Tomomura, J., Physiol Rev. 58, 1-79 (1978).
9. Ikemoto, N., Ann. Rev. Physiol. 44, 297-317 (1982).
10. Endo, M., Physiol. Rev. 57, 71-108 (1977).
11. Winegrad, S., Ann. Rev. Physiol. 44, 451-462 (1982).
12. de Meis, L. and Inesi, G., in Membrane Transport of Ca^{2+} (E. Carafoli, editor) Academic Press, London and New York, pp 141-181 (1982).
13. Moller, J. V., Anderson, J. P., LeMaire, M., Mol. Cell. Biochem. 42, 83-107 (1982).
14. Tanford, C., Ann. Rev. Biochem. 52, 379-409 (1983).
15. Fleischer, S., Wang, C-T., Saito, A., Pilarska, M. and McIntyre, J.O., in Cation Flux Across Biomembranes. (Y. Mukohata and L. Packer, Eds.) Academic Press, pp 193-205 (1979).
16. Fleischer, S., Wang, C-T., Hymel, L., Seelig, J., Brown, M.F., Herbette, L., Scarpa, A., McLaughlin, A.C. and Blasie, J.K., in "Function and Molecular Aspects of Biomembrane Transport". E. Quagliariello, F. Palmieri, S. Papa and M. Klingenberg, editors. Elsevier/North Holland, pp. 465-485. (1979).
17. Meissner, G., Conner, G.E. and Fleischer, S., Biochim. Biophys. Acta 298, 246-269 (1973).
18. MacLennan, D.H., Yip, C.C., Iles, G.H. and Seeman, P., Cold Spring Harbor Symp. Quart. Biol. 37, 469-177, New York (1972).
19. MacLennan, D.H. and Wong, P.T.S.. Proc. Natl. Acad. Sci. U.S. 68, 1231-1235 (1971).
20. Meissner G. and Fleischer, S., Biochim. Biophys. Acta 241, 356-378 (1971).

21. Meissner, G., Biochim. Biophys. Acta 389, 51-68 (1975).
22. Meissner, G., Biochim. Biophys. Acta 298, 906-926 (1973).
23. Meissner, G. and Fleischer, S., J. Biol. Chem. 249, 302-309 (1974).
24. Racker, E., J. Biol. Chem. 247, 8198-8200 (1972).
25. Inesi, G., Nakamoto, R., Hymel, L. and Fleischer, S., J. Biol. Chem. 258, 14804-14809 (1983).
26. Hymel, L. and Fleischer, S., Fed. Proc. 42, 2243a (1983).
27. Wang, C-T., Saito, A. and Fleischer, S.. J. Biol. Chem., 254, 9209-9219 (1979).
28. Seelig, J., Tamm, L., Hymel L., and Fleischer, S., Biochemistry 20, 3922-3932 (1981).
29. Martonosi, A., Donley, J. and Halpin, R.A., J. Biol. Chem. 243, 61-70 (1968).
30. Meissner, G. and Fleischer, S., Biochim. Biophys. Acta 255, 19-33 (1972).
31. Dean, W.L. and Tanford, C., Biochemistry 17, 1683-1690 (1978).
32. Lunardi, J., DeFoor, P. and Fleischer, S., manuscript in preparation.
33. Bloj, B. and Zilversmit, D.B., Methods in Enzymol. 98, 574-581 (1983).
34. Metcalfe, J.C. and Warren, B.B., in "International Cell Biology" (Brinkley, B.R. and Porter, K.R., Eds.) Rockefeller University Press, pp. 15-23 (1977).
35. Hymel, L., Maurer, A., Berenski, C., Jung, C. and Fleischer, S.,(This volume).
36. Blasie, J.K., Herbette, L., Pierce, D.H., Pascolini, D., Skita, V., Scarpa, A. and Fleischer, S.,(This volume).
37. McLaughlin, A.C., Blasie, J.K., Herbette, L., Wang, C-T., Hymel, L. and Fleischer, S., Biochim. Biophys. Acta, 643, 1-16 (1981).
38. Deese, A.J., Hymel, L., Fleischer, S. and Dratz, E.A., Biophys. J. 37, 207-216, (1982).
39. Fleischer, S. and McIntyre, J.O., in Ann. N.Y. Acad. Sci. (E. Carafoli and A. Scarpa, eds.) 402, pp. 558-560 (1982).
40. Kasai, M., Nunogaki, K., Nagasaki, K. and Tanifugi, M., (This volume).
41. Ikemoto, N. and Kim, D.H., (This volume).
42. Ohnishi, S.T., J. Biochem. 86, 1147-1150 (1979).
43. Seiler, S. and Fleischer, S., J. Biol. Chem. 257, 13862-13871 (1982).
44. Saito, A., Seiler, S. and Fleischer S., Studies in progress.
45. Mitchell, R., Palade, P. and Fleischer, S., J. Cell Biol. 96, 1008-1016 (1983).

46. Mitchell, R., Saito, A., Palade, P. and Fleischer, S., J. Cell Biol. 96, 1017-1029 (1983).
47. Mitchell, R.D., Volpe, P., Palade P. and Fleischer, S., J. Biol. Chem. 258, 9867-9877 (1983).
48. Saito, A., Seiler, S., Chu, A. and Fleischer, S., J. Cell Biol. 99, 875-885, (1984).
49. Miyamoto, H. and Racker, E., J. Membr. Biol. 66, 193-201 (1982).
50. Chu. A., Volpe, P., Costello, B. and Fleischer, S., Manuscript in preparation.

A NEW PROCEDURE FOR THE RECONSTITUTION OF SARCOPLASMIC RETICULUM VESICLES BY SALTING-OUT METHOD

Takahisa Taguchi
Michiki Kasai

Department of Biophysical Engineering
Faculty of Engineering Science
Osaka University
Toyonaka, Osaka, Japan

I. INTRODUCTION

Both ion channels and ion transporters exist in membranes of sarcoplasmic reticulum. In order to study the structure and the function of them, reconstitution of the transport systems is indispensable. In the last decade, many successful reconstitution experiments have been reported. Concerning sarcoplasmic reticulum, Meissner and Fleischer (1), Zimniak and Racker (2) and Chiesi,et al. (3) reported reconstitution of Ca^{2+}-ATPase in sarcoplasmic reticulum membrane. However,we can find only a few cases of the reconstitution of transport systems with Triton X-100 even though it is a non-ionic "mild" surfactant and is convenient to solubilize and purify them. This is because the formation of vesicles is made difficult by the large micelle size and the low critical micelle concentration of Triton X-100.

As mentioned by Maclay (4), non-ionic surfactants such as Triton X series have an ethylene oxide chain as a hydrophilic portion. The high water solubility of the polyethylene oxide chain is due to the hydrogen bonding between the solvent and the ether oxygen atoms in the chain. Since the hydrogen bonding is a temperature sensitive phenomenon, for each non-ionic surfactant there exists a temperature at which the degree of hydration of the hydrophilic portion becomes insufficient to solubilize the remaining hydrocarbon fragment. Consequently, the surfactant forms large aggregation and turbidity results. There is generally a well-defined temperature and is called the "cloud

ISBN 0-12-260380-X

point". Recently, Bordier (5) reported that the phase separation in solutions of the non-ionic surfactant, Triton X-114, separated integral membrane proteins from hydrophilic proteins. The low cloud point at about 20°C of Triton X-114 made this separation possible. We investigated the possibility of utilizing Triton X-114 for the formation of membrane vesicles.

In this report, we describe a new method to form membrane vesicles. It was found that an addition of salt to the solution containing Triton X-114 and lipid forms membrane vesicles and this phenomenon can be utilized to reconstitute sarcoplasmic reticulum vesicles.

II. CLOUD POINT OF TRITON X-114 SOLUTION IN VARIOUS CONDITIONS

A solution of 1%(v/v) of Triton X-114, 5 mM sodium Hepes (pH 7.4) has a cloud point at about 20°C. But, the cloud point disappeared when lipid was added, as shown in FIGURE 1(A). An

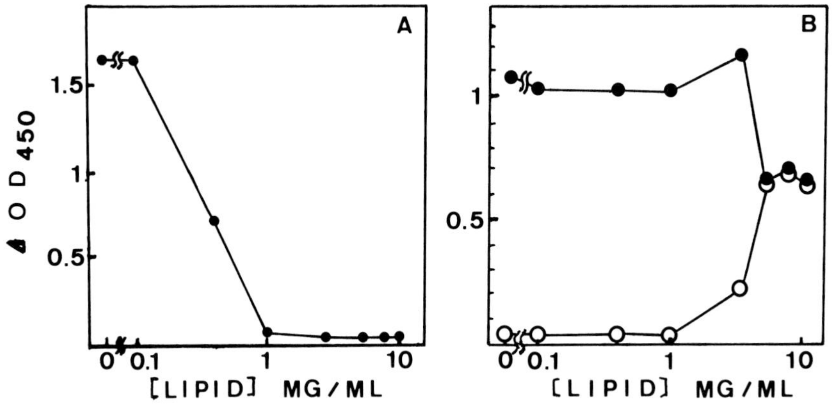

FIGURE 1. (A) Effect of lipid on the cloud point of the solution of 1%(v/v) Triton X-114. The increments of optical density at 450 nm were measured when the temperature in the solutions containing 1%(v/v) Triton X-114, different concentrations of asolectin and 5 mM sodium Hepes (pH 7.4) were changed from 20°C to 30°C. The mixture of asolectin and Triton X-114 was stirred by voltex mixer. (B) Effect of lipid on the turbidity change of the solution of Triton X-114 and NaCl. closed circles (●); optical density measured at 20°C after the addition of 200 mM (final) NaCl into the same solutions in (A). open circles (○); optical density of the supernatant measured at 20°C after the centrifugation (1,300 x g, for 10 min). (from Taguchi and Kasai (6))

addition of 200 mM(final) NaCl made the solutions of Triton X-114 and various concentrations of lipid turbid at 20°C. But the appearances of the solutions were different depending on the lipid concentration. The solutions containing lipid at less than 3 mg/ml were light gray and become clear with the decrease of temperature in the ice bath. On the other hand, those containing lipid at more than 5 mg/ml were ivory-white and remained turbid with the decrease of temperature. As shown in FIGURE 1(B) (open circles), the supernatants obtained by the low speed centrifugation of the solutions containing lipid at less than 3 mg/ml were clear, but those of the solutions of more than 5 mg lipid/ml were still turbid. Furthermore, the former supernatants did not contain Triton X-114 and the latter contained about 90% of Triton X-114. These results indicate that the material obtained by the addition of salt into the Triton X-114 solution with lipid at more than 5 mg/ml was not the aggregation of Triton X-114 but that of lipid.

In the next step, we tried to separate the aggregated lipid from Triton X-114. Since the aggregated lipid sediments by centrifugation (100,000 x g, for 60 min), washing is possible. For better elimination of Triton X-114,however, hydrophobic chromatography was used with ω-amino-n-octyl derivative of Sepharose 4B. This procedure could eliminate more than 99% of Triton X-114 within 15 min.

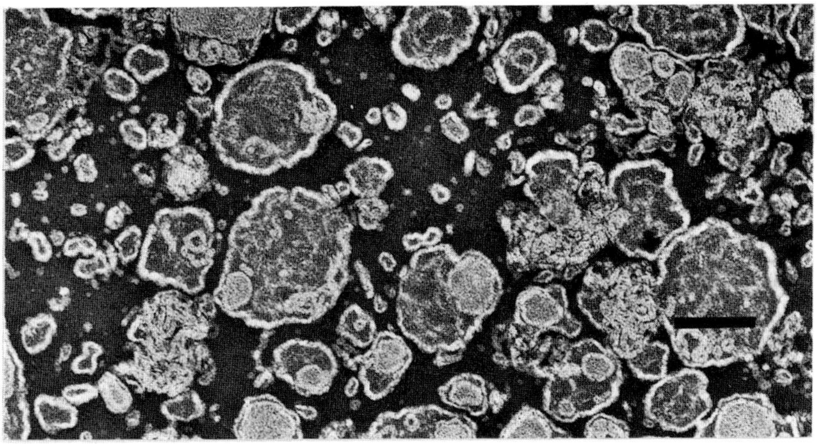

FIGURE 2.Electron micrograph of vesicles formed by the salting-out method. The vesicles were formed by the addition of 200 mM (final) KCl into the solution containing 8 mg/ml asolectin, 1% (v/v) Triton X-114 and 5 mM sodium Hepes (pH 7.4). Triton X-114 in vesicle solution was eliminated by hydrophobic chromatography and then washed three times with 5 mM sodium Hepes (pH 7.4) according to Taguchi and Kasai (6). Sample was negatively stained with uranyl acetate. A bar represents 200 nm.

III. FORMATION OF CLOSED VESICLES BY THE ADDITION OF SALT

The structure of the aggregated lipid, which was formed by the addition of 200 mM KCl, was confirmed to be closed membrane vesicles by electron microscopy (FIGURE 2). The mean diameter of the vesicles was 110 nm $\pm$ 69 nm(S.D.).

As shown in FIGURE 3, the membrane of these vesicles was semipermeable. According to Kometani and Kasai (7), the increase of scattered light intensity corresponds to the shrinkage of the vesicles due to the outflow of water caused by the osmotic pressure difference across the membrane, and the following decrease corresponds to their swelling due to the water inflow accompanied with the inflow of molecules or ions. FIGURE 3, therefore, indicates that the membrane of the vesicles has no permeability to KCl, glycine and glucose, but has high permeability to glycerol, ethyleneglycol and water.

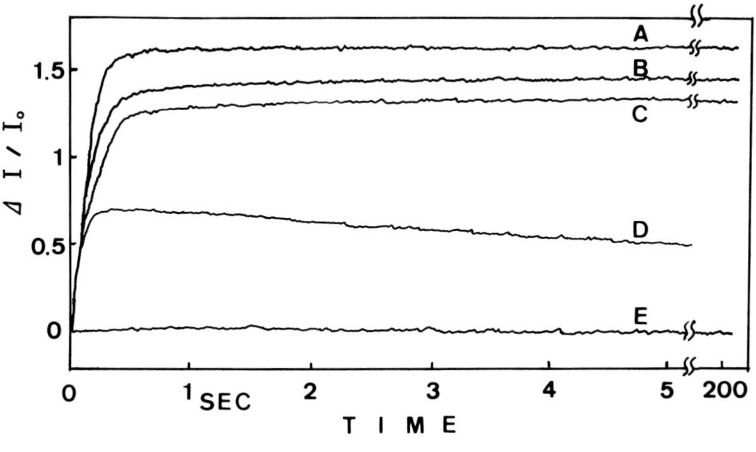

FIGURE 3. Osmotic volume change of vesicles. The vesicles were formed as described in the legend to FIGURE 2 and suspended in the same buffer overnight at 4°C. These vesicle suspensions at 23°C were mixed with an equal volume of 5 mM sodium Hepes (pH 7.4) containing various solute. Then the changes in the intensity of the scattered light on rapid mixing were measured at 450 nm by a stopped-flow apparatus. The ordinate indicates the ratio of the increment of the intensity of scattered light (ΔI) to the intensity before the shrinkage of the vesicles (I_0); (A) 50 mM(final) KCl, (B) 100 mM glycine, (C) 100 mM glucose, (D) 100 mM glycerol, (E) 100 mM ethyleneglycol. (from Taguchi and Kasai, (6)).

IV. FORMATION OF VESICLES BY THE ADDITION OF SALT UNDER VARIOUS CONDITIONS

The dependence of vesicle formation on NaCl concentration is shown in FIGURE 4. More than 100 mM NaCl was required to get maximal increment of optical density. Similar experiments were carried out by using various kinds of salts. The values of maximal increments of optical density at 450 nm can be arranged with regard to anions and to cations, as follows:

anion (cation; K):
Oxalate > F > SO_4 > methanesulfonate > Cl > Br > NO_3 > I > SCN
cation (anion; Cl):
Cs > Li > K > Na > choline > Tris

These orders are similar to the lyotropic series, especially for anions. Neutral molecules, such as sucrose and urea, were not effective on the vesicle formation. We, therefore, name this new method of forming vesicles by addition of salt, "Salting-Out Method".

The dependences of vesicle formation on pH and on temperature are shown in FIGURE 5. The yield of vesicles was virtually invariant in the range of biochemically important pH, but increased with temperature from 0°C to 30°C.

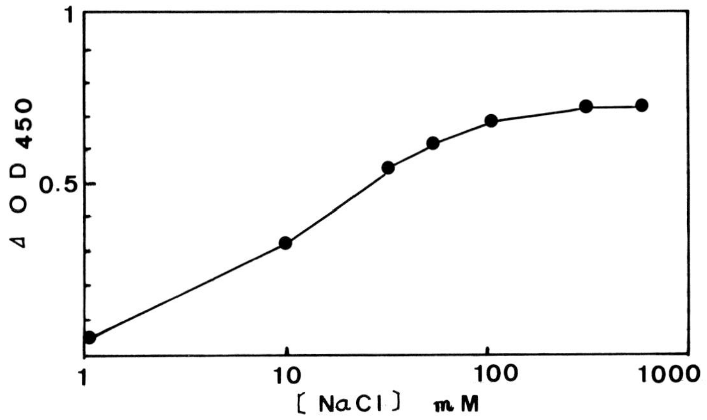

FIGURE 4. Vesicle formation at various concentrations of NaCl. The increments of optical density (450 nm) of the solution of 8 mg/ml asolectin, 1%(v/v) Triton X-114 and 5 mM sodium Hepes (pH 7.4) were measured at 20°C after the addition of NaCl. (from Taguchi and Kasai (6)).

Effects of the contamination of other surfactants upon the formation of vesicles by the salting-out method were shown in FIGURE 6. High concentration of the second surfactants prevent the vesicle formation. Among the surfactants studied, cholate was most effective and the increment of optical density became negative at the concentration of cholate more than 0.4%(w/v). This is because the critical micelle concentration of the ionic surfactant such as cholate was decreased by the addition of salt and then the solution with Triton X-114 and lipid became clearer. These data are important in the reconstitution experiments for the selection of the suitable surfactant and of its concentration to solubilize and to purify the membrane-bound proteins.

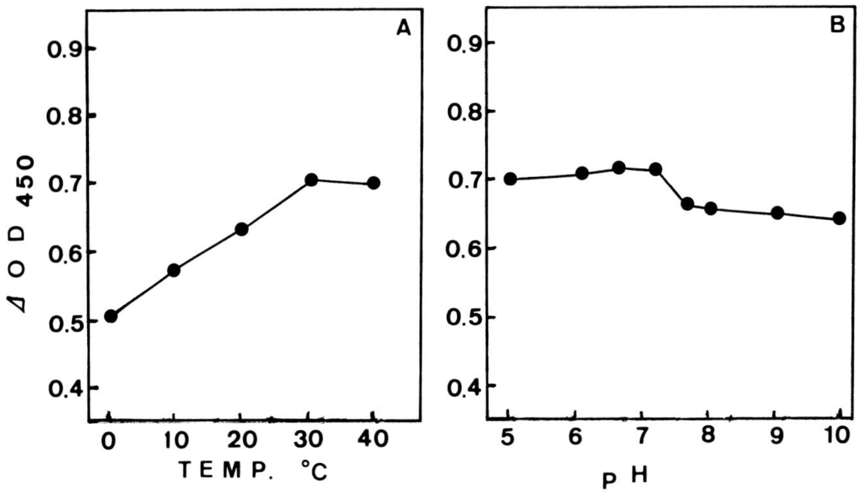

FIGURE 5. Effects of temperature and pH on the vesicle formation by the salting-out method. (A) The change of optical density at 450 nm was measured as soon as 200 mM(final) NaCl was added into the temperature-controlled solution containing 8 mg/ml asolectin, 1%(v/v) Triton X-114 and 5 mM sodium Hepes (pH 7.4). (B) The change of optical density at 450 nm was measured at 20°C when 200 mM(final) NaCl was added into the solution containing 8 mg/ml asolectin, 1%(v/v) Triton X-114, and 5 mM sodium Hepes (pH 7.4).

V. THE APPLICATION OF SALTING-OUT METHOD TO THE RECONSTITUTION OF SARCOPLASMIC RETICULUM MEMBRANE

The reconstitution of sarcoplasmic reticulum vesicles by the salting-out method was performed as follows. The suspension of sarcoplasmic reticulum vesicles (12 mg/ml), 1%(v/v) Triton X-100 and 5 mM sodium Hepes (pH 7.4), and the solution of 8 mg/ml asolectin, 1%(v/v) Triton X-114 and 5 mM sodium Hepes (pH 7.4) were centrifuged (100,000 x g, for 60 min), separately at 4°C to remove unsolubilized materials. The supernatant of the former solution (1.5 ml) and that of the latter solution (13.0 ml) were mixed at 20°C and then concentrated KCl (or oxalate) was added into the mixture (final 200 mM).

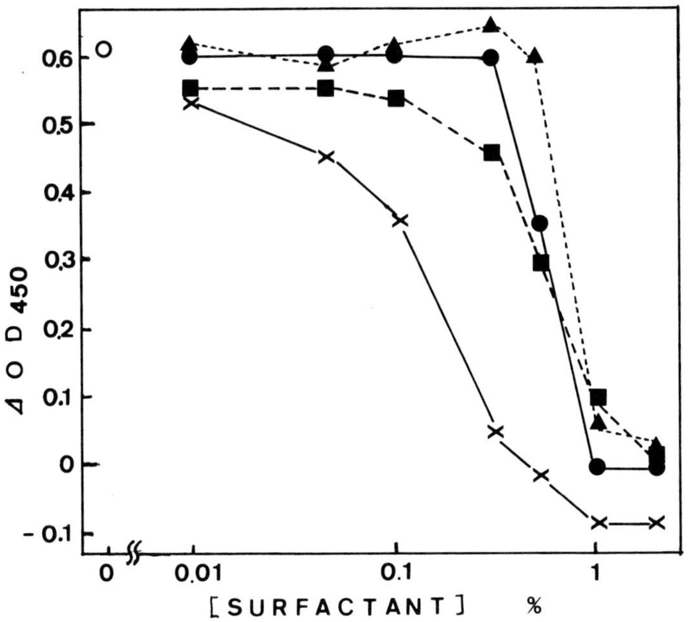

FIGURE 6. Effects of the second surfactants of the vesicle formation by the salting-out method. The change of optical density at 450 nm was measured at 20°C when 200 mM(final)NaCl was added into the solution containing 8 mg/ml asolectin, 1%(v/v) Triton X-114,5 mM sodium Hepes (pH 7.4) and various concentration of the second surfactants: (▲)Nonidet P-40, (●) Triton X-100, (■) Lubrol PX, (✕)Na-cholate. The abscissa indicates the concentration of the second surfactants. The unit of the concentration is %(v/v) for the non-ionic surfactants (Nonidet P-40, Triton X-100 and Lubrol PX) and %(w/v) for ionic one (Na-cholate). (from Taguchi and Kasai (6)).

This turbid mixture with salt was applied to a column (2.2 x 30 cm) containing ω-amino-n-octyl derivative of Sepharose 4B which had been washed with 200 mM KCl (or K_2oxalate) and 5 mM sodium Hepes (pH 7.4), and then eluted with the same solution. The turbid fractions were collected and washed 3 times with suitable buffer solution by centrifugation (100,000 x g, for 60 min) to remove the residual Triton X-114. We got 40 to 60%(w/w) recovery of sarcoplasmic reticulum protein in the reconstituted membrane.

SDS polyacrylamide gel electrophoresis represented that Ca^{2+}-ATPase was incorporated into the membrane of the reconstituted vesicles but calsequestrin was not and that some proteins of molecular weight more than 100,000 were also incorporated into the membrane. The reconstituted membrane vesicles could hydrolyze ATP (363 nmole ATP/min/mg protein) and take up Ca^{2+} (115 nmole Ca^{2+}/min/mg protein).

The new method, "salting-out", to form vesicles has some advantages to the reconstitution experiments of transport systems in biomembranes as follows. (1) Completion of the reconstitution process takes place in a short time. The vesicle formation is completed within 1 sec after addition of salt. This is favorable for incorporation of unstable proteins into membrane vesicles. (2) Availability of non-ionic surfactants to solubilize the membrane protein. (3) High yield and good reproducibility. This method has potential for application to reconstitution along with other methods including the dialysis method, the freeze-thaw method, the dilution method and the sonication method.

REFERENCES

1.Meissner,G., and Fleischer,S.,J.Biol.Chem. 249:302 (1974)
2.Zimniak,P., and Racker,E., J.Biol.Chem. 253:631 (1978)
3.Chiesi,M.,Peterson,S.W.,and Acuto,O., Arch.Biochem.Biophys. 189:132 (1978)
4.Maclay,W.N., J.Colloid Sci., 11:272 (1956)
5.Bordier,C., J.Biol.Chem. 256:1604 (1981)
6.Taguchi,T.,and Kasai,M., Biochim.Biophys.Acta in the press (1983)
7.Kometani,T.,and Kasai,M., J.Memb.Biol. 41:295 (1978)

TARGET SIZE OF CALCIUM PUMP PROTEIN FROM SKELETAL MUSCLE SARCOPLASMIC RETICULUM[1]

Lin Hymel, Andreas Maurer, Charles Berenski*, Chan Jung* and Sidney Fleischer

Department of Molecular Biology, Vanderbilt University, Nashville, Tennessee

*Biophysics Laboratory, Veterans Administration Medical Center State University of New York at Buffalo, Buffalo, NY

I. INTRODUCTION

The quaternary structure of the calcium pump protein (CPP) within the sarcoplasmic reticulum (SR) membrane has been the subject of many investigations. An early study (1) using classical target analysis of lyophilized SR membranes estimated a molecular weight of 190,000 daltons for the Ca^{++}-ATPase activity. This was even before the calcium pump protein was identified as a band of the monomeric size of 115,000 to 119,000 daltons (2). Using freeze-fracture electron microscopy (3) or labeling studies with fluorescein isothiocyanate (4) the size of the CPP was estimated to be a dimer or a larger oligomer. Several other studies have suggested a variety of sizes, from monomer (5,6,7) to tetramer (8). The tendency of detergent-solubilized CPP to associate into a variety of multimeric forms is well documented (9). The oligomeric structure of the CPP has remained at issue. In the present work, we used the radiation-inactivation technique (10) to measure the molecular size of CPP in normal and reconstituted SR (12) membrane vesicles.

II. MATERIALS AND METHODS

Highly purified sarcoplasmic reticulum was prepared as previously described (11). Reconstitution of SR was carried essentially according to Meissner and Fleischer (12).

Target irradiation procedure. SR or RSR vesicles were diluted to 0.75-2.9 mg protein/ml in 0.3 M sucrose, 0.1 M KCl and 5 mM HEPES pH 7.1 at 0°C. Glucose-6-phosphate dehydrogenase from Leuconostoc mesenteroides (Sigma, St. Louis, MO) was added to the vesicle suspension at 5 unit/ml for use as an internal standard for target size. The samples were then frozen on aluminum trays as described in (13) and maintained at liquid nitrogen temperature during transit for irradiation between Nashville, TN. and Buffalo, NY. The samples were

[1]Supported in part by NIH AM 14632 to SF and AM 13376 to CJ.

ISBN 0-12-260380-X

irradiated with a Van der Graff generator producing a 0.5 or 0.6 mA beam of 1.5 MeV electrons. The irradiation dose was measured using the transmittance change of blue cellophane calibrated against a chemical dosimeter (14). The dose for each sample was controlled by varying the number of passes through the electron beam from 0 to 10 at approximately 0.5 MRad/pass. The irradiation chamber was cooled with liquid nitrogen. Control samples had no loss of activity when subjected to the same manipulations but with the electron beam turned off. The percentage of enzymic activity or intact CPP remaining after irradiation was plotted on a logarithmic scale versus the measured radiation dose. Linear regression analysis was then performed to obtain the best straight line, including the extrapolated 100% activity, determined from all the points. The target size molecular weight (M_r) was calculated according to the formula of Kepner and Macey (15), Mr = 6.4 x 10"/D_{37}. D_{37} is the dose in Rad required to reduce the activity or polypeptide to 37% of the extrapolated 100% value. Similar analysis were made for the remaining glucose-6-phosphate dehydrogenase activity. In most cases the target size of this enzyme was within ± 15% of the expected value of 103,700 (16). Since this is approximately the typical variation between repeat molecular weight determinations, no correction was made.

Protein was determined according to Lowry, et al. using bovine serum albumin as standard (17). Phosphorus was measured using a modification (18) of Chen et al. (19). Calcium loading and ATPase rates were measured at 25°C in a medium containing 0.3 M Sucrose, 0.1 M KCl, 100 µM CaCl, 60 µM EGTA, 5.5 mM $MgCl_2$ and 5 mM NA_2 ATP. Calcium loading was determined using ^{45}Ca and filtration through 0.22 µ millipore filters. The ATPase assays were performed with 10 µg protein/ml and 1 µg/ml A 23187 (Calbiochem, San Diego, CA) in order to obtain maximal results.

The amount of CPP remaining in samples after irradiation was quantitated on SDS-polyacrylamide gels after silver-staining (20) or staining with Coomassie brilliant blue. SDS-PAGE was carried out according to Laemmli (21), using 10 or 11% polyacrylamide slab gels. The gels were then scanned at 557 nm (for Coomassie brilliant blue R) or 622 nm (for silver staining) with an E-C densitometer (E-C Apparatus Corporation, St. Petersburg, FL). The areas were calculated as the peak hight times the width at half hight. For each set of samples at least 2 gels were prepared and the average values of the peak areas were used for calculations. For calibration, different amounts of SR were electrophoresed and analyzed as described above. The absorbance of the CPP band was linear for silver-staining in the range of 0.25 to 2.5 µg and for Coomassie brilliant blue staining in the range of 1 to 10 µg of SR loaded on the gel.

III. RESULTS

The irradiation of sarcoplasmic reticulum vesicles resulted in a simple exponential loss of Ca^{++} loading activity, Ca^{++}-ATP'ase (Fig. 1a) and the CPP band observed by densitometry (Fig. 1b) as a function of radiation dose. The lines drawn in Figure 1 were determined by least square fit of the averaged data. The slope of these lines then provided the D_{37} value from which the average molecular weights of the target were calculated. The target size for both the Ca^{++} loading and Ca^{++}-ATPase ranged from 213,000 to 250,000. Since the molecular weight of the monomeric CPP is 115,000 to 119,000 (2), these target sizes are consistent with a dimer. The results are summarized in Table 1.

The Ca^{++}-ATPase was measured in the presence of ionophore, so that the integrity of the membrane did not limit the activity, and the decline of the activity was linear to 75% of

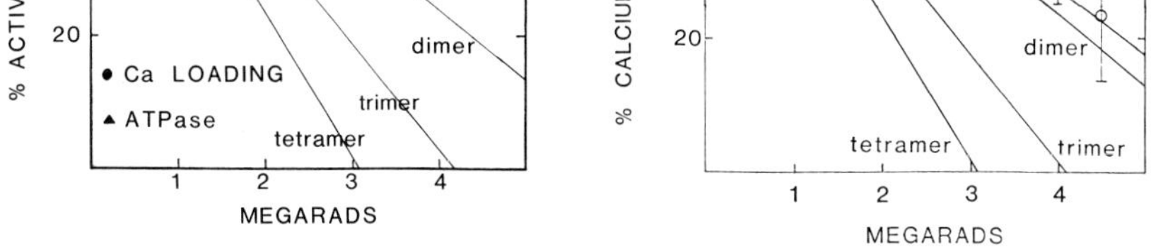

FIGURE 1. Samples were irradiated in the frozen state at approximately -40°C (Fig. 1a). Calcium loading and Ca-ATP'ase rates (average of 2 experiments) were measured in the presence of 5 mM oxalate at 25°C; 100% = 2.2 μmoles Ca^{++}/min/mg protein. Samples which received more than about 2.5 megarads were essentially unable to accumulate calcium (not shown). ATP'ase rates (average of 4 experiments) were measured in the presence of 1 μg/ml A23187 to give maximal activities. 100% = 4.0 μmoles Pi/min/mg protein (Fig. 1b). Fragmentation of CPP polypeptide was determined by densitometry after SDS-polyacylamide gel electrophoresis (average of 6 gels), cf Figure 2. Bars indicate standard deviations. The calculated slopes for monomer, dimer, trimer, and tetramer are indicated.

loss of initial activity and greater (Fig. 1a). However, for Ca^{++} pumping the membranes must remain intact. The decline of activity was linear to only 60% loss of initial activity. Samples irradiated with more than 2.95 Megarads showed a sharp decrease in Ca^{++} loading due to leakiness of the membrane. Such a sharp decline is typical of free radical catalyzed lipid peroxidation induced by the high energy electron bombardment.

The integrity of the polypeptide chain of the calcium pump protein before and after irradiation was followed by SDS-PAGE (Fig. 2). A decline in the staining intensity of the CPP band

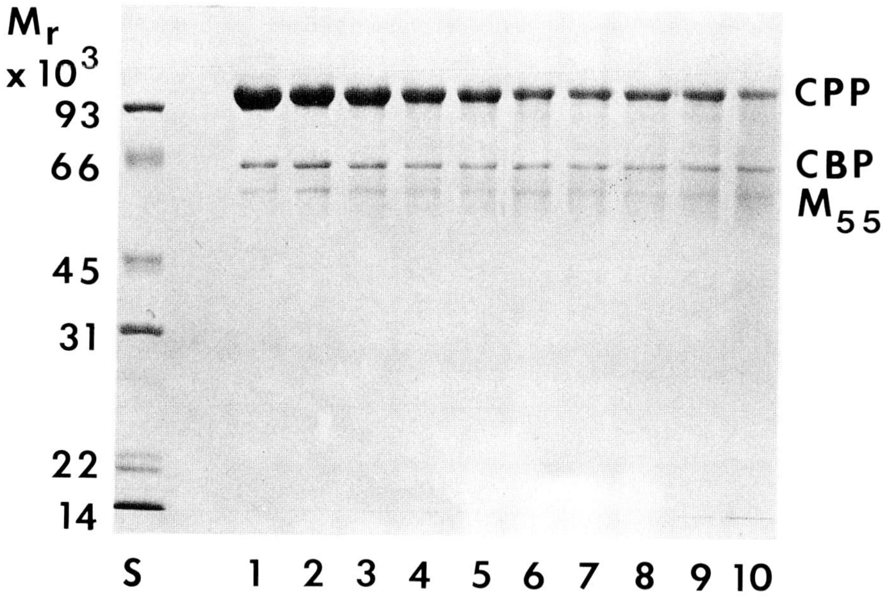

FIGURE 2. SDS-gel electrophoresis of normal sarcoplasmic reticulum reticulum using 11% polyacrylamide slab gels. Staining was with Coomassie brilliant blue R. The samples were exposed to 1-9 passes of irradiation with a dose of 0.49 Mrad per pass. Sample 1 (control) was not irradiated. Samples 2-10 were irradiated with 1-9 passes, respectively, (0.49-4.41 Mrad). S refers to the molecular weight standards. The bands referable to the calcium pump protein (CPP), the calcium binding protein (CBP) or calsequestin, and M_{55} are indicated to the right of the gel. Ten μg of each sample were applied to the gel. Staining of CPP was found to be linear from 1-10 μg. The loss of band intensity of the calcium pump protein and the increase in degradation products with higher irradiation are clearly discernable.

TABLE I. Target Size of Calcium Pump Protein (CPP)[a]

	TARGET SIZE MOLECULAR WEIGHT (KILODALTONS)		
SR MEMBRANE	Ca^{++} loading	ATPase	CPP
NORMAL	251±35 (6)	214±14 (4)	213±31 (6)
RECONSTITUTED	-------c	242±45 (3)	227±40 (3)

[a]Target size was calculated from the three characteristics as indicated (cf text). Values are the mean of the data ± standard deviation. The number of experiments is indicated in parenthesis.

[b]The lipid/protein molar ratio of the RSR preparations used for this experiment ranged from 52 to 105 moles phospholipid/mole calcium pump protein, compared with 115 for the normal SR membrane. The target size did not vary within this range of lipid/protein ratio.

[c]The calcium loading activity of irradiated reconstituted SR decreased rapidly and was non linear with respect to radiation dose on semilog plot. Reconstituted SR is even more susceptible to radiation damage than normal SR.

with increasing irradiation was observed both with silver staining and Coomassie brilliant blue staining and a concomitant increase of smaller polypeptides was observed. Like Ca^{++}-ATPase activity, the fragmentation of CPP polypeptide was linear in the range studied to 75% loss of the initial value or to irradiation of up to 4.95 Megarads. The lines in Figure 1 show theoretical slopes for monomer, dimer, trimer, and tetramer. The slope of the intact CPP polypeptide chain remaining exhibits a target size of 220,000 daltons and demonstrates clearly that the CPP is a dimer. Similar studies were carried out for reconstituted SR. The CPP in the reconstituted SR membrane is also a dimer.

IV. DISCUSSION

Classical target theory analysis of radiation inactivation data leads to a definitive conclusion regarding the oligomeric size of the calcium pump protein in the SR membrane. The CPP in the sarcoplasmic reticulum membrane is a dimer.

The methodology for target inactivation studies is now vastly improved. No longer are lyophilized samples required. Instead the sample can be irradiated as a suspension of vesicles in the frozen state. Therefore, Ca^{++} pumping can also be measured and used for the molecular weight determination.

In addition to loss of enzymic function, the fragmentation of the CPP can be directly correlated with the irradiation dose. An internal standard, glucose-6-phosphate dehydrogenase, has been included to insure quantitation of oligomeric size. Radiation inactivation analysis is a simple relatively non-perturbing method that is less prone to the kind of artifacts that may result from crosslinking and chemical labeling studies. It is well suited for determination of molecular size in the native membrane. The properties of the CPP were readily analyzed after radiation in the frozen state. In the present work, loss of Ca^{++}-ATPase, Ca^{++} loading activity and CPP-polypeptide remaining, as a function of the radiation dose yielded similar results for both normal and reconstituted SR. The CPP is a dimer in the membrane.

The CPP of highly purified cardiac SR was also found to be a dimer by radiation inactivation analysis using the same methodology and criteria (23).

Target inactivation analysis studies have been reported in which the size obtained for a known oligomer gave a partial molecular weight; with other enzymes the whole oligomeric size was obtained (24). Most of these studies were carried out on soluble enzymes. Recently, a number of membrane proteins have been found to give an oligomeric size by target inactivation, including the CPP from skeletal muscle (this study) and heart muscle SR (23), D-β-hydroxybutyrate dehydrogenase (13), a lipid requiring enzyme (25), the glucose carrier of the red blood cell (26) and the (K^+, H^+) pump of the gastric mucosa (27). Obtaining an oligomeric size by target inactivation analysis can be explained by: 1) a single hit can destroy an essential subunit which would result in complete loss of function of the oligomeric complex; or 2) the energy of the irradiating particle is absorbed by the oligomeric unit and is redistributed so that all the subunits in the oligomer become fragmented. This would explain the loss of activity of all the subunits in the complex, even if the monomers would otherwise be active. In this context, obtaining an oligomeric size reflects an oligomeric structure whereas a monomer molecular weight by target analysis would not preclude oligomeric quaternary structure.

No significant loss of particles or change in vesicle ultrastructure was observed by freeze-fracture electron microscopy indicating that the gross morphology of the vesicles was largely unchanged and that the particle integrity of the pump is apparently unaltered even though the CPP molecule has been fragmented (22).

In summary, the oligomeric size of the calcium pump protein (CPP) in sarcoplasmic reticulum (SR) and reconstituted SR membranes was determined using classical target theory analysis of radiation inactivation data. The loss of Ca^{++}-ATPase and Ca^{++}-loading activity, with increasing irradiation dose paralleled one another as did the decrease in CPP measured from the staining intensity of the CPP band on polyacrylamide gels run in sodium dodecyl sulfate. The molecular size from this analysis ranged between 210,000 to 250,000 daltons, which is consistent with a dimer. Similar values were obtained for reconstituted SR (RSR).

REFERENCES

1. Vegh, K., Spiegler, P., Chamberlain, C., and Mommaerts, W.F.H.M. (1968). Biochim. Biophys. Acta 163, 266-268.
2. Rizzolo, L.J., le Maire, M., Reynolds, J.A. and Tanford, C. (1976) Biochemistry 15, 3433-3437.
3. Wang, C.T., Saito, A., and Fleischer, S. (1979) J. Biol. Chem. 254, 9209-9219.
4. Pick, U., and Karlish, S.J.D. (1980) Biochim. Biophys. Acta 626, 255-261.
5. Andersen, J.P., Moller, J.V., and Jorgensen, P.L. (1982) J. Biol. Chem. 257, 8300-8307.
6. Moller, J.V., Mahrous, T.S., and Andersen, J.P. (1982) Z. Naturforsch. 37, 517-521.
7. Anderson, J.P., Skriver, E., Mahrous, T.S. and Moller, J.V. (1983) Biochim. Biophys. Acta 728, 1010 (1983).
8. Vanderkooi, J.M., Ierokomas, A., Nakamura, H., and Martonosi, A. (1977) Biochemistry 16, 1262-1267.
9. Moller, J.V., Andersen, J.P., and le Maire, M. (1982) Mol. and Cell. Biochem. 42, 83-107.
10. Pollard, E.C. (1953) Adv. Biol. Med. Phys. 3, 153-189.
11. Meissner, G., Conner, G.E., and Fleischer, S. (1973) Biochim. Biophys. Acta 298, 246-259.
12. Meissner, G., and Fleischer, S. (1974) J. Biol. Chem. 249, 302-309.
13. McIntyre, J.O., Churchill, P., Maurer, A., Berensky, C., Jung, C.Y. and Fleischer, S. (1983) J. Biol. Chem. 258, 953-959.
14. Fricke, H. and Hart, E.J. (1966) In Radiation Dosimetry Aftix, F.H. and Roech, W.C., eds.) Vol. 11, pp. 167-234.
15. Kepner, G.R., and Macey, R.I. (1968) Biochim. Biophys. Acta 163, 188-203.
16. Olive, C., and Levy, H.R. (1971) J. Biol. Chem. 246, 2043-2046.
17. Lowry, O.H., Rosebrough, N.J., Faar, A.L., and Randall, R.J. (1951) J. Biol. Chem. 193, 265-275.
18. Rouser, G., and Fleischer, S. (1967) Methods Enzymol. 10, 385-406.
19. Chen, P.S., Toribara, T., and Warner, H. (1956) Anal. Chem. 28, 1756-1758.

20. Merril, C., Goldmann, D., and van Keuren, M.L. (1983) Methods Enzymol., Vol. 96, in press.
21. Laemmli, U.K. (1970) Nature 227, 680-685.
22. Hymel, L., Maurer, A., Jung, C., Berenski, C., and Fleischer, S. (1983, in preparation).
23. Chamberlain, B.K., Berenski, C.J., Jung, C.Y. and Fleischer, S., (1983) J. Biol. Chem. (in press).
24. Kempner, E.S., and Schlegel, W. (1979) Analytical Biochem. 92, 2-10 (1979).
25. Fleischer, S., Bock, H-G. and Gazzotti, P. (1974) in "Membrane Proteins in Transport and Phosphorylation" (G.F. Azzone, M. Klingenberg, E. Quagliariello and N. Siliprandi) North-Holland Pub. Co., Amsterdam, 125-136.
26. Jung. C.Y., Hsu, T.L., Hah, J.S., Cha, C. and Haas, M.N. (1980) J. Biol. Chem. 255, 361-364.
27. Saccomani, G., Sachs, G., Cuppoletti, J. and Jung, C.Y. (1981) J. Biol.Chem. 256, 7727-7729.

OLIGOMERIC FORM OF THE Ca^{2+}, Mg^{2+}–DEPENDENT ATPase IN SARCOPLASMIC RETICULUM

Taibo Yamamoto
Robert E. Yantorno[1]
Yuji Tonomura[2]

Department of Biology, Faculty of Science
Osaka University
Toyonaka, Osaka, Japan

I. INTRODUCTION

Ca^{2+}, Mg^{2+}-dependent ATPase is a major component of the sarcoplasmic reticulum membrane which accounts for more than 70% of the total protein. It serves as a translocator for Ca^{2+} as well as an energy transducer. However, the manner in which the movement of the enzyme molecule is coupled to the Ca^{2+} translocation across the membrane and associated with ATP hydrolysis is not clearly understood.

To better understand the molecular basis of the active transport of Ca^{2+}, information is needed concerning the structural aspects of the Ca^{2+}, Mg^{2+}-dependent ATPase molecules in the sarcoplasmic reticulum membrane.

Information from a number of sources, e.g., cross-linking

---

[1]Present address: 7027 Radbourne Road, Upper Darby, PA 19082

[2]This investigation was supported grants from Muscular Dystrophy Association, Inc. and the Ministry of Education, Science and Culture of Japan.

Abbreviations: $C_{12}E_8$, octaethyleneglycoldodecylmonoether; DACM, N-(7-dimethylamino-4-methyl-3-coumarinyl)maleimide; N-1-P, N-(1-pyrene)-maleimide; TES, N-Tris(hydroxymethyl)methyl-2-aminomethanesulfonate

ISBN 0-12-260380-X

(Murphy, 1976; Chyn & Martonosi, 1977; Baskin & Hanna, 1979; Balin, 1980; Pick & Karlish, 1980) and negative staining (Scales & Inesi, 1976) and freeze fracture of fragmented sarcoplasmic reticulum (Jilka et al., 1975; Wang et al., 1979) indicate that Ca^{2+}, Mg^{2+}-dependent ATPase exists in the sarcoplasmic reticulum membrane in an oligomeric form. Use of a Ca^{2+}, Mg^{2+}-dependent ATPase inhibitor has led to the suggestion of a possible dimer structure for this pump (Pick & Bassilian, 1981). Recent kinetic evidence also indicates that cooperativity exists between subunits of the Ca^{2+}, Mg^{2+}-dependent ATPase (Ikemoto et al., 1981). Non-ionically solubilized Ca^{2+}, Mg^{2+}-dependent ATPase has been used to show that this protein tends to form oligomers of various sizes (Le Maire et al., 1978). The fluorescence energy tansfer data of Vanderkooi et al. (1977) is considered good evidence for the oligomeric structure of the Ca^{2+}, Mg^{2+}-dependent ATPase. However, their data did not provide information on either the number of subunits or subunit interaction.

II. REACTION MECHANISM OF SARCOPLASMIC RETICULUM ATPase IN THE PRESENCE OF $C_{12}E_8$

Recently Dean and Tanford (1977, 1978) reported that ATPase exists as a monomer when the sarcoplasmic reticulum is solubilized in a high concentration of a nonionic detergent, such as $C_{12}E_8$. They observed that the monomeric ATPase is fully active in the presence of Ca^{2+} and Mg^{2+}, and the basic aspects of the ATPase reaction of the monomeric and the membranous forms are similar. These findings suggest that most of the catalytic functions of sarcoplasmic reticulum may be performed by a monomeric ATPase. To test this possibility, we (Yamamoto and Tonomura, 1982 a,b) investigated the kinetic properties of the reaction steps in the cycle of ATP hydrolysis under the condition where the sarcoplasmic reticulum ATPase was solubilized into a monomeric form by the addition of a sufficiently high concentration of $C_{12}E_8$. When the sarcoplasmic reticulum ATPase was phosphorylated by $AT^{32}P$ in the presence of various concentrations of Mg^{2+} then chased by nonradioactive ATP, the radioactivity in the phosphoenzyme, EP, decreased with a rate constant of about 0.3 s^{-1} at 0°C. The rate of EP decomposition was independent of Mg^{2+} in the range from 0 to 10 mM. On the other hand, the value of the apparent rate constant of EP decomposition calculated from the ratio of ATPase activity to the amount of EP at steady state increased from about 0.3 to 1.5 s^{-1} when the Mg^{2+}

concentration was raised from 0 to 10 mM. Furthermore, no activity was observed in the absence of Ca^{2+}. These results are summarized as the following reaction scheme.

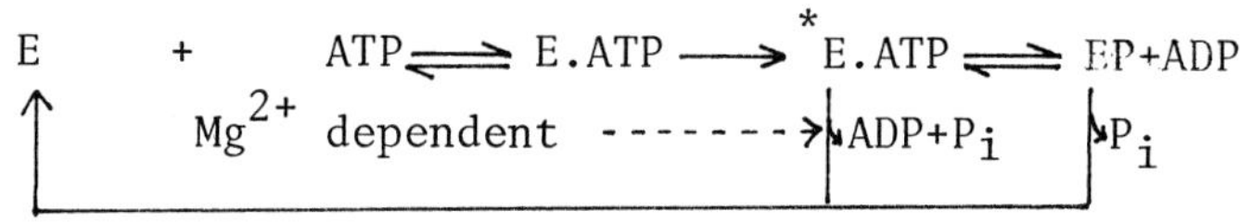

Most of the EP formed under these conditions exists in the ADP-sensitive form, E_1P. In the presence of Mg^{2+}, most of the ATP added can be hydrolyzed into ADP and P_i via the enzyme·substrate complex. The existence of the Mg^{2+}-dependent uncoupled reaction was also suggested by the fact that a high concentration of Mg^{2+} severely inhibits ATP formation by backward reaction from EP and ADP without preventing the ADP-stimulated decay of EP.

One of the most important findings in the kinetic studies on the Ca^{2+}, Mg^{2+}-dependent ATPase reaction in the presence of $C_{12}E_8$ is that the effect of $C_{12}E_8$ on the catalytic function of the sarcoplasmic reticulum ATPase is entirely reversible; when $C_{12}E_8$ was removed from the ATPase by incubation with about 0.2 g/ml Bio-Beads SM-2 for longer than 20 min at room temperature, the Mg^{2+} sensitivity of EP decomposition was fully restored to that of the intact sarcoplasmic reticulum ATPase.

Recently, further evidence for the reversible effect of $C_{12}E_8$ on the ATPase reaction has been obtained using the sarcoplasmic reticulum treated with N-ethylmaleimide, NEM. In the experiment of Fig. 1, sarcoplasmic reticulum vesicles were allowed to react with 1 mM NEM essentially under the same condition as described by Kawakita et al. (1980). The sarcoplasmic reticulum membranes were washed by centrifugation and then suspended in the reaction medium containing 0.5 M KCl, 20% glycerol, 50 mM TES (pH 7.5), 5 mM $CaCl_2$, 5 mM $MgCl_2$, and 20 μM A23187. The sarcoplasmic reticulum ATPase pretreated with NEM was phosphorylated by 50 uM ATP, then the reaction of EP formation was stopped by the addition of 15 mM EGTA and EP decomposition was measured as a function of time. As shown in Fig. 1, the amount of EP did not decrease with the addition of EGTA even in the presence of high concentrations of $MgCl_2$ and Ca^{2+}-ionophore. Complete inhibition of EP decomposition is considered to be caused by preventing the conversion step of E_1P to E_2P (Takisawa & Tonomura, 1979). In the case where sarcoplasmic reticulum membranes were solubilized in the high concentration of $C_{12}E_8$, the EP level decreased rapidly upon addition of EGTA. Inhibition of EP decomposition by NEM was observed again when $C_{12}E_8$ was removed from the reaction medium

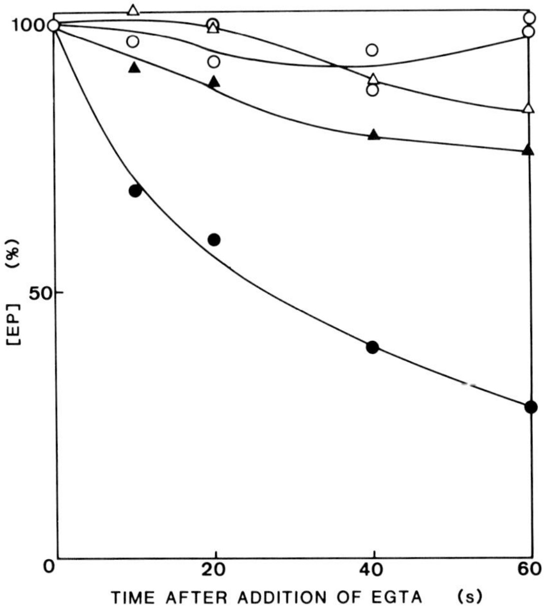

FIGURE 1. Effect of NEM treatment on EP decomposition in the presence or absence of $C_{12}E_8$. ○, control; ●, + $C_{12}E_8$; △ and ▲, + $C_{12}E_8$ + Bio-Beads SM-2.

by incubation with Bio-Beads at room temperature for longer than 20 min.

These results suggest that ATP can be hydrolyzed by the sarcoplasmic reticulum ATPase solubilized in $C_{12}E_8$ without going through the reaction step from E_1P to E_2P which is an essential step for Ca^{2+} transport across the membrane. Since decomposition of EP and liberation of the corresponding amount of P_i can be observed (data not shown) even when Mg^{2+} is removed by EDTA to prevent direct hydrolysis of ATP through the Michaelis complex as described above, E_1P may be directly hydrolyzed into E and P_i.

III. ENERGY TRANSFER BETWEEN FLUORESCENT DYES LOCATED ON ATPase MOLECULES UPON REMOVAL OF $C_{12}E_8$

The reversible effect of $C_{12}E_8$ on the reaction mechanism of ATPase may be correlated with the conversion of the enzyme states between oligomeric and monomeric forms. To examine this hypothesis, we used fluorescence energy transfer and fluorescence polarization decay measurements. Fluorescence energy transfer can yield information on the molecular level concerning the distance between separately labeled molecules or the way in which molecules interact as a function of changes in distance (Forster, 1959; Stryer, 1978; Fairclough & Cantor, 1978). Only the proper choice of probes can yield this type of information. We considered the possibility of monitoring the subunit interaction with respect to changes in enzyme state or the process of the subunits coming together during oligomerization. We chose N-1-P, as the donor and DACM as the acceptor. The emission and excitation spectra of N-1-P- and DACM-protein are shown in Fig. 2. Note the large overlap between the fluorescence emission of the N-1-P-protein and the excitation of the DACM-protein. This is one good criterion for energy transfer (Fairclough & Cantor, 1978). The excitation maximum of N-1-P is 343 nm and the emission maximum of DACM-protein is 465 nm.

The sarcoplasmic reticulum protein was labeled with the fluorescent probes in 0.5 M KCl, 20% glycerol, 20 mM TES (pH 7.5), 5 mM $CaCl_2$, 10 μM fluorescent dyes at 0°C. Both probes, DACM (Yamamoto et al., 1977) and N-1-P (Weltman et al., 1973), have been used for covalent binding to sulfhydryl groups.

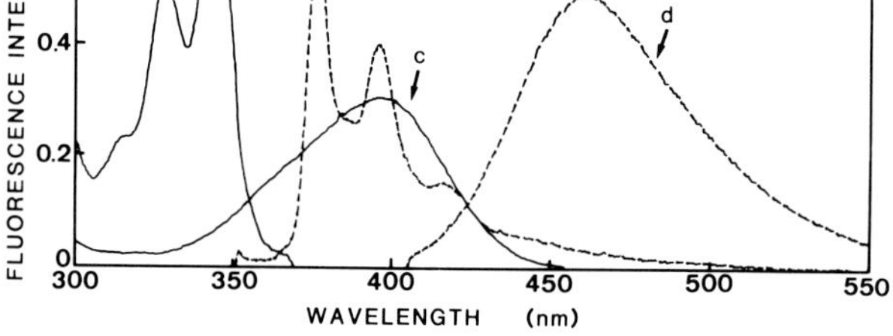

FIGURE 2. Emission and excitation spectra of N-1-P-protein and DACM-protein. N-1-P-protein emission (b) maximum, 375 nm; excitation (a) maximum, 343 nm. DACM-protein emission (d) maximum, 465 nm; excitation (c) maximum 398 nm.

Diagram 1 shows how the sarcoplasmic reticulum ATPase is first solubilized in a high concentration of $C_{12}E_8$ so that the protein is in the monomeric form. Next, the ATPase protein is separately labeled with either N-1-P or DACM. After a prescribed interval of incubation, β-mercaptoethanol in large excess was added to stop the reaction. The two labeled ATPases were combined, and Bio-Beads SM-2 were added to the mixture to remove $C_{12}E_8$.

1. MIX AFTER

```
N-1-P-protein in C12E8 — Bio-Beads — Filtration ┐
                                                 ├→
DACM-protein  in C12E8 — Bio-Beads — Filtration ┘
```

2. MIX BEFORE

```
N-1-P-protein in C12E8 ┐
                       ├— Bio-Beads — Filtration →
DACM-protein  in C12E8 ┘
```

DIAGRAM 1. Procedure for labeling sarcoplasmic reticulum ATPase protein with both N-1-P and DACM.

In the reaction of sarcoplasmic reticulum ATPase with either of these probes, about one mole of probe was incorporated into 10^5g protein within 20 min at room temperature, and the kinetic properties of both the formation and decomposition of EP in the presence of $C_{12}E_8$, were unaffected as shown in Fig. 3.

As can be seen in Fig. 4, when the labeled N-1-P was excited at 343 nm, there was a very large increase in the fluoresecence of the "MIX BEFORE" sample at the emission maximum of DACM of 465 nm, while there was no increase in the fluorescence of the "MIX AFTER" sample. There was also a large decrease in fluorescence at the emission maximum of N-1-P of 375 nm.

Energy transfer using steady state fluorescence can be observed as either a decrease in the fluorescence emission of the donor or an increase in the excitation spectra of the donor when the emission of the acceptor is monitored (Stryer, 1978; Fairclough & Cantor, 1978). This type of experiment (Fig. 5) showed a very large increase in the fluorescence of the "MIX BEFORE" condition in the range of the excitation maximum of N-1-P-protein (343nm).

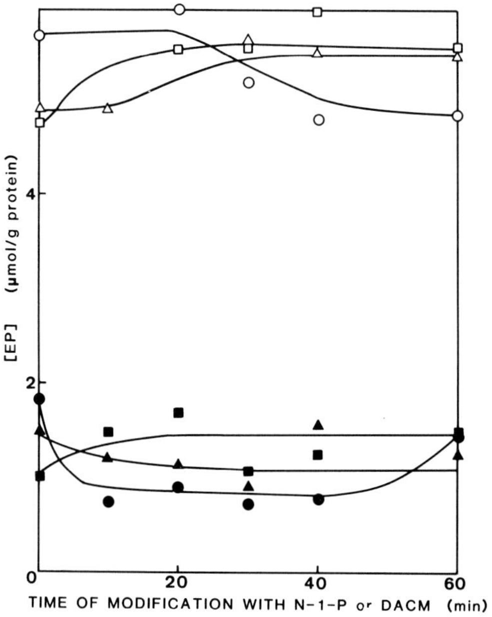

FIGURE 3. Effect of modification with N-1-P or DACM on EP formation and decomposition. Sarcoplasmic reticulum (0.2 mg/ml) was incubated with 10 μM N-1-P (□,■) or DACM (△,▲) at 20°C. EP formation (●,▲,■) and decomposition (○,△,□) were measured as described in the text.

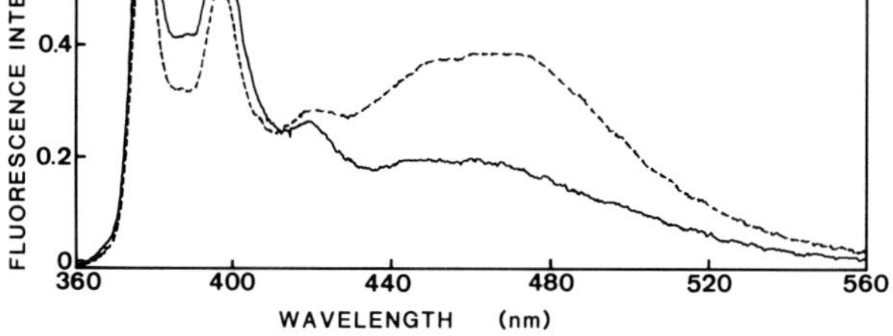

FIGURE 4. Emission spectra of N-1-P- and DACM-proteins for the "MIX BEFORE" (----) and "MIX AFTER" (——) samples.

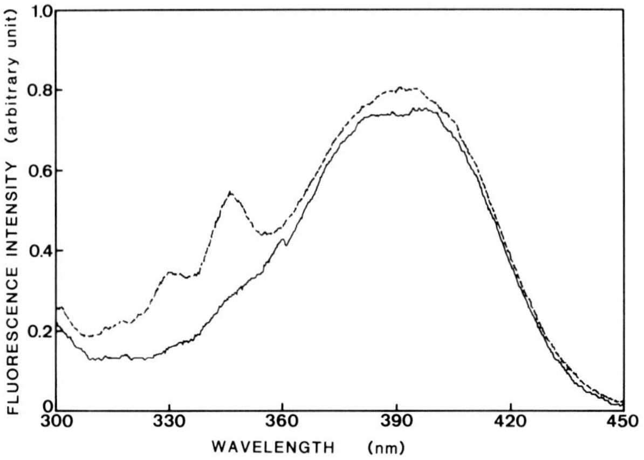

FIGURE 5. Excitation spectra of N-1-P- and DACM-protein for the "MIX BEFORE" (upper curve) and "MIX AFTER" (lower curve) conditions. Emission maximum 465 nm.

Based on energy transfer efficiency, the emission and excitation data agree well. However, this is not conclusive data for nonradiative energy transfer. Thus, the lifetime of the energy donor in the presence and absence of the energy acceptor were measured. Figure 6 clearly shows a decrease in the lifetime of the fluorescence of the donor, N-1-P, in the presence of the acceptor, DACM, from 150 to 120 ns.

As discussed above, fluorescence energy transfer is a good indicator of the close proximity of separately labeled molecules, and the energy transfer data should provide information concerning the phenomenon of oligomerization of the Ca^{2+}, Mg^{2+}-dependent ATPase monomers. This type of information is also shown in Fig. 7. Fig. 7b is the difference of the fluorescence excitation spectra of the "MIX BEFORE" sample minus the fluorescence excitation spectra of the "MIX AFTER" one (see Fig. 5 & DIAGRAM 1). The resultant value is the amount of fluorescence due to energy transfer. As can be seen from Fig. 7b, the amount of fluorescence increased with the incubation time with Bio-Beads. This data is plotted in Fig. 7a along with data showing restoration of the Mg^{2+} sensitivity of EP decomposition and $C_{12}E_8$ concentration as a function of the incubation time with Bio-Beads. These data suggests that interaction between the ATPase molecules is essential for the ATPase reaction coupled to the Ca^{2+} transport.

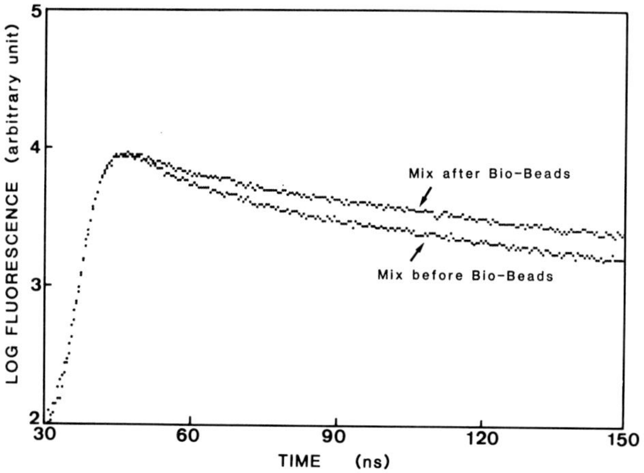

FIGURE 6. Fluorescent lifetime of N-1-P-protein in close proximity with DACM-protein (lower curve) and N-1-P-protein at a distance great enough for no energy transfer to the DACM-protein (upper curve). Time-dependent fluorescence lifetime measurements were performed with an Applied Photophysics, Ltd. Model 9200 System Nanosecond Spectoscopy Fluorimeter.

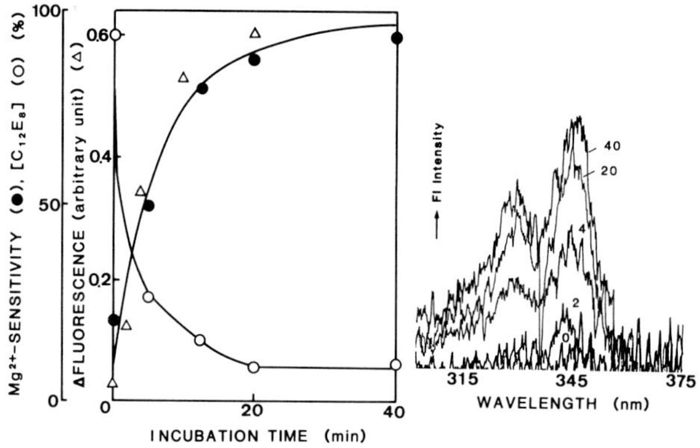

FIGURE 7. Comparison of fluorescence energy transfer data and Mg^{2+} sensitivity of EP as a function of incubation time with Bio-Beads. b) difference of the "MIX BEFORE" and "MIX AFTER" (see Fig. 5) fluorescence excitation spectra for various incubation times. a) a graph of data from b) as well as data for the Mg^{2+} sensitivity of EP and $C_{12}E_8$ concentration in the reaction medium. The Mg^{2+} sensitivity of EP is defined as the ratio of the EP level 40 s after addition of EDTA to the initial level of EP.

IV. SEPARATION OF MONOMERIC AND OLIGOMERIC ATPase BY GEL FILTRATION

Le Maire et al. (1978) have shown that by varying the weight ratio of $C_{12}E_8$ to the sarcoplasmic reticulum protein applied to a column equilibrated with Tween 80, different elution profiles can be obtained and each peak of the profiles is associated with a different oligomeric form of the Ca^{2+}, Mg^{2+}-dependent ATPase. They showed that both the activity and structure of their oligomeric ATPase are stable for a long time after the elution. We performed gel filtration by a method essentially similar to that of Le Maire et al. to separate oligomeric ATPase of different sizes.

In the experiment of Fig. 8, N-1-P-labeled sarcoplasmic reticulum protein was injected onto the Sephacryl S-300 column and eluted with solution containing 0.5M KCl, 20% glycerol, 20 mM TES (pH 7.5), 5 mM $CaCl_2$, and 1 mg/ml $C_{12}E_8$ and the fluorescence of each elution volume was measured. Here we assumed that a good correlation exists between the fluorescence and the protein concentration. Fluorescence scanning of an SDS gel electrophoretic pattern of the labeled protein showed only a single band at about the 100-Kdalton level.

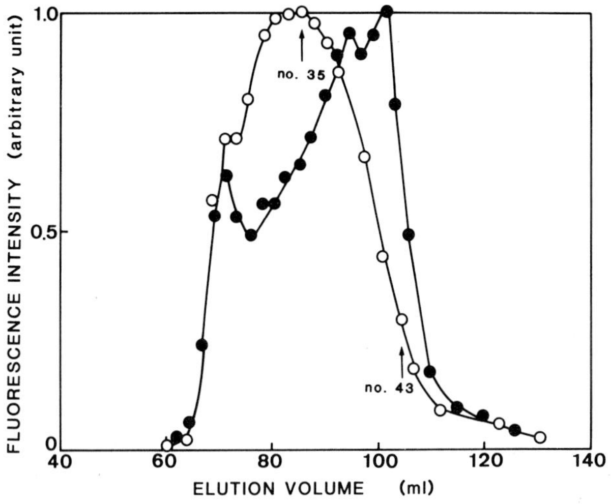

FIGURE 8. Elution profiles N-1-P labeled sarcoplasmic reticulum protein on Sephacryl S-300 column. 5 ml of 0.25 (●) and 0.5 (○) mg/ml protein in 2 mg/ml $C_{12}E_8$ were injected onto the column. Arrows indicate fractions selected for assay.

As Le Maire et al. (1978) suggested, accompanying an increase in the weight ratio of $C_{12}E_8$ to sarcoplasmic reticulum protein is an increase in the distribution of smaller sized oligomers. We randomly selected two chromatographic fractions for comparison of physical and kinetic properties; fraction number 35 near the void volume and fraction number 43 far from the void volume (Fig. 8).

Measurements of fluorescence polarization decay can provide information concerning the molecular dimensions and sometimes the molecular configuration of proteins using either intrinsic or extrinsic fluorescence probes. For fluorescence polarization decay measurements of large proteins, e.g., 100 K to 500 Kdaltons, a fluorescence system of lifetime longer than 20 ns is necessary. We chose N-1-P as the fluorescence indicator as it has a very long fluorescent lifetime when attached to proteins (Weltman et al., 1973).

Fluorescent polarization decay measurements were performed on two selected fractions (no. 35 and no. 43) from the Fig. 8 chromatography and also on a sample of N-1-P-protein, which had not been chromatographed. The results are shown in Fig. 9.

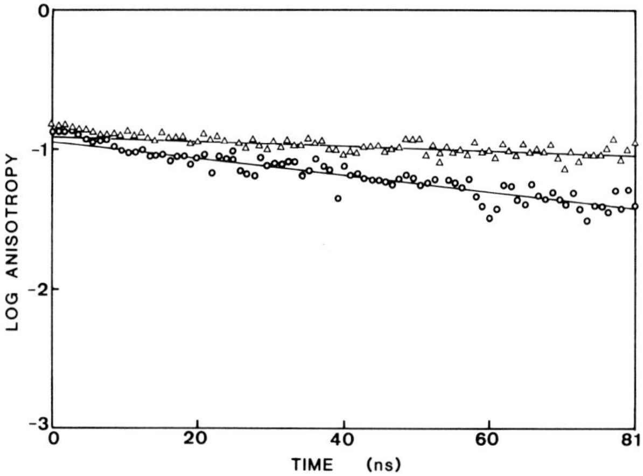

FIGURE 9. Time-resolved anisotropy data of fractions no. 35 (△) and no. 43 (○) of the column chromatography of Fig. 8. Linear least squares fit of the data for the rotational correlation time gave 251 ns for fraction no. 35 and 74 ns for fraction no. 43. Measurements were performed with the same apparatus as those shown in Fig. 6.

From the data of Fig. 9, the rotational correlation time were estimated to be 251 ns for no. 35 and 74 ns for no. 43. The rotational correlation time for the sarcoplasmic reticulum protein solubilized in $C_{12}E_8$ and not chromatographed, was close to that of no. 43 (data not shown).

As shown in Fig. 10, we estimated the values of the Stokes radius and molecular weight for the two fractions by measuring the elution profiles of several standard proteins under the conditions used to separate the different sized oligomeric ATPases. To determine whether $C_{12}E_8$ could change the column characteristics, we applied BSA to the column and found no difference in its elution profiles with and without $C_{12}E_8$.

From the data shown in Fig. 10A the Stokes radius can be estimated to be 8.3 nm and 5 nm for fraction no. 35 and 43, respectively. The molecular weight is about 240 K and 150 K-daltons for fraction no. 35 and 43, respectively, according to the column calibration curve in Fig. 10B.

The results of Figs. 9 and 10 suggest that the main component of fraction no. 35 is a dimeric ATPase, while that of fraction no. 43 is a monomeric ATPase.

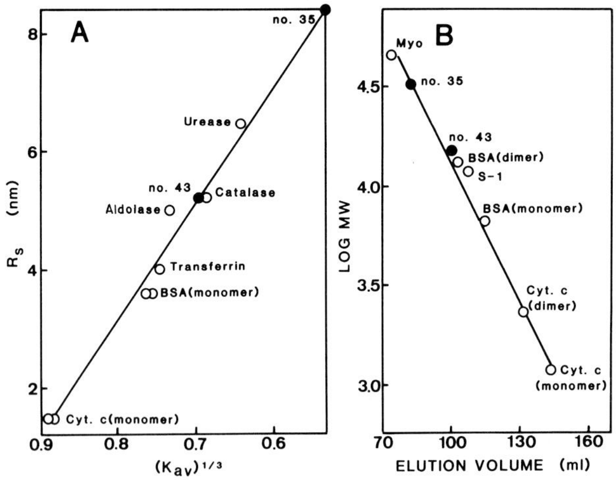

FIGURE 10. Estimation of Stokes radius (A) and molecular weight (B) of two fractions from chromatography on Sephacryl S-300 using standard proteins of known Stokes radius or molecular weight. Column chromatography was performed as described in Fig. 8.

We performed gel filtration under essentially the same conditions as described in Fig. 8, except that non-labeled ATPase was used, and selected two fractions corresponding to monomeric and dimeric ATPases. We examined the Mg^{2+} sensitivity of these fractions by measuring the time course of EP decomposition after addition of EDTA (Fig. 11).

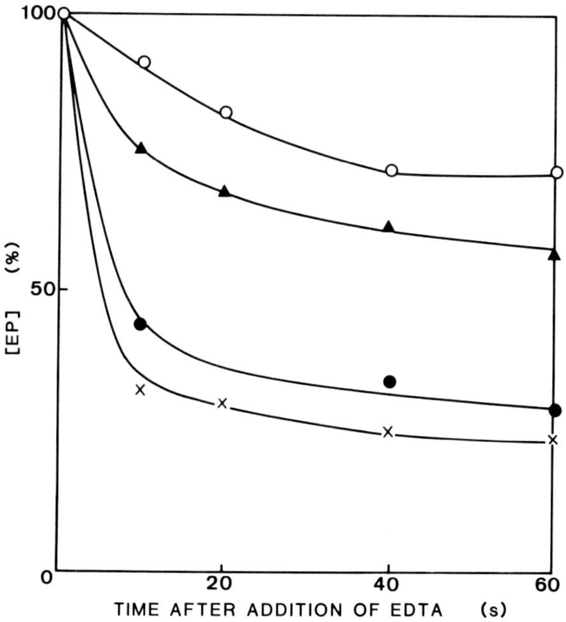

FIGURE 11. Time courses of EP decomposition after addition of EDTA. Smaller (●) and larger (▲) sizes of the ATPase protein fraction were eluted as described in Fig. 8 and phosphorylated by 50 μM ATP. The reaction was stopped by addition of 15 mM EDTA. In the control, the sarcoplasmic reticulum was solubilized by $C_{12}E_8$ at the protein/$C_{12}E_8$ ratios of 0.1 (x) and 0.5 (○).

This figure shows that the smaller ATPase fraction from column chromatography had a much faster rate of EP decomposition than that of the larger size fraction. The smaller-sized fraction which contains $C_{12}E_8$ at the protein/$C_{12}E_8$ weight ratio of 0.4, showed EP decomposition at a rate similar to that of the sarcoplasmic reticulum which was solubilized in $C_{12}E_8$ at the weight ratio of protein to $C_{12}E_8$ of 0.1, but the rate was much higher than that at the protein/$C_{12}E_8$ ratio of 0.5. These results suggest that the rate of EP decomposition is dependent on the size of the ATPase aggregation but independent of the concentration ratio of the sarcoplasmic reticulum protein to $C_{12}E_8$ in the assay medium, although we can not exclude the possibility that the difference in the rate of EP decomposition between these fractions is due to the difference in the amount of $C_{12}E_8$ which is tightly bound to the sarcoplasmic reticulum protein.

On the basis of our findings, we conclude that the oligomeric structure of the ATPase in the sarcoplasmic reticulum membrane is essential for the ATPase reaction which is coupled with the active transport of calcium across the membrane. However, the functional role of the interaction between the ATPase molecules in the oligomeric structure during the ATP hydrolysis or Ca^{2+} transport remains unclarified. Studies are under way at our laboratory to elucidate this important issue.

REFERENCE

Balin G. (1980). Biochim. Biophys. Acta 624:511.

Baskin, R.J. and Hanna, S. (1979). Biochim. Biophys. Acta 576:61.

Chyn, T. and Martonosi, A. (1977). Biochim. Biophys. Acta 468:114.

Dean, W.L. and Tanford, C. (1977). J. Biol. Chem. 252:3551.

Dean, W.L. and Tanford, C. (1978). Biochemistry 17:1683

Fairclough, R.H. and Cantor, C.R. (1978) "Methods in Enzymology" vol. XLVIII, part F.

Forster, T. (1959). Disc. Fard. Soc. 27:7.

Ikemoto, N., Miyao, A., and Kurobe, Y. (1981). J. Biol. Chem. 265:10809.

Jilka, R.L., Martonosi, A.N., and Tillack, T.W. (1975). J. Biol. Chem. 250:7511.

Kawakita, M., Yasuoka K., and Kaziro, Y. (1980). J. Biochem. 87:609.

Le Maire, M., Lind, K.E., Jørgensen, K.E., Roigaad, H., and Møller, J.V. (1978). J. Biol. Chem. 253:7051.

Murphy, A.J. (1976) Biochem. Biophys. Res. Comm. 70:160.

Pick, V. and Bassilian, S. (1981). FEBS Lett. 123:127.

Pick, V. and Karlish, S.J.D. (1980). Biochim. Biophys. Acta 626:255.

Scales, D. and Inesi, G. (1976). Biophys. J. 16:735.

Stryer, L. (1978). Ann. Rev. Biochem. 47:819.

Takisawa, H. and Tonomura Y. (1979). J. Biochem. 86:425.

Vanderkooi, J.M., Ierokomas, A., Nakamura, H., and Martonosi, A. (1977). Biochemistry 16:1262.

Wang, T., Saito, A., and Fleischer, S. (1979). J. Biol. Chem. 254:9209.

Weltman, J.K., Szaro, R.P. Frackelton, Jr., A.R., Dowben, B.J.R., and Cathou, R.E. (1973). J. Biol. Chem. 248:3173.

Yamamoto, K., Sekine, T., Kanaoka, Y. (1977). Anal. Biochem. 79:83.

Yamamoto, T. and Tonomura, Y.(1982a). J. Biochem. 91:477.

Yamamoto, T. and Tonomura, Y.(1982b). Membranes and Transport 1:573

Section III

A. Kinetics and Thermodynamics of the Calcium Pumping Mechanism: Thermodynamic and Kinetic Considerations

BINDING OF LIGANDS TO CALCIUM ATPase OF THE SARCOPLASMIC RETICULUM[1]

Yoichi Nakamura[2] and Yuji Tonomura

Department of Biology, Faculty of Science
Osaka University
Toyonaka, Osaka, Japan

I. INTRODUCTION

The kinetic properties of Ca^{2+}-ATPase in the sarcoplasmic reticulum have been studied extensively by many workers [Tonomura, 1972; Tada *et al.*, 1978; Yamamoto *et al.*, 1979; de Meis and Vianna, 1979; Inesi, 1979], and the following reaction scheme has been proposed:

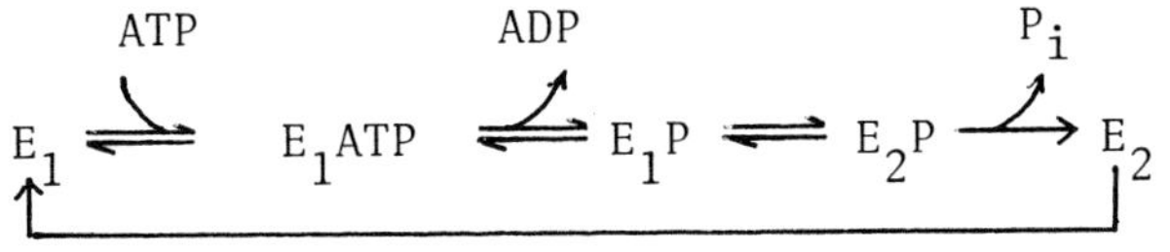

In this reaction scheme, E_1P and E_2P are ADP-sensitive and ADP-insensitive phosphoenzymes, respectively. Although

---

[1]Supported by grants from the Ministry of Education, Science and Culture, the Ministry of Health and Welfare of Japan, and Muscular Dystrophy Association, Inc.

[2]Present address: Department of Biological Chemistry, University of Maryland, Baltimore, Maryland, U.S.A.

Abbreviations: SR, sarcoplasmic reticulum; EP, phosphoenzyme; E_1P, ADP-sensitive EP; E_2P, ADP-insensitive EP; E*ATP, slowly exchanging enzyme-ATP complex; DOC, deoxycholate; EGTA, glycoletherdiamine-N,N,N',N'-tetraacetic acid; AMPPNP, adenyl-5'-yl-imidodiphosphate; CK, creatine kinase; CP, creatine phosphate.

ISBN 0-12-260380-X

kinetics is a very powerful tool for studying the reaction mechanism, the conclusions derived by kinetics are indirect, and thus direct measurements of ligand binding are also required to elucidate the coupling mechanism between the Ca^{2+}-transport and the ATPase reaction.

We have recently measured the amounts of Ca^{2+} and ATP bound to the enzyme in various enzyme states using the double membrane filtration method devised by Yamaguchi and Tonomura [1979]. Our studies on the binding of Ca^{2+} to SR ATPase showed that (1) 2 mol of Ca^{2+} binds to 1 mol of the ATPase with high affinity, (2) Ca^{2+} bound to the enzyme takes an "occluded" form when E_1P is formed, and (3) Ca^{2+} is released by the conversion of E_1P into E_2P [Nakamura and Tonomura, 1982a]. Furthermore, our measurements of ATP binding indicated two classes of ATP binding sites on the ATPase, i.e., a catalytic site with high affinity and a regulatory site with low affinity [Nakamura and Tonomura, 1982b]. Upon addition of Ca^{2+}, the ATP bound at the catalytic site was converted to EP + ADP, which was in equilibrium with a slowly exchanging enzyme-ATP complex, E^*ATP. The equilibrium was shifted toward EP by the ATP binding to the regulatory site.

II. METHODS

We measured the binding of ligands to leaky SR vesicles, that is, DOC-treated SR ATPase by the method of double membrane filtration. The reaction mixture contained the enzyme, ^{45}Ca or [α-^{32}P]ATP, and [^3H]glucose. About 30 μl of the reaction mixture was applied to a set of two membrane filters (upper, Diaflow XM-100A; lower, Millipore filter), and suction was applied for 10-20 sec. Almost all the reaction mixture remained on the upper filter, and only 0.5 to 1.0 μl of the filtrate was trapped by the lower Millipore filter. The concentrations of Ca^{2+} and ATP in the filtrate were determined as the ratios of $^{45}Ca/^3H$ and $^{32}P/^3H$, respectively. The amount of active site was determined as the maximum amount of EP produced by the reaction with ATP.

III. RESULTS AND DISCUSSION

A. Ca^{2+} Binding to ATPase

We found that the dependence of the amount of Ca^{2+} bound to the enzyme on the concentration of Ca^{2+} can be explained by assuming that 2 mol of Ca^{2+} binds to 1 mol of the enzyme cooperatively with a Hill coefficient of 2. The dissociation constants for Ca^{2+} in the presence of 0.2 and 20 mM $MgCl_2$ were 0.12 and 0.3 µM, respectively (data not shown). These findings are consistent with the conclusion previously derived from kinetic studies [Yamada and Tonomura, 1972] that the affinity of the Ca^{2+} binding site outside the SR membrane for Ca^{2+} is 10^5-fold that for Mg^{2+}.

The amount of bound Ca^{2+} in the presence of 20 mM $MgCl_2$ was unaffected by the addition of 0.2 mM AMPPNP, and the bound Ca^{2+} was rapidly removed by the addition of EGTA both in the presence and absence of AMPPNP (data not shown). These findings are consistent with the suggestion made by Kanazawa <u>et al</u>. [1971] that the binding of 2 mol of Ca^{2+} and 1 mol of substrate, in this case AMPPNP, occurs in a random sequence.

B. Occlusion of Ca^{2+} in SR ATPase upon Formation of E_1P

Figure 1 shows the time course of the change in the amount of bound Ca^{2+} after the addition of 0.2 mM ATP in the presence of 75 mM KCl and 20 mM $MgCl_2$ at pH 7.0 and 0°C. Under these conditions, more than 90% of the EP formed was E_1P. The amount of bound Ca^{2+} increased upon addition of ATP, and the increase was observed even under the conditions where the high affinity Ca^{2+} binding sites were completely saturated by Ca^{2+}, and it was unaffected by the addition of a Ca^{2+} ionophore, A-23187. Therefore, we can reasonably conclude that the affinity of the Ca^{2+} binding site(s) other than the high affinity site increases with the formation of E_1P.

Before the addition of ATP, almost all the bound $^{45}Ca^{2+}$ was displaced very rapidly by the addition of 1 mM unlabeled Ca^{2+} and 1.04 mM EGTA to keep the free Ca^{2+} concentration at 1 µM. However, when E_1P was formed by the addition of ATP, the bound $^{45}Ca^{2+}$ was displaced biphasically (Fig. 1). The amount of $^{45}Ca^{2+}$ displaced during the first rapid phase was equal to the amount of Ca^{2+} that increased upon the addition

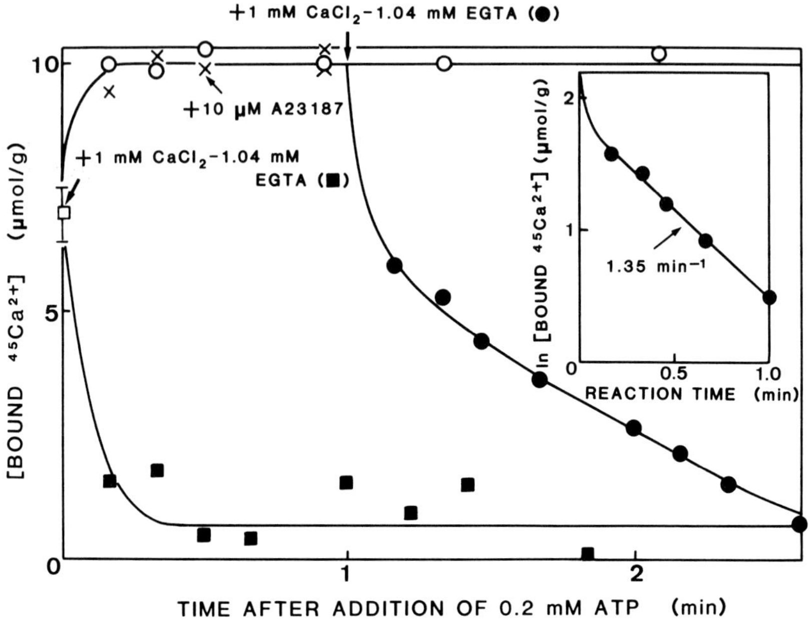

FIGURE 1. Time courses of the increase in the amount of bound $^{45}Ca^{2+}$ on addition of ATP and the decrease on further addition of unlabeled $CaCl_2$. The reaction was started by adding 200 µM ATP (○) to 2.5 mg/ml DOC-treated SR ATPase in the presence of 58 µM $^{45}Ca^{2+}$, 50 µM EGTA, 2 mg/ml CK, 10 mM CP, 20 mM $MgCl_2$, 75 mM KCl, 100 mM [^{3}H]glucose, 10% glycerol, and 50 mM Tris-maleate at pH 7.0 and 0°C. The time course in the presence of 10 µM A-23187 and 0.2% ethanol (×) is also shown. At 0 (■) or 1 min (●), 1 mM unlabeled $CaCl_2$ + 1.04 mM EGTA was added, and the time course of the displacement of bound $^{45}Ca^{2+}$ was measured. The latter time course is replotted logarithmically in the inset.

of ATP. The time course of the second slow phase, that is the displacement of the remaining 2 mol of bound $^{45}Ca^{2+}$, obeyed first-order kinetics, and the apparent rate constant was 1.35 min^{-1}. This rate constant was almost equal to the turnover rate of the ATPase reaction under the conditions used. These findings indicate that the Ca^{2+} bound to the high affinity site in E_1P is unexchangeable with medium Ca^{2+}, as suggested by Takakuwa and Kanazawa [1979] from their kinetic studies on E_1P. Thus, upon formation of E_1P,

Ca^{2+} bound to the enzyme takes an occluded form, which corresponds to the step for Ca^{2+} translocation in the SR membrane.

C. Release of Ca^{2+} from ATPase upon Conversion of E_1P into E_2P

Figure 2 shows the time course of the change in the amount of bound Ca^{2+} after addition of ATP in the presence of 75 mM KCl and 20 mM $MgCl_2$ at pH 8.8. Under these conditions, more than 90% of the EP formed was E_2P. Upon addition of ATP, the amount of bound Ca^{2+} increased slightly, then decreased slowly to reach the steady state level. We measured both the amount of E_2P and the amount of bound Ca^{2+} at steady state in the presence of various concentrations of ATP. The inset shows that the amount of E_2P plotted against that of bound Ca^{2+} gave a straight line with a slope of -1. These findings strongly indicate that 1 mol of the 2 mol of bound Ca^{2+} is released from the enzyme with the formation of 1 mol of E_2P. The enzyme is known to react with P_i to form E_2P in the presence of Mg^{2+} and the absence of Ca^{2+}. Therefore, we concluded that there are two kinds of E_2P, i.e., one with 1 mol of bound Ca^{2+} per mol and another without it.

D. Binding of ATP to SR ATPase

Figure 3A shows the amount of ATP bound to the enzyme as a function of the ATP concentration in the presence of 20 mM $MgCl_2$ and 75 mM KCl. Excess amounts of CK and CP were used as an ATP-regenerating system. In the absence of Ca^{2+}, the dependence could be explained by assuming that 1 mol of ATP was bound to a high affinity site with a dissociation constant of 7 μM and another 1 mol of ATP was bound to a low affinity site with a dissociation constant of 80-300 μM. Upon addition of Ca^{2+}, the high affinity ATP-binding disappeared and only the low affinity ATP-binding was observed. On the other hand, the amount of EP was saturated at about 10 μM ATP and the ATP concentration for half saturation of the EP level was less than 1 μM. These findings indicate that the high affinity site is the catalytic site, and that the equilibrium between an enzyme-ATP complex and EP + ADP shifts extremely toward EP in the presence of Ca^{2+} at a high KCl concentration.

At low KCl concentrations, the reaction mechanism of the SR ATPase is modified at several reaction steps, and the rate

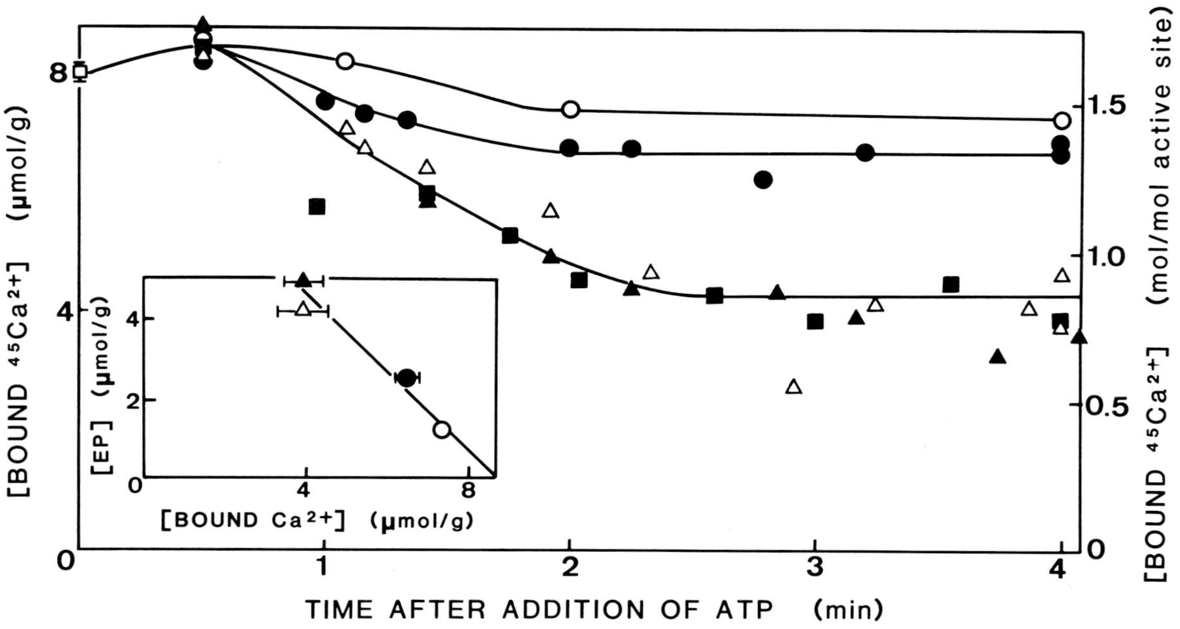

FIGURE 2. Decrease in the amount of bound Ca^{2+} after addition of various concentrations of ATP in the presence of 20 mM $MgCl_2$ at pH 8.8. The reaction was started by the addition of 5 (○), 10 (●), 30 (△), 100 (▲), or 300 μM ATP (□) to a solution containing 2.5 mg/ml DOC-enzyme, 58 μM $^{45}Ca^{2+}$, 30 μM EGTA, 20 mM $MgCl_2$, 75 mM KCl, 100 mM [^{3}H]glucose, 10% glycerol, and 75 mM Tris-HCl at pH 8.8 and 0°C. The amounts of EP were measured under the same conditions as those for Ca^{2+}-binding measurements, and plotted in the inset against the amounts of bound Ca^{2+} at steady state.

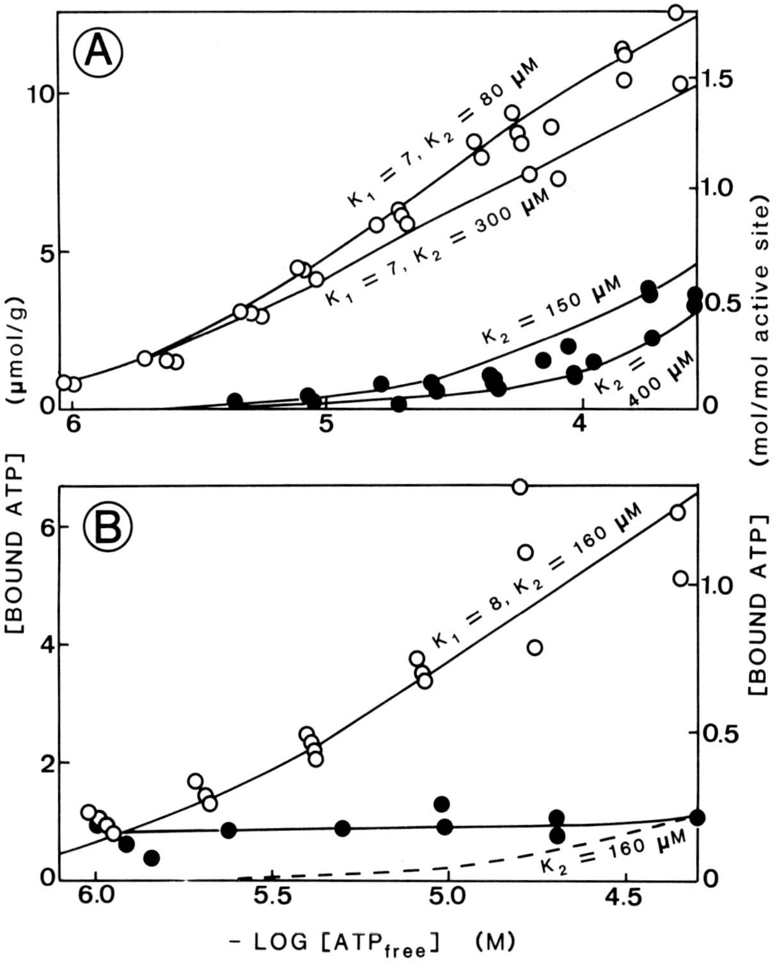

FIGURE 3. Dependence of the amount of bound ATP on the concentration of ATP in the presence and absence of Ca^{2+} at 75 mM (A) or 2 mM KCl (B). The amount of [α-^{32}P]ATP bound to the enzyme was measured in the presence of 20 mM CP and 100 mM [^{3}H]glucose at 0°C. A: 5 mg/ml enzyme, 1 mM EGTA (○) or 1 mM $CaCl_2$ (●), 2 mg/ml CK, 20 mM $MgCl_2$, 75 mM KCl, 75 mM Tris-HCl, pH 8.8. B: 0.5 mg/ml enzyme, 2 mM EGTA (○) or 5 mM $CaCl_2$ (●), 1 mM $MgCl_2$, 2 mM KCl, 60 mM imidazole-HCl, pH 7.0. The curves were obtained by calculation according to the following equation with the constants indicated in the figures.

$$\frac{[\text{bound ATP}]}{\varepsilon} = \frac{1}{1 + K_1/[\text{ATP}]} + \frac{1}{1 + K_2/[\text{ATP}]}$$

of EP decomposition into E + P_i is extremely small [Sigekawa and Pearl, 1976; Shigekawa et al., 1978; Shigekawa and Dougherty, 1978a,b]. Figure 3B shows the dependence of the amount of bound ATP on the concentration of ATP in the presence of 2 mM KCl. In the absence of Ca^{2+}, the profile of ATP binding was essentially equal to that in the presence of 75 mM KCl (Fig. 3A). Upon addition of Ca^{2+}, the amount of bound ATP decreased, as in the presence of 75 mM KCl. However, a significant amount of bound ATP remained and was apparently more than the amount obtained by assuming that only the low affinity ATP-binding remained. Therefore, we concluded that a part of the bound ATP at the catalytic site remained even when EP was formed in the presence of Ca^{2+}.

E. Slowly Exchanging Enzyme-ATP Complex

We found that when ADP is added to EP formed at low KCl concentrations in the presence of 5 mM $CaCl_2$, the amount of EP decreases and an equal amount of ATP is formed. Furthermore, the amount of EP increased when ADP was removed using an ATP-regenerating system (data not shown). These findings suggest that the EP formed is E_1P and is in equilibrium with an enzyme-ATP complex.

We examined the effect of ATP on the equilibrium at a fixed concentration of ADP (Fig. 4). The enzyme was phosphorylated with 8 μM [γ-^{32}P]ATP for 20 sec in the presence of 1 μM ADP, then 20 volumes of the reaction mixture containing 1 μM ADP and various concentrations of unlabeled ATP were added. When the reaction mixture was diluted with a solution without ATP, the amount of E_1P was unaffected. When the reaction mixture was diluted with the solution containing 1 μM ADP and 8 μM unlabeled ATP, the amount of ^{32}P-labeled E_1P decreased slowly according to first-order kinetics. These findings indicate that the rate of decomposition under the conditions used is very small as previously reported by Shigekawa et al. [1978], and that the ATP in the enzyme-ATP complex, which is in equilibrium with EP + ADP, is exchanged slowly with the medium unlabeled ATP. When the concentration of unlabeled ATP increased, the apparent rate constant of ^{32}P-labeled E_1P decay decreased. Furthermore, the amount of ^{32}P-labeled E_1P estimated by extrapolating to the time of the dilution increased with an increase in the concentration of unlabeled ATP. These findings clearly indicate that the binding of ATP at a site other than the catalytic site shifts the equilibrium between E_1P + ADP and the enzyme-ATP complex toward E_1P. Therefore, we suggest that the low affinity ATP-binding site is a regulatory site of the ATPase.

sition, indicating that the Ca^{2+} bound to ADP-sensitive EP takes an "occluded" form. When ADP-sensitive EP was converted into ADP-insensitive EP, 1 mol in 2 mol of the bound Ca^{2+} was released. In the absence of Ca^{2+}, two kinds of ATP binding sites were observed on the ATPase, i.e., 1 mol of the catalytic site and 1 mol of the regulatory site per mol of active site. Upon addition of Ca^{2+}, all the ATP bound to the catalytic site disappeared at 75 mM KCl, while a significant amount of ATP remained at 0-2 mM KCl. The latter binding was found to be due to the formation of a slowly exchanging enzyme-ATP complex, which is in equilibrium with EP + ADP. The ATP binding to the regulatory site shifted the above equilibrium toward EP.

REFERENCES

de Meis, L. (1981). The Sarcoplasmic Reticulum: Transport and Energy Transduction. John Wiley & Sons, Inc., New York.

de Meis, L. and Vianna, A. (1979). Ann. Rev. Biochem. 48, 275-292.

Inesi, G. (1979). In Membrane Transport in Biology (Giebisch, G., Tosteson, D., and Ussing, H., eds.), Springer-Verlag, West Berlin and Heidelberg, pp. 357-393.

Kanazawa, T., Yamada, S., Yamamoto, T., and Tonomura, Y. (1971). J. Biochem. 70, 95-123.

Nakamura, Y. and Tonomura, Y. (1982a). J. Biochem. 91, 449-461.

Nakamura, Y. and Tonomura, Y. (1982b). J. Bioenerg. Biomem. 14, 21-32.

Shigekawa, M. and Dougherty, J.P. (1978a). J. Biol. Chem. 253, 1451-1457.

Shigekawa, M. and Dougherty, J.P. (1978b). J. Biol. Chem. 253, 1458-1464.

Shigekawa, M., Dougherty, J.P., and Katz, A.M. (1978). J. Biol. Chem. 253, 1442-1450.

Shigekawa, M. and Pearl, L.J. (1976). J. Biol. Chem. 251, 6947-6952.

Tada, M., Yamamoto, T., and Tonomura, Y. (1978). Physiol. Rev. 58, 1-79.

Takakuwa, Y. and Kanazawa, T. (1979). Biochem. Biophys. Res. Commun. 88, 1209-1216.

Tonomura, Y. (1972). In Muscle Proteins, Muscle Contraction and Cation Transport Chapter 11, Univ. Tokyo Press and Univ. Park Press, Tokyo and Baltimore.

Yamada, S. and Tonomura, Y. (1972). J. Biochem. 72, 417-425.

Yamaguchi, M. and Tonomura, Y. (1979). J. Biochem. 86, 509-523.

Yamamoto, T., Takisawa, H., and Tonomura, Y. (1979). In Current Topics in Bioenergetics (Sanadi, R., ed.) Vol. 9, pp. 179-236, Academic Press, New York.

EQUILIBRIUM AND KINETIC CORRELATIONS OF THE PARTIAL REACTIONS OF THE Ca^{2+} TRANSPORT CYCLE[1]

Giuseppe Inesi
Mark Kurzmack
David Lewis

Department of Biological Chemistry
University of Maryland School of Medicine
Baltimore, Maryland, USA

I. INTRODUCTION

The specific functional feature of Sarcoplasmic Reticulum (SR) vesicles is active transport of Ca^{2+}, which is coupled to utilization of ATP (1-3). The Ca^{2+}-dependent ATPase of SR vesicles is the operator of this function, and utilization of ATP includes formation of a phosphorylated enzyme intermediate (4,5) due to transfer of the ATP terminal phosphate to an aspartyl residue at the catalytic site (6,7).

The transducing mechanism of SR ATPase is based on a tight interdependence of catalytic and calcium binding sites, whereby occupancy of the binding sites by calcium permits utilization of ATP by the catalytic site, and phosphorylation of the catalytic site by ATP produces dissociation of Ca^{2+} from the binding sites. The vectorial character of the calcium pump requires that the change in affinity of the calcium sites be associated with a change in orientation from "outside" to "inside" with respect to the vesicles delimited by the SR membrane. A reaction sequence for the ATPase cycle, which is commonly found in the literature is outlined

[1]Supported by NIH grant HL 27867 and by the Muscular Dystrophy Association of America.

ISBN 0-12-260380-X

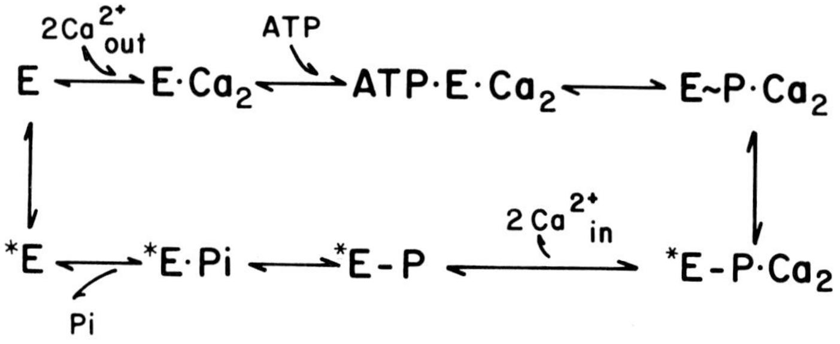

FIGURE 1. A reaction scheme for the catalytic and transport cycle of SR ATPase (8).

in Fig. 1, as proposed originally by deMeis and Vianna (8). Characterization of such a mechanism requires measurement of phosphorylation and calcium binding stoichiometry, as well as determination of kinetic and equilibrium parameters pertinent to the partial reactions of the catalytic and transport cycle.

II. CALCIUM BINDING SITES

The Ca^{2+} dependence of SR ATPase and the established ratio of two calcium ions transported per enzyme cycle, are an obvious indication of calcium binding to the enzyme. It should be understood however that the shape of a curve representing the enzyme dependence on Ca^{2+} concentration is not necessarily identical to that representing actual calcium binding. If enzyme activation is dependent on occupancy of two calcium sites, the relationship between the curves representing binding saturation and enzyme activation is determined by the comparative magnitudes of the intrinsic binding constants of the two calcium sites as shown in Fig. 2. Only in the case of a sequential calcium binding mechanism, with the affinity of the second site higher than that of the first, is positive cooperativity present in both binding and activation curves, as we found experimentally by determining the Ca^{2+} concentration dependence of calcium binding at equilibrium, and of ATP fluxes in steady state (Fig. 3). It is of interest that sequential binding accounts for positive cooperativity in equilibrium conditions, and predicts resolution of two binding components of equivalent magnitude if access to the second site is kinetically hindered (Fig. 4). If established experimentally, such a kinetic resolution would be a further demonstration of sequential binding. As for the stoichiometry of calcium binding, we find 8-10 nanomoles of high affinity sites (apparent $K_a \cong 10^6\ M^{-1}$)/mg protein, in our preparation of SR vesicles.

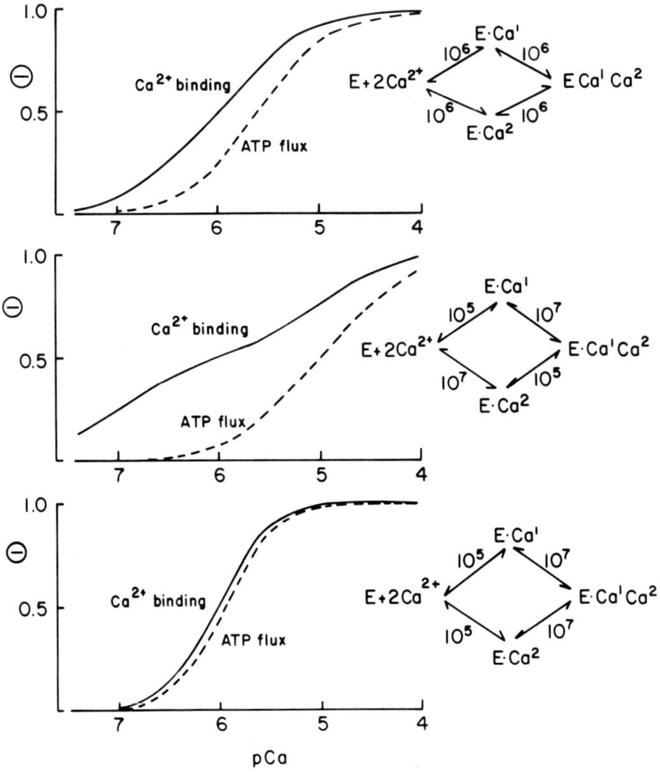

FIGURE 2. Simulated saturation of binding sites and ATP flux based on a simple mechanism whereby one catalytic enzyme site is activated only when two binding sites are occupied by Ca^{2+}. The upper scheme assumes random binding to two identical sites; the middle scheme requires sequential binding to a higher and then to a lower affinity site; the lower scheme requires sequential binding to a lower affinity and then to a higher affinity site. The constants are expressed as M^{-1}.

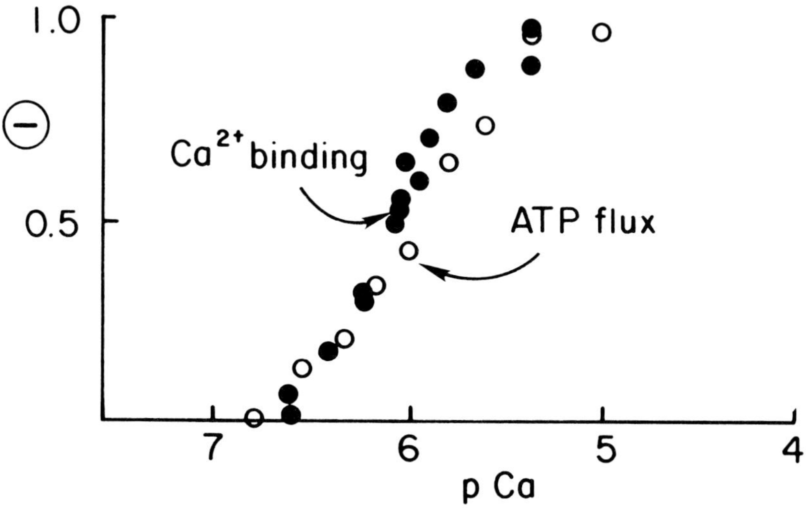

FIGURE 3. Experimental determination of calcium binding to, and ATP utilization by SR ATPase, as functions of Ca^{2+} concentration. For experimental methods see ref. (9).

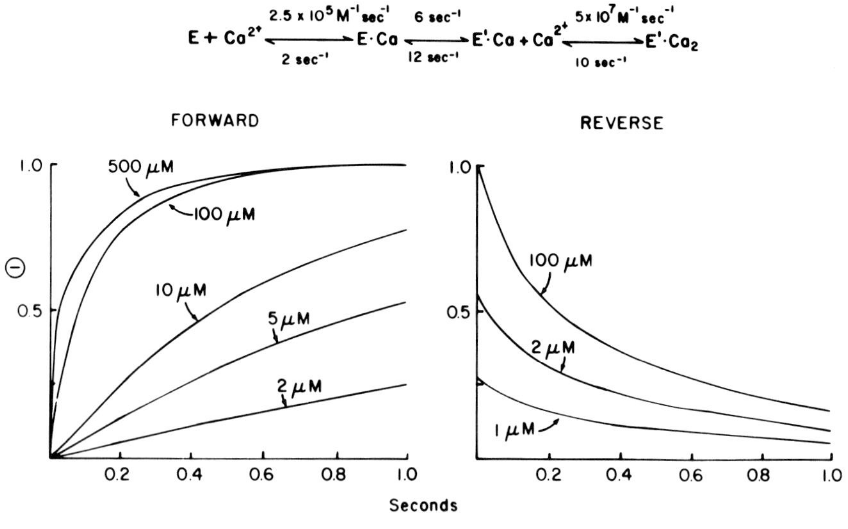

FIGURE 4. Simulated kinetics of calcium binding (forward) and dissociation (reverse) when Ca^{2+} (concentrations on Fig.) is added to E, or EGTA (1mM) is added to E preincubated with Ca^{2+} (concentrations on Fig.). The simulation assumes reactions and rate constants on the Figure.

III. PHOSPHORYLATION SITES

One of the uncertainties that we have encountered is related to the stoichiometry of the ATPase catalytic (phosphorylation) sites with respect to the 116,000 MW polypeptide chains identified with the ATPase enzyme. Electrophoretic analysis (Fig. 5) indicates that the 116,000 MW band accounts for 60-70% of the total protein in our preparation of SR vesicles, corresponding to approximately 7 nanomoles of 116,000 MW chains/mg total protein. Yet, we find that the maximal level of phosphorylated intermediate obtained following addition of ATP are 4-5 nanomoles/mg protein, in conditions inhibiting progression of the cycle to hydrolytic cleavage. (Higher values which are occasionally obtained with GTP or other analogues are due to partial degradation and unreliable radiolabeling of these substrates). It is likely that this discrepancy is due to the inadequacy of electrophoretic gel densitometry for accurate determination of non-specific protein components.

An alternative approach to determination of catalytic site stoichiometry is the phosphorylation reaction with Pi in the absence of Ca^{2+} (10,11), which is the reverse of the hydrolytic phosphoenzyme cleavage completing the ATPase cycle in Fig. 1. Most commonly, this reaction is studied at pH 6.0, 25°C temperature and in the absence of K^+, a condition favoring phosphorylation. Based on the currently accepted reaction scheme (Ref. 12 and Fig. 5), phosphoenzyme levels obtained in the presence of increasing Pi and Mg^{2+} concentrations can be extrapolated (Fig. 6) to yield a maximal level of phosphoenzyme ($E\text{-}P_{max}$) which we find to be 4-5 nanomoles/mg protein in our SR vesicles. This value is

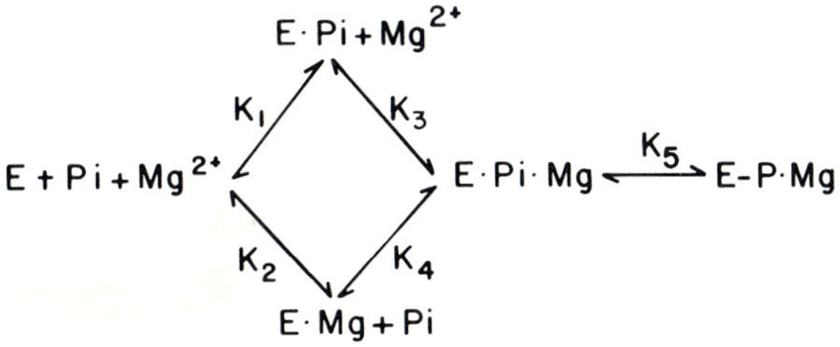

FIGURE 5. Reaction scheme for the Pi reaction with SR ATPase in the absence of Ca^{2+} (12).

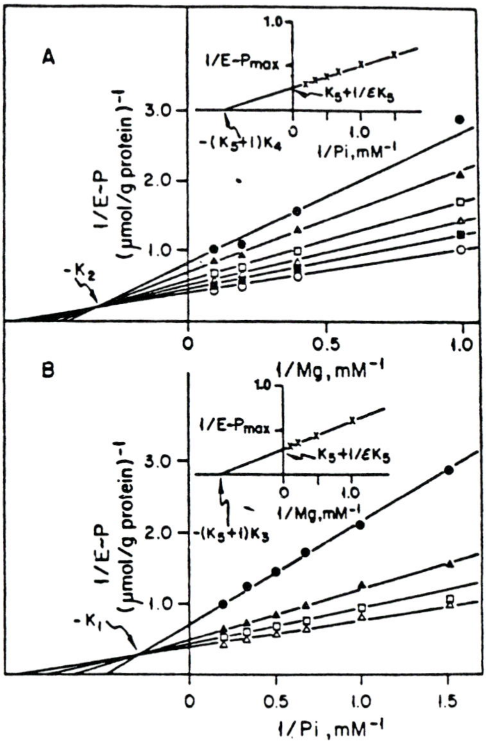

FIGURE 6. Equilibrium levels of phosphoenzyme in the presence of various Pi and Mg^{2+} concentrations.

related to the total number of reactive site (ε) by the equilibrium constant K_5 which is unknown. An estimate of K_5 has been obtained by assuming ε = number of 116,000 MW chains determined by electrophoresis analysis, and then $K_5 = E\text{-}P_{max}/\varepsilon$ (12,13). On the other hand, K_5 can be estimated without assumptions by independently measuring both the forward and reverse rate constants for the phosphorylation reaction, thereby deriving K_5

$$\left(K_5 = \frac{k_{phosph}}{k_{hyd}}\right) \text{ and } \varepsilon \left(\varepsilon = \frac{(1 + K_5)\ E\text{-}P_{max}}{K_5}\right)$$

The hydrolytic rate constant (k_{hyd}) is obtained in kinetic studies in which, following equilibration of enzyme with (^{32}P)-Pi, a large excess of non-radioactive Pi is added, and the time dependence of (^{32}P) phosphoenzyme decay is followed by rapid quench methods (Fig. 7). Since formation

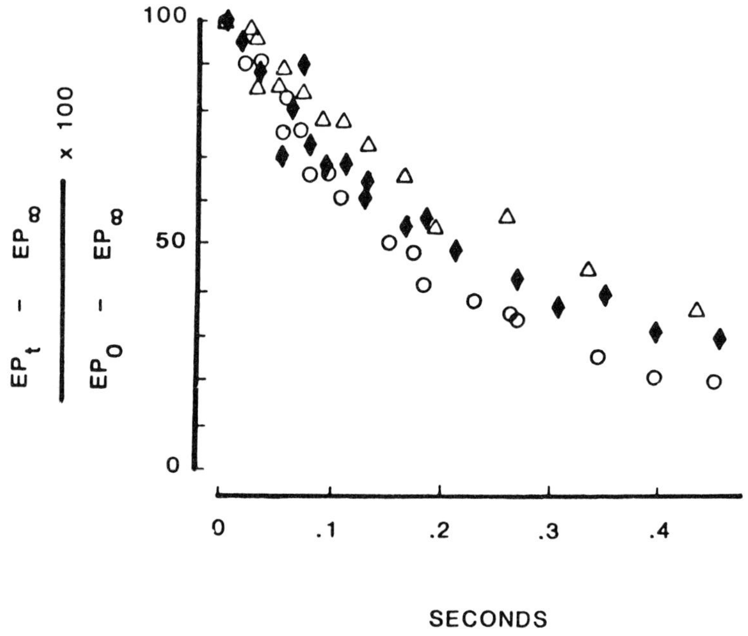

FIGURE 7. Decay of (^{32}P)-phosphoenzyme following 12-fold dilution with non-radioactive Pi (25°C).

of (^{32}P)-phosphoenzyme is negligible due to the large dilution of (^{32}P)-Pi with non-radioactive Pi, the observed rate constant is nearly equal to K_{hyd}, and is found to be 3-4 sec^{-1}.

In parallel rapid quench experiments, the time dependence of ^{32}P-phosphoenzyme formation can be measured following addition of (^{32}P)-Pi. In this case, extrapolation of the observed rate constants to infinite Pi concentration (Fig. 8) yields a value > 50 sec^{-1}. Since $k_{obs} = k_{phosph} + k_{hyd}$, and we know that k_{hyd} is only 3-4 sec^{-1}, k_{phosph} may be considered to be > 50 sec^{-1}. The low value for k_{hyd} is independently verified by extrapolation of k_{obs} to zero Pi in Fig. 8 (i.e., hydrolysis in the absence of phosphorylation), whereby a value of 3-4 sec^{-1} is found in agreement with the dilution experiments (Fig. 7).

It should be pointed out that significantly different values are obtained for the phosphorylation equilibrium constant (K_5), as well as for the association constants for Mg^{2+} and Pi secondary binding (K_3 and K_4), when the estimates

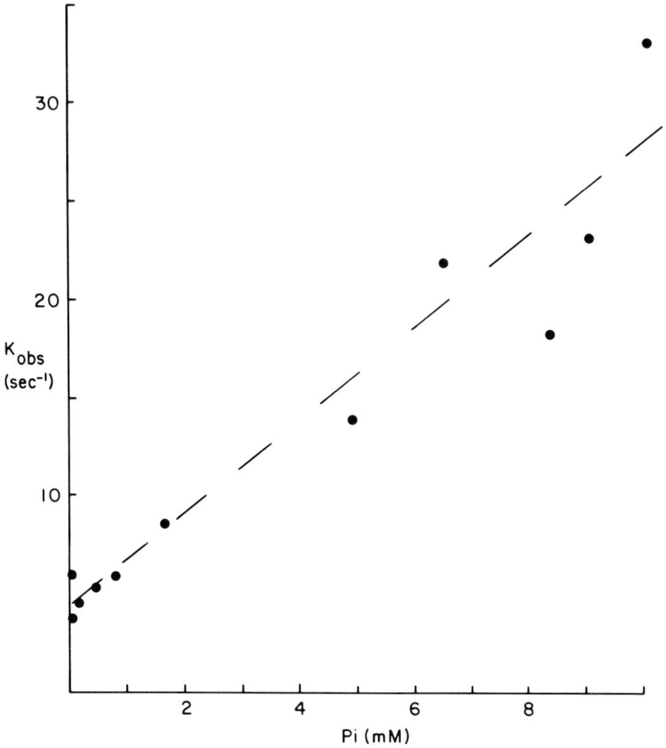

FIGURE 8. Observed rate constants of enzyme phosphorylation with various concentrations of Pi (25°C). Higher Pi concentrations cannot be used owing to technical limits of the experimentation.

are based on the combined equilibrium and kinetic measurements without assumptions on the number of reactive sites (ε), as compared to estimates based on a number of active sites (ε) assumed to be equivalent to the number of 116,000 MW chains (Table I). In the former case K_5 is much higher (favoring phosphorylation), while K_3 and K_4 are significantly lower. If $K_5 >> 10$, then E is nearly equal to $E\text{-}P_{max}$ derived from equilibrium studies (Fig. 6), i.e., 4-5 nanomoles/mg protein. An identical value is obtained when enzyme and Pi are equilibrated in the presence of 30% dimethylsulfoxide which is known to produce nearly complete phosphorylation of the available sites (14). These values are consistent with those obtained from phosphorylation studies with ATP and

TABLE I. Number of sites, equilibrium constants and rate constants in the reaction of Pi with SR ATPase in the presence of Mg^{2+} (Fig. 5), pH 6.0, absence of Ca^{2+} and K^{+} (25°C). The upper figures were obtained only by equilibrium studies assuming the number of sites, the lower figures were derived from combined equilibrium and kinetic studies without assumptions (see text).

Method	E $(\frac{nmoles}{mg})$	K_1 (M^{-1})	K_2 (M^{-1})	K_3 (M^{-1})	K_4 (M^{-1})	K_5 (M^{-1})	k_{ph} (sec^{-1})	k_{hy} (sec^{-1})
Eq.	8 (assumed)	110	120	744	35[illegible]	~1	-	-
Eq. + kinet.	4-5 $E\text{-}P_{max}$.	110	120	<100	<100	>10	>50	4

again lower than those expected from the number of 116,000 MW chains (Table I). We find identical phosphorylation levels even following solubilization of 116,000 MW monomers with the detergent $C_{12}E_8$. This demonstrates that availability of sites is not limited by inside-out vesicles or aggregation of chains. Therefore, it is apparent that densitometric analysis of electrophoretic gels is not a suitable criterion for accurate determination of ATPase stoichiometry. Based on measurements of phosphorylation sites, it appears that Ca^{2+} dependent ATPase accounts for 50-60% of the total protein in our preparation of SR vesicles, rather than 60-70% as suggested by SDS gels.

The measurements described above indicate that one ATPase unit, most likely corresponding to a single polypeptide chain, comprises two calcium binding sites and one phosphorylation site.

Finally, with regard to the enzyme reaction with Pi, it must be noted that the rate and equilibrium constants listed in Table I were obtained at pH 6.0 and in the absence of K^{+}. We have also determined these parameters in conditions more commonly used for studies of the ATPase reaction (i.e., pH 6.8 and 80 mM KCl). These values are listed in Fig. 11 together with other experimentally measured constants, as they apply apparently to the reaction scheme proposed by de Meis and Vianna (8).

IV. EXPRESSION OF THE TRANSDUCTION MECHANISM. EFFECTS OF ATP, Pi AND VANADATE

Experimental determination of the apparent association constant for binding of calcium to the ATPase before and after phosphorylation, yields the evidence required to demonstrate that the mechanism of free energy transduction is operated through direct coupling between calcium binding sites and phosphorylation sites. Calcium binding to the enzyme before phosphorylation can be measured in equilibrium conditions as explained above, and found to occur with an apparent $K_a \cong 10^6\ M^{-1}$. This constant remains unchanged in the presence of ATP analogues (i.e., AMP-PNP) which do not yield significant levels of phosphorylation (9). On the other hand, following phosphorylation of the catalytic site by ATP, calcium is internalized by the ATPase protein (a step leading to its transfer across the membrane) and then released from the binding sites (15-19).

Therefore, it is apparent that the binding sites undergo a reduction in their affinity for calcium as a consequence of phosphorylation.

While the enzyme can be phosphorylated by ATP in the presence of Ca^{2+}, phosphorylation of the enzyme with Pi requires prior removal of Ca^{2+} from the high affinity sites (10). The phosphoenzyme so formed reacts with ADP to form ATP (reversal of the ATPase cycle) only if high concentrations (9mM) of Ca^{2+} are added (20,21). This indicates that the binding sites have a low affinity ($K_a = 10^3\ M^{-1}$) for Ca^{2+} even if the enzyme is phosphorylated with Pi. If Ca^{2+} is added (even at μM concentrations) in the absence of ADP, the phosphoenzyme formed with Pi reverts entirely to non-phosphorylated form displaying high affinity for Ca^{2+} (10). It is then clear that in a reaction cycle

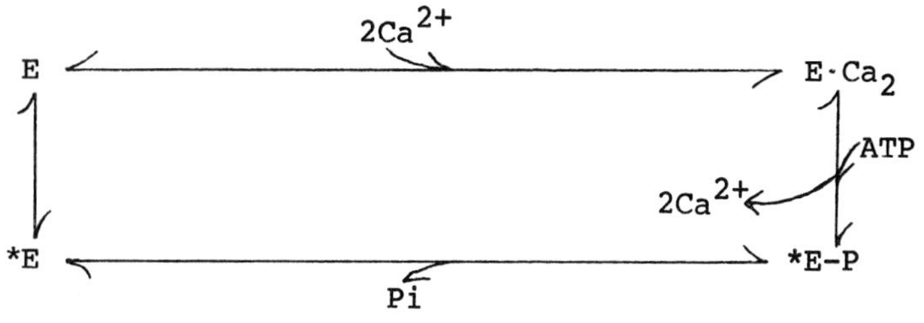

in which the enzyme may reside alternatively in a high (E) or low (*E) affinity state for Ca^{2+}, $E \cdot Ca_2$ is in a low free energy state from which it can be removed only by a substrate of high chemical potential such as ATP. On the other hand, no net free energy input is required to obtain the low affinity phosphoenzyme (*E-P) from the free enzyme and Pi, on conditions that the binding sites are free of Ca^{2+}.

A special and very interesting case is presented by vanadate. Following studies on the Na^{+}-K^{+} ATPase (22-24), it was found that vanadate inhibits the Ca^{2+} ATPase as well, by acting as an analogue of phosphate (25-27). We now find that vanadate inhibits high affinity calcium binding. This effect is analogous to that of ATP; however, while the effect of ATP can be studied only in steady state owing to the tendency of ATP to undergo hydrolysis, vanadate is stable and its effect can be studied by equilibrium experimentation (Fig. 9). It is likely that vanadate functions as a stable stereo-analogue of a phosphate transition state of the phosphorylation reaction, owing to its trigonal bipyramidal structure. Stabilization of vanadate at the site may also be gained by

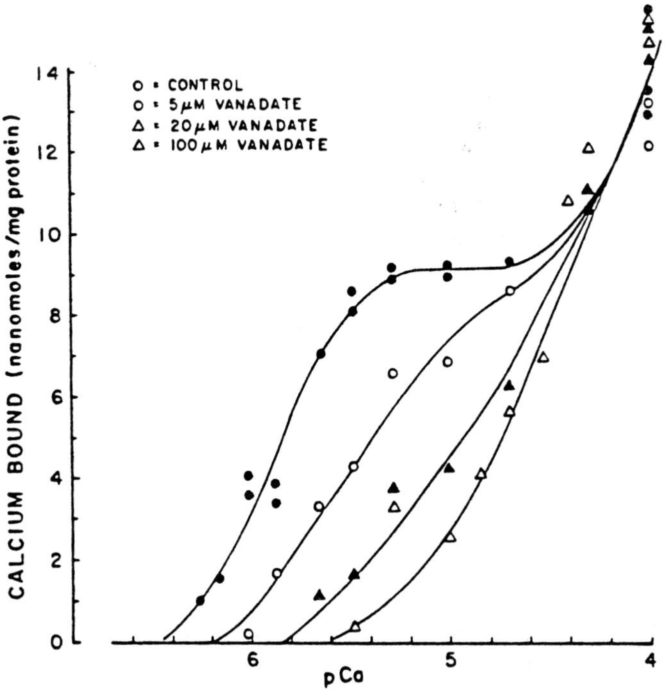

FIGURE 9. Vanadate inhibition of calcium binding to SR ATPase. The inhibition is specific for the high affinity sites, and does not involve the lower affinity, non-specific sites. Binding was measured at equilibrium in the absence of ATP, as previously described (9).

acceptance of electrons from neighboring oxygens into its d orbitals. Thereby, vanadate is able to overcome the energy barrier presented by the calcium bound enzyme ($E \cdot Ca_2$). The resulting inhibition of high affinity calcium binding demonstrates that the effective mechanism is the interaction of a phosphate moiety (or its analogue) with the catalytic site, rather than the hydrolytic cleavage of a phosphorylated substrate. In the case of ATP, the phosphorylation reaction is a means for reacting the phosphate moiety with the catalytic site in the presence of calcium, whereby free energy transduction is manifested by a decrease in affinity of the binding sites, and dissociation of Ca^{2+} against a concentration gradient. The subsequent phosphohydrolase reaction is simply a means to allow the enzyme to recycle.

V. KINETIC CONSTRAINTS AND ENZYME TURNOVER

The accumulation of high levels of phosphorylated enzyme intermediate during steady state utilization of ATP, suggests that the most important rate limit for the ATPase cycle follows enzyme phosphorylation and precedes hydrolytic cleavage of the phosphoenzyme. This is clearly evident on inspection of the experiments shown in Fig. 10, in which following addition of ATP, early bursts of enzyme phosphorylation and calcium translocation are followed by steady state

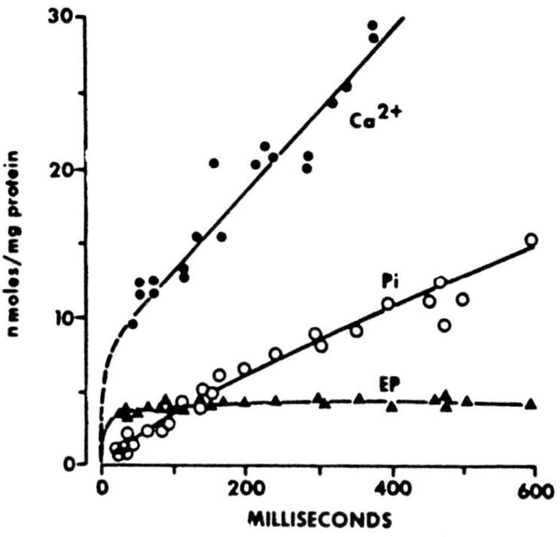

FIGURE 10. Time resolution of phosphoenzyme formation, calcium translocation and Pi production following addition of ATP.

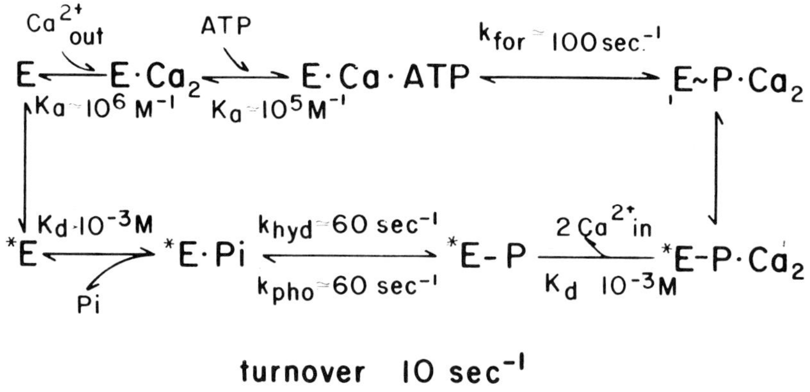

FIGURE 11. Experimentally determined equilibrium and rate constants (see text) as they apparently apply to the reaction scheme proposed by de Meis and Vianna (8).

Pi production (28). In fact, the rate constant for enzyme phosphorylation by ATP was found to be approximately 100 sec^{-1} (29,30). On the other hand, the rate constant for hydrolysis of the phosphorylated enzyme intermediate (pH 6.8, 80 mM K^+, 5 mM Mg^{2+}) is approximately 60 sec^{-1} (Figs. 1 and 11). It is then clear that partial reactions related to calcium dissociation and evolution of the phosphoenzyme to hydrolytic cleavage, impose kinetic constraints on the enzyme turnover which is known to be only 10 sec^{-1} (31) even in the absence of back-inhibition by accumulated Ca^{2+} (Fig. 11).

VI. CONFORMATIONAL CHANGES

Several lines of evidence indicate that protein conformational changes are operative in at least two states of the transport mechanism: (1) cooperative calcium binding and activation of the enzyme, and (2) internalization of bound calcium following enzyme phosphorylation by ATP. Our own evidence for the calcium effect (32,33) is based on spectral changes produced by calcium binding on spin-labeled SR (Fig. 12). With regard to the phosphorylation effects, we have recently observed that enzyme phosphorylation is followed by reversible disassociation of the ATPase chains in the presence of subliminal detergent concentrations, as demonstrated by a reduction in light scattering and in

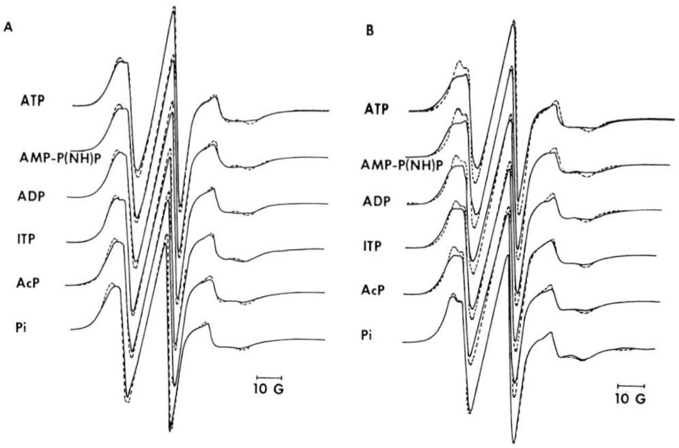

FIGURE 12. EPR spectra of covalently labeled SR ATPase in the absence (dashed lines) and in the presence (solid lines) of various substrates: (A) in the absence of Ca^{2+} (EGTA present); or (B) in the presence of Ca^{2+} (33).

fluorescence energy transfer (34,35, see paper by Watanabe and Inesi in this volume). This indicates that phosphorylation produces a rather extensive structural change, favoring interaction of the detergent with the ATPase hydrophobic segments which are known to intrude into the membrane bilayer. These structural changes are likely to be operative in the displacement of the calcium binding sites, as required for vectorial transport of the divalent cation.

VII. CONCLUSIONS

Explanation of bioenergetic transductions is generally formulated in terms of transmembrane potentials, covalent interactions, or conformational changes. It is clear that all these factors are pertinent to the calcium pump of SR vesicles. In the first place, the Ca^{2+} concentration transmembrane potential can be expressed by equations defining its thermodynamic relationship with the chemical potential of ATP (36). In addition, experimental characterization of intermediate steps is necessary to clarify the mechanism of enzyme catalysis coupled to active transport. For instance, direct measurements of calcium binding reveal a change in orientation and a reduction in the affinity of the binding sites for calcium following enzyme phosphorylation. This demonstrates a primary and self-contained mechanism of transport within the ATPase protein, as well as the relationship between the covalent interaction of the enzyme with ATP and its ability to bind or dissociate Ca^{2+}. The observed reduction of the calcium association constant requires a free energy input consistent with that derived from utilization of ATP, and with that required to overcome the maximal Ca^{2+} gradient generated by the pump. Furthermore, spectroscopic experimentation indicates that intermediate steps such as the enzyme activation by calcium binding and the phosphorylation reaction, are accompanied by changes in the conformation of the ATPase protein. It is likely that these conformational changes are operative in the catalytic and transport mechanism.

It should be understood that the high and low affinity states of the calcium binding sites correspond to two discrete energy levels of the enzyme, and their interconversion is permitted by a corresponding free energy exchange. However, the ratio between the binding constants of the two states is fixed by the natural limits of the two discrete energy levels of the enzyme, and cannot be varied in proportion to the free energy of the substrate. This means that the pump is made to concentrate Ca^{2+} from 10^{-7} -10^{-6} M to 10^{-3} M (for the purpose of regulating muscle contraction), and the maximal gradient that can be formed is limited by the dissociation constants acquired by the calcium binding sites in the non-phosphorylated and phosphorylated states of the enzyme.

REFERENCES

1. Hasselbach, W., Makinose, M., Biochem. Zeitschrift 333: 518 (1961).
2. Ebashi, S., Lipman, F., J. Cell Biology 14:389 (1962).
3. Hasselbach, W., Makinose, M., Biochem. Zeitschrift 339: 94 (1963).
4. Yamamoto, T., Tonomura, Y., J. Biochem. (Tokyo) 62:558 (1967).
5. Makinose, M., Europ. J. Biochem. 10:74 (1969).
6. Degani, C., Boyer, P., J. Biol. Chem. 248:8222 (1973).
7. Bastide, F., Meissner, G., Fleischer, S., Post, R. L., J. Biol. Chem. 248:8385 (1973).
8. deMeis, L., Vianna, A., Ann. Rev. Biochemistry 48:275 (1979).
9. Inesi, G., Kurzmack, M., Coan, C., Lewis, D., J. Biol. Chem. 255:3025 (1980).

10. Masuda, H., deMeis, L., Biochemistry 12:4581 (1973).
11. Kanazawa, T., Boyer, P., J. Biol. Chem. 248:3163 (1973).
12. Punzengruber, C., Prager, R., Kolassa, N., Winkler, F., Suko, J., Europ. J. Biochem. 92:349 (1978).
13. Martin, D., Tanford, C., Biochemistry 20:4597 (1981).
14. deMeis, L., Otero, A., Martins, O., Alves, E., Inesi, G., Nakamoto, R., J. Biol. Chem. 257:4993 (1982).
15. Kurzmack, M., Verjovski-Almeida, S., Inesi, G., Biochem. Biophys. Res. Commun. 78:772 (1977).
16. Dupont, Y., Europ. J. Biochem. 109:231 (1980).
17. Takakuwa, Y., Kanazawa, T., Biochem. Biophys. Res. Commun. 88:1209 (1979).
18. Ikemoto, N., J. Biol. Chem. 251:7275 (1976).
19. Watanabe, T., Lewis, D., Nakamoto, R., Kurzmack, M., Fronticelli, C., Inesi, G., Biochemistry 20:6617 (1981).
20. Knowles, A., Racker, E., J. Biol. Chem. 250:1949 (1975).
21. deMeis, L., Tume, R., Biochemistry 16:4455 (1977).
22. Cantley, L., Josephson, L., Warner, R., Yanagisawa, M., Lechene, C., Guidotti, G., J. Biol. Chem. 252:7421 (1977).
23. Cantley, L., Cantley, L., Josephson, L., J. Biol. Chem. 253:7361 (1978).
24. Smith, R., Zinn, K., Cantley, L., J. Biol. Chem. 255:9852 (1980).
25. Inesi, G., Kurzmack, M., Nakamoto, R., deMeis, L., Bernhard, S., J. Biol. Chem. 255:6040 (1980).
26. Dupont, Y., Bennett, N., FEBS Letters 139:237 (1982).
27. Pick, U., J. Biol. Chem. 257:6111 (1982).
28. Inesi, G., "The sarcoplasmic reticulum of skeletal and cardiac muscle," in Cell and Muscle Motility (R. Dowben and J. Shay, eds.), Plenum Publishing Corporation, New York 1:63 (1981).
29. Froelich, J., Taylor, E., J. Biol. Chem. 250:2013 (1975).
30. Verjovski-Almeida, S., Kurzmack, M., Inesi, G., Biochemistry 17:5006 (1978).
31. Inesi, G., Scarpa, A., Biochemistry 11:356 (1972).
32. Coan, C., Inesi, G., J. Biol. Chem. 252:3044 (1977).
33. Inesi, G., Watanabe, T., Coan, C., Murphy, A., Ann. N.Y. Acad. Sci. accepted.
34. Watanabe, T., Inesi, G., Biochemistry 21:3254 (1982).
35. Watanabe, T., Inesi, G., J. Biol. Chem. 257:11510 (1982).
36. Tanford, C., Proc. Nat. Acad. Sci. 79:6527 (1982).

KINETIC AND THERMODYNAMIC ASPECTS OF ATP-DEPENDENT CA^{2+} TRANSPORT IN SARCOPLASMIC RETICULUM USING ATPßS DIASTEREOMERS AS SUBSTRATES

Antonio Scarpa
Elizabeth Pintado
Mildred Cohn

Department of Biochemistry & Biophysics
University of Pennsylvania School of Medicine
Philadelphia, Pennsylvania

Use of ATPßS Diastereomers

It has often been suggested that the Ca^{2+} concentration gradient across the sarcoplasmic reticulum (SR) membrane of a resting skeletal muscle cell is close to what would be predicted for the equilibrium state of the ATP-driven Ca^{2+} pump (1, 2, 3). Recently Tanford (1) has suggested that unlike the (Na^+, K^+)-ATPase system of either erythrocyte plasma membranes or squid axons, where the free energy of the ATP hydrolyzed greatly exceeds the free energy required for the ion concentration gradient, for the ATP-driven Ca^{2+} pump of SR, the Ca^{2+} concentration gradient across the sarcoplasmic reticulum membrane is very close to that predicted for the equilibrium state of the pump.

This hypothesis can be quantitatively addressed using analogues of ATP which can be used as substrates by the ATPase of the SR, but whose energy of hydrolysis is markedly different from that of ATP. In this study we took advantage of the availability of two diastereomers of adenosine 5'-0-(2-thiotriphosphate), ATPßS(R) or ATPßS(S), whose stereochemistry has been recently determined and which can be synthesized in large amounts and great purity (4, 5).

Using a variety of phosphorylable substrates, it was found that the free energy of hydrolysis of ATPßS to form ATPßS and P_i is approximately 2.5 kcal greater than that associated with the hydrolysis of ATP to form ADP and P_i.

ISBN 0-12-260380-X

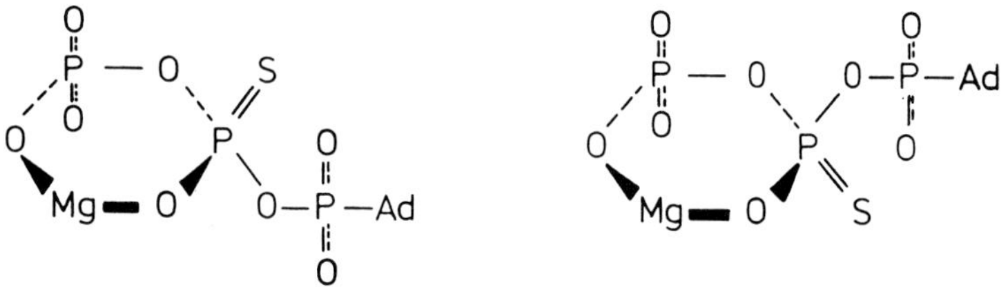

FIGURE 1. Structure of the ß, γ-bidentate magnesium complexes of the diastereomers of ATPßS(S) and ATPßS(R). For more details see ref. 4.

(6). These findings are summarized in Table I. Some of these measurements were carried out using ^{31}P NMR, which is a convenient means for simultaneously measuring the concentrations of several species in the equilibrium mixtures of the reactions catalyzed by arginine kinase and creatine kinase.

$$\text{MgATP} + \text{X} = \text{MgADP} + \text{XP},$$

where X = arginine or creatine and XP = P-arginine or P-creatine. The free energy of phosphorylation of various metabolites by adenosine 5'-O-(2-thiotriphosphate) at pH 8.0 and 30°C is more exergonic than the corresponding phosphorylations by ATP by about 2.5 kcal/mol, resulting in a displacement of the equilibrium toward the nucleoside diphosphates by a factor of approximately 60. Since this factor does not depend on the nature of the metabolite, the equilibrium constants of thionucleotide reactions have also been used to determine the equilibrium constants of corresponding oxynucleotide reactions which lie too far toward ATP.

The availability of ATPßS(S) then permits one to address a quantitative aspect of the coupling in the energy transduction process in ionmotive ATPases, i.e., whether the ion concentration gradient across the membrane is close to the value predicted for the equilibrium state of the ATP-driven ion pump. Consequently, the SR ion-transporting system was chosen to be investigated with ATPßS as substrate by simultaneously monitoring Ca^{2+} uptake and the concomitant nucleoside triphosphate hydrolysis. Since it was found

TABLE I

Apparent equilibrium constants for kinase reactions at pH 8.0 and 30°C

Initial concentrations: 200 mM Hepes, 30 mM total nucleotides, 30 mM total acceptor and phosphorylated acceptor ($[X] + [XP]$), 38 mM Mg^{2+}, 5 mM EDTA, 20% D_2O

$$K_{oxy} = \frac{[\text{MgADP}][XP]}{[\text{MgATP}][X]}$$

$$K_{thio} = \frac{[\text{MgADP}\beta\text{S}][XP]}{[\text{MgATP}\beta\text{S}][X]}$$

X	K_{oxy}	K_{thio}	K_{thio}/K_{oxy}
	Observed	Observed	
Arginine	0.17 ± 0.02	11 ± 2	65
Creatine	0.053 ± 0.002	3.1 ± 1	58
	Calculated[a]	Observed	
Pyruvate	3.1×10^{-4}	0.019 ± 0.004	
3-P-glycerate	2.9×10^{-4}	0.018 ± 0.003	

[a] Calculated from K_{thio} using the average value of $K_{thio}/K_{oxy} = 62$

that both diastereomers of ATPßS, (R) and (S), unlike other substrates investigated (7), were hydrolyzed at the same rate as ATP and with similar values of K_M, the interpretation of the experiments of the coupling between hydrolysis with Ca^{2+} transport were not limited by kinetic considerations.

Hydrolysis of ATP and ATPßS

Figure 2 shows that the rates of "uncoupled" Ca^{2+} dependent hydrolysis are very similar for ATP, ATPßS(R) and ATPßS(S), with a K_M of about 100 µM for all three substrates. Comparative rates could not be followed at concentrations of nucleotides below 10 µM, due to the inability of using the necessary regenerating system at these low concentrations with the ATPßS diastereomers. This complete lack of stereospecificity of an enzyme for the two diastereomers has not been published previously. Furthermore, the property of the Ca-dependent SR ATPase, that the rate of hydrolysis of both diastereomers of ATPßS is as high as that of ATP itself, is not shared by other ATPases, as exemplified by myosin (8) and the ATPases investigated in this report (cf. below).

Figure 3 shows that the basal ATPase, i.e., the hydrolysis observed in the absence of Ca^{2+} and in the presence

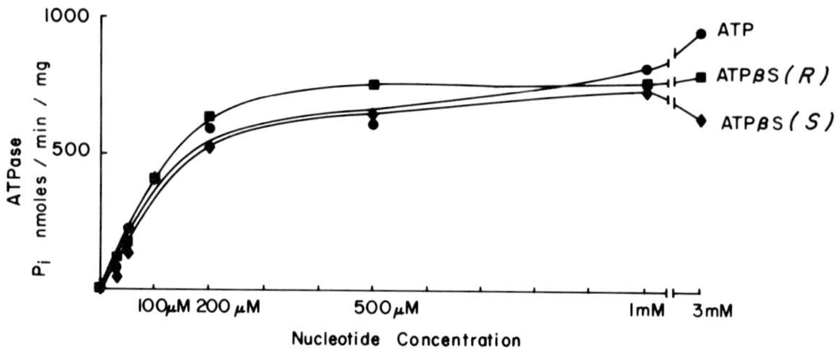

FIGURE 2. Calcium-dependent ATPase activity of sarcoplasmic reticulum as a function of nucleotide concentration with ATP (●), ATPβS(R) (■), and ATPβS(S) (◆) as substrates. The reaction mixture contained 20 mM Tris-maleate, pH 6.8, 10 mM $MgCl_2$, 50 mM KCl, 50 μM $CaCl_2$, and the concentrations of ATP, ATPβS(R) or ATPβS(S) indicated in the figure (temperature, 23°C). The reaction was begun with the addition of 0.46 mg protein/ml and was stopped with 2% SDS. The basal ATPase activity measured in a calcium-free medium, with 500 μM EGTA added, was subtracted for each point. Both basal and Ca-dependent ATPase activities were determined in the presence of ionophore A23187, 20 μM, which effectively prevented intravesicular Ca^{2+} accumulation.

of EGTA, is almost absent for the ATPβS diastereomers. This finding is entirely consistent with recent evidence reported in the literature concluding that the basal ATPase of SR is not an intrinsic enzymic activity of the SR membrane, but originates from a variable amount of contaminant, mostly mitochondria, in SR preparations (9). As a fringe benefit of this investigation, a criterion for the purity of the Ca-dependent SR ATPase has become available, namely how closely the ratios of hydrolytic activities of ATP to ATPβS(R) and to ATPβS(S) approach 1. Furthermore, using ATPβS as substrate, Ca^{2+} transport by SR can be selectively studied even in the fraction of SR isolated <u>in vitro</u> is highly contaminated (i.e., cardiac and smooth muscle SR fractions) with mitochondria and other soluble or organelle-bound ATPases which do not use either ATPβS diastereomer effectively as substrate.

For measurements of both nucleoside triphosphate hydrolysis (Fig. 3) and synthesis (see below) HPLC separa-

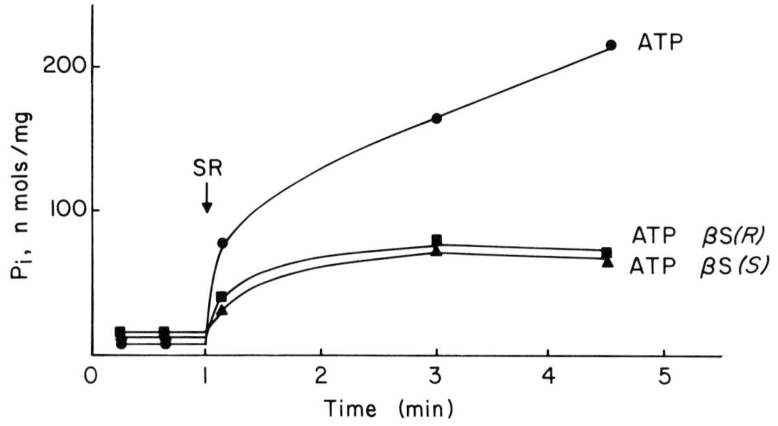

FIGURE 3. "Basic" SR ATPase activity with ATP or ATPβS(R) or ATPβS(S) as substrate. The reaction mixture contained 50 mM KCl, 20 mM Tris-maleate, pH 6.8, 2 mM EGTA, 10 mM $MgCl_2$, and 500 μM nucleoside triphosphate. The reaction was initiated by the addition of 0.4 mg/ml of SR protein. At successive intervals a 50 μl sample was withdrawn and the reaction was stopped with 200 μl of 14% perchloric acid at 0°C. After centrifugation of insoluble material and neutralization of the sample, nucleotide concentrations were measured by HPLC, as shown in Figure 4 (from ref. 16).

tion and analysis was used. Figure 4 shows the elution profile of various nucleoside diphosphates and triphosphates.

ATPβS(S) and Ca^{2+} Uptake

Ca^{2+} uptake by SR vesicles was measured either by determining the decrease of Ca^{2+} concentration in the medium spectrophotometrically with the dye arsenazo III, as shown in Figure 5, or by uptake of $^{45}Ca^{2+}$ at various time intervals, as shown in Fig. 6A. Both the rate of disappearance of Ca^{2+} from the medium and the steady state level attained are the same with ATP and the two diastereomers of ATPβS.

To determine the stoichiometry between moles of ATP hydrolyzed and moles of Ca^{2+} transported, $^{45}Ca^{2+}$ uptake and P_i production were measured in the same reaction mixture as shown in Fig. 6. In order to obtain linear rates for both Ca^{2+} uptake and ATP hydrolysis, intravesicular

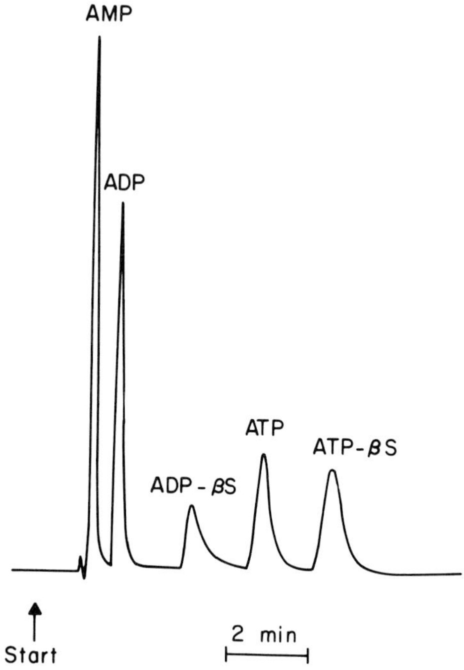

FIGURE 4. HPLC elution profile of AMP, ADP, ADPßS, ATP, and ATPßS after single injection. The 10 µl injection consisted of an aqueous standard mixture containing 20 µM each of AMP, ADP, ATP, and ATPßS, and 10 µM of ADPßS. The chromatographic separation of ATP, ATPßS, ADP, ADPßS, and AMP was achieved at 20°C on a Partisil 10 SAX, 20 µ particle size ionic exchange column, 250 mm long, 4.6 mm i.d., supplemented with a small precolumn (30 X 2.9 mm i.d.) filled with similar but coarser particulate material, using a Waters 600 solvent delivery system with a model U6K sample injector and a model 441 absorbance detector. The mobile phase, consisting of 500 mM ammonium phosphate (pH 4.4), was pumped at a flow rate of 2 ml/min (1800 lb/square inch). Nucleotide absorbance was recorded at 214 nm (0.01 absorbance full scale).

free Ca^{2+} gradients must be avoided, and such gradients were effectively prevented by the presence of oxalate (10). Within experimental error, the rates of Ca^{2+} uptake measured by this method are again the same for all three substances as in Fig. 5. The rates of nucleoside triphosphate hydrolysis are also the same for all three substrates.

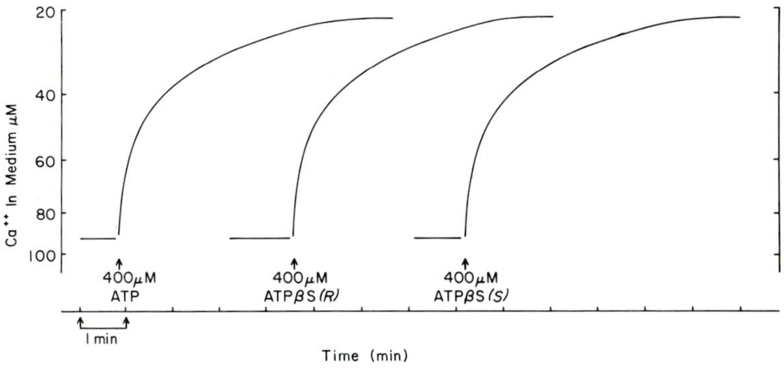

FIGURE 5. Ca^{2+} uptake by sarcoplasmic reticulum vesicles in the presence of ATP and of ATPßS(R) or ATPßS(S). The reaction mixture contained 50 μM arsenazo III, 100 mM KCl, 10 mM $MgCl_2$, 100 μM total Ca^{2+}, 20 mM Tris-maleate, pH 6.9, and 0.5 mg/ml of SR protein. The reaction was initiated with the addition of nucleoside triphosphate (100 μM-1 mM), and the changes in absorbance of arsenazo III (675-685 nm) were measured continuously as a function of Ca^{2+} concentration in the medium, as described previously (17). Quantitatively identical steady state levels of Ca^{2+} accumulation were obtained by measuring Ca^{2+} disappearance from the medium with a Ca^{2+} selective electrode (18), as described previously (19), and by using a variety of Ca^{2+}, SR, and ATP concentration ratios (from ref. 16).

Furthermore, the stoichiometry, i.e., the ratio of moles of Ca^{2+} transported to moles of nucleoside triphosphate hydrolyzed, is the same for ATP, ATPßS(R), and ATPßS(S).

The rates in the experiments described above were determined on a time scale of seconds, and it was possible that differences in the initial rates of Ca^{2+} transport might be observable at much shorter times. Therefore, using $^{45}Ca^{2+}$, the time course of Ca^{2+} uptake over the time range 20-120 ms was determined with ATP and ATPßS(S) as substrates in a rapid mixing and quenching apparatus. As shown in Fig. 7, in spite of considerable scatter in the data, it is very clear that within experimental error the rates of Ca^{2+} uptake, 56 nmol/s/mg of protein, are the same whether ATP or ATPßS(S) are hydrolyzed and agree well with values reported in the literature (11, 12). However, the initial burst of Ca^{2+} uptake (not resolved on the time scale of this experiment) is about 20% higher for ATPßS than for ATP.

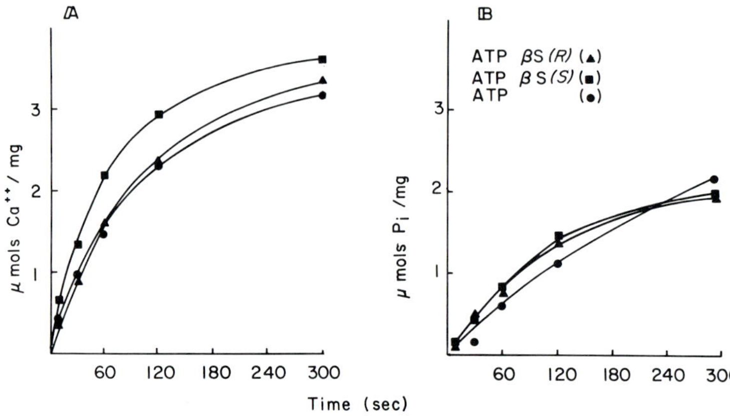

FIGURE 6. $^{45}Ca^{2+}$ uptake (A) and ATP, ATPßS(R), or ATPßS(S) hydrolysis (B) by sarcoplasmic reticulum in the presence of oxalate. The reaction mixture contained 20 mM Tris-maleate, pH 6.5, 5 mM $MgCl_2$, 10 mM sodium oxalate, 1 mM $CaCl_2$ ($^{45}Ca^{2+}$ = 0.3 μCi/ml), 0.69 mM EGTA, and 1 mM ATP, ATPßS(R) or ATPßS(S), and the reaction was started with the addition of SR at a concentration of 0.1 mg of protein/ml (temperature 21°C). For $^{45}Ca^{2+}$ uptake, A, 500 μl aliquots were withdrawn at the times indicated in the figure and the SR was separated from the medium by filtration through Millipore filters which were then counted for $^{45}Ca^{2+}$ radioactivity. A small correction for the nucleoside triphosphate-independent binding was made at each experimental point by subtracting the amount of $^{45}Ca^{2+}$ present in the SR in the absence of nucleoside triphosphate. For ATP hydrolysis, B, the concentration of P_i liberated was measured as described in Fig. 2 after withdrawing aliquots at the times indicated and transferring them to 2% ice-cold SDS to quench the reaction (from refs. 16 & 27).

The question posed initially, whether the coupling of ATPßS hydrolysis to Ca^{2+} uptake would result in a larger Ca^{2+} gradient relative to ATP because the free energy of ATPßS hydrolysis exceeds that of ATP by approximately 2.5 kcal, has been answered unequivocally. Both the rate and extent of Ca^{2+} uptake is the same for both diastereomers of ATPßS and ATP. It is fortunate that the rates of hydrolysis for ATP and ATPßS are the same, so that the interpretation is straightforward and not limited by kinetic differences, as it would be, for example, with acetylphosphate (13, 14) as substrate. The energy available from the hydro-

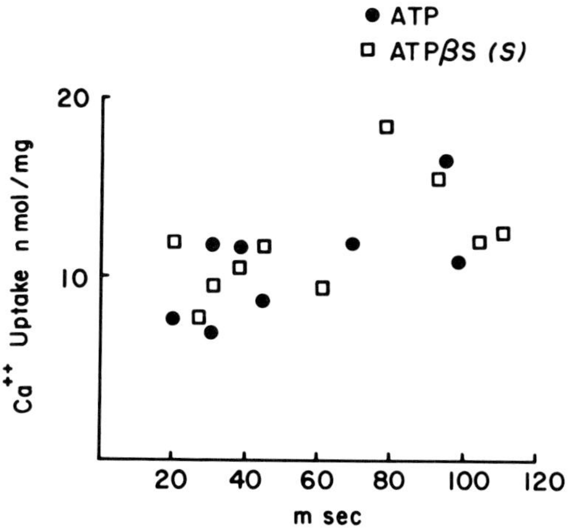

FIGURE 7. Initial velocities of Ca^{2+} uptake in sarcoplasmic reticulum. Ca^{2+} uptake was measured with the EGTA rapid quenching technique in a reaction mixture containing (after the mixing) 80 mM KCl, 5 mM $MgCl_2$, 20 mM MOPS (pH 6.8), 50 µM $^{45}CaCl_2$, 0.25 µCi/ml, 100 µM nucleoside triphosphate, and 0.5 mg of SR protein/ml (temperature, 25°C). The reaction was quenched at the time indicated by the addition of EGTA, 5 mM final concentration. ATP (●) and ATPßS(S) (■). More details on this technique in ref. 12.

lysis of ATP is both necessary and sufficient to maintain the Ca^{2+} gradient, and any excess available is apparently not utilized to yield a larger Ca^{2+} gradient.

Phosphoenzyme Formation Using ATPßS

The first step in the hydrolysis of ATP catalyzed by SR ATPase is the transfer of the γ-phosphoryl group to an aspartyl residue of the enzyme (20), and it has been established that the rate of the first step is much faster than the rate of the overall reaction (21, 22). The apparent complete lack of specificity and stereoselectivity observed with the Ca-dependent SR ATPase for ATP, ATPßS(R), and ATPßS(S) could result from the occurrence of the rate-determining step in all cases at a step in the reaction sequence beyond the formation of the E-P intermediate.

In order to establish whether SR ATPase does indeed discriminate between ATP and ATPßS and between the diastereomers in the formation of E-P, the rate of formation of E-P was measured with all three substrates. Figure 8 shows the results of an experiment where the rate of formation and the steady state concentration of E-^{32}P formed from γ-^{32}P-labeled nucleoside triphosphate substrates was investigated in a rapid quenching apparatus.

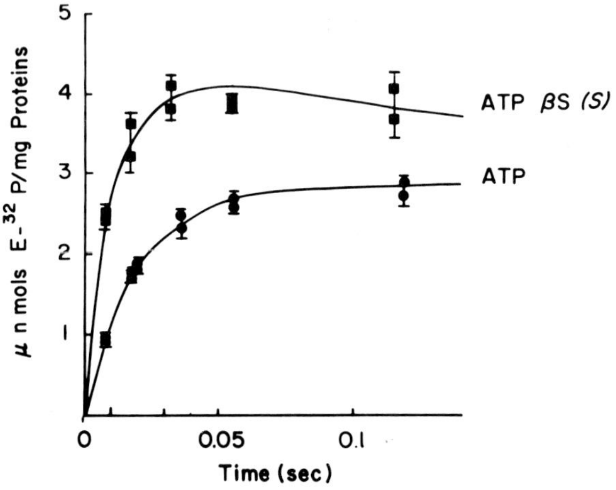

FIGURE 8. Kinetics of phosphoenzyme formation after rapid mixing of [γ-^{32}P]ATP or [γ-^{32}P]ATPβS(S) followed by rapid quenching. Content of the two syringes and experimental design are described in refs. 16 and 20. The final concentrations of protein and nucleotide after mixing were 0.5 mg/ml and 100 μM, respectively. The reaction was quenched with 2.5% perchloric acid solution containing 1 mM nonradioactive nucleotide, and 0.25 mM P_i (temperature, 0°C). [γ-^{32}P]ATPβS(S) was synthesized according to ref. 16.

A comparison of E-^{32}P formation in the time range from 0 to 120 ms with ATP and ATPβS(S) as substrates is shown in the figure. The rate of E-^{32}P formation appears somewhat higher for ATPβS than for ATP, but the major difference is that the steady state concentration of E-^{32}P for ATP is about 2/3 that for ATPβS(S). The rate of E-^{32}P formation and the steady level attained from ATPβS(R) relative to ATP were measured in another experiment and the behavior was similar for both diastereomers. The ratios of the levels of E-^{32}P attained with ATPβS(S) and ATPβS(R) as substrates relative to the level for ATP are 1.3 ± 0.22 (6 experiments), and 1.8 ± 0.26 (11 experiments), respectively.

The rate of formation of E-P is approximately 2-3 times greater for the thioanalogues than for ATP, but both diastereomers have approximately the same rate. Thus, all substrates have a rate-determining step beyond E-P formation, but the enzyme surprisingly exhibits practically no stereospecificity between Δ and Λ configurations of MgATPßS in the rate of formation of E-P. For nucleotides other than ATP, such as ITP and GTP, both the overall rate of hydrolysis (23) and the rate of E-P formation (20, 23) are lower than those for ATP, particularly at Ca^{2+} concentrations below 1 mM.

Reversal of Hydrolysis, Net Nucleoside Triphosphate Synthesis

Net synthesis of ATP from ADP and P_i in a Ca^{2+} free medium has been observed (24, 25) accompanying efflux of Ca^{2+} from SR vesicles previously loaded with Ca^{2+}. In these experiments, only initial velocities of ATP (or ATPßS) synthesis and Ca^{2+} efflux can be compared, because the unavailability of a trapping system equally effective for both ATP and ATPßS and the consequent omission of a trapping system makes the interpretation of steady state levels questionable. As shown in Fig. 9B, if ADPßS is substituted for ADP, net synthesis of ATPßS ensues. However, both the rate and extent of ATPßS formation is considerably lower than that of ATP. The concomitant efflux of Ca^{2+} coupled to nucleoside triphosphate production is correspondingly lower in both rate and extent for ADPßS than for ADP (Fig. 9A), thus establishing that the stoichiometry of transport to synthesis remains constant for the two nucleotides. The difference in rates cannot be ascribed to a difference in K_M values for ADP and ADPßS, as evidenced by the persistence of the difference in nucleoside diphosphate-dependent Ca^{2+} efflux rates over a wide range of concentrations (1 µM to 1 mM) for each nucleotide (not shown). The ratio of the rates for ADP to ADPßS was also found to remain constant over the pH range of 5.5 to 8.

These results show that net synthesis of ATPßS from ADPßS and P_i can be achieved accompanying Ca^{2+} efflux. However, both the rate of synthesis and the concomitant rate of Ca^{2+} efflux are reduced about 6-fold. The apparent extent of nucleoside triphosphate formation as well as the Ca^{2+} efflux are reduced about 2-3-fold with ADPßS, but the stoichiometry between synthesis and Ca^{2+} efflux remains the same as with ADP. Thus, as for the forward reaction, the quantitative aspect of the coupling between Ca^{2+}

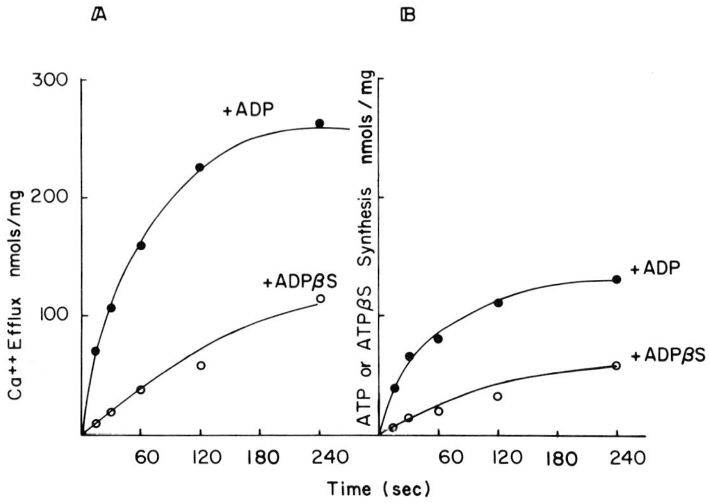

FIGURE 9. Ca^{2+} efflux (A) and ATP or ATPßS synthesis (B). SR was loaded with Ca^{2+} for 3-5 min in a medium containing 50 mM KCl, 20 mM Tris-maleate, pH 6.5, 5 mM $MgCl_2$, 1 mM $CaCl_2$ (0.3 µCi/ml of $^{45}Ca^{2+}$), 0.69 mM potassium EGTA, 20 mM P_i, and 1 mM ATP. After centrifuging the Ca^{2+}-loaded vesicles, the pellet was washed and resuspended at a concentration of 0.2 mg of protein/ml in a medium containing 20 mM Tris-maleate, pH 6.5, 10 mM $MgCl_2$, 5 mM EGTA, 100 mM KCl, and 10 mM P_i. The reaction was initiated with the addition of 1 mM ADP or ADPßS (temperature, 25°C). For measurements of $^{45}Ca^{2+}$ efflux as a function of time (A), aliquots were withdrawn at times indicated and filtered through Millipore filters. Each point was corrected for leakage of Ca^{2+} by subtracting the amount of $^{45}Ca^{2+}$ efflux in the absence of ADP or ADPßS. For the measurement of net ATP or ATPßS synthesis as a function of time (B), aliquots of the reaction mixture were withdrawn at the times indicated and added to ice-cold 10% perchloric acid. Nucleoside triphosphate synthesis was measured in the extract by HPLC (from ref. 16).

transport and nucleoside triphosphate breakdown or synthesis remains the same for ATP and ATPßS in the reverse reaction. IDP, which also reacts with P_i to yield ITP accompanying Ca^{2+} efflux (23), has an apparent K_M for the synthetic reaction 3 orders of magnitude greater than the K_M for ADP, but the rate of the reverse reaction and of Ca^{2+} efflux are the same for both nucleotides. However, with

ADPßS, the K_M value is of the same order of magnitude but the rate of both processes, reversal of hydrolysis and rate of Ca^{2+} efflux, are reduced.

Nucleoside Diphosphate-Nucleoside Triphosphate Exchange Reaction

To address the question of which step in the sequence of nucleoside triphosphate synthesis is responsible for the slower rate with ADPßS compared to ADP, the rates of the exchange reaction between nucleoside triphosphate and diphosphate were compared for ADP and ADPßS. The NDP-NTP exchange reaction, followed by the appearance of radioactivity in nucleoside triphosphate originating in ^{14}C-linked nucleoside diphosphate, represents a partial reaction in the overall pathway of hydrolysis or synthesis of nucleoside triphosphates. The exchange reaction should be useful in dissecting the reaction pathway and in establishing which step in the pathway is responsible for the discrimination between ADP and ADPßS in the synthesis of nucleoside triphosphate. The time course of the exchange reactions shown in Fig. 10 reveals that the exchange rate for the thioanalogues is approximately 10-fold slower than for the oxynucleotides. Thus, both the rate of synthesis of ATPßS and the rate of its exchange with ADPßS are about an order of magnitude less than for ATP. Since both reactions have rather complex kinetics, inasmuch as the rates are dependent on the functional state of the SR, the calcium concentration, and the concentration of inorganic phosphate, the data presented in this investigation are insufficient to determine uniquely the step in the sequence which leads to a slower rate in the reversal of hydrolysis and in the exchange reaction for ADPßS compared to ADP as substrate. The rate-determining step may be the rate of association of ADPßS with E-P, the actual conversion rate to ATPßS, the rate of dissociation of ATPßS from the enzyme, or a shift in the ratio between the various species of E-P. Further experiments are needed to differentiate the individual rates.

ATPßS and Activities of Other Ionmotive ATPases

To determine the generality of the behavior of SR ATPase with the ATPßS diastereomers for ion transporting ATPases, ATP, ATPßS(S), and ATPßS(R) were compared as substrates for ATPases coupled to ion transport other than SR, including (Na^+, K^+)-ATPase and two proton transporting

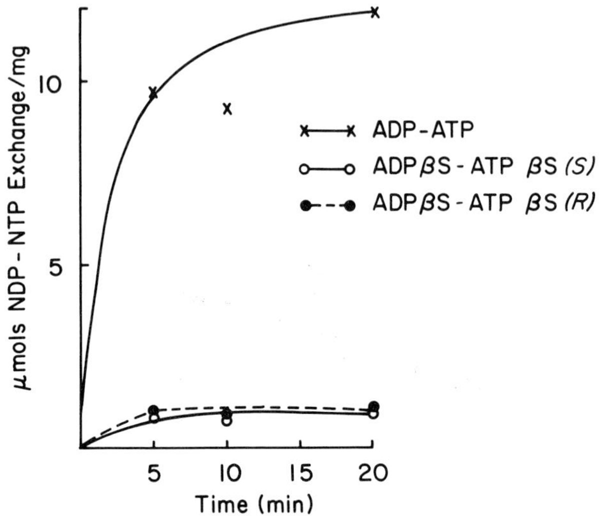

FIGURE 10. Comparison of nucleoside diphosphate-nucleoside triphosphate exchange for oxynucleotides and thionucleotides as a function of time. x——x, ADP-ATP; o——o, ADPβS-ATPβS(S); •——•, ADPβS-ATPβS(R). The reaction mixture contained 20 mM histidine buffer, pH 7.0, 0.1 mM $CaCl_2$, 40 mM KCl, 0.1 mM Ap_5A, 1 mM NTP, 2 mM NDP uniformly labeled with ^{14}C (0.57 μCi/μmol), 4 mM $MgCl_2$ (temperature, 25°C). The reaction was initiated by the addition of SR, 0.02 mg of protein/ml, and stopped by adding aliquots of the reaction mixture to ice-cold perchloric acid (final concentration, 5%).

systems, submitochondrial particles and chromaffin ghosts. The results of some exploratory experiments are presented in Table II. The SR ATPase is unique among these enzymes because all three substrates are hydrolyzed at the same rate. For the (Na^+, K^+)-ATPase, the rate for both thioanalogues is the same, but it is 50% of the rate for ATP. For the proton-transporting systems, activity could be detected only with the R isomer, and the V_{max} values for hydrolysis relative to that of ATP were 14 and 2% for chromaffin ghosts and submitochondrial particles, respectively.

TABLE II

RELATIVE RATES OF HYDROLYSIS

OF ATPβS DIASTEREOMERS

(ATP HYDROLYSIS = 100)

ATPase	V_{max} (S) %	V_{max} (R) %	V_S/V_R
Sarcoplasmic Reticulum	100	100	1.0
Na^+-K^+-ATPase	*50	*50	1.0
Chromaffin ghost	n.d.	14	<0.14
Submitochondrial particles	n.d.	2	<0.05

*Rate at 1 mM Nucleotides

(from ref. 16, where experimental details are given)

Of the four ATPases which have been investigated in addition to SR ATPase, namely, the (Na^+, K^+)-ATPase, mitochondrial ATPase, chromaffin ghost ATPase, and myosin ATPase (26), only the (Na^+, K^+)-ATPase forms an E-P intermediate. Although the rate of hydrolysis with ATP as substrate is reduced to 50% with ATPβS as substrate in the (Na^+, K^+)-ATPase, both diastereomers of the thioanalogue react at the same rate. For those ATPases for which cogent evidence exists that they do not form an E-P intermediate, there is high stereoselectivity and often considerably reduced rates even for the more active isomer.

In conclusion, ATPβS and ADPβS have proven to be unique in their behavior patterns in the reactions catalyzed by SR ATPase relative to other substrate analogues. Because of the affinity for the enzyme and the rate of hydrolysis for ATPβS and ATP, it was possible to conclusively establish that both the rates and the extent of Ca^{2+} uptake were the same for both substrates in spite of a large difference in free energy of hydrolysis between ATP and ATPβS.

ACKNOWLEDGMENTS

Many thanks are due to Dr. D. Trentham, University of Pennsylvania, and to Dr. G. Inesi, University of Maryland, for their encouragement and help in carrying out the transient kinetic experiments. This work was supported by grants HL-18708 and AA-05662 from the United States Public Health Service.

REFERENCES

1. Tanford, C., J. Gen. Physiol. 77:223 (1981).
2. Inesi, G., Science (Wash, D.C.) 171:901 (1971).
3. Hasselbach, W., Biochim. Biophys. Acta 515:23 (1978).
4. Jaffe, E.K., and Cohn, M., J. Biol. Chem. 253:4823 (1978).
5. Burgers, P.M.J., and Eckstein, F., J. Biol. Chem. 254:6889 (1979).
6. Lerman, C.L., and Cohn, M., J. Biol. Chem. 255:8756 (1980).
7. deMeis, L., "The Sarcoplasmic Reticulum," John Wiley and Sons, New York, 1981.
8. Goody, R.S., and Hofmann, W., J. Muscle Res. Cell Motil. 1:101 (1980).
9. Fernandez, J.L., Rosenblatt, M., and Hidalgo, C., Biochim. Biophys. Acta 599:552 (1980).
10. Weber, A., Herz, R., and Reiss, I., Biochem Z. 345:329 (1966).
11. Inesi, G., and Scarpa, A., Biochemistry 11:356 (1972).
12. Kurzmack, M., Verjovski-Almeida, S., and Inesi, G., Biochem. Biophys. Res. Comm. 78:772 (1977).
13. deMeis, L., J. Biol. Chem. 244:3733 (1969).
14. Pucell, A., and Martonosi, A., J. Biol. Chem. 246:3389 (1971).
15. Fiske, C.H., and Subbarow, Y., J. Biol. Chem. 66:375 (1925).
16. Pintado, E., Cohn, M., and Scarpa, A., J. Biol. Chem. 257:11346 (1982).
17. Scarpa, A., Methods in Enzymology 56:301 (1979).
18. Simon, W., Ammann, D., Oehme, M., and Morf, W.E., Ann. N.Y. Acad. Sci. 307:52 (1978).
19. Dubyak, G., and Scarpa, A., J. Muscle Res. Cell Motil. 3:87 (1982).
20. Degani, C., and Boyer, P.D., J. Biol. Chem. 248:8222 (1973).
21. Hasselbach, W., and Suko, J., Biochem. Soc. Spec. Publ. 4:159 (1974).

22. Martonosi, A., Lagwinska, E., and Oliver, M., Ann. N.Y. Acad. Sci. 227:549 (1974).
23. Carvalho, M.G.C., deSouza, D.G., and deMeis, L., J. Biol. Chem. 251:3629 (1976).
24. Makinose, M., and Hasselbach, W., FEBS Lett. 12:271 (1970).
25. deMeis, L., J. Biol. Chem. 251:2055 (1976).
26. deMeis, L., and Fialho deMello, M.C., J. Biol. Chem. 248:3691 (1973).
27. Pierce, D., Scarpa, A., Topp, M., and Blasie, J.K., Biochemistry, submitted (1983).

CONFORMATIONAL CHANGES OF THE SARCOPLASMIC RETICULUM Ca-ATPase INDUCED BY SUBSTRATE BINDING AND PHOSPHORYLATION

Yves Dupont[1], Nelly Bennett[1], Richard Pougeois

Département de Recherches Fondamentales
Centre d'Etudes Nucléaires
Grenoble, France

Jean-Jacques Lacapere

Département de Biologie
Centre d'Etudes Nucléaires
Saclay, France

CONTENTS

1- Supported by CNRS grant: ATP - Conversion de l'énergie dans les membranes biologiques

ISBN 0-12-260380-X

I INTRODUCTION

Understanding the functioning of the sarcoplasmic reticulum Ca-ATPase means being able to answer precise questions like: how does this enzyme catalyse the transfer of calcium ions through the sarcoplasmic reticulum membrane and how is this transport coupled to the phosphate transfer reactions which supply energy to the enzyme? We know little about the structure of this protein: the low resolution view emerging from the X-rays diffraction studies (1,2) only indicates that it is asymmetrically inserted into the lipid bilayer with an important fraction of the enzyme being immersed into the myoplasm. This view does not provide any indication on the location of the active sites. It follows that the present image of the Ca-ATPase functioning is essentially derived from studies using more standard enzymological approaches. The concept of enzyme conformational changes plays an important role in the models advocated by researchers in this field; firstly the simplest intuitive scheme of transport involves an alternate sites mechanism where the transport sites are alternatively exposed to both sides of the membrane, with a simultaneous change in binding constant for the ion transported; secondly in most of the schemes published the enzyme conformation provides the link between the ions translocation reaction and the phosphate transfer reactions (3-5). These two key roles of the enzyme conformation change are the basis for the so-called $E \rightleftharpoons E^*$ scheme of the Ca-ATPase (3) and the closely related $E1 \rightleftharpoons E2$ scheme of the (Na-K)-ATPase(4,5).

Attempts to detect these putative conformational changes of the Ca-ATPase and (Na-K)-ATPase were numerous: measurements of the reactivity of active sites, chemical modifications, susceptibility of SH groups or to tryptic attack, use of optical properties of extrinsic probes or measurement of the protein intrinsic fluorescence changes. In this article we will review the results which have been obtained with this latter technique. Combined with accurate static and time resolved titration of active sites with labeled substrates, the recording of the intrinsic fluorescence changes provides one of the best method presently available to study the events in the life of the Ca-ATPase enzyme.

Since the first report (6) of an intrinsic fluorescence change induced by calcium binding, nearly all other known substrates have been reported to affect the fluorescence of the Ca-ATPase (see Table I for references):

TABLE I. Ca-ATPase Fluorescence

- Calcium binding	--------(6-10)
- Magnesium binding	--------(11,12)
- ATP binding	--------(13)
- Phosphorylation by ATP	---(7,13)
- Phosphorylation by Pi	
gradient dependent	---(14)
gradient independent	---(15,16)
- Vanadate binding	--------(17)

In the following chapters we will successively describe how the enzyme responds to binding of its main substrates: calcium, ATP and phosphate. At the end of this paper we wish to propose and discuss a general interpretation of these fluorescence changes. The scheme which will be proposed could, at least, be a simple unification of these data in terms of the minimal number of conformational states of the protein. Such an attempt will, moreover, be also justified by the fact that it gives at the same time a plausible framework to explain the cooperative calcium binding and the half-of-the-sites reactivity for ATP binding or phosphorylation. The scheme proposed is in many aspects significantly different from the $E \rightleftharpoons E^*$ scheme used by most workers in this field.

II EXPERIMENTAL TECHNIQUES

Most of the experiments presented here have been performed with already published techniques.

Substrates binding assays were performed using Millipore filters and a thermostated filtration apparatus as described in (18). Static fluorescence levels were measured with a fluorimeter described in previous reports (9,10); this fluorimeter was especially designed for measuring the small fluorescence changes reported in this work. Improvements over conventional designs were mainly centered on lamp stability, mixing device and efficiency of light detection using interference filter. As shown in figure 1, this results in a very good signal to noise ratio and little fluorescence decay due to photolysis (compare with experiments reported by us in (6)).

Transient fluorescence changes were measured with an original stopped-flow apparatus fitted in the beam of the fluorimeter mentionned above. Sample and substrate syringes are driven by electronicaly controled stepping motors; mixer is a Berger type ball-mixer (19), the observation cuvette is vertical with no stop-syringe, dead time is around 5 msec.

In some experiments which will be described, a fast filtration device using Millipore filters was used (dead time 20 to 50 msec. depending on experimental conditions). This fast filtration process will be described extensively elsewhere.

All experiments reported in this work were performed with sarcoplasmic reticulum vesicles obtained as described in (20). Measurements of phosphoenzyme concentration were performed with a procedure identical to that described in (21).

III RESULTS WITH EACH SUBSTRATES

A) Calcium

- Equilibrium binding. Equilibrium binding of calcium ions to the high affinity sites of the Ca-ATPase was measured by Millipore filtration. We have previously reported an affinity around 0.5 µM (pH 7.2) and a number of sites of 6 to 7 nmol/mg. The affinity of these calcium binding sites is strongly pH dependent (22) as confirmed recently by Hill and Inesi (23). Although temperature has been reported to affect the number of measurable high affinity sites (24,25), we have been unable to reproduce this temperature effect. In our hands the calcium binding capacity of the Ca-ATPase remains unchanged from 30°C to -11°C (in 30% glycerol) (18); this insensitity to temperature was later confirmed by others (22).

Binding of calcium ions affects the intrinsic fluorescence of the Ca-ATPase; this was first reported in 1976 (6) and then reproduced and used in a large number of studies (6-10,26-28).

This fluorescence change is also observed with the solubilized protein (9,27,28) clearly disproving a contribution of light scattering to the change observed. Calcium binding increases the fluorescence of the Ca-ATPase (Fig. 1) and this change is accompanied by a slight blue shift of the emission spectrum (11). At 340 nm emission wavelength and 20°C, the fluorescence increase is around 5% of the total fluorescence.

The calcium ion concentration dependence of the fluorescence change indicates an affinity for calcium equivalent to that found by direct titration of the calcium binding sites with 45Ca.

The fact that calcium ions affect the intrinsic fluorescence of the Ca-ATPase is a very good indication that binding of this ion is associated with a conformation change of the protein. Yet, studies at equilibrium are insufficient to disprove an alternative interpretation: the fluorescence changes could be produced by a close interaction between calcium ions and one

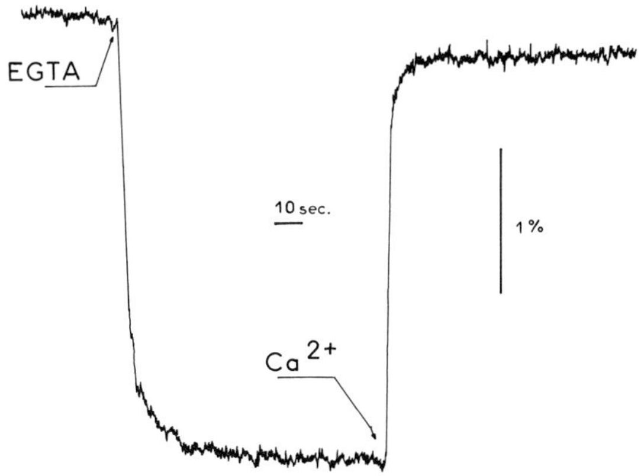

FIGURE 1. Calcium induced intrinsic fluorescence change of the sarcoplasmic reticulum ATPase. Experimental conditions: 10 mM Tris-Mops, 100mM KCl, 5mM Mg and where indicated: 200μM EGTA or 300 μM Ca, 0°C and pH 7.2. Excitation: 295 nm, emission: 350 nm.

or more tryptophanyl residue located in the vicinity of the ion binding site. This kind of uncertainty is common to all other techniques used so far in the search for Ca-ATPase conformation changes. However, fluorescence changes can in principle be resolved down to the millisecond range where good tests on the origin of the fluorescence change can be applied. This point will be discussed further in this article.

- Transient state studies. The $E \rightleftharpoons E^*$ scheme published by Carvalho et al. (3) has been one of the most extensively used so far in the mechanistic studies of the Ca-ATPase. This scheme postulates the existence of two distinct functional conformations of the enzyme (E and E*) which are differentiated mainly by their phosphorylating reaction: E is phosphorylated by ATP and catalyses an ATP-ADP exchange, while the E* form can be phosphorylated directly from Pi and catalyses a water-phosphate oxygen exchange. An other important property of the enzymatic state proposed by Carvalho et al. concerns the sidedness of the calcium binding sites: it was proposed that the sites are accessible from the cytoplasmic side of the vesicles in the E form, while they are facing the internal space in the E* form. Guillain et al. (15) have clearly demonstrated that the calcium induced fluorescence change and conversion into a form phosphorylable by ATP occured at the same rate. It is therefore very likely that the calcium induced fluorescence change is directly related to the $E \rightleftharpoons E^*$ isomerization proposed by Carvalho et al. The simple two-state model described by these authors can lead to very precise predictions for the rate constant of transition between the two conformers. This is schematically described in figure 2 (scheme I). In this figure are also described alternative two-state schemes where calcium binding proceeds through quite different mechanisms. It is obvious that investigation of the calcium concentration dependence of the observed rate constants of isomerization might lead to a clear choice among these various schemes.

The first kinetic studies published (7,8) were in clear disagreement with schemes I and II (Fig. 2). The basic point was that the calcium concentration dependence of the observed rate constant conclusively exhibited a very steep decrease above 100 nM calcium followed by an increase above 5 μM calcium. These results indicated that, if the enzyme has two conformation states E and E*, calcium ions are able to bind directly to the E form with a dissociation constant of around 0.1 μM but, surprisingly, also to the E* form with a non negligible dissociation constant of 10 to 20 μM. This process can be accounted for by two of the schemes presented in figure 2 (III and IV).

Distinction between schemes III and IV (Fig. 2) was attempted in a study at low temperature published recently (10). At -10°C and in the presence of 30% glycerol the fluorescence change induced by calcium occurs with a very slow rate (k=0.03/sec); it was therefore possible to measure under the same conditions

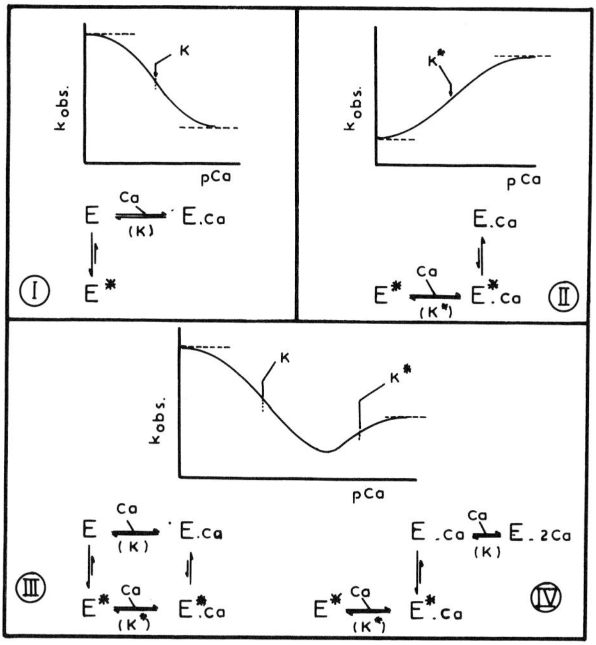

FIGURE 2. Two-state E = E* schemes. Schematic representation of the observed rate constants for various sequences of calcium binding. Details of calculation may be found in ref. 10.

the time course of ^{45}Ca binding by filtration. These results are summarized below and some are reproduced in figure 3. 1) The calcium concentration dependence of the observed rate constants exhibits the same decreasing and increasing phases that were found at higher temperatures, 2) A fast accessible ^{45}Ca binding phase ($k_{obs} \gg 0.2$/sec) is found corresponding to a binding site of apparent dissociation constant K_d = 25 μM, 3)A slow ^{45}Ca binding phase follows with a rate constant equal to that measured in fluorescence (k_{obs} = 0.015/sec in 100 μM calcium), 4) The maximal amplitude of the fast phase is about 3 nmol/mg i.e. one half the calcium binding capacity of the enzyme, 5) Binding of calcium to the rapidly accessible site

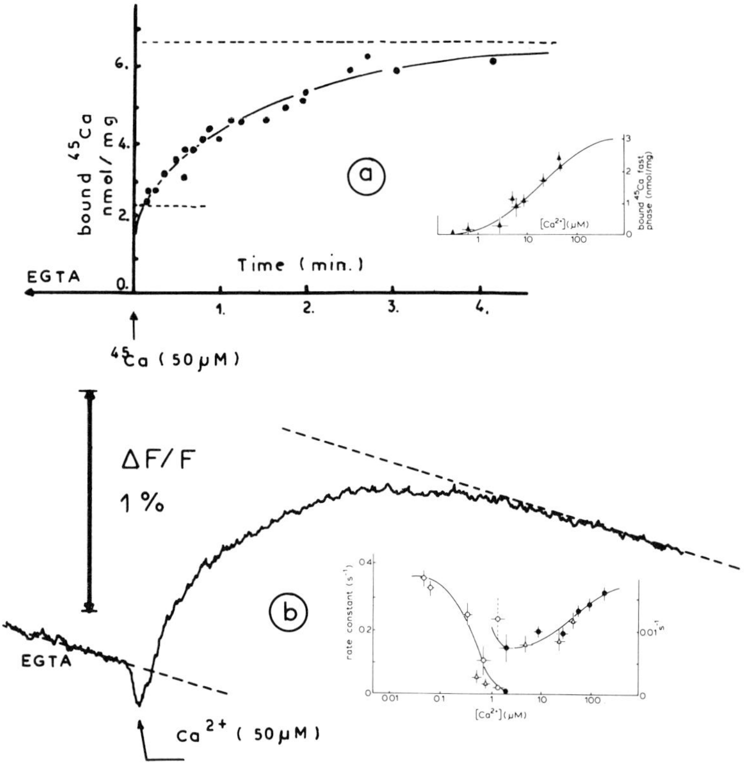

FIGURE 3. Time resolved measurements of the calcium induced transition at −10°C. (a) Measurement of calcium binding by filtration on Millipore filters (0.45 μm). Inset: amplitude of the fast phase of calcium binding as a function of free calcium concentration. (b) Measurement of intrinsic fluorescence change in similar conditions. Inset: rate constant of the slow phase of calcium binding and of the fluorescence change. Time scale is identical in a) and b). Experimental details may be found in ref. 10.

increases the observed rate constant of the slow phase of calcium binding and of the fluorescence change (see insets in figure 3).

These characteristics are those of a sequential binding process and correspond to the scheme IV described in figure 2. Interestingly it is the only one among the four schemes which explicitly includes the fact that the Ca-ATPase has two ion binding sites with cooperative interaction (27,29). In fact the data at low temperature are better fitted by modifying slighty the scheme IV and adding a secondary route for binding and dissociation to the rapidly accessible calcium site (10).

One embarrassing result of these studies is that one of the two calcium sites is rapidly accessible from the outer face of the vesicles with an intermediate affinity (Kd' = 25 μM). The rate of access to this site, its affinity and the fact that the vesicles were tightly closed safely indicates that this site has nothing to do with an internal form of the calcium binding site. This result is in disagreement with one of the basic proposal of the $E \rightleftharpoons E^*$ scheme in which the calcium sites are accessible from the inner face of the vesicles when the enzyme is in the E^* form. As this proposal is supported by a considerable amount of work, we have tentatively proposed an important modification of the kinetic scheme by including a third functional state of the enzyme. In this case the fast calcium binding phase is dependent on an initial fast isomerization of the enzyme. This scheme published in (10) is reproduced below (scheme V). A two-state $E \rightleftharpoons E^*$ scheme is also reproduced for comparison (Scheme VI). On this figure the state of the calcium sites is schematically represented: □ for low affinity and inwards facing, ○ for high affinity and exposed to the cytoplasmic side of the vesicle. In the case of a two-state scheme (VI), cooperative binding to the two calcium sites may be accounted for by a concerted modification of the two sites. This is obviously different from the sequence proposed by scheme V which includes the transient formation of an hybrid intermediate E21 .

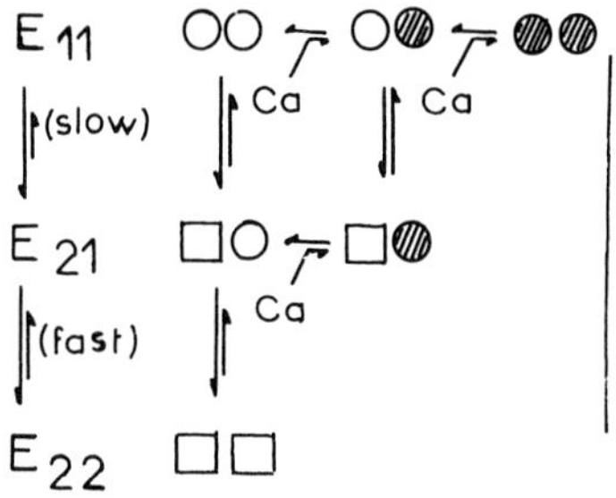

SCHEME V

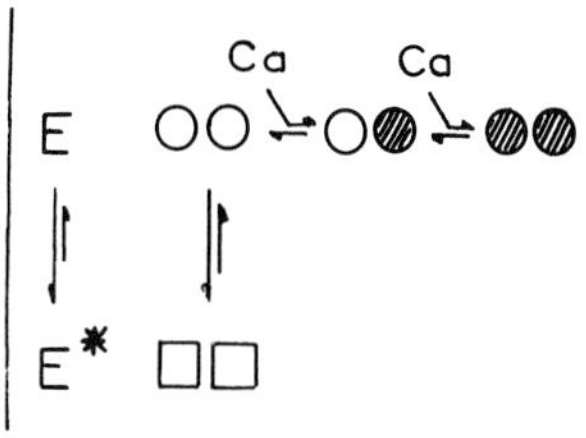

SCHEME VI

Only scheme V can explain the biphasic ^{45}Ca binding, the relatively low apparent affinity of the rapidly accessible site being accounted for by the existence of the first E22 $\rightleftharpoons$ E21 unfavoured transition.

Two important implications of this three-state model are:
i) In the E22 low fluorescence form, no calcium site should be accessible from the cytoplasmic side of the vesicle,
ii) addition of calcium ions to an enzyme incubated in EGTA should induce a two-step isomerization according to the sequence:

$$E22 \longrightarrow E21 \longrightarrow E11.$$

These two predictions have recently received some support:
i) It has been demonstrated that the low fluorescence form of the enzyme (the present E22 state) can be stabilized by low concentrations of vanadate (17). Therefore calcium binding to the E22 form can be assayed without simultaneous induction of conformational transition. Results shown in figure 4(b) indicate conclusively that below 100 µM no calcium binding can be detected on the E22 form stabilized by vanadate. A similar result can be found in a recent article by Pick (30).
ii) In a very careful analysis of the calcium induced fluorescence changes, Champeil et al. (12) have clearly observed and analysed the predicted biphasic transition. They have also evidenced the important role of magnesium and pH on the fast E22 $\longrightarrow$ E21 transition.

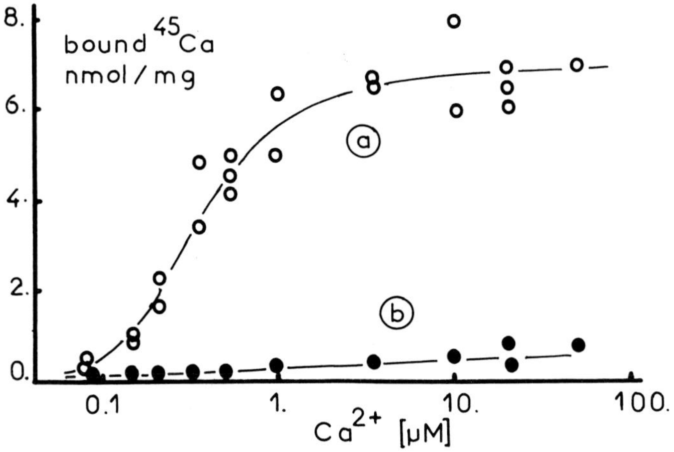

FIGURE 4.Measurement of calcium binding to the Ca-ATPase. Conditions were: 20mM Tris-Mops pH 7.2, 100mM KCl, 5mM Mg, 50 µM total Ca and 0 to 200 µM EGTA. Vesicles were preincubated 10 min in 200 µM EGTA (○) or in 200 µM EGTA + 100 µM VO4 (●). Measurements were performed using filtration on Millipore filters as described in (18).

In the absence of magnesium or at acid pH this transition can become rate limiting. An experiment showing a biphasic fluorescence transition is shown in the figure 5.

- Non equivalence of the calcium binding sites. In the scheme described above (scheme V) the two calcium binding sites do not play equivalent roles. We have indeed reported an experiment which shows that, even when saturated by calcium ions, the two sites remained different. This was evidenced by measuring the rate of $^{45}Ca \rightleftharpoons Ca$ exchange at low temperature (-10°C) (10). Since these experiments were reported, we have improved the time resolution of the filtration technique and the same

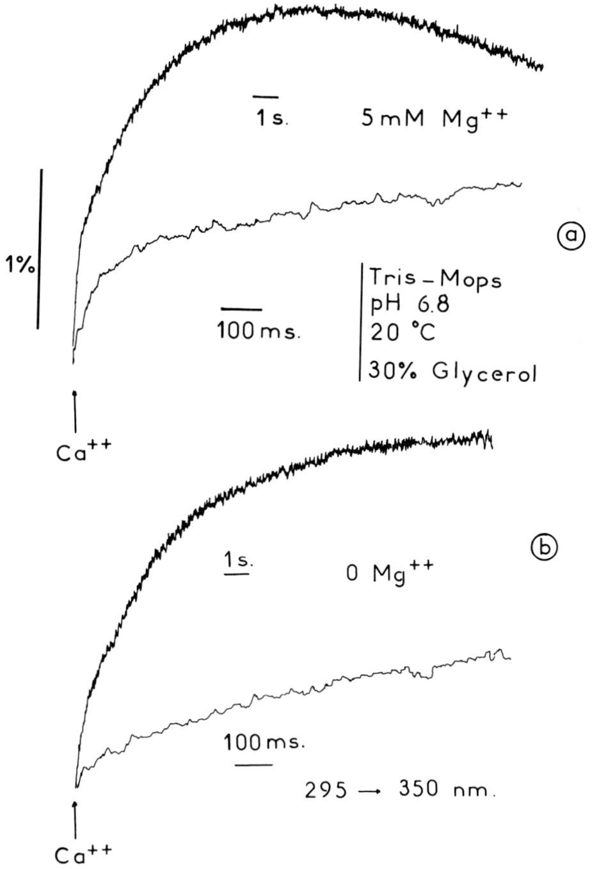

FIGURE 5. Time resolved fluorescence changes induced by calcium in the presence (a) or in the absence (b) of magnesium. Vesicles preincubated in 0.5mM EGTA were mixed with calcium to give a final concentration of 0.5mM calcium. The traces shown are average of four stopped-flow records. Two traces with different time scale are shown for each condition.

process can now be measured at room temperature. Figure 6 displays this experiment which confirms that the two calcium binding sites present very different rates of calcium exchange: k1 ⩾ 50/sec, k2 = 0.8/sec at 20°C and 50 μM calcium. The same concept of two families of non equivalent calcium binding sites has been developed in a work by Ikemoto et al. (31).

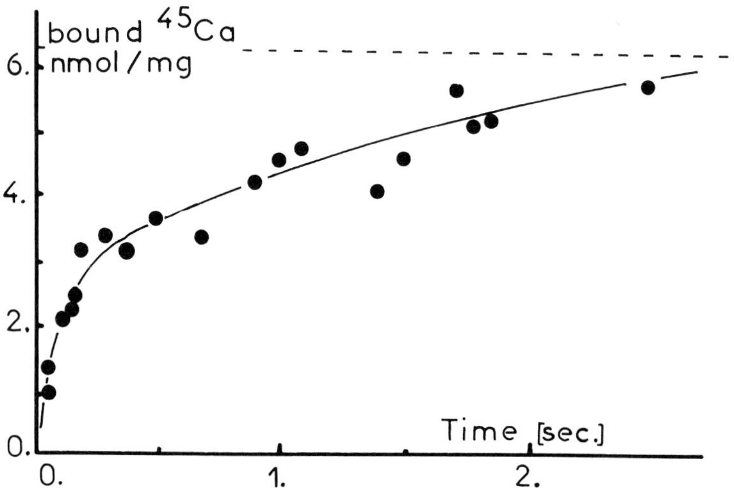

FIGURE 6. Isotopic exchange of calcium ions at equilibrium. Vesicles incubated a few minutes in 50 μM cold calcium were layered on a Millipore filter and washed during the time indicated with 50 μM radioactive calcium. Measurements were performed using an original rapid filtration equipment to be described elsewhere[1]. Temperature 20°C and pH 7.2.

-Conclusion for calcium binding

- Existence of two interacting calcium binding sites (6 to 7 nmol/mg)
- Calcium binding is sequential and is associated to two successive conformational changes of the protein:
  E22 ⟶ E21 ⟶ E11
- The intermediate state E21 is an hybrid conformation with only one accessible high affinity calcium binding site (3.0 nmol/mg)
- The two calcium sites are not equivalent.

---

1. Patented process, manuscript in preparation.

B) ATP

- Binding in the absence of calcium. In the absence of calcium (EGTA containing buffer) ATP binding can be easily assayed with radioactive ATP: around 3.0 to 3.3 nmol/mg of binding sites were reproducibly found (13,18,20). As the number of calcium sites, the number of ATP sites is independent of temperature. The affinity for ATP binding to these sites is around 3.0 uM at 20°C and pH 7.2. Figure 7(a) represents a typical ATP binding curve. At ATP concentrations higher than 100 µM further binding can be observed. This fact was reported in an earlier study (20), and has been attributed to a secondary and possibly regulatory site (20,33). Figure 7b also shows the strong inhibitory effect of vanadate ions on ATP binding.

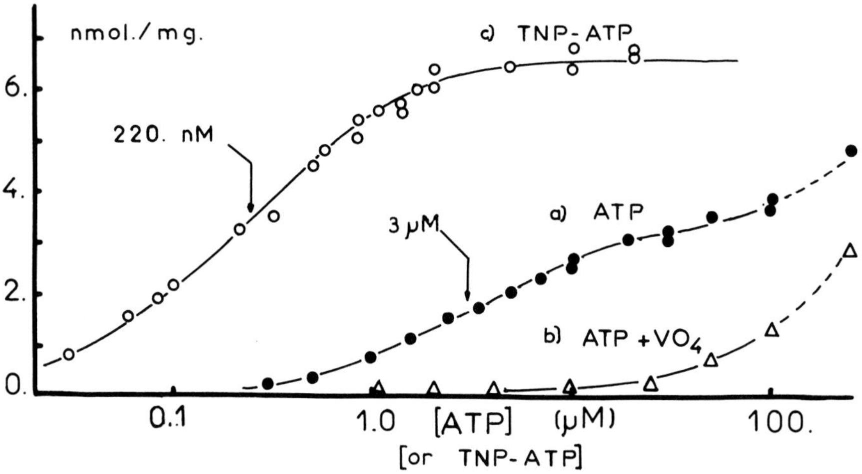

FIGURE 7. Nucleotides binding to the Ca-ATPase. (a) Mg-ATP binding measured with ^{14}C-ATP by filtration in 20mM Tris-Mops pH 7.2, 100mM KCl, 5mM Mg and 0.5mM EGTA at 20°C. (b) Same conditions as in a) except that vesicles were incubated in 10 µM vanadate prior to binding assays. (c) TNP-ATP binding measured as in a) except for the omission of Mg and addition of 1mM EDTA to prevent TNP-ATP hydrolysis which may affect significantly the measurements at the low concentration of substrate used. The curve drawn in c) is a fit calculated for one single family of site of concentration 6.7 nmol/mg and dissociation constant Kd = 220 nM. Assuming two families of equal concentration, the largest difference in affinity which can yield a reasonable fit is: Kd1 = 100 nM, Kd2 = 400 nM. Other experimental details may be found in (32).

ATP (or ADP) binding in the absence of calcium produces an important fluorescence increase (ref.13 and figure 8). The amplitude of this change is, however, smaller than that induced by calcium binding (about 60% of the latter under the conditions described in figure 8). The apparent affinity for ATP (or ADP) to produce this fluorescence change is equal to the dissociation constant measured with labelled nucleotides (compare figure 7a and inset in figure 8).

- Phosphorylation by ATP. In the presence of calcium, ATP induces the very fast formation of about 3.0 nmol/mg of phosphoenzyme (ADP sensitive)(18). This value is very similar to that found for the number of ATP or ADP binding sites. At 0°C in sarcoplasmic reticulum vesicles the concentration of ATP giving half-maximum saturation of the phosphoenzyme is around 100 nM (18). This low value is mainly due to a slow dephosphorylation phase k = 0.035 /sec in low ATP. Measurement of the intrinsic fluorescence under the same conditions reveals that ATP induces a very fast decrease of the fluorescence (from the E.Ca level, see figure 8). Fluorescence remains at an intermediate level as long as ATP is present. When all added ATP has been exhausted the initial E.Ca level is restored. At 0°C a clear distinction between the decreasing and relaxation phases can be made so that accurate measurement of the amplitude of the fluorescence drop can be obtained. These measurements (fig. 8) indicate a very high apparent affinity for ATP to produce this effect. The (ATP)1/2 for this effect is indeed much smaller than the enzyme concentration and therefore cannot be evaluated with accuracy; this implies, however, that it is possible to determine directly the active site concentration from the fluorescence data (inset of figure 8). The number found (3.2 nmol/mg) is in very good agreement with that found in similar conditions with (γ-32P)-ATP.

The rate constant of the relaxation phase of fluorescence (0.035/sec) is identical to that of dephosphorylation (18). Altogether this indicates that the fluorescence decrease observed on adding ATP in the presence of calcium is due to the formation of the ADP sensitive phosphoenzyme.

Figure 8 also shows an important result: the intermediate fluorescence levels obtained either after ATP binding (in EGTA) or after ATP dependent phosphorylation (in the presence of calcium) cannot be distinguished. This implies that in the presence of saturating amounts of ATP calcium binding induces no fluorescence change.

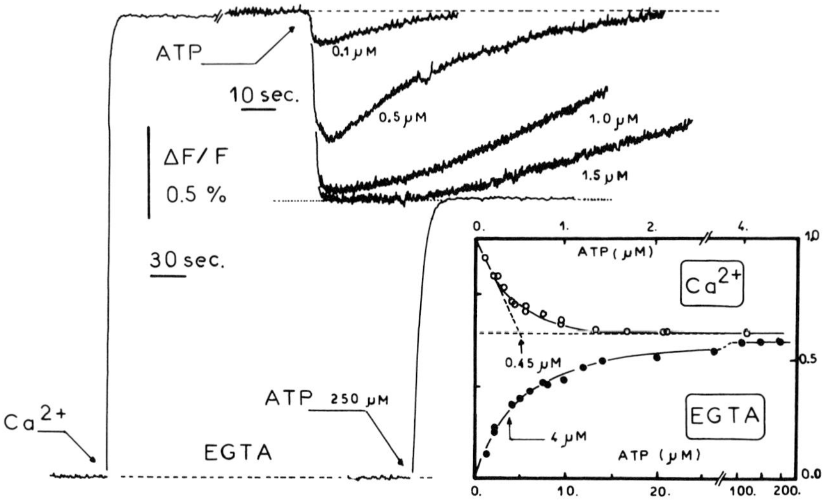

FIGURE 8. Intrinsic fluorescence changes induced by ATP in the presence or absence of calcium. Protein concentration was 140 μg/ml in 20mM Tris-Mops pH 7.2, 100mM KCl, 5mM Mg and 0°C. Excitation: 295 nm, emission: 350 nm. Inset: titration of the amplitude of the fluorescence change induced by ATP, expressed as fraction of the amplitude of the calcium induced fluorescence change (note the different scales on the abscissa); 0.45 μM of active sites titrated from the response in the presence of calcium correspond to 3.2 nmol/mg.

- What is the true number of ATP binding sites? As described above, phosphorylation by ATP is saturated at very low concentrations of ATP and yields no more than about 3.0 to 3.2 nmol/mg of active sites i.e. one half the number of calcium sites. This ratio is also in very good agreement with the coupling ratio of the transport. For the actual number of ATP sites the situation is not so clear-cut since multiple effects of ATP have been reported suggesting the existence of more than one single class of binding sites (20,33-35). In addition to the high affinity sites titrated with labelled ATP, there are some indications of further binding to a secondary class of sites at high ATP concentrations (see fig. 7 and ref. 20). Titration of these low affinity sites is difficult because of the lower accuracy of the measurements above 100 μM ATP. We have recently, in some way, overcome this difficulty by using

TNP-ATP[1], a fluorescent analogue of ATP exhibiting a much higher affinity for the Ca-ATPase ATP sites than ATP (32). Measurements of TNP-ATP binding by fluorescence, optical density changes and direct titration with labelled TNP-ATP have clearly revealed the existence of 6 to 7 nmol/mg of binding sites for this analogue. This number is twice that obtained for the ATP high affinity sites (fig.7). These TNP-ATP sites are specific, as indicated firstly by the clear titration (no further binding can be observed from 2 μM up to at least 100 μM TNP-ATP, while the affinity measured lies between 100 and 500 nM), secondly by the fact that all TNP-ATP can be chased by millimolar amounts of ATP, so that ATP can certainly bind to all the sites revealed by TNP-ATP. It should be mentionned here that an identical study has been reported recently by Watanabe and Inesi (36): the number of TNP-ATP binding sites titrated is similar to ours, but there are important differences concerning the competition with ATP and the affinity for TNP-ATP binding.

The simplest conclusion from the discussion above is that the Ca-ATPase possesses the same number of sites for ATP as for calcium (6 to 7 nmol/mg) but that only half of these are able to bind ATP with high affinity or to be phosphorylated. ATP binding or phosphorylation of this site results in the stabilization of the same intermediate fluorescence state.

C) Phosphorylation by Pi

- Titration of the phosphorylation sites. In the absence of calcium, at acid pH and in the absence of a calcium concentration gradient, the sarcoplasmic reticulum Ca-ATPase can directly be phosphorylated by Pi (37). This important reaction was observed and studied by many groups so that we will only discuss here some recent results.

The phosphorylation by Pi apparently proceeds through a two-step reaction as follows:

$$E + P_i + Mg^{++} \rightleftharpoons E.P.Mg \underset{(K_5)}{\rightleftharpoons} E{-}P.Mg$$

It has been proposed (38-40) that Pi and magnesium can bind independently to the Ca-ATPase to form the ternary complex: E.P.Mg; this fast association is followed by a slower step (K5) leading to the formation of E-P.Mg, an ADP-insensitive covalent phosphoenzyme. About 3 nmol/mg of phosphoenzyme were

---

1. TNP-ATP: 3'-O-(2,4,6-trinitrophenyl) adenosine 5'-triphosphate.

reproducibly found in many recent reports (16,21,39-42). It has been argued in some reports that in sarcoplasmic reticulum vesicles the actual number of phosphorylation sites is in fact 5 to 6 nmol/mg and that the incomplete saturation of these sites for phosphorylation is due to an even equilibrium between the ternary complex and the covalent intermediate (21,40,43). This idea of an unfavored phosphorylation reaction was supported by kinetic measurements of the rate of formation and decay of E-P in which a value of K5 close to one was reported (21,43). These data were, however, in contradiction with those of Lacapere et al. (16) who measured the rate of phosphorylation by Pi from the increase of fluorescence that they observed in the absence of a calcium concentration gradient. The fluorescence measurements of Lacapere et al. were extended over a larger range of phosphate and magnesium concentrations than those using acid-quench techniques and the equilibrium constant found was largely higher than 1 ($K5 \geqslant 20$). This indicates that phosphorylation is a favorable reaction and the very fact that it induces a fluorescence increase, moreover, reveals that E-P formation is probably associated with a conformational change of the protein (see figure 9). It is interesting to note that Inesi et al. (41) have recently confirmed the evaluation of K5 obtained by Lacapere et al.(16).

The fact that the phosphorylation by Pi is fully saturated in excess magnesium and Pi indicates that the actual number of phosphorylation sites at pH 6.0 can hardly be larger than 3 to 4 nmol/mg, a value identical to the number of phosphorylation

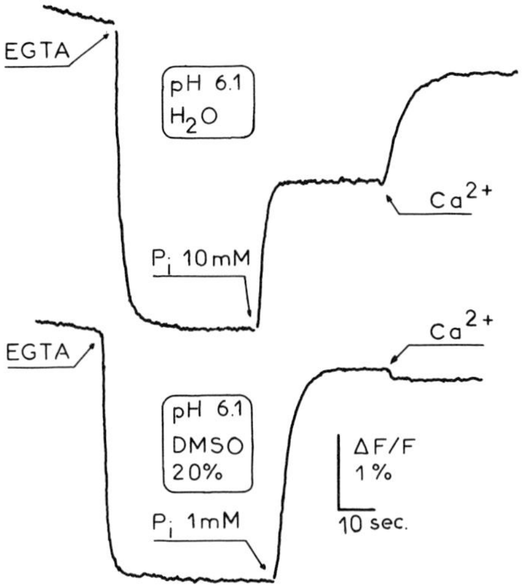

FIGURE 9. Intrinsic fluorescence changes induced by Pi in the absence of calcium. Conditions were: 100mM Mes-Tris pH 6.1, 20mM Mg and 25°C. Protein concentration was 200 μg/ml.

obtained with ATP or to the number of high affinity ATP binding sites. This analogy is strengthened by the fact that the fluorescence level measured after phosphorylation by Pi is identical to the intermediate level stabilized by ATP binding or after phosphorylation by ATP (compare figures 8,9 and 10).

- Effect of organic solvents. deMeis and coworkers (21,42,43) have recently reported that the E-P level at saturation and the apparent affinity for phosphorylation by Pi were significantly increased in the presence of organic solvents like DMSO, DMFA or glycerol. They have attributed this effect to a reduction of phosphate solvation energy related to lower water activity. Figure 9 indicates that we have been able to confirm this activating effect of DMSO by following the Pi induced intrinsic fluorescence change. We have reproduced the very important increase of the apparent affinity for Pi of two orders of magnitude in the presence of 20% DMSO (Fig 10), (concentrations higher than 20% greatly increase the turbidity of the vesicular suspension making the fluorescence measurements inaccurate). In addition to this we have observed a significant increase of the fluorescence level at saturation of Pi: in 20% DMSO the fluorescence increase induced by excess phosphate reaches the maximum level obtained in the presence of calcium (Fig. 9 and 10). This point is corroborated by a measurement of E-32P which clearly indicates an increase of the concentration of phosphorylated enzyme at saturation reaching nearly 5.0 nmol/mg (Table II). That organic solvents induce an increase of E-P concentration at saturation was previously

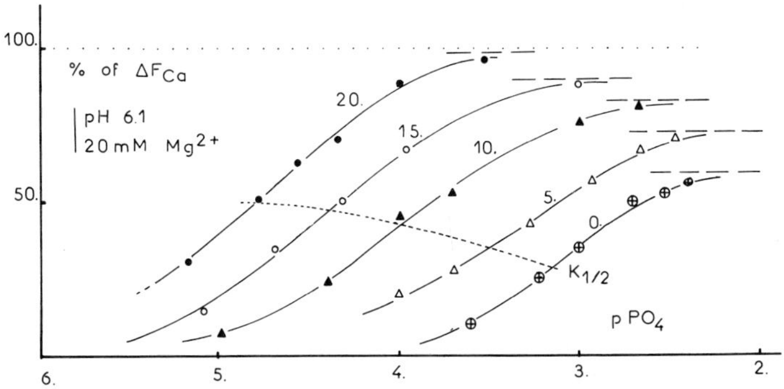

FIGURE 10. Phosphate concentration dependence of the phosphate induced fluorescence changes described in figure 9. DMSO concentration was adjusted to the percentage (v/v) indicated on each curve. The fluorescence changes amplitudes are expressed as fraction of the total amplitude of the calcium induced fluorescence increase.

reported by deMeis et al. (21,42,43) (mean values 3.1 nmol/mg in water and 5.5 nmol/mg in 40% DMSO). deMeis et al., however, have interpreted this effect as being due to a change in the equilibrium constant K5. As discussed above, we think that this interpretation is not valid and we prefer an alternative explanation according to which this effect is due to an increase of the number of sites available for phosphorylation. The mechanism by which DMSO induces this effect has to be further investigated.

TABLE II. Maximum level of phosphorylation

DMSO (%)	Pi (mM)	E-P (nmol/mg)
0.	10.0	3.0 to 3.2
20.	5.0	4.2 to 5.0
40.	0.5	4.5

- Conclusion for ATP and E-P

- Under standard conditions the number of ATP (high affinity) and phosphorylation sites (from ATP or from Pi) is 3.0 to 3.3 nmol/mg in sarcoplasmic reticulum vesicles.
- Binding of ATP or phosphorylation results in the stabilization of the same intermediate level of fluorescence.
- Another fraction of ATP sites can be revealed indicating a total number of ATP sites of around 6.0 to 7.0 nmol/mg a number identical to that found for the high affinity calcium sites. These secondary ATP sites are revealed with TNP-ATP.
- More phosphorylation sites may be revealed in the presence of organic solvents like DMSO; in this case phosphorylation by Pi increases the fluorescence to its maximum.

IV - GENERAL DISCUSSION

In the preceding chapters we have described separately the reactivity of the Ca-ATPase with its main substrates: calcium, ATP and Pi. We have described our present knowledge of the titration of the active sites for these effectors and their interaction with the enzyme leading to what we believe to be important conformational changes of the protein. This latter assertion is supported by various facts: i) the fluorescence changes induced by the various substrates are not additive, ii)the rate constants do not obey simple first-order kinetics for substrate concentration as it should be the case for a simple interaction between the substrate and a tryptophan located in or near the active site, iii) when they can be compared, the rate constants of the fluorescence changes are found to be identical to those derived from measurements using direct labelling techniques.

It, therefore, emerges from these studies not only that reorientation of the transport calcium sites is probably produced by a structural transition of the Ca-ATPase, but also that ATP binding and phosphorylation are also associated with significant conformation changes of the protein.

We may now consider this problem with a more general view and ask whether the various changes observed could be explained by a common and comprehensive scheme. This kind of unification is also justified by the idea that the protein conformation is the source of the coupling between the ion translocation and the phosphate transfer reactions. This implies that each change observed might, at the same time, affect the reactivity of the calcium binding sites and that of the ATP or phosphorylation sites or, if we consider this problem in reverse, that the same conformational transition might be induced by different substrates.

Following these ideas we have attempted to reassemble the fluorescence changes described in this article; this is schematically represented in figure 11.

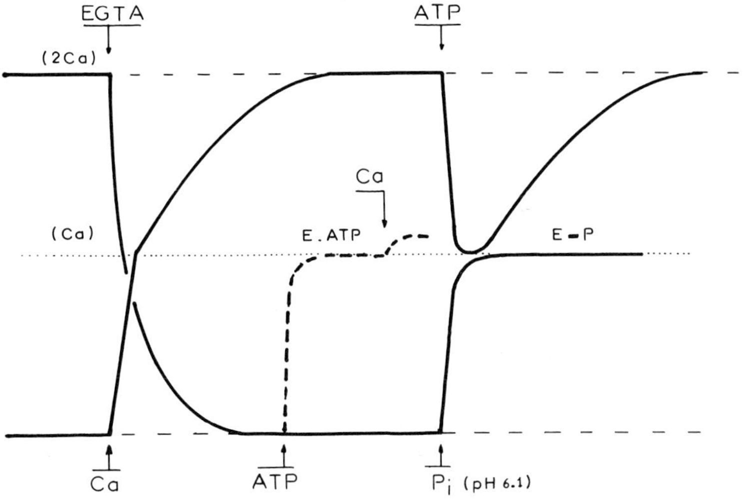

FIGURE 11. Schematic representation of the fluorescence changes described in this article.

It is apparent that the number of fluorescence levels described can be reduced to three. From the nomenclature defined in chapter III-A, they will be termed E22 ⇌ E21 ⇌ E11. This means that we propose the hybrid E21 state, whose existence was postulated in the calcium binding studies, to be the same as that stabilized by ATP binding or phosphorylation. Assuming this, we are forced to conclude that a striking symmetry relation exists between the accessibility of the active sites for the various substrates and the fluorescence levels, as shown below:

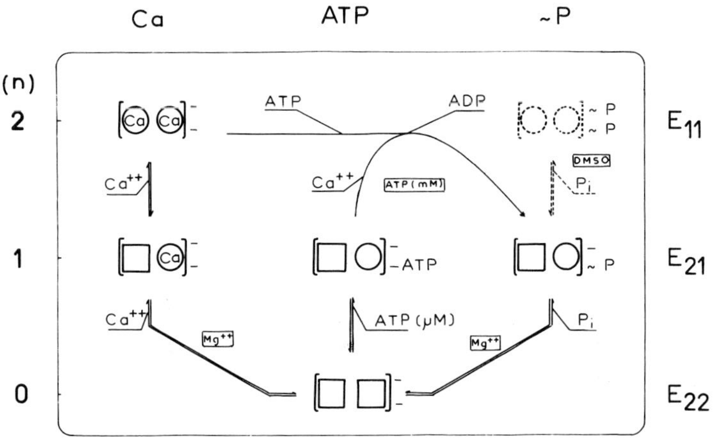

FIGURE 12. The proposed enzymatic states are represented with respect to their reactivity for calcium, ATP and phosphate. (n) is the number of titrable sites on each state. [Mg], [ATP] and [DMSO] represent an activating effect on some of the transitions represented.

The main ideas proposed in this scheme are:

- The Ca-ATPase has three main conformational states: E22,E21 and E11.
- The Ca-ATPase has two sets of sites for calcium, ATP and possibly for phosphorylation.
- These sites are probably not equivalent.

Our present ideas on the three proposed enzymatic states are the following:

-E22. This state is probably identical to the E* state of Carvalho et al.(3). It is stabilized by vanadate and has no high affinity calcium binding sites accessible from the cytoplasmic side of the vesicles. No ATP (nor ADP) binding can be observed on the E22-Vanadate complex (see figure 7); this could, however, reflect the occupancy of the nucleotide site by vanadate and not necessarily a low affinity for ATP when the enzyme is in the E22 state. We have, indeed, little information about ATP binding on this form of enzyme since ATP addition induces the transition to the intermediate E21 form.

-E21. The characterization of this state is the most interesting result derived from the data presented here. The simplest interpretation of these data is that the E21 form is an half-site-reactive enzyme for calcium, ATP and for phosphorylation. This state is only transiently produced during calcium binding since calcium binding to E21 produces further isomerization and unmasking of a second high affinity site (positive cooperativity). This can be opposed to the effect of phosphorylating substrates like ATP and Pi which induce a stabilization of E21 preventing further isomerization or binding (half-of-the-sites reactivity).

We have been forced, in our scheme, to postulate that phosphorylation of the enzyme always leads to the stabilization of the same E21-P state and this independently of the phosphorylating reaction pathway (from ATP or from Pi). This might indicate that the differences between the ADP-sensitive and ADP-insensitive forms of the phosphoenzyme are more subtle than expected in previous schemes (3) and are not necessarily linked to an important change of conformation of the enzyme.

-E11. This form of the enzyme has two high affinity cytoplasmic calcium sites. The affinity for ATP and the number of binding sites has not been presently evaluated since ATP addition induces a very fast ATP splitting and enzyme phosphorylation with transition to the E21 state.

In scheme V and in figure 12, as well as in ref.10 we have proposed other symbols for the Ca-ATPase conformational states:

$$\bigcirc\bigcirc \rightleftharpoons \bigcirc\square \rightleftharpoons \square\square$$

These symbols were used to described more explicitly the reactivity of the calcium sites. This representation is an obvious suggestion that the three states proposed may result from an association between two subunits, each of which has only two main conformations: $\square$ and $\bigcirc$. This could clearly be a way to reconcile our view with the E $\rightleftharpoons$ E* model by assuming that each of the putative subunits is able to undergo the deMeis's E $\rightleftharpoons$ E* two-state scheme. Since the intermediate fluorescence state is stabilized during turn-over (see figure 8), we propose that the enzyme is composed of two subunits working out of phase, the hybrid intermediate E.E* being the most probable steady state intermediate (flip-flop mechanism).

More structural work is needed to confirm without ambiguities the idea that the functional Ca-ATPase is an oligomer. In any case, this will not exclude the possibility that the monomer may also support calcium activated ATPase activity and transport but regulation of the activity and coupling ratio would be completely different.

In the case of a dimeric association, the observation of an hybrid intermediate, the clear non-equivalence of the two calcium and of the two ATP sites suggest a possible asymmetric (heterologous) association between two ATPase molecules. It is likely that magnesium plays an important role into this association as suggested by its effect on the kinetics of calcium binding.

AKNOWLEDGEMENTS

The authors wish to thank L. deMeis, F. Guillain and P. Champeil for stimulating discussions and communication of unpublished results.

REFERENCES

1. Dupont,Y., Harrison,S.C. and Hasselbach,W. (1973) Nature 244,555-558
2. Herbette,L., Marquardt,J., Scarpa,A. and Blasie,J.K. (1977) Biophys. J. 20,245-272
3. Carvalho,M.G.C., deSouza,D.G. and deMeis,L. (1976) J. Biol. Chem. 251,3629-3636
4. Post,R.L., Hegevary,C. and Kume,S. (1972) J. Biol. Chem. 247,6530-6540
5. Albers,J.W. (1976) in: The Enzymes of Biological Membranes (Martonosi,A.N.ed.) Vol.III, pp.283-301, Plenum Press, New York
6. Dupont,Y. (1976) Biochem. Biophys. Res. Commun. 71,544-550
7. Dupont,Y. and Leigh,J.B. (1978) Nature 273,396-398
8. Guillain,F., Gingold,M.P., Buschlen,S. and Champeil,P. (1980) J. Biol. Chem. 255,2072-2075
9. Dupont,Y. and Le Maire,M. (1980) FEBS Lett. 115,247-252
10. Dupont,Y. (1982) Biochim. Biophys. Acta 688,75-87
11. Guillain,F., Gingold,M.P. and Champeil,P. (1982) J. Biol. Chem. 257,7366-7371
12. Champeil,P., Gingold,M.P., Guillain,F. and Inesi,G. (1983) J. Biol. Chem. in the press

13. Dupont,Y., Bennett,N. and Lacapere,J.J. (1983) Ann. of N.Y. Acad. Sci. In the press
14. Dupont,Y. (1978) Biochem. Biophys. Res. Commun. 82,893-900
15. Guillain,F., Champeil,P. and Lacapere, J.J. (1981) J. Biol. Chem. 256,6140-6147
16. Lacapere,J.J., Gingold,M.P., Champeil,P. and Guillain,F. (1981) J. Biol. Chem. 256,2303-2306
17. Dupont,Y. and Bennett,N. (1982) FEBS Lett. 139,237-240
18. Dupont,Y. (1980) Eur. J. Biochem. 109,231-238
19. Berger,R.L., Balko,B. and Chapman,H.F. (1968) The Rev. of Sci. Inst. 39,493-498
20. Dupont,Y. (1977) Eur. J. Biochem. 72,185-190
21. deMeis,L., deSouza Otero,A., Martins,O.B., Alves,E.W., Inesi,G. and Nakamoto,R. (1982) J. Biol. Chem. 257, 4993-4998
22. Watanabe,T., Lewis,D., Nakamoto,R., Kurzmack,C., Fronticelli,C. and Inesi,G. (1981) Biochemistry 20,6617-6625
23. Hill,T.L. and Inesi,G. (1982) Proc. Natl. Acad. Sci. USA 79,3978-3982
24. Ikemoto,N. (1975) J. Biol. Chem. 250,7219-7224
25. Sumida,M. and Tonomura,Y. (1974) J. Biochem. (Tokyo) 75,283-297
26. Verjovsky-Almeida, S. (1981) J. Biol. Chem. 256, 2662-2668
27. Verjovsky-Almeida,S. and Silva,L. (1981) J. Biol. Chem. 256,2940-2944
28. Dean,W.L. and Gray,R.D. (1980) J. Biol. Chem. 255,7514-7516
29. Inesi,G., Kurzmack,M., Coan,C., and Lewis,D.E. (1980) J. Biol. Chem. 255,3025-3031
30. Pick,U. (1982) J. Biol. Chem. 257,6111-6119
31. Ikemoto,N., Garcia,A.M., Kurobe,Y. and Scott,T.L. (1981) J.Biol. Chem. 256,8593-8601
32. Dupont,Y., Chapron,Y. and Pougeois,R. (1982) Biochem. Biophys. Res. Commun. 106,1272-1279
33. Taylor,J.S. and Hattan,D. (1979) J. Biol. Chem. 54, 4402-4407
34. Inesi,G., Goodman,J.J. and Watanabe,S. (1967) J. Biol. Chem. 242,4637-4643
35. Yamamoto,T. and Tonomura,Y. (1967) J. Biochem. (Tokyo) 62,558-575
36. Watanabe,T. and Inesi,G. (1982) J. Biol. Chem. 257, 11510-11516
37. Masuda,H. and deMeis,L. (1973) Biochemistry 12,4881-4885
38. Kolassa,N., Punzengruber,C., Suko,J. and Makinose,M. (1979) FEBS Lett. 108,495-500

39. Suko,J., Plank,B., Preis,P., Kolassa,N., Hellman,G. and Conca,W. (1981) Eur. J. Biochem. 119,225-236
40. Martin,D.W. and Tanford,C. (1981) Biochemistry 20,4597-4602
41. Inesi,G., Watanabe,T., Coan,C. and Murphy,A. (1983) Ann. of N.Y. Acad. Sci. In the press
42. deMeis,L., Martins,O.B. and Alves,E.W. (1980) Biochemistry 19,4252-4261
43. deSouza Otero,A. and deMeis,L. (1982) Z. Naturforsch. 376,527-531

CONFORMATIONAL CHANGE OF SR-ATPase INDUCED BY PHOSPHORYLATION[1]

TAKAHIDE WATANABE[2]
GIUSEPPE INESI

Department of Biological Chemistry
University of Maryland, School of Medicine
Baltimore, Maryland USA

The sarcoplasmic reticulum(SR) calcium pump utilizes ATP for vectorial transport of calcium ions against a concentration gradient(1-3). A number of sequential reactions have been proposed to describe the catalytic and transport cycle of the SR-ATPase(4-6). Conformational change in response to Ca were reported by various investigators(7-11). Furthermore in order to produce a change in orientation and reduction in affinity of calcium binding sites, structural transition must follow utilization of ATP. We have observed reversible structural transition of SR-ATPase(including Ca affinity change) upon phosphoenzyme formation in the following conditions. (1) In the presence of a small amount of neutral detergent, octaetyleneglycol monododecylether($C_{12}E_8$)(12). (2) Using the ATP analogue, TNP-ATP, as a reporter of conformational change upon phosphorylation by ATP(13). (3) In the presence of dimethylsulfoxide which is known to inhibit hydrolysis of the ADP-insensitive phosphoenzyme with low affinity for Ca^{2+}(14).

---

[1]Supported by NIH grant HL-16607 and a grant from the Muscular Dystrophy Association.

[2]Present address: Department of Molecular Physiology, National Cardiovascular Center Research Institute, Suita, Osaka, Japan.

ISBN 0-12-260380-X

RESULTS AND DISCUSSION

Conformational change of SR-ATPase induced by ATP utilization.--- Addition of ATP to a suspension of SR in the absence or in the presence of saturating detergent concentrations does not produce any significant turbidity change. However, if ATP is added in the presence of low $C_{12}E_8$ concentrations, which produce only partial solubilization of SR, a pronounced reduction in turbidity is observed. It is of great interest that such an effect of ATP is reversible in the presence of 20 % glycerol, as shown by an increase in turbidity after ATP hydrolysis(Fig 1). The duration of the ATP effect is directly proportional to the concentrations of ATP used. In fact it can be demonstrated that reversal occurs as soon as added ATP is hydrolyzed. This clearing effect of ATP is Ca^{2+} dependent and utilization of ATP by the ATPase is necessary, AMP-P(NH)P and ADP are not effective. ITP and acetyl phosphate can also produce the clearing effect, although at a lower rate and to a lesser extent than ATP, in analogy to their slow utilization as substrates for the enzyme. To explore further the clearing effect of ATP with respect to dissociation and reassociation of ATPase chains, we measured fluorescence energy transfer among chains

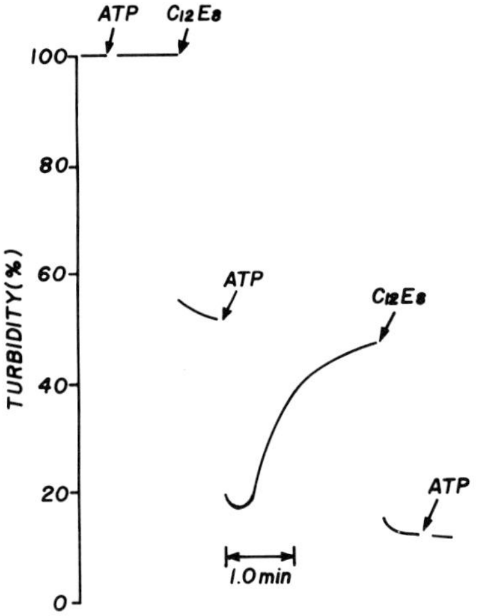

FIGURE 1. Turbidity change induced by $C_{12}E_8$ and ATP. 0.2 mg of SR/ml was suspended in a medium which contains 20 mM Tris-HCl, pH 8.0, 80 mM KCl, 5 mM $MgCl_2$, 50 µM $CaCl_2$, and 20 % glycerol. The concentration of ATP and $C_{12}E_8$ added each time was 16.7 µM and 0.67 mM, respectively. Temperature was 25°(12).

using the technique of Vanderkooi et al(15). We labeled SR vesicles with either 1,5-IAEDANS(donor) or IAF (acceptor) separately for two hours at pH 8 at room temperature. In this condition, approximately 2 mol of IAEDANS or 1 mol of IAF per mol of ATPase chains was incorporated. The enzyme activity of the labeled sample remained essentially the same as control vesicles. We chose 470 nm wavelength to monitor changes in the fluorescence emission of the IAEDANS label, as an expression of the energy transfer and interaction of ATPase chains. We found that when IAEDANS labeled SR and IAF labeled SR are mixed in the absence of detergent, there is no significant exchange of labeled chains among different vesicles, and the two fluorophores remain at distance precluding energy transfer.

However, addition of nonsaturating concentrations of $C_{12}E_8$ to this SR suspension produced slow reduction in IAEDANS fluorescence and addition of ATP accelerated its reduction(Fig 2A). A parallel reduction in the donor

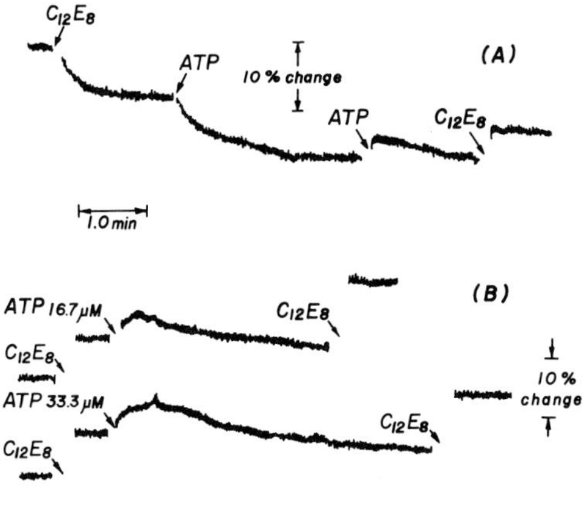

FIGURE 2. Effects of $C_{12}E_8$ and ATP on donor(IAEDANS) fluorescence. A. The sample contains a 3:7 mixture of IAEDANS- and IAF-labeled SR vesicles which were not treated with DOC for randomization of chains. The reaction medium contained 0.15 mg of protein/ml, 20 mM Tris-HCl, pH 8.0, 80 mM KCl, 5 mM $MgCl_2$, 50 μM $CaCl_2$, and 20 % glycerol. $C_{12}E_8$ was added to yield a 0.5 mM concentration each time and ATP to yield 66.7 μM the first and 33.3 μM the second time. The excitation 340 nm; emission 470 nm. B. A 3:7 mixture of IAEDANS- and IAF-labeled SR vesicles were treated with DOC for randomization of chains. The concentrations of added $C_{12}E_8$ were 0.5 and 0.67 mM the first and second time respectively for each run(12).

fluorescence lifetime was also observed. This effect may reflect the exchange of chains between the two populations of vesicles and randomization of chains labeled with either of the two fluorophores. Secondary addition of ATP produced a transient rise of the donor fluorescence intensity which may reflect the dissociation of randomized chains by the formation of phosphoenzyme intermediate. Addition of excess $C_{12}E_8$ produced chain dissociation and loss of energy transfer, manifested by an increase in IAEDANS fluorescence intensity.

In the case of reconstituted vesicles with randomized chains, prepared by mixing IAEDANS labeled SR and IAF labeled SR in the presence of 1.0 mg of DOC/mg of protein and removing DOC by column chromatography, addition of low concentration of $C_{12}E_8$ produced a rise of the donor fluorescence intensity, consistent with an increase in the average distance among chains and a reduction in the energy transfer (Fig 2B). When ATP was added in the presence of such low detergent concentrations, a transient increase in fluorescence intensity and lifetime was noted. The duration of this fluorescence increase was proportional to the concentration of ATP used. These results indicate that the average distance between different fluorophore increases by the formation of phosphoenzyme intermediate in the presence of small amount of $C_{12}E_8$. Although low detergent concentrations are required to observe dissociation of chains upon substrate utilization, our observations indicate that a structural perturbation is produced by enzyme phosphorylation at the active site, and that this perturbation plays a role in the mechanism of catalysis associated with vectorial transport.

<u>TNP-ATP, a useful reporter of conformational change of SR-ATPase upon phosphorylation by ATP</u>---The ATP analogue, 2',3'-O-(2,4,6-trinitrophenyl)adenosine 5'-triphosphate,

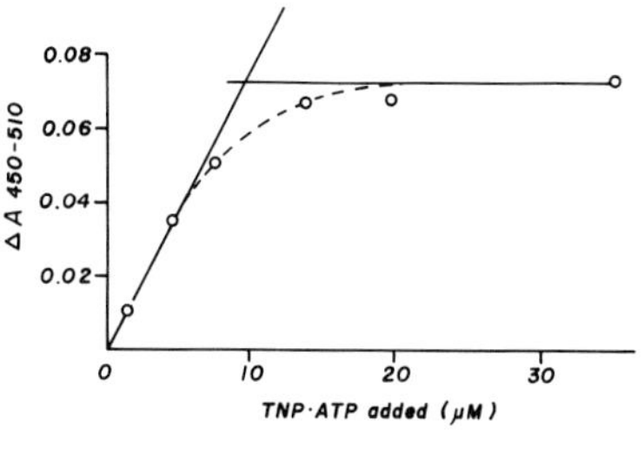

FIGURE 3. Titration of SR ATPase with TNP-ATP. Reaction mixture; 20 mM Tris-maleate, pH 8.0, 80 mM KCl, 5 mM $MgCl_2$, 50 μM $CaCl_2$ and 20 % glycerol. The SR concentration was 1.37 mg of protein/ml. The absorption difference between the trough at 450 and the peak at 510 nm was plotted as a function of TNP-ATP added(13).

(TNP-ATP), which was first synthesized by Hiratsuka and Uchida(16), binds to SR ATPase and produces changes in its absorption spectrum in the visible region with a trough at approximately 445 nm and peaks at 415 and 510 nm. The stoichiometry of TNP-ATP binding was estimated from plots of the absorbance difference as a function of the concentration of TNP-ATP added. Based on the protein concentration in the mixture, the plots indicate binding of 7-8 nmol of TNP-ATP/mg of SR protein(Fig 3). TNP-(^{32}P)-ATP binding was also measured directly by the centrifuge column method of Penefsky (17). By this method, the maximum levels of nucleotide bound is estimated to be 6-7 nmol of TNP-ATP/mg of SR protein. Dupont et al also (18) reported that the number of TNP-nucleotide binding site of SR ATPase is 6.0-7.0 nmol/mg of protein. The addition of 100 μM ATP reduced the TNP-ATP differential spectra and also TNP-ATP bound to a minimum of about one-half the original value(Fig 4). Therefore, a maximum of 4 nmol of TNP-ATP/mg of protein was displaced by ATP. This is in agreement with the number of sites phosphorylated by ATP, which is also 4 nmol/mg of protein in our preparation. The TNP-ATP displacement was transient, and the absorption returned to nearly the original level after hydrolysis of the added ATP. In order to totally replace the bound TNP-ATP, it was necessary to use ATP concentrations as high as 10 mM, indicating that part of the sites occupied by TNP-ATP possess a low affinity for ATP. TNP-ATP is very weakly fluorescent in aqueous solution and its intensity is slightly increased by the binding to SR-ATPase. We expected

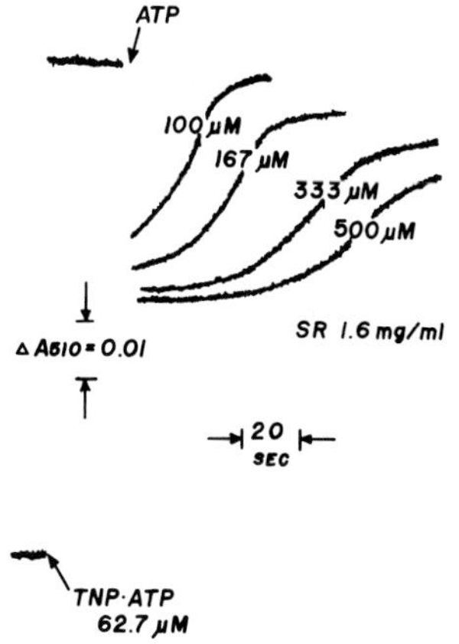

FIGURE 4. Change in differential absorption of bound TNP-ATP(510 nm) following addition of ATP. The concentration of SR was 1.6 mg/ml and the medium was as described in figure 3 (13).

that the fluorescence enhancement would be partially reversed upon addition of 100 μM ATP as a consequence of partial analogue displacement. On the contrary, a strong enhancement of the fluorescence intensity of the bound TNP-ATP was observed when 100 μM ATP was added(Fig 5). This fluorescence increase may reflect a conformational change in the protein. High concentration of ATP(10 mM), as mentioned above, displaced the bound analogue , and consequently reduced the fluorescence enhancement.

These experiments indicate that the effect can be observed only when TNP-ATP and ATP are present at concentrations permitting occupancy of part of the sites by TNP-ATP and part by ATP. The fluorescence enhancement is reversible upon exhaustion of added ATP. This ATP induced TNP-ATP fluorescence enhancement is Ca dependent. These observations suggest that ATP utilization by the Ca^{2+} dependent ATPase is obligatory for the fluorescence enhancement to occur. A large fluorescence enhancement by ATP is also observed in the presence of detergent, so that this rules out the possibility that this fluorescence enhancement is due to a light scattering change or effects of Ca^{2+} and H^+ gradient across the membrane as a consequence of ATP utilization. It is concluded that TNP-ATP acts as a reporter of an ATPase conformational change following enzyme phosphorylation of the catalytic site by ATP and that the observed conformational change is operative in the mechanism of calcium site translocation for active transport. Furthermore, the sensitivity of the bound analogue to the phosphorylation reaction is likely to be related to nucleotide regulation of enzyme turnover.

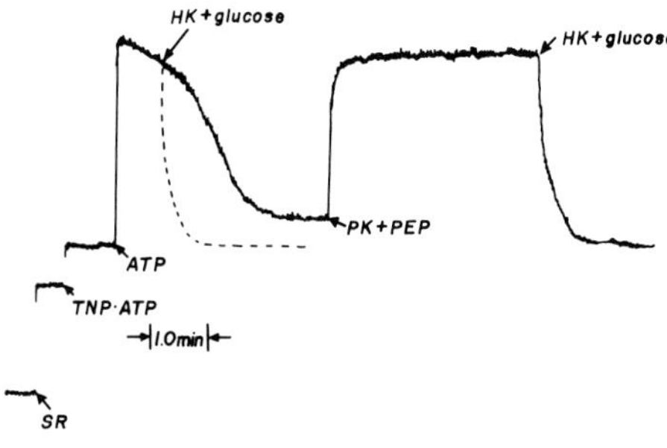

FIGURE 5. Fluorescence changes induced by ATP. The concentrations of SR protein and TNP-ATP were 0.25 mg/ml and 5.0 uM, respectively. ATP(40 μM), glucose(2.5 mM), hexokinase(HK)(32.3 μg/ml), pyruvate kinase(PK)(10.3 μg/ml), phosphoenol pyruvate(PEP)(0.125 mM) were added at where indicated. Excitation 410 nm; emission 515 nm.

Ca^{2+} release from SR ATPase upon phosphorylation in the presence of dimethylsulfoxide(Me_2SO))---Ikemoto(19,20) observed a release of bound calcium following ATPase phosphorylation with ATP, using an enzyme prepared by solubilization of native vesicles with Triton X-100. On the other hand, Takakuwa and Kanazawa(21,22), and Dupont(23) showed that the bound calcium is not released immediately upon phosphorylation but rather retained in an occluded form which is not accessible from the outer or the inner side of the membrane. Therefore it is reasonable to suppose that the occluded state would be prevalent in steady state at neutral pH and in the presence of 100 mM KCl and 5 mM $MgCl_2$. The and Hasselbach (24) reported that the ATPase hydrolytic activity is inhibited by Me_2SO, while the amount of phosphorylated enzyme intermediate formed is maintained at maximum levels. Shigekawa and Akowitz(25) also reported that Me_2SO specifically inhibits hydrolysis of ADP-insensitive phosphoenzyme, and de Meis et al(26) demonstrated that the equilibrium of Pi reaction with the enzyme is shifted greatly in favor of phosphoenzyme formation. Therefore we reasoned that , in the presence of Me_2SO, Ca^{2+} might be released upon phosphorylation with ATP. As expected, significant calcium release was detected upon ATP addition in the presence of Me_2SO(Fig 6). Released calcium binds again to the ATPase when ATP is hydrolyzed. The duration of release is proportional to the concentration of ATP added. It is clear that release and rebinding of calcium are due to changes in affinity of the binding site, related to phosphorylation and dephosphorylation of the enzyme.

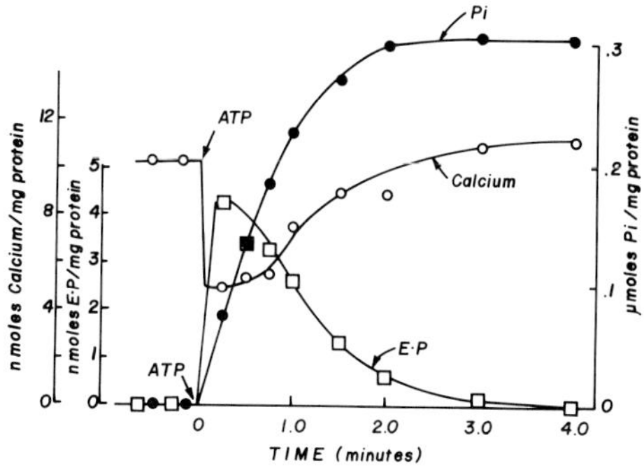

FIGURE 6. Ca release, formation of phosphorylated enzyme intermediate, and Pi liberation following addition of ATP in the presence of 30% Me_2SO. Calcium bound to purified ATPase(27) was measured by the filtration method. ATP(200 uM) was added at zero time. The reaction mixture consisted of 20 mM MES pH 6.1, 80 mM KCl, 5 mM $MgCl_2$, 50 μM $^{45}CaCl_2$, 10 mM (^3H)-glucose, and 0.5 mg of purified ATPase/ml(14).

Besides ATP, other substrates known to phosphorylate ATPase, such as ITP, p-nitrophenyl phosphate, and acetyl phosphate, cause calcium release. The release phenomenon is also observed upon ATP addition with native SR in the presence of the ionophore A-23187 and Me_2SO. Ca^{2+} release from the enzyme upon addition of ATP can be also observed spectrophotometrically using arsenazo-III only in the presence of Me_2SO(Fig 7). These results clearly show that the phosphoenzyme formed in the presence of Me_2SO has low affinity to calcium. Nakamura and Tonomura(28) also observed Ca^{2+} release from SR ATPase at alkaline pH(8.8) and in the presence of high concentration of $MgCl_2$. Other organic solvent such as ethanol and dimethylformamide are also effective for detection of the Ca release.

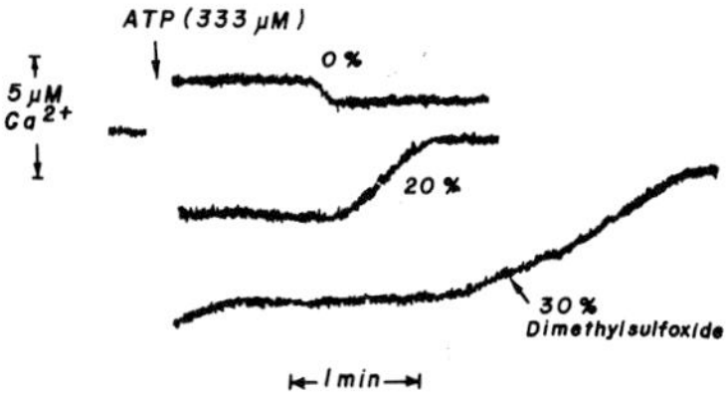

FIGURE 7. Spectrophotometric determination of calcium release from ATPase following addition of ATP. The reaction mixture consisted of 0.4 mg of ATPase per ml, 20 mM MES, pH 6.1, 45 μM $CaCl_2$, 80 mM KCl, 5 mM $MgCl_2$, 50 μM arsenazo-III and various concentrations of Me_2SO. ATP(333 μM) was added at time zero. Aminco double-wavelength spectrophotometer(585 vs 575 nm) was used(14).

REFERENCES

1. Hasselbach, W., & Makinose, M. (1961) Biochem. Z. 333, 518.
2. Hasselbach, W., & Makinose, M. (1963) Biochem. Z. 339, 94.
3. Ebashi, S., & Lipman, F. (1962) J. Cell Biol. 14, 389.
4. de Meis, L., & Vianna, A. (1979) Annu. Rev. Biochem. 48, 275.
5. Inesi, G. (1979) in Membrane Transport in Biology (Giebisch, G., Tosteson, D., & Ussing, H., Eds.) p357, Springer-Verlag, West Berlin and Heidelberg.
6. Yamamoto, T., Takisawa, H., & Tonomura, Y. (1979) Curr. Top. Bioenerg. 9, 179.
7. Coan, C., & Inesi, G. (1977) J. Biol. Chem. 252, 3044.

8. Champeil, P., Buschlen-Boucly, S., Bastide, F., & Gary-Bobo, C. (1978) J. Biol. Chem. 253, 1179.
9. Ikemoto, N., Morgan, T., & Yamada, S. (1978) J. Biol. Chem. 253, 8027.
10. Murphy, A. (1978) J. Biol. Chem. 253, 385.
11. Dupont, Y., & Leigh, J. (1978) Nature (London) 273, 396.
12. Watanabe, T., & Inesi, G. (1982) Biochemistry 21, 3254.
13. Watanabe, T., & Inesi, G. (1982) J. Biol. Chem. 257, 11510
14. Watanabe, T., Lewis, D., Nakamoto, R., Kurzmack, M., Fronticelli, C. & Inesi, G. (1981) Biochemistry 20, 6617.
15. Vanderkooi, J., Ierokomas, A., Nakamura, H., & Martonosi, A. (1977) Biochemistry 16, 1262.
16. Hiratsuka, T., & Uchida, K. (1973) Biochim. Biophys. Acta 320, 635
17. Penefsky, H. S. (1977) J. Biol. Chem. 252, 2891.
18. Dupont, Y., Chapron, Y., & Pougeois, R. (1982) Biochem. Biophys. Res. Commun. 106, 1272.
19. Ikemoto, N. (1975) J. Biol. Chem. 250, 7219.
20. Ikemoto, N. (1976) J. Biol. Chem. 251, 7275.
21. Takakuwa, Y., & Kanazawa, T. (1979) Biochem. Biophys. Res. Commun. 88, 1209.
22. Takakuwa, Y., & Kanazawa, T. (1981) J. Biol. Chem. 256, 2691.
23. Dupont, Y. (1980) Eur. J. Biochem. 109, 231.
24. The, R., & Hasselbach, W. (1977) Eur. J. Biochem. 74, 611.
25. Shigekawa, M., & Akowitz, A. (1979) J. Biol. Chem. 254, 4726.
26. de Meis, L., Martins, O., & Alves, E. (1980) Biochemistry 19, 4252.
27. MacLennan, D. (1970) J. Biol. Chem. 245, 4508.
28. Nakamura, Y., & Tonomura, Y. (1982) J. Biochem. 91 449.

MECHANISM FOR THE INTERCONVERSION OF CHEMICAL AND OSMOTIC FREE ENERGY[1]

Charles Tanford[2]

Department of Physiology
Duke University Medical Center
Durham, North Carolina

I. INTRODUCTION

This paper deals with the mechanism of active transport at the level of the transport protein molecule. It is not concerned with the fine details of the reaction sequence of the catalytic cycle, but rather with a physical picture of how free energy transfer to or from the transported ion is probably accomplished.

As other papers of this symposium testify, the sequence of steps that constitute the reaction cycle for the sarcoplasmic reticulum Ca^{2+} pump is by no means completely established, and the same statement applies to other active transport systems. Even if a reaction sequence were established - if all steps of the sequence had been identified and the rate constants for each step had been measured - it would not constitute adequate information for the elucidation of the mechanism as here defined. For the sarcoplasmic reticulum Ca^{2+} pump, for example, the central problem is how free energy is transferred from ATP to the Ca^{2+} ions, which means that we need to be able to keep track of what is happening separately to each substrate during its processing

---

[1]Supported by NSF grant PCM-79-20676 and NIH grant AM-04576.
[2]Research Career Awardee, National Institutes of Health.

ISBN 0-12-260380-X

by the transport protein. The usual kind of kinetic analysis does not provide this information. Stepwise rate constants can do no more than yield values for the total free energy change for each step, representing the sum of what is happening to the separate substrates. Steps in which the chemical potential of bound Ca^{2+} ions goes up at the expense of a free energy decrease in bound ATP or phosphate might have an overall free energy difference close to zero, and would not be identifiable on the basis of conventional kinetic analysis (1, 2).

The alternative approach to be used in this paper is to ask what essential steps are theoretically required for chemi-osmotic free energy transduction in active transport, or in ATP synthesis resulting from reversal of active transport. This approach has been previously employed by Jencks (3), and the present paper in part reiterates and confirms the "rules" formulated by Jencks. It also, however, extends them, both in terms of additional theoretical ideas, and, more particularly, by focusing on protein structural features that would be required to put the "rules" into practice.

When one begins to define the framework for a theoretical approach, it becomes immediately evident that the problem is more difficult for active transport than for a typical biochemical reaction in which a chemical transformation occurs. In the typical biochemical reaction, matter is transferred between interacting substrates. For example, in creatine kinase, a phosphate group is exchanged between creatine and ADP, and we can easily specify a plausible physical picture of the catalytic role of the protein, and don't need a detailed pathway or rate constants to do so. Catalysis is likely to involve binding of the two substrates close to each other, properly oriented so that exchange of a phosphoryl group between them can readily occur. [For creatine kinase a mechanism of this kind has in fact been established (4)] In active transport the situation is quite different. One of the substrates, the transported ion, remains chemically unchanged. There is no transfer of matter between ATP and Ca^{2+} ions in the example of a Ca^{2+} pump; only free energy is transferred, and this makes it much more difficult to imagine how the process might work.

One consequence of this difference is that there is no longer any obvious advantage (in any active transport process) to binding the interacting species close to each other. In the case of a Ca^{2+} pump, it would in fact be a disadvantage because Ca^{2+} and ATP interact to form a stable compound with an overall decrease in free energy, e.g., for $ATP^{-4} + Ca^{2+} \rightleftharpoons CaATP^{-2}$, we have ΔG^{o} = -7 kcal/mol. The overall pump reaction, however, involves an exchange of free energy, and at the physiological steady state of a resting cell involves virtually no <u>net</u> free energy change (5). Approximate numerical data for the overall process, at typical levels of ATP, ADP, and P_i, are

$$ATP \rightleftharpoons ADP+P_i \qquad \Delta G=-14kcal/mol \tag{1}$$

$$2Ca^{2+}(cyto),40nM \rightleftharpoons 2Ca^{2+}(SR),2mM \qquad 2\Delta\mu_{Ca}=+13.3kcal/mol \tag{2}$$

and the sum of the two is essentially zero. Complexation between Ca^{2+} and ATP would make it more difficult to raise the chemical potential of Ca^{2+} to its ultimate level. In discussing the reaction mechanism, therefore, I shall consider it reasonable to assume that the sites on the protein for Ca^{2+} (or other transported ions) and for ATP processing are not close together. There is experimental evidence to support this assumption, both for the sarcoplasmic reticulum Ca^{2+} pump and for ATP-driven Na^{+},K^{+} pumps (6, 7).

II. DISSECTION OF THE PROBLEM

The overall problem of how the transport protein might catalyze free energy transfer coupled to ion translocation can be dissected into component parts as follows:

(1) How does the protein provide a passage across the membrane through which ion translocation can occur?

(2) How can a protein change the chemical potential of an ion, regardless of where the energy is derived?

(3) How can the change in chemical potential be linked to ion translocation?

(4) How is ATP processed? The problem here is how _uncoupled_ ATP hydrolysis can be prevented.

(5) How are these separate processes, assumed to occur in different parts of the protein molecule, rigidly coupled to each other, so that one cannot occur without the other?

III. ION TRANSLOCATION

In active transport, ion translocation in a _specified direction_ has to be coupled to ATP hydrolysis or other source of free energy. It is easy to show that a trans-membrane channel that is simultaneously open to both sides of the membrane cannot accomplish this, even if the channel is gated so that it is open only part of the time, and even if energy is required to open the gate. All realistic models for active transport that have ever been suggested involve _alternating access_ (3, 8-10). The protein alternates in each reaction cycle between two conformations. In one of them a site for binding the translocated ion is accessible

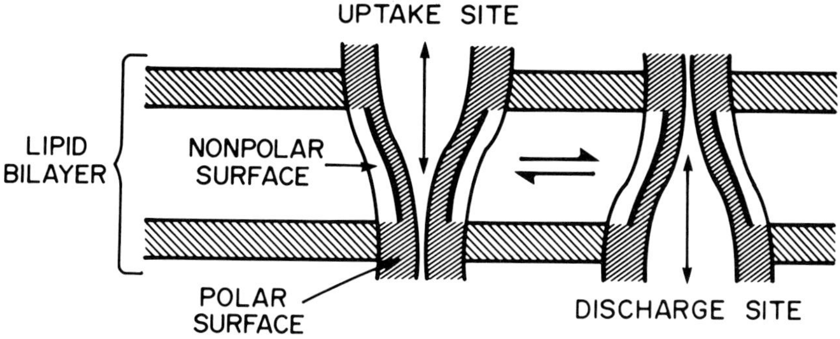

FIGURE 1. Schematic diagram illustrating the alternating access model for ion translocation. The transport protein remains fixed in the membrane during the conformational transition, with no change in the location of external hydrophilic and hydrophobic regions. (Only the trans-membrane portion of the protein is shown. Access to the alternating site could be via access channels through the part of the protein that is not shown.)

from one side of the membrane only and in the other conformation it is accessible only from the other side. (Fig. 1 provides a schematic sketch). Intermediate states are of course possible, provided that the ion is occluded and accessible to neither side of the membrane. The binding sites need not be directly open to the bulk solution, as in Fig. 1, but they could be connected to the outside by access channels in which the ion might be transiently bound to other sites on its way to or from the critical alternating site. The proteolipid of the F_0 complex of mitochondrial or chloroplast H^+-linked ATPase might represent such an access channel, since it can be separated from the rest of the ATPase molecule and then exhibits the properties of an open trans-membrane channel (11). (The intact ATPase molecule of course does not exhibit this property).

IV. CHANGE IN CHEMICAL POTENTIAL OF THE TRANSPORTED ION

It has been intuitively believed for a long time that a change in the chemical potential of the transported ion, occurring in synchrony with translocation, arises from a change in binding affinity. The binding constant for the ion is thought to be high in the left-hand conformation of Fig. 1, so that ion uptake can occur in spite of a very low concentration in the adjacent medium. The binding constant in the other conformation is thought to be low, so that the ion can be released into a medium of high concentration. This attractive hypothesis is not a thermodynamic necessity, as a number of authors have pointed out (3, 12), but can in fact be shown to be a practical necessity to meet physiological and kinetic requirements (13).

A difference in binding constant is not thermodynamically essential because the equilibrium state towards which a system strives is determined by the stoichiometry of the overall reaction, and is independent of its mechanism. Thus any reaction sequence that leads to the overall reaction

$$2Ca^{2+}(cyto) + ATP \rightleftharpoons 2Ca^{2+}(out) + ADP + P_i \quad (3)$$

(where "out" refers to the sarcoplasmic reticulum lumen or extracellular space) will necessarily approach an equilibrium state defined by

$$\frac{[Ca(out)]^2}{[Ca(cyto)]^2} = K_{ATP} e^{\Delta\psi F/RT} \frac{[ATP]}{[ADP][P_i]} \tag{4}$$

where K_{ATP} is the equilibrium constant for hydrolysis of ATP in aqueous solution, $\Delta\psi$ is the difference in electrical potential (inside - outside) across the membrane, and F is Faraday's constant. With $K_{ATP} = 2 \times 10^5$ M and the concentrations of ATP, ADP, and P_i at typical physiological levels (14), the right-hand side of equation 4 is >>1 at any reasonable membrane potential, so that the reaction represented by equation 3 will strive for the establishment of a high ion concentration gradient regardless of mechanism or any consideration other than that the stoichiometry has to be the stoichiometry given by the equation. A similar argument applies to any other active transport system, and it is therefore evident that such systems will generally seek to establish a high ion concentration gradient even if the binding constants for the ion are the same on the two sides of the membrane. The product of all stepwise equilibrium constants for a reaction cycle has to be consistent with the equilibrium state of the overall reaction (in the present example the product has to be equal to K_{ATP}), but there are no constraints on individual K's (12).

Formal thermodynamics is, however, not a sufficient criterion for making mechanistic choices. It is not sufficient that the thermodynamic driving force is in the right direction. An ion pump must also work at an adequate rate to counter leaks when a cell is at rest, and must respond with an increased rate when challenged by changes in the environment, e.g., the sarcoplasmic reticulum Ca^{2+} pump must be able to extrude Ca^{2+} rapidly to relax a muscle after contraction. Reversibility is another requirement for many transport systems, and for most others (e.g., the sarcoplasmic reticulum Ca^{2+} pump) the ability to reverse direction (to synthesize ATP when the thermodynamic driving force demands it) is an experimental fact (15). These functional requirements must be met by a mechanism in which rate constants have physically reasonable values. For example, the rate constant for a bimolecular binding step has a rigorously defined upper limit, corresponding to diffusion-controlled rate, which it is physically impossible to exceed (13).

I recently carried out a computer simulation for an ATP-driven Ca^{2+} pump, using a conventional reaction sequence of the type proposed by de Meis and coworkers (16, 17). I allowed all rate constants to have a range of values within reasonable physical limits, which generated 20,000,000 different sets of kinetic parameters. It was found that only about 7,000 of these sets satisfied the physiological criteria of the preceding paragraph, and every one of these "valid" sets had the high and low equilibrium constants for uptake and discharge that is intuitively expected, e.g., the equilibrium constant for binding of Ca^{2+} from the cytoplasmic side, obtained from the data as the ratio of the on and off rate constants, was found to exceed 2×10^5 M^{-1} for each site in more than 95% of the "valid" combinations. Actual experimental data bear out this verification of the intuitive result, not only for the sarcoplasmic reticulum Ca^{2+} pump (18-21) but also for Na^+,K^+ pumps (22, 23) and for the F_0F_1 H^+-linked ATPase of chloroplasts (24). Some recent data are given in Table I.

Table I. Binding Constants for Transported Ions[a]

Protein	Ion	Binding constant, K_{eq} (M^{-1})	
		Uptake side	Discharge side
SR Ca^{2+}pump	Ca^{2+}	10^7-10^8	300
Na^+,K^+ pumps	Na^+	4×10^3	<20
	K^+	2×10^4	<10
Chloroplast F_0F_1	H^+	$>10^8$	$<10^6$

[a]Uptake and discharge sides are defined for direction of uphill transport of the ion for which data are given. See text for references.

With these results, the problem of finding a mechanism for the change in chemical potential of the transported ion reduces to the problem of how the protein can change its binding affinity for the ion, with the understanding that the ion must remain bound to the protein while this happens. It is easy to postulate a reasonable answer to this question in the case of Ca^{2+} because of the existence of a lot of experimental data for model compounds and for Ca^{2+} binding proteins with known three-dimensional structure (10, 25). These data show that chelation with 6 coordinating groups (usually including several COO^- groups) is required to generate a high affinity binding site. Three or four chelating groups, with other coordination positions filled by H_2O molecules, would suffice for a low affinity site. A conformational change leading to a change in binding constant can therefore be expected to be accompanied by the movement

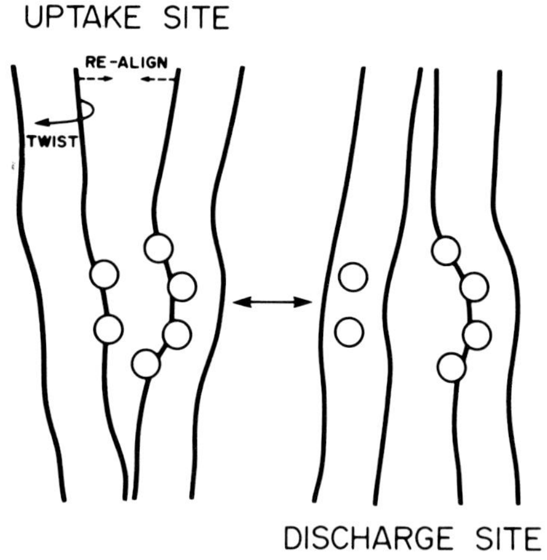

FIGURE 2. Schematic illustration of how a change in binding affinity might be coupled to a change in accessibility of the type illustrated by Fig. 1. The circles represent amino acid side chains and peptide groups that constitute the binding site for the ion, and they are visualized as attached to helical transmembrane segments of the protein molecule. Only two segments are shown, but the overall structure is assumed to be an assembly of six or more such segments, as observed in bacteriorhodopsin (28).

of chelating amino acid side chains in and out of the binding site. For example, the structure of the Ca^{2+} binding site of the cytoplasmic Ca^{2+} binding protein parvalbumin is known from x-ray structure analysis to consist of appropriately arranged chelating groups contributed by 3 aspartyl residues, 1 glutamyl residue, 1 seryl residue and one peptide carbonyl group (26). The binding constant for Ca^{2+} is about $10^8 M^{-1}$, so that this site is a reasonable model for the high affinity site of a Ca^{2+} pump. Two of the chelating groups (one aspartyl, one glutamyl) in this structure are located on a separate segment of polypeptide chain from the others, and can easily be visualized as twisting away from the ion, which would remain attached to the protein by virtue of the other four groups. The result would be a decrease in the binding constant, which would be equivalent to a large increase in the chemical potential of the bound Ca^{2+} ion. Parvalbumin itself does not undergo a conformational transition in which this kind of change occurs, but many well-known examples of the same kind of process exist, e.g. hexokinase exists in two conformational states with different binding affinity for glucose, brought about by movement of a hinged domain of the protein towards or away from the glucose binding site (27).

Both in hexokinase, and in the hypothetical example based on the parvalbumin structure, the binding site is exposed to the same solution in the two protein conformations. In active transport we are dealing with a trans-membrane protein, and the affinity change, to be functionally meaningful, has to be structurally coupled to translocation of the binding site from one side of the membrane to the other. With nothing known about the three-dimensional structure of any Ca^{2+} pump protein, any proposal as to how this coupling might be accomplished has to be purely speculative. I have chosen to base a speculative model on the structure of the only trans-membrane protein for which even low resolution structural data are available, namely the protein bacteriorhodopsin. This protein is itself an ion pump (a light-driven proton pump) so that its use as a source for a speculative model is not unrealistic.

The trans-membrane portion of the bacteriorhodopsin structure consists of seven roughly parallel helical polypeptide segments, joined outside the membrane by short and probably flexible segments of the chain (28, 29). The entire structure is contained within a single polypeptide chain with a molecular weight of only 26,000. There are cavities between the helices that extend part way into the membrane domain from either side (but not all the way through) and an ion binding site could be located in such a cavity, as illustrated by Fig. 2. The polypeptide segments between helices could then serve as hinges or ball-joints, capable of producing a concerted twisting and rocking motion that would achieve the desired result of moving chelating groups away from the binding site at the same time as the direction of access is changed from one side of the membrane to the other (10). Fig. 2 illustrates this possibility schematically. In a multi-helical structure, several ions could be moved simultaneously in a structural transition of this kind.

While the foregoing considerations were discussed with Ca^{2+} in mind as the transported ion, the same general principles may be applicable to most ion transport processes. A conformational transition of the kind illustrated by Fig. 2 could, for example, be responsible for a change in pK of the protons transported by bacteriorhodopsin itself.

V. ATP PROCESSING

In many active transport systems, including the sarcoplasmic reticulum Ca^{2+} pump, the free energy for uphill transport is derived from the exergonic hydrolysis of ATP (equation 1). The problem of how ATP should be processed to permit the appropriate free energy exchange to occur is conceptually simpler than the problem of devising a model for the chemical potential change of a transported ion, because the reaction occurs entirely on one side of the membrane. Apart from this difference, the basic principles are the same: (1) It is necessary to have two distinct conformational states of the protein (3). In one of these states the protein should be capable of binding ATP from the solution, but incapable of carrying its processing of the substrate to the point of release of both ADP and P_i. In the other state, the protein should be able to exchange ADP, P_i, or both with the adjacent medium, but should be incapable of completing the reaction ADP + $P_i \rightleftharpoons$ ATP to the point of ATP release into the aqueous medium. Without this proviso, the protein could catalyse uncoupled ATP hydrolyze. (2) The same kind of theoretical considerations and model calculations that were used to show that the binding constant for

Ca^{2+} has to be different in the two conformations of an alternating access model for ion transport can be used to demonstrate that the standard chemical potential of bound substrate (in any of a variety of possible forms) has to be different in the two conformational states employed for ATP processing.

Both conditions are satisfied by all known ATP-linked ion pumps. The sarcoplasmic reticulum Ca^{2+} pump, and other pumps that operate via phosphoenzyme intermediates, have two forms of this intermediate, E~P and E'-P, which can be interconverted only if the protein conformation is interconverted between states E and E'. In E~P the chemical potential of the bound phosphoryl group is high and the group can react with ADP to synthesize ATP. In E'-P the chemical potential of the bound phosphoryl group is much lower (by about 12 kcal/mol!) (1, 30), and it is unreactive towards ADP. The E'-P form can be made by reaction of E' with inorganic phosphate, or can be hydrolyzed to release P_i into the aqueous medium (16, 17, 23). Thermodynamically similar states, involving "tightly bound" and "loosely bound" ATP, occur in the reaction cycle of proton-linked ATPases of bacteria, mitochondria and chloroplasts, where a covalent phosphoenzyme derivative is not involved (31-33).

Although the existence of these states has been well established for a long time, there does not exist even a speculative structural model for them, such as I have presented in Fig. 2 for the chemical potential change of bound Ca^{2+}. The reason for this is that much less information is available about the energetics of protein-phosphate interaction than there is about protein-Ca^{2+} interaction, so that there are no obvious model systems that could serve as prototypes. A plausible qualitative suggestion to account for the different thermodynamic properties of the E~P and E'-P states has been proposed by de Meis (17, 34).

VI. FREE ENERGY COUPLING

The final question posed at the beginning of this paper (and posed pictorially in Fig. 3) is the question of how ion translocation (with its attendant chemical potential change) and ATP processing, probably taking place some distance from each other, can be coupled so that neither can occur without the other. Given what has been said about the two separate processes, the most likely answer to the question is virtually self-evident. Each separate function requires two distinct conformational states, differing in accessibility of the ion binding sites in the case of ion translocation, and differing in the chemical reactivity of bound phosphate or ATP in the case of ATP processing. The obvious way to achieve functional coupling is then by means of conformational coupling. If the conformational change at the ATP site in Fig. 3 cannot occur separately from the conformational change that is associated with translocation, then functional coupling will automatically be assured.

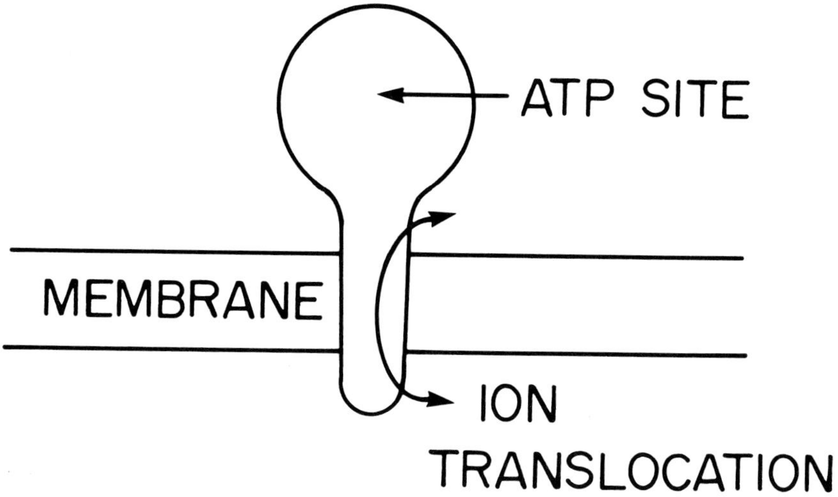

FIGURE 3. Conformational coupling. If the essential conformational changes at the translocation site (Figs. 1 and 2) and at the ATP site are coupled through the protein structure, so that one cannot occur without the other, then functional coupling is automatically assured. The ATP site is shown to be far from the ion translocation part of the molecule because there would be no obvious advantage for free energy transduction if they were close together. Having the sites close together, however, would have no effect on the principle involved in conformational coupling.

Such structural coupling is a normal phenomenon in protein chemistry. Ordered structures result from cooperative interactions that involve the entire protein molecule, and it is a common occurrence that structural changes at different sites within the same molecule are incapable of taking place independently. Hemoglobin is the prototype example. Each of its four subunits can exist in two states with high and low binding constants for O_2 at the heme binding sites. The four sites are about 25 Å apart and there is no direct interaction between them, but nevertheless one cannot change the affinity at a single site by itself. The protein as a whole exists predominantly in one of two alternative states - one in which all four O_2 binding sites have high affinity and the other in which all four have low affinity. Structural changes at the four sites are coupled to each other through the fabric of the protein molecule by constraints on the kind of overall conformational transition that is allowed (35).

Conformational coupling still permits two possibilities. If the two protein conformations have very different intrinsic free energies, two-step coupling could occur. In one step the free energy derived from ATP hydrolysis could be used to drive the protein into its thermodynamically unfavorable state, and in a second step the free energy released by return of the protein to its more stable state could be used to drive the chemical potential increase of the transported ions. This is essentially the "rack" mechanism of Lumry, proposed in a more general context some years ago (36).

The second possibility is that the two protein conformations have close to the same free energy, so that the protein itself is energetically inert. In this case there would have to be <u>direct exchange</u> of free energy between the free energy donor and the transported ions (1), i.e., the conformational change (with bound ligands attached) would be accompanied by the <u>release</u> of free energy from bound ATP (or equivalent E~P complex) and the <u>uptake</u> of free energy by the transported ions, the two processes taking place <u>simultaneously</u>. Return of the protein to its original conformation (now with no ligands attached) would not contribute significantly to the free energy balance.

It is known that the second possibility applies to the sarcoplasmic reticulum Ca^{2+} pump. The two conformational states have nearly the same intrinsic free energy (21, 37, 38), i.e., ΔG^{o} for the interconversion is much smaller than the 13-14 kcal/mol that is transferred between ATP and the transported ions in each reaction cycle (equations 1 and 2). Free energy transduction is accomplished virtually entirely in a single transition, $Ca_2E{\sim}P \rightleftharpoons Ca_2E'{-}P$, of the conventional pump cycle (16, 17).

VII. DISCUSSION

The mechanism for free energy transduction that has emerged from the theoretical analysis that I have presented is of course essentially the same as the mechanism of free energy transduction proposed by de Meis for the sarcoplasmic reticulum Ca^{2+} pump on the basis of an experimentally deduced reaction scheme (16, 17), and it has not therefore produced a new and different view of how the sarcoplasmic reticulum Ca^{2+} pump works. The importance of the theoretical analysis is that it is not based on this or any other reaction scheme. Some of the work reported in the recent literature and at this meeting suggests that the de Meis scheme may require modification, e.g., it has been suggested that there is an "occluded" state for Ca^{2+} in between the two states accessible from the two sides of the membrane (39, 40), and that there may be only one Ca^{2+} binding site per catalytic polypeptide chain instead of two (41). The theoretical analysis would not be affected by such modifications, and the general principles I have discussed would need to continue to apply. In summary, these principles are

1. Ion translocation has to occur by movement of the ion binding site with bound ion attached, which is accomplished by means of a conformational change in the protein. Movement of the ion between <u>fixed</u> sites facing the two sides of the membrane cannot lead to active transport.

2. The ion translocation step (regardless of whether there is an intervening "occluded" step) has to be coupled to a change in the standard chemical potential of the bound ion, which is equivalent to a change in binding constant. This is not a thermodynamic requirement, but results from kinetic limitations.

3. ATP processing must likewise involve two states of the protein, with different standard chemicals potential for bound substrate in some form, e.g., bound ATP or phosphoenzyme intermediate.

4. Free energy transduction steps have to be thermodynamically coupled so that the total free energy change in each step is small (2, 3). When the chemical potential of the bound ion changes, there has to be a compensating change in the same step, either in the chemical substrate, or else in the conformational free energy of the protein itself. Again, this is a kinetic and not thermodynamic requirement.

5. The first alternative (direct exchange) applies to the sarcoplasmic reticulum Ca^{2+} pump. The protein conformational transition itself contributes little to free energy bookkeeping. The exchange per se, $\Delta\mu^{o}_{p} \sim -12$kcal/mol, $\Delta\mu^{o}_{Ca} \sim +6$kcal/mol ($2\Delta\mu^{o}_{Ca} \sim +12$ kcal/mol), probably occurs in a single step of the reaction sequence.

A final justification for the theoretical analysis of this paper is that it is in principle quite general, and should not be unique for the transport of Ca^{2+}. The same principles should apply to all ATP-driven ion pumps or to ATP-synthesizing transport systems working in the opposite direction. Speculative mechanisms proposed for these systems (42) are more diverse that the mechanisms that have been proposed for the sarcoplasmic reticulum Ca^{2+} pump, and the general principles deduced in this paper can be expected to lead to rejection of some of them (43).

REFERENCES

1. Tanford, C., Proc. Natl. Acad. Sci. USA 78: 270-73 (1981).
2. Tanford, C., Proc. Natl. Acad. Sci. USA 79: 6527-31 (1982).
3. Jencks, W.P., Adv. Enzymol. 51: 75-106 (1980).
4. Watts, D.C., in The Enzymes, (Boyer, P. D. ed.), 3rd ed., vol. 8, ch. 12, Academic Press, New York, 1973.
5. Tanford, C., J. Gen. Physiol. 77: 223-29 (1981).

6. Miki, K., Scott, T.L., and Ikemoto, N., J. Biol. Chem. 256: 9382-85 (1981).
7. Carilli, C.T., Farley, R.A., Perlman, D.M., and Cantley, L.C., J. Biol. Chem. 257: 5601-6 (1982).
8. Jardetsky, O., Nature 211: 969-70 (1966).
9. Dutton, A., Rees, E.D., and Singer, S.J., Proc. Natl. Acad. Sci. USA 73: 1532-36 (1976).
10. Tanford, C., Proc. Natl. Acad. Sci. USA 79: 2882-84 (1982).
11. Sebald, W., and Hoppe, J., Curr. Top. Bioenerget. 12: 1-64 (1981).
12. Hill, T.L., Free Energy Transduction in Biology, p. 229. Academic Press, New York, 1977.
13. Tanford, C., Proc. Natl. Acad. Sci. USA 79: 6161-65 (1982).
14. Veech, R.L., Lawson, J.W.R., Cornell, N.W., and Krebs, H.A., J. Biol. Chem. 254: 6538-47 (1979).
15. Hasselbach, W., Biochim. Biophys. Acta. 515: 23-53 (1978).
16. de Meis, L., and Vianna, A.L., Ann. Rev. Biochem. 48: 275-92 (1979).
17. de Meis, L., "The Sarcoplasmic Reticulum" p. 163 Wiley, New York, 1981.
18. Inesi, G., Kurzmack, M., Coan, C., and Lewis, D.E., J. Biol. Chem. 255: 3025-31 (1980).
19. Ikemoto, N., J. Biol. Chem. 250: 7219-24 (1975).
20. de Meis, L., and Tume, R.K., Biochemistry 16: 4455-63 (1977).
21. Tanford, C., and Martin, D.W., Z. Naturforsch. 37c: 522-6 (1982).
22. Robinson, J.D., and Flashner, M.S., Biochim. Biophys. Acta 546: 146-76 (1979).
23. Jørgensen, P.L., Biochim. Biophys. Acta 694: 27-68 (1982).
24. Dewey, T.G., and Hammes G.G., J. Biol. Chem. 256: 8941-46 (1981).
25. Williams, R.J.P. In Calcium Binding Proteins and Calcium Function, (R.H. Wasserman et al, ed.), p. 3-12 1977.
26. Kretsinger, R.H., Ann. Rev. Biochem. 45: 239-66 (1976).
27. Bennett, W.S., Jr., and Steitz, T.A., Proc. Natl. Acad. Sci. USA 75: 4848-52 (1978).
28 Henderson, R., and Unwin, P.N.T., Nature 257: 28-32 (1975).

29. Engelman, D.M., and Zaccai, G., Proc. Natl. Acad. Sci. USA 77: 5894-98 (1980).
30. Taniguchi, K., and Post, R.L., J. Biol. Chem. 250: 3010-18 (1975).
31. Boyer, P.D. Ann. Rev. Biochem. 46: 957-66 (1977).
32. Boyer, P.D. 1979. In Membrane Bioenergetics, (C.P. Lee, G. Schatz, L. Ernster, ed.), p. 461-79. Addison-Wesley, Reading, Mass, 1979.
33. Hammes, G.G., Proc. Natl. Acad. Sci. 79: 6881-84 (1982).
34. de Meis, L., Otero, A. de S., Martins, O.B., Alves, E.W., Inesi, G., and Nakamoto, R., J. Biol. Chem. 257: 4993-98 (1982).
35. Wyman, J., Adv. Protein Chem., 19: 223-86 (1964).
36. Lumry, R., Ann. N.Y. Acad. Sci. 227: 46-73 (1974).
37. Loomis, C.R., Martin, D.W., McCaslin, D.R., and Tanford, C., Biochemistry 21: 151-56 (1982).
38. Pick, U., and Karlish, S.J.D., J. Biol. Chem. 257: 6120-26 (1982).
39. Dupont, Y., Eur. J. Biochem. 109: 231-38 (1980).
40. Takisawa, H., and Makinose, M., Nature 290: 271-73 (1981).
41. Watanabe, T., Lewis, D., Nakamoto, R., Kurzmack, M., Fronticelli, C., and Inesi, G., Biochemistry 20: 6617-25 (1981).
42. Boyer, P.D., Chance, B., Ernster, L., Mitchell, P., Racker, E., and Slater, E.C., Ann. Rev. Biochem. 46: 955-1026 (1977).
43. Tanford, C., Ann. Rev. Biochem. 52: in press.

B. Kinetic Studies of the Calcium Pumping Cycle

KINETICS OF ACCUMULATION OF THE ADP-SENSITIVE AND ADP-INSENSITIVE PHOSPHOENZYMES AND THEIR RELATIONSHIP TO Ca^{2+} TRANSLOCATION IN SARCOPLASMIC RETICULUM

Jeffrey P. Froehlich
Phillip F. Heller

National Institute on Aging, National Institutes of Health,
Gerontology Research Center, Baltimore City Hospitals,
Baltimore, Maryland 21224

I. INTRODUCTION

Fragmented membrane vesicles of sarcoplasmic reticulum prepared from skeletal muscle will actively accumulate Ca^{2+} coupled to ATP hydrolysis by a mechanism which is believed to include the following steps (1,2):

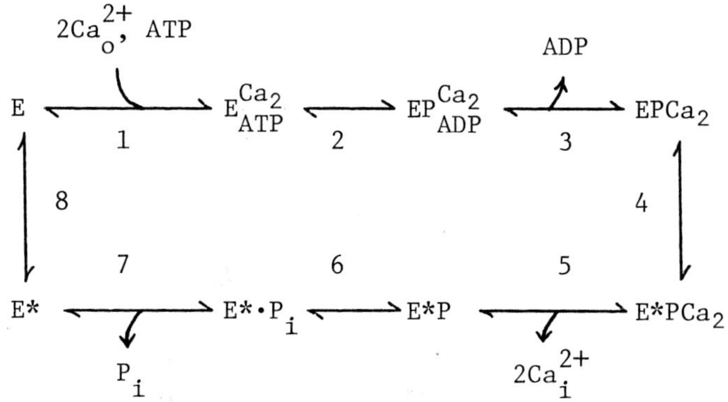

MECHANISM I

In this scheme, transfer of gamma phosphate from ATP to an aspartyl residue on the ATPase (3) results in the formation of an acid-stable phosphorylated intermediate, the ADP-sensitive phosphoenzyme ($EPCa_2$). The latter is referred to as ADP-sensitive because it will react rapidly with added ADP to form ATP by reversal of phosphorylation (4,5). The ADP-sensitive

ISBN 0-12-260380-X

phosphoenzyme is converted through a change in conformation to a second acid-stable phosphorylated intermediate, the ADP-insensitive phosphoenzyme (E^*PCa_2). This species will not react with ADP, but is hydrolysed to inorganic phosphate (P_i) by a reaction that is activated by K^+ and Mg^{2+} ions (4,6,7). During the conversion of the ADP-sensitive phosphoenzyme to the ADP-insensitive form the enzyme loses its high affinity for Ca^{2+} which subsequently dissociates from the transport site (8). The enzyme regains its high affinity for Ca^{2+} by undergoing a second major conformational change which takes place in the final step of the reaction cycle (2,9,10).

Phosphate bound to the ADP-insensitive phosphoenzyme formed from P_i in the absence of a Ca^{2+} gradient (11,12) will react with ADP to form ATP following an increase in the Ca^{2+} concentration of the medium (13). This occurs as a result of occupation of low affinity Ca^{2+} binding sites on E*P which activate its conversion to the ADP-sensitive phosphoenzyme.

Although it is generally believed that the outside-to-inside carrier transition is coupled to the phosphoenzyme conversion reaction (1,2,14,15), the sequence of these events has not been experimentally determined. In order to resolve this problem, we have undertaken measurements of the time course of formation of the Ca^{2+}-depleted, ADP-insensitive phosphoenzyme (E*P) using the acid quench-flow technique (10). By adding ADP to the phosphoenzyme and measuring the amplitude of the slow phase of decay at various times after the addition of ATP, we were able to obtain an estimate of the time course of accumulation of the ADP-insensitive phosphoenzyme. The secondary phase of phosphoenzyme decay following the addition of ADP may result from dephosphorylation of any one of the following intermediates: $EP^{Ca_2}_{ADP}$ (16), E^*PCa_2 or E*P. Only the latter, however, will show an increased rate of decomposition in response to a jump in the Ca^{2+} concentration during dephosphorylation following the addition of ADP. This will take place as a result of activation by Ca^{2+} of the E*P to EP conversion followed by reversal of the phosphorylation reaction. In the present study, we have attempted to demonstrate the formation of this species by adding high Ca^{2+} to the enzyme together with ADP after a brief interval of phosphorylation. The time course of E*P formation determined in this fashion was compared to that of intravesicular $^{45}Ca^{2+}$ accumulation measured by rapid quenching with EGTA + ADP (17).

II. METHODS

Sarcoplasmic reticulum membrane vesicles were prepared from rabbit skeletal muscle as previously described (18). The vesicles were either used immediately after preparation for measurements of Ca^{2+} uptake or were quick frozen in liquid nitrogen and stored at -80°. Treatment with the Ca^{2+} ionophore A23187 was carried out by adding the ionophore in 95% ethanol to a suspension of the vesicles (40 μg A23187/mg SR protein) with slow stirring. Ionophore-treated vesicles were unable to accumulate Ca^{2+} but showed no impairment of phosphorylation or ATPase activity.

Rapid mixing experiments were carried out using the multi-mixing apparatus of Froehlich *et al.* (19). Unless otherwise indicated, SR vesicles were phosphorylated in a medium containing 10 μM [γ-^{32}P]ATP, 100 mM KCl, 3 mM $MgCl_2$, 0.1 mM $CaCl_2$, 0.1 mM EGTA and 20 mM Tris-maleate at 21°. In the dephosphorylation experiments, 1.33 mM ADP or 1.33 mM ADP + 5 mM $CaCl_2$ (final concentrations) was added to vesicles that had been previously exposed to ATP for a constant time interval (usually 6 or 116 msec). The incubation was allowed to proceed for an additional variable time interval (0-542 msec) after which 9% perchloric acid (2.25% final concentration) was added to terminate the reaction. For determination of intravesicular Ca^{2+} accumulation, the reaction medium contained [^{45}Ca] $CaCl_2$ and 10 μM ATP and the quench medium contained 6 mM EGTA + 3 mM ADP (2 mM EGTA + 1 mM ADP after mixing). The vesicles were collected on a Millipore filter (0.45 μ) and washed with a medium containing 1 mM EGTA, 100 mM KCl and 20 mM Tris maleate, pH 6.8. Assays for protein, [^{32}P]phosphoenzyme, and [^{32}P]P_i were carried out as previously described (20).

III. RESULTS

In the experiment shown in Fig. 1, SR vesicles were phosphorylated for 116 msec in a medium containing 10 μM [γ-^{32}P]ATP and approximately 10 μM free Ca^{2+}, and then dephosphorylated by the addition of 1.33 mM ADP. The resulting time course of phosphoenzyme decomposition (closed circles, Fig. 1) can be resolved into three components (inset, Fig. 1): a very fast component, a slow component, and a very slow component accounting for roughly 40%, 50%, and 10% of the total phosphoenzyme, respectively. Inorganic phosphate (P_i) production (open circles), which was not observed during the rapid phase of dephosphorylation, had a rate that was similar to the slow phase (∿30 s^{-1}). These results suggest that the fast and slow components correspond to the ADP-sensitive (EP) and ADP-insensitive (E*P) phosphoenzymes, respectively. The very slow residual component, which decomposes at a rate at least 10 times slower than the ADP-insensitive phosphoenzyme, probably arises by a separate (parellel) pathway.

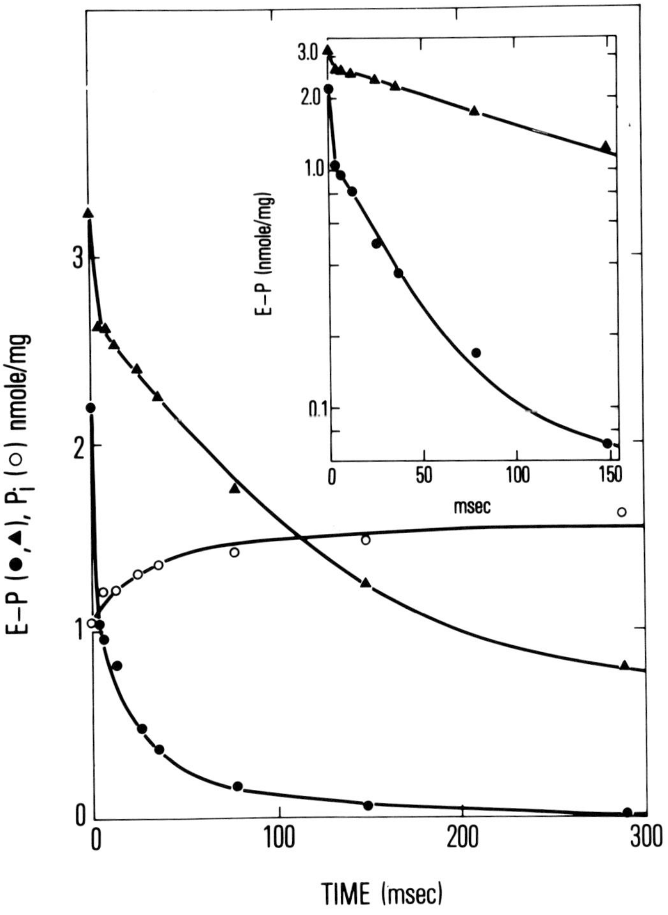

Fig. 1. Phosphoenzyme decomposition and P_i release following the addition of ADP. Sarcoplasmic reticulum vesicles (0.5 mg/ml) were suspended in a medium containing 3 mM $MgCl_2$, 0.1 mM each $CaCl_2$ and EGTA, 20 mM Tris maleate pH 6.8, and either 100 mM KCl (●,○) or 100 mM choline chloride (▲) and mixed with an equal volume of the same medium containing 20 [γ-^{32}P]ATP. After 116 msec, a solution containing 5 mM ADP (final concentration 1.33 mM) was added to the reaction mixture and dephosphorylation was allowed to proceed for the indicated times prior to the addition of acid (2.25% perchloric acid + 2 mM P_i). (●,▲) phosphoenzyme decomposition; (o) P_i release. Inset shows the logarithm of the phosphoenzyme concentration vs time after the addition of ADP.

Potassium ions have been shown to activate decomposition of the ADP-insensitive phosphoenzyme (7,21-23). In the dephosphorylation experiment represented by the closed triangles in Fig. 1, KCl in the reaction medium was replaced by 100 mM choline chloride resulting in a slower rate of decomposition of the slow component. This lends additional support to the conclusion that the slow phase is due to the ADP-insensitive phosphoenzyme.

Although P_i production and dephosphorylation of the slow component following the addition of ADP have similar rates, the amplitude of the former reaction is only about one-third that of the latter. This behavior can be explained by assuming that during dephosphorylation, the ADP-insensitive intermediate can break down in both the forward and reverse directions as depicted below:

$$EP \leftrightarrow E^{*}P \leftrightarrow E^{*}\cdot P_i$$

By forming a complex with EP, ADP will prevent the further conversion of EP to E*P and pull the first reaction to the left, allowing some of the phosphate that was originally bound to E*P to be transferred to ADP to form ATP. If, in addition, the hydrolytic reaction is essentially irreversible, then the rate of disappearance of E*P will equal the sum of the rate constants controlling its breakdown in the forward and reverse directions ($\sim 30\ s^{-1}$). From our earlier studies (20), the rate constant for dephosphorylation of E*P is about $10\ s^{-1}$; consequently, the rate of conversion of E*P to EP is $\sim 20\ s^{-1}$. These values for the forward and reverse constants for E*P decay are consistent with the observation that about two-thirds of the phosphate bound to the latter is converted to a form that can be removed from the supernatant with activated charcoal.

The fact that a significant fraction of the ADP-insensitive phosphoenzyme decays back in the direction of EP implies that sufficient Ca^{2+} must be available on the inside of the vesicle to occupy the low affinity transport sites on E*P (Mechanism I). Although the time interval during which Ca^{2+} transport occurs is relatively short (116 msec), the small excluded volume (~ 5 μl/mg; ref. 24) and high density of transport sites (18) might allow the accumulation of sufficient intravesicular Ca^{2+} to occupy the low affinity transport sites. To determine if preventing the formation of the transport gradient would inhibit the reverse transition and increase P_i production, we repeated the experiment in Fig. 1 using vesicles that had been treated with the Ca^{2+} ionophore A23187. When these vesicles were dephosphorylated by ADP, the level of P_i production did not increase nor did the rate of dephosphorylation decrease indicating that Ca^{2+} which activates the reverse transition is not in the intravesicular space but in a separate compartment that is inaccessible to the ionophore.

From the similarity in the steady state levels of the ADP-sensitive and ADP-insensitive phosphoenzymes in Fig. 1, we might expect the forward rate and reverse rate constants for the conversion reaction to have similar values. If this prediction is correct, then most of the phosphoenzyme present after a brief exposure to ATP should be in the ADP-sensitive form. To test this hypothesis, we measured dephosphorylation produced by adding ADP to vesicles that had been phosphorylated with ATP for only 6 msec. The results of this experiment, which are plotted in Fig. 2, show that the concentration ratio of the two forms of the phosphorylated enzyme is very similar to that found in the dephosphorylation experiment carried out at 116 msec. Inorganic phosphate production (not shown) was again equal to about one-third of the amplitude of the slow phase of phosphoenzyme decay despite the presence of ionophore and the short incubation time.

According to mechanism I, the rate of decay of E*P in the reverse direction will depend on the level of saturation of the low affinity sites and on the rate constant for the conversion of E*P to EP. By raising the [Ca^{2+}] during dephosphorylation, the degree of saturation of these sites will increase causing E*P to decompose more rapidly in the direction of EP and P_i production to decrease. When the Ca^{2+} concentration was raised to 5 mM during dephosphorylation after allowing the phosphoenzyme to accumulate for 6 msec at a Ca^{2+} level of ∿10 μM, a dramatic increase in the rate of decay of the slow component was observed (open circles, Fig. 2). Assuming that Ca^{2+} and ADP form a stoichiometric complex, the Ca^{2+} available to the pump sites following the jump was slightly greater than 3.5 mM. In addition, P_i production associated with the decay of the slow component was virtually eliminated. From this we conclude that the slow phase of dephosphorylation at 6 msec is due to the Ca^{2+}-depleted form of the ADP-insensitive phosphoenzyme (E*P) and not to EP^{Ca}_{ADP}, since the latter will not show an accelerated rate of decomposition in response to raising the Ca^{2+} concentration.

The Ca^{2+} site involved in activating the dephosphorylation of E*P was further characterized by measuring the extent of activation at different Ca^{2+} concentrations (inset, Fig. 2). Phosphoenzyme was allowed to accumulate for 6 msec at ∿10 μM free Ca^{2+}; then ADP was added together with additional Ca^{2+} to activate the conversion of E*P to EP. In order to increase the accessibility of the transport sites to Ca^{2+}, the vesicles were treated with A23187 beforehand. As seen in the figure, the amount of phosphoenzyme remaining after 6 msec decreased

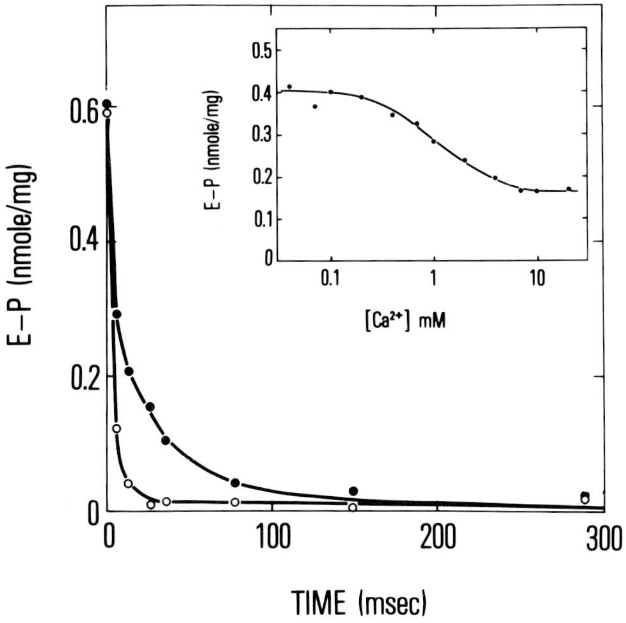

Fig. 2. Phosphoenzyme decomposition resulting from the addition of ADP or ADP + Ca^{2+} (Ca^{2+} jump) after 6 msec of phosphorylation. Sarcoplasmic reticulum vesicles (0.5 mg/ml) were mixed with A23187 in 95% ethanol (40 µg ionophore/mg SR protein) and allowed to incubate for 10 minutes. The ionophore-treated vesicles were phosphorylated according to the conditions described in the legend to Fig. 1 for 6 msec after which 1.33 mM ADP (●) or 1.33 mM ADP + 5 mM $CaCl_2$ (o) (concentrations after mixing) were added to initiate dephosphorylation. The residual phosphoenyzme, which accounts for about 15% of the total phosphoenzyme, was subtracted from these data. Inset: phosphoenzyme remaining after 6 msec of dephosphorylation with ADP + Ca^{2+} *vs* [Ca^{2+}]. The enzyme was pretreated with ionophore and phosphorylated as described above except that the Ca^{2+} concentration was 20 µM and unbuffered. After 6 msec, 1.33 mM ADP and sufficient Ca^{2+} to bring the total concentration to the indicated level was added and the reaction allowed to proceed for an additional 6 msec before the addition of acid.

with increasing [Ca^{2+}] such that between 1 and 2 mM Ca^{2+} the extent of activation of dephosphorylation was half-maximal. Assuming that the decline in the phosphoenzyme level is proportional to the initial velocity of E*P decomposition, then the maximum rate of the back transition occurs between 8 and 10 mM Ca^{2+}. The Hill coefficient for the dependence of the E*P level on the Ca^{2+} concentration was 1.73.

The apparent affinity and stoichiometry of the Ca^{2+} sites involved in activating dephosphorylation of the ADP-insensitive phosphoenzyme strongly suggest that these sites are the transport sites. According to the generally accepted model for Ca^{2+} transport (Mechanism I), the transport sites should face the interior of the vesicle when the carrier is phosphorylated and return to the external surface after dephosphorylation. If the system obeys this sequence of events, then leaving ionophore out of the reaction mixture should prevent the activation of dephosphorylation of E*P produced by a Ca^{2+} jump. When freshly prepared vesicles exhibiting high levels of transport activity were phosphorylated for 6 msec and then dephosphorylated in the presence of 5 mM (total) Ca^{2+}, the amount of phosphoenzyme remaining after 13 msec was .129 $\pm$ .015 nmole/mg protein as compared to .267 $\pm$ .024 nmole/mg for an experiment in which only ADP was added during the second stage of mixing. The corresponding values for ionophore-treated vesicles were .088 $\pm$ .007 nmole/mg and .233 $\pm$ .003 nmole/mg, respectively. These results indicate that a high percentage of the low affinity (transport) sites on the ADP-insensitive phosphoenzyme have an external orientation and that consequently, the inside-to-outside carrier transition takes place prior to dephosphorylation.

The results of the experiment shown in Fig. 2 suggest that formation of the ADP-insensitive phosphoenzyme from the ADP-sensitive precursor takes place very rapidly during the initial turnover of the enzyme. Since it was unnecessary to add the ionophore to generate a response to the Ca^{2+} jump, it is apparent that the conformational transitions involved in flipping the carrier back and forth across the membrane must also be very rapid. If ATP binding and phosphorylation are treated as a single irreversible reaction with a rate of 65 s^{-1} (k_1[ATP] $\simeq$ 100 s^{-1}; k_2 = 180 s^{-1}) and if the conversion reaction and the two carrier transitions are also combined into a single step, viz.

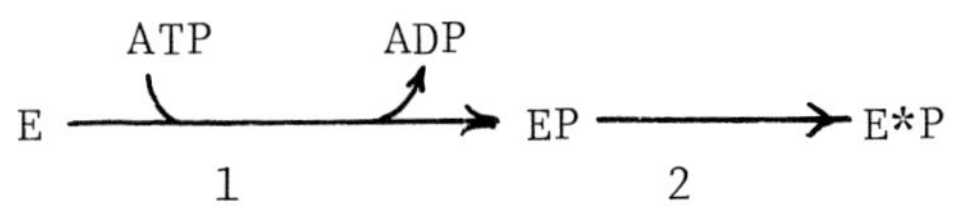

then the rate of formation of the ADP-insensitive phosphoenzyme can be estimated from the time-dependent solution for the above scheme and the ratio of EP to E*P present after 6 msec of phosphorylation. The time-dependent equation for EP is (25):

$$EP(t) = \frac{E_o k_1}{k_2 - k_1} \left(e^{-k_1 t} - e^{-k_2 t}\right)$$

while the equation for E*P is:

$$E^*P(t) = E_o + \frac{E_o}{k_1 - k_2} \left(k_2 e^{-k_1 t} - k_1 e^{-k_2 t}\right)$$

It can be shown that for $k_1 = 65\ s^{-1}$ and E*P/EP = 0.8 at t = .006 sec, k_2 equals 295 s^{-1}. Since the rate of reversal of the second step is only about 20 s^{-1} in the absence of a Ca^{2+} jump, the concentratration ratio of the two phosphoenzymes in the steady state should strongly favor the ADP-insensitive form. This is in contrast to the observed ratio which is close to 1:1 (Fig. 1).

The level of accumulation of the ADP-insensitive phosphoenzyme was determined in a series of double mixing experiments carried out between 6 and 116 msec. The results of these experiments, summarized in Table I, reveal that the amount of phosphoenzyme present as E*P remains fairly close to 50% throughout the entire time course of phosphorylation. When the fraction of E*P present at time t was multiplied by the corresponding total phosphoenzyme level measured in a separate single addition experiment (closed triangles, Fig. 3), the resulting time course (dashed line, Fig. 3) gave a reasonably good fit to a single exponential with a rate of 42 s^{-1}. The similarity between this rate and the apparent rate of phosphorylation is consistent with the view that conversion of the ADP-sensitive phosphoenzyme to the ADP-insensitive form takes place very rapidly after the addition of ATP.

A question raised by the less-than-stoichiometric P_i release in Fig. 1 is whether during translocation Ca^{2+} is being trapped in an ionophore inaccessible compartment prior to its release inside of the vesicle. In the simplest case, the translocation sequence will include the following steps:

$$EP(Ca_2)_o \rightleftharpoons EP(Ca_2)_i \overset{2Ca_i^{2+}}{\rightleftharpoons} E^*P_i \rightleftharpoons E^*P_o$$

where the subscripts i and o refer to carrier states in which the Ca^{2+} transport sites face the inner and the outer surface of the membrane, respectively. A simple prediction based on

TABLE I. Time dependence of the ADP-insensitive phosphoenzyme.

The conditions for phosphorylation of the enzyme were identical to those given in the legend to Fig. 1. At time t (column 1), 1.33 mM ADP (final concentration) was added to dephosphorylate the enzyme. The level of ADP-insensitive phosphoenzyme (E*P), obtained by extrapolation of the slow phase of dephosphorylation to the time at which ADP was added, is reported as a fraction of the total phosphoenzyme in column 2. The total phosphoenzyme (EP_{Total}), measured at time t in a single addition experiment (Fig. 3), is given is column 3. The expected level of E*P at time t for the experiment shown in Fig. 3 (column 4) was calculated from the product of the fractional level of E*P (column 1) and the total phosphoenzyme level (column 3).

Time msec	$E{*}P/EP_{Total}$	EP_{Total} nmole/mg	Predicted E*P nmole/mg
6	.49	1.0	.49
10	.50	1.5	.75
20	.35	2.4	.84
25	.34	2.7	.92
43	.41	3.1	1.27
59	.48	3.0	1.44
116	.58	2.60	1.51

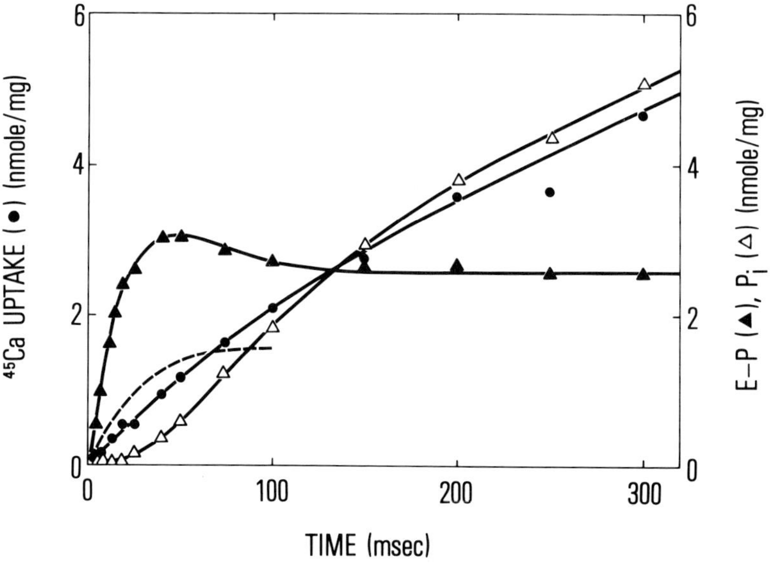

Fig. 3. Time courses of phosphoenzyme formation, P_i liberation and Ca^{2+} accumulation in sarcoplasmic reticulum vesicles. Conditions identical to those described in the legend to Fig. 1 were used for determining phosphoenzyme formation (▲) and P_i release (△). In a similar experiment using the same preparation, ATP-dependent intravesicular $^{45}Ca^{2+}$ accumulation (●) was determined by mixing the vesicles with a solution containing 10 μM unlabelled ATP and 0.1 mM [^{45}Ca]$CaCl_2$ and allowing them to incubate for the indicated times prior to adding 2 mM EGTA+1 mM ADP (final concentrations) to terminate the reaction. The quenched reaction mixture was immediately passed through a Millipore filter (0.45 μ) and washed with a solution containing 1 mM EGTA, 100 mM KCl, 3 mM $MgCl_2$, and 20 mM Tris maleate, pH 6.8. The dashed line represents the approximate time course of formation of the Ca^{2+}-depleted, ADP-insensitive phosphoenzyme obtained from the data in Table I.

this scheme is that Ca^{2+} will begin to accumulate on the inside of the vesicle before the unoccupied transport site appears at the external surface. Since the ADP-insensitive phosphoenzyme becomes sensitive to a Ca^{2+} jump almost as soon as it is formed (Fig. 2), we shall assume that the final step in the sequence (step 3) is very rapid and that essentially all of the ADP-insensitive phosphoenzyme is in the E^*P_o state. This assumption allows us to estimate the concentration of unoccupied transport sites at the external surface from the level of the ADP-sensitive phosphoenzyme reported in Table I.

As noted previously, the apparent rate of accumulation of the ADP-insensitive phosphoenzyme at 10 μM ATP is 42 s^{-1}. If the system behaves according to the proposed mechanism, then the initial rate of intravesicular Ca^{2+} accumulation should be comparable to or faster than the rate of formation of the ADP-insensitive phosphoenzyme. To test this hypothesis, we measured intravesicular $^{45}Ca^{2+}$ accumulation using the technique of quenching with EGTA + ADP (17) and compared it to the time course of E*P formation computed from the values in Table I. The amount of Ca^{2+} sequestered by the system will consist of two fractions, one of which is bound to the membrane in association with EP and E*P and the other which is present inside of the vesicle. Because we were only interested in the latter fraction, ADP was included in the quench to remove Ca^{2+} bound to EP by reversal of phosphorylation. Calcium that is bound to E*P, on the other hand, will either be returned to the external medium as consequence of reversal of the phosphoenzyme conversion and phosphorylation reactions or it will be pumped to the inside of the vesicle coupled to the hydrolysis of E*P. Thus, the measured amount of intravesicular $^{45}Ca^{2+}$ will equal that which was present at the time the quench was added plus whatever accumulates after the addition of the quench. This is plotted in Fig. 3 (closed circles) together with the time courses of phosphoenzyme formation (closed triangles) and P_i release (open triangles) measured in a similar experiment in which acid was used to terminate the reaction. As seen in the figure, $^{45}Ca^{2+}$ accumulation is biphasic and is delayed with respect to the time course of the ADP-insensitive phosphoenzyme (dashed line). The initial rate of $^{45}Ca^{2+}$ accumulation, obtained by subtracting the steady phase of the reaction from the initial time course and plotting the difference semilogarithmically, was 15 s^{-1}. This behavior does not conform to the sequence of events predicted by the simple translocation model.

IV. DISCUSSION

The results of the present study can be summarized as

follows:

(1) Decomposition of the phosphoenzyme resulting from the addition of ADP can be resolved into rapid and slow phases corresponding to the ADP-sensitive (EP) and ADP insensitive (E*P) phosphoenzymes, respectively.
(2) The distribution of ADP-sensitive and ADP-insensitive phosphoenzymes is approximately 1:1 in the steady state. This ratio is rapidly achieved following phosphorylation with ATP.
(3) Decomposition of the ADP-insensitive phosphoenzyme liberates a less-than-stoichiometric amount of P_i. This discrepancy is observed in ionophore-treated vesicles at short reaction times and becomes more pronounced when the external Ca^{2+} concentration is raised during dephosphorylation.
(4) Simultaneous addition of ADP and high (millimolar) Ca^{2+} activates the breakdown of the ADP-insensitive phosphoenzyme in the direction of the ADP-sensitive form. The site involved in this effect faces the extravesicular surface, has a low affinity for Ca^{2+} ($K_{0.5} \simeq$ 1-2 mM), and has a Hill coefficient greater than 1 ($n \cong 1.73$).
(5) Accumulation of intravesicular $^{45}Ca^{2+}$ in the presteady state occurs at a slower rate than the formation of the Ca^{2+} jump-sensitive, ADP-insensitive phosphoenzyme.

A necessary condition for activation of reversal of the phosphoenzyme conversion reaction is that the low affinity sites on the ADP-insensitive phosphoenzyme bind Ca^{2+}. This will occur when the Ca^{2+} level reaches the millimolar concentration range either as a result of transport or due to an artificial elevation in the Ca^{2+} concentration (Ca^{2+} jump). At much lower Ca^{2+} levels (*i.e.* micromolar), breakdown of the ADP-insensitive phosphoenzyme is expected to occur predominantly in the forward direction resulting in stoichiometric levels of P_i production. The fact that we observed less-than-stoichiometric P_i release following the addition of ADP under conditions favoring desaturation of the low affinity sites suggests that during translocation Ca^{2+} may be temporarily detained in an internal cavity or a compartment that is distinct from the intravesicular space. The level of Ca^{2+} present within this compartment during transport will presumably be higher than the level found in the bulk medium surrounding the membrane when the ionophore is present. This would enable Ca^{2+} to rebind to the phosphorylated carrier following re-exposure of the transport sites to the internal cavity during ADP-activated dephoshorylation. After releasing Ca^{2+}, the translocator must return to the external membrane surface where it undergoes a conformation at change increasing its affinity for Ca^{2+}. Since breakdown of the ADP-insensitive phosphoenzyme can be accelerated by a Ca^{2+} jump in the absence of ionophore, it follows that re-orientation of the carrier from the cavity to the external surface must take place prior to dephosphorylation. A mechanism depicting these events is shown in Fig. 4.

Additional support for the internal cavity model comes from the observation that accumulation of intravesicular $^{45}Ca^{2+}$ measured by the ADP + EGTA quench technique appears to be delayed with respect to formation of the Ca^{2+} jump-sensitive, ADP-insensitive phosphoenzyme (Fig. 3). Had Ca^{2+} bound to the ADP-insensitive phosphoenzyme been released directly to the intravesicular compartment upon dissociation from the enzyme, then the time course of E*P formation would either have coincided with or been delayed with respect to the initial phase of intravesicular Ca^{2+} accumulation. Comparison of the initial behavior of these reactions is complicated by the possibility that some additional $^{45}Ca^{2+}$ flux may occur after the addition of the quench thereby increasing the observed level of intravesicular $^{45}Ca^{2+}$. In the case where the membrane contains Ca^{2+} trapped within an internal cavity as well as Ca^{2+} bound to EP, addition of the quench will cause rapid release of the latter while Ca^{2+} remaining in the cavity will either be slowly released to the external surface by reversal of steps 1-4 (Mechanism I) or enter the lumen coupled to the production of P_i. When this occurs, the measured level of intravesicular $^{45}Ca^{2+}$ will be greater than that which was present at the instant the quench was added. By knowing how much P_i is released during dephosphorylation and the pump stoichiometry (∿1Ca/ATP at 10 μM ATP), it is possible to estimate how much Ca^{2+} was originally trapped inside the vesicle. This type of calculation yields a result which suggests that dephosphorylation of the carrier either triggers or immediately follows the release of internal Ca^{2+}, in agreement with an earlier observation by DuPont (26). It is apparent that more direct information about the time course of intravesicular Ca^{2+} accumulation is needed before this issue can be fully resolved.

Although the scheme in Fig. 4 depicts Ca^{2+} release coupled to dephosphorylation of the carrier, there is evidence which suggests that a conformational change preceding hydrolytic cleavage of E*P is responsible for triggering internal Ca^{2+} release. For example, Watanabe *et al.* (27) have shown that when sarcoplasmic reticulum vesicles are treated with dimethyl sulfoxide which inhibits E*P hydrolysis, Ca^{2+} dissociates from the enzyme before P_i is released. Recent quench-flow studies of E*P hydrolysis (28) indicate that the breakdown rate is

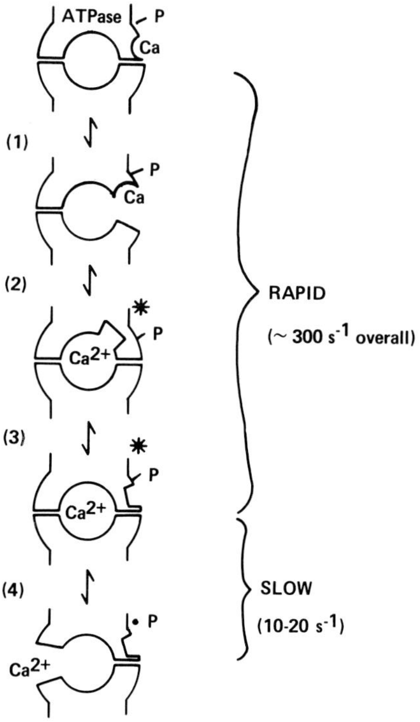

Fig. 4. Internal cavity model for active Ca^{2+} transport. The carrier is in a bilayer whose horizontal plane vertically bisects the figure. The compartment, which lies between the inner and outer surfaces of the membrane, is surrounded by the protein of the ATPase. The steps of the translocation sequence are as follows: (1) translocation of Ca^{2+} bound to the ADP-sensitive phosphoenzyme from the external membrane surface to the internal cavity; (2) conversion of the ADP-sensitive phosphoenzyme to the ADP-insensitive phosphoenzyme (denoted by asterisk) and Ca^{2+} dissociation; (3) reorientation of the low affinity site from the internal cavity to the external surface; (4) dephosphorylation and Ca^{2+} release on the inside of the vesicle.

about 60 s^{-1}, which is about 4 times faster than the initial rate of P_i liberation in Fig. 3. These results favor a mechanism in which a step preceding dephosphorylation of E*P is rate-limiting in the translocation sequence.

From the estimated forward and reverse rate constants for the phosphoenzyme conversion step, the ratio of EP to E*P should be less than 0.1 in the steady state. By contrast, the observed ratio was close to 0.5 and remained fairly constant throughout the entire time course of phosphorylation. The apparent stability of the observed ratio cannot be attributed to a rapid equilibrium between EP and E*P or to a similarity in the rate constants governing the formation and breakdown of E*P since either situation would lead tc a much faster rate of E*P decomposition than was actually measured. This raises the possibility that some constraint in the system prevents EP and E*P from achieving their expected distribution after the first partial turnover has been completed. This constraint may arise from an interaction between subunits which are coupled by a mechanism which prevents them from simultaneously existing in the ADP-sensitive or ADP-insensitive phosphorylated states. A mechanism which exhibits this behavior was recently proposed by Ikemoto and coworkers (29,30) based on observations of the phosphorylation reaction in the transient state. They obtained evidence for two distinct classes of Ca^{2+} binding sites which they assumed were present on separate subunits of a dimer. According to their scheme, one subunit of each pair is one step ahead of its neighbor in completing the reactions of the translocation cycle. If the system obeys this type of coupling relationship, it will produce a doubly phosphorylated state in which one half of the dimer is ADP-sensitive and the other half ADP-insensitive (Fig. 5). In order for this mixed state to accumulate, it is necessary that the hydrolysis of E*P to E*·P_i take place much more slowly than the reactions preceeding it. This will enable Ca^{2+} bound to the left-hand subunit in Fig. 5 to be rapidly transferred from the external surface to the internal cavity where it is detained until dephosphorylation or a conformational change occurs and triggers its release to the inside. Calcium entering the cavity from the right-hand subunit will remain bound to the protein until the Ca^{2+} that has entered initially leaves the cavity. Since the combined rates for steps 6 and 7 is smaller than the rate for step 5, Ca^{2+} entering from the right-hand subunit will exit more slowly than Ca^{2+} entering initially. This could account for the biphasic appearance of the time course of intravesicular $^{45}Ca^{2+}$ accumulation in Fig. 3.

Some additional assumptions about the coupling relationship are necessary to explain the observed pattern of

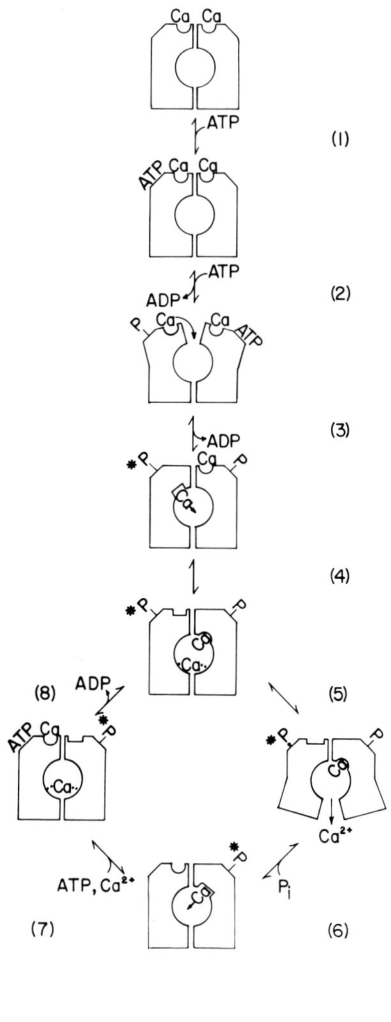

Fig. 5. ATPase dimer Ca^{2+} translocation mechanism. The transport system consists of two 10^5 dalton subunits embedded in the bilayer with an internal compartment located in the plane between the subunits. The catalytic site coupling mechanism allows the left-hand subunit to complete the reaction cycle one step ahead of the right-hand subunit (29-31). The left-hand and right-hand subunits switch positions at reaction 8, so that the reaction sequence completed by the dimer during the initial cycle of Ca^{2+} transport is 1234567856. The ADP-insensitive phosphoenzyme is denoted by an asterisk.

dephosphorylation following the addition of ADP. Unless the ADP-sensitive and ADP-insensitive phosphorylated subunits uncouple and dephosphorylate independently, both will transfer bound phosphate to ADP to form ATP. This would lead to a situation in which no P_i is liberated during the slow phase of dephosphorylation. The fact that a significant amount of P_i is released during dephosphorylation by ADP suggests that the coupling mechanism may vary depending on which ligands are present in the reaction medium. It is also possible that the degree of coupling varies so that some steps are more tightly coupled than others. Examination of the properties of the dephosphorylation reaction in preparations treated with detergents that produce active monomers (27,28,32) may be helpful in identifying the steps in the reaction mechanism that are strongly affected by the presence of subunit interactions.

REFERENCES

1. Tada, M., Yamamoto, T., and Tonomura, Y., Physiol. Rev. 58:1-79 (1978).
2. de Meis, L. in "Transport in the Life Sciences" (E. Bittar, ed.), Vol II, pp. 1-163, John Wiley and Sons, New York, 1981.
3. Bastide, F., Meissner, G., Fleischer, S., and Post, R.L., J. Biol. Chem. 248:8385-8391 (1973).
4. Kanazawa, T., Yamada, S., Yamamoto, T., and Tonomura, Y., J. Biochem. (Tokyo) 70:95-123 (1971).
5. Shigekawa, M., Dougherty, J.P., and Katz, A.M., J. Biol. Chem. 253:1442-1450 (1978).
6. Inesi, G., Maring, E., Murphy, A.J., and McFarland, B.H. Arch. Biochem. Biophys. 138:285-294 (1970).
7. Duggan, P. F., J. Biol. Chem. 252:1620-1627 (1977).
8. Yamada, S., and Tonomura, Y., J. Biochem. (Tokyo) 71:417-425 (1972).
9. Carvahlo, M.G.C., Souza, D.O., and de Meis, L., J. Biol. Chem. 251:3629-3636 (1976).
10. Sumida, M., Wang, T., Mandel, F., Froehlich, J. P., and Schwartz, A., J. Biol. Chem. 253:8722-8777 (1978).
11. Kanazawa, T., and Boyer, P. D., J. Biol. Chem. 248:3163-3172 (1973).
12. Kanazawa, T., J. Biol. Chem. 250:113-119 (1975).
13. Knowles, A. A., and Racker, E., J. Biol. Chem. 250:3538-3544 (1975).
14. Ikemoto, N., J. Biol. Chem. 251:7275-7277 (1976).
15. Yamada, S., and Ikemoto, N., J. Biol. Chem. 255:3108-3119 (1980).

16. Pickart, C. M. and Jencks, W. P., J. Biol. Chem. 257: 5319-5322 (1982).
17. Sumida, M., and Tonomura, Y., J. Biochem. (Tokyo) 75:283-297 (1974).
18. Froehlich, J. P., and Taylor, E. W., J. Biol. Chem. 250: 2013-2021 (1975).
19. Froehlich, J. P., Sullivan, J. V., and Berger, R. L., Anal. Biochem. 73:331-341 (1976).
20. Sumida, M., Wang, T., Schwartz, A., Younkin, C., and Froehlich, J. P., J. Biol. Chem. 255:1497-1503 (1980).
21. Shigekawa, M., Dougherty, J.P., and Katz, A.M., J. Biol. Chem. 253:1442-1450 (1978).
22. Shigekawa, M., and Dougherty, J. P., J. Biol. Chem. 253: 1458-1464 (1978).
23. Shigekawa, M., and Akowitz, A. A., J. Biol. Chem. 254: 4726-4730 (1979).
24. Duggan, P.F., and Martonosi, A., J. Gen. Physiol. 56:147-167 (1970).
25. Frost, A. A., and Pearson, R. G., in "Kinetics and Mechanism" (1st edition), pp. 1-405, John Wiley and Sons, New York, 1956.
26. Dupont, Y., Eur. J. Biochem. 109:231-238 (1980).
27. Watanabe, T., Lewis, D., Nakamoto, R., Kurzmack, M., Fonticelli, C., and Inesi, G., Biochemistry 20:6617-6625 (1981).
28. Inesi, G., Kurzmack, M., Kosk-Kosicka, D., Lewis, D., Scofano, H., and Guimaraes-Motta, H., Z. Naturforsch. 376:685-691 (1982).
29. Ikemoto, N., Garcia, A. M., Kurobe, Y., and Scott, T. L., J. Biol. Chem. 256:8593-8601 (1981).
30. Ikemoto, N., Miyao, A., and Kurobe, Y., J. Biol. Chem. 256:10809-10814 (1981).
31. Froehlich, J. P., and Taylor, E. W., J. Biol. Chem. 251: 2307-2315 (1976).
32. Dean, W. and Tanford, C., Biochemistry 17:1683-1690 (1978).

PRE-STEADY STATE KINETICS OF E~P FORMATION AND DECOMPOSITION BY Ca^{2+},Mg^{2+}-ATPase IN BOVINE AORTA MICROSOMES

Michihiro Sumida[1]
Hiromichi Okuda
Minoru Hamada

Department of Medical Biochemistry
Ehime University
Ehime 791-02 Japan

Hitoshi Takenaka[2]
James M. Watras[2]
J. George Sarmiento[2]

Department of Medicine
University of Connecticut
Farmington, Connecticut U.S.A.

Jeffrey P. Froehlich

Laboratory of Molecular Aging
National Institute of Aging
National Institute of Health
Baltimore, Maryland, U.S.A.

Microsomal fractions were isolated from bovine aorta smooth muscle tissue and the transient-kinetic properties of an acid stable phosphorylated intermediate (E~P) of the

[1]Supported in the part by a grants from the Ministry of Education, Science and Culture of Japan.
[2]Supported in part by NIH grants HL 22135 and HL 21812, and American Heart Association Connecticut Affiliate.

ISBN 0-12-260380-X

Ca^{2+},Mg^{2+}-dependent ATPase of the fraction were studied. Although the steady state amount of E~P formation was less than 2 % and 10 % of that present in rabbit skeletal and dog cardiac muscle, respectively, the rate of E~P formation was fast (its half time = 13 ms) and similar to that of rabbit skeletal and dog cardiac SR under similar experimental conditions (10 μM ATP, 21 °C). The time course of E~P decomposition following the addition of 5mM EGTA was shown to be the first order (k_d=2s^{-1}) and was accelerated by 1mM ATP addition (k_d=51s^{-1}).

I. INTRODUCTION

It has been observed that vascular smooth muscle contains intracellular organelles (1), including sarcoplasmic reticulum (SR) which could regulate the cytosolic Ca^{2+} concentration that activates the contractile apparatus (2,3). Microsomal fractions isolated from the vascular smooth muscle have been shown to take up Ca^{2+} ions coupled to ATP hydrolysis (3,4,5); thus, these fractions have been suggested to be the counterparts of SR found in both skeletal and cardiac muscles. We have recently been able to demonstrate the Ca^{2+},Mg^{2+}-ATPase and its acid-stable phosphorylated intermediate (E~P) in microsomal fractions isolated from bovine aorta smooth muscle tissue (5). The calcium dependency of E~P formation was the same as that of the rabbit skeletal SR, although the amount of the E~P was less than 2% ofthat of the skeletal SR, possibly due to contamination by proteins of low molecular weight (5).

In this paper we further characterized the rates of the E~P formation and decomposition using the rapid acid quench technique and demonstrated presteady state profiles of phosphorylation identical to those found in skeletal and cardiac SR.

II. METHODS

Fresh bovine aorta tissue was chilled in ice-cold saline solution and the endothelium, the subendothelium and the elastic exterior were completely removed. The treated tissue (100g) was minced and homogenized vigorously with an omni-mixer for 2 min in 3 volumes of 0.1 M KCl and 10 mM Bis-tris propane (BTP) at pH 6.8. The homogenate was centrifuged 4,000 xg for 10 min and the resultant supernatant passed through 4 layers of cheesecloth and centrifuged at

15,000 xg for 30 min to pellet the mitochondria. The supernatant was passed through 4 layers of cheesecloth and centrifuged at 100,000 xg for 30 min. The pellet was homogenized in 50 ml of 0.1 M KCl and 10 mM BTP (pH 6.8) and placed on a discontinuous sucrose gradient (15, 30 and 60 % sucrose) containing 10 mM BTP (pH 6.8) and centrifuged in a Beckman SW40 rotor at 29,000 rpm for 90 min. The band present at the 15-30 % sucrose interface was collected and suspended in 10 volumes of 0.6 M KCl and 10 mM BTP (pH 6.8) and centrifuged at 100,000 xg for 30 min. The pellet was homogenized in 5ml of 0.1 M KCl and 10 mM BTP (pH6.8). The average yield of the microsomal protein was approximately 0.5 mg/g tissue (wet weight). The microsomes were frozen in liquid nitrogen and stored at -80.C before use.

The transient state measurements of E~P formation and decomposition were performed using the quench-flow apparatus previously described (Froehlich et al.,(6), Sumida et al.,(7)).

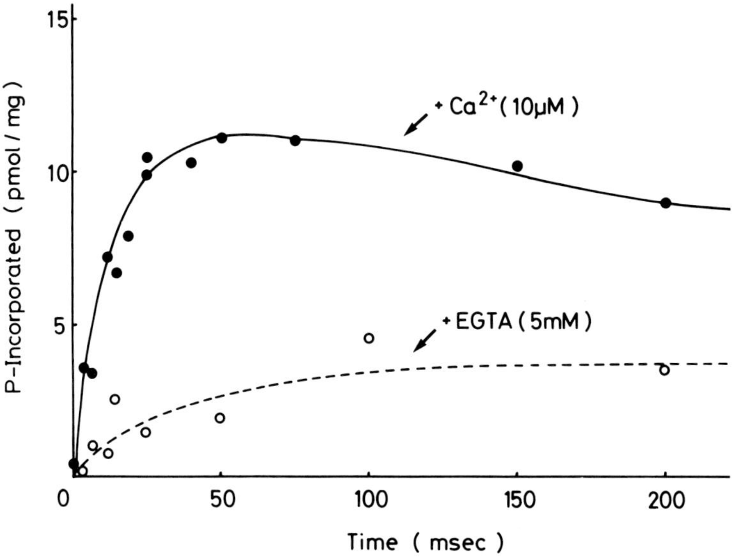

FIGURE 1. Time course of phosphoenzyme formation in aorta smooth muscle microsomes --- The reaction was performed at 21 °C by mixing the microsomes (0.86 mg/ml) with an equal volume of substrate medium containing 20 μM ATP (including $5x10^{13}$ cpm/mol γ-^{32}P ATP). Both media contained 0.1M KCl, 0.5 mM $MgCl_2$, 5 mM NaN_3 20 mM BTP and 0.1mM $CaCl_2$ + 0.1mM EGTA(●) or 5 mM (o) EGTA. The reactions were quenched by the addition of 3% perchloric acid at the indicated times.

III. RESULTS AND DISCUSSION

E~P formation --- The time course of E~P formation in bovine aorta microsomes measured at 21.C in the presence of 10 μM ATP, 0.5 mM $MgCl_2$, and 10 μM (free) Ca^{2+} is shown in Fig.1. The rising phase of E~P formation of the aorta microsomes was much like that found in skeletal and cardiac SR (cf. Fig. 1 in ref. 7) and the half time , $t_{1/2}$, which is defined as the time between the start of the reaction and 50 % of maximal phosphorylation (11 pmol/mg), was $13s^{-1}$. In the presence of 5 mM EGTA, the preparation exhibited Ca^{2+}-independent phosphorylation (dashed line, Fig. 1) which rose to a maximum level (3.5 pmol/mg) equal to about one-third of that observed in the presence of Ca^{2+} In the latter case, the phosphoenzyme declined slightly after 75 msec, similar to the behavior found in skeletal and cardiac SR (7). When the Mg^{2+} concentration was increased from 0.1 mM to 3 mM, the steady state level of the Ca^{2+}-dependent phosphoenzyme rose from 11 to 40 pmol/mg. Raising the Mg^{2+} concentration did not affect the level of the Ca2+-independent phosphoenzyme.

Inorganic phosphate liberation during the initial 200 msec appeared to increase linearly although scatter in the data precluded an accurate determination of the rate. The velocity of P_i liberation was estimated to be about 60 pmol/mg/sec.

E~P-decomposition --- As shown in Fig. 2, E~P decomposition was observed when 5 mM EGTA was added to microsomes that had been phosphorylated for 116 msec in a medium containing 10 μM ATP, 10 μM free Ca^{2+}, and 3 mM Mg^{2+}. Compared to skeletal and cardiac SR, the initial rate of E~P decomposition was slower (k_d - 2 s^{-1} vs 10 s^{-1}), in agreement with the slow decay of the E~P overshoot seen in Fig. 1. Addition of 1 mM unlabelled ATP together with EGTA produced a marked acceleration of the initial phase of decay, suggesting that the system may contain a regulatory site for ATP. Under these conditions, the time course of E~P decay could be resolved into rapid (k_d = 51 s^{-1}) and slow (k_d = 6 s^{-1}) phases. Since P_i production occurring during dephosphorylation was not measured in these experiments, the possibility that contaminating ADP was responsible for the acceleration of E~P decomposition could not be excluded.

Reaction Mechanism --- The present results indicate that smooth muscle aortic microsomes catalyze a Ca^{2+}-dependent rate of phosphorylation that is indistinguishable from that found in skeletal and cardiac SR microsomes under comparable conditions. In skeletal SR, enzyme phosphorylation was shown to be an obligatory reaction by demonstration of the presence of a lag in P_i production during the initial rapid phase of

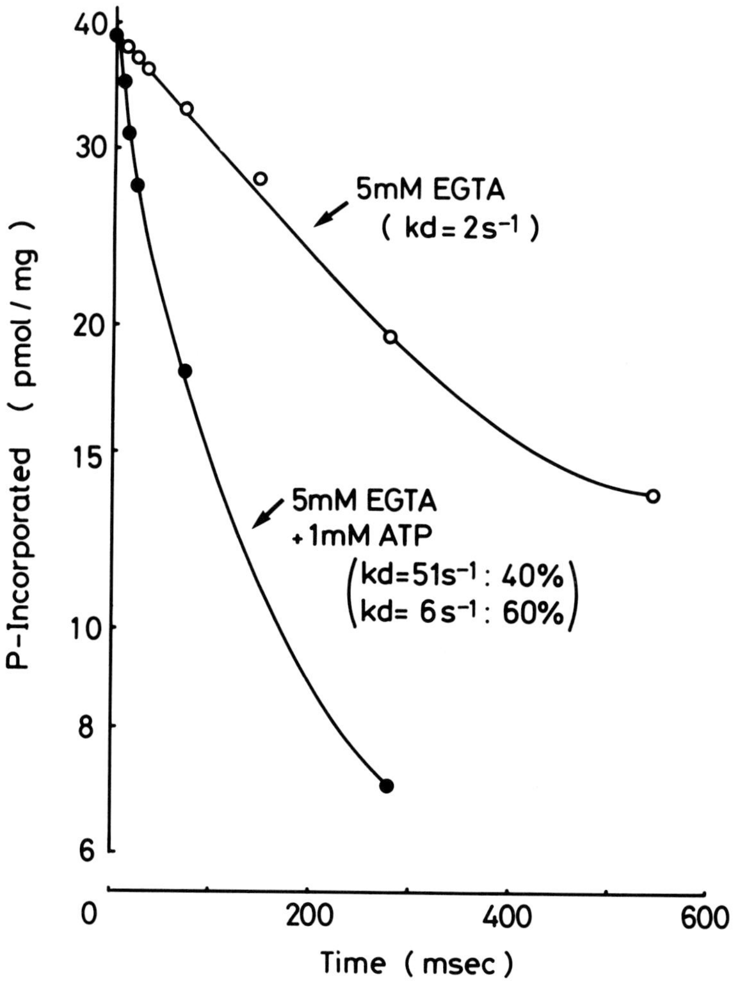

FIGURE 2. Time course of phosphoenzyme decomposition in aorta smooth muscle microsomes --- The reaction condition used to form E~P was the same as that in Fig. 1, except 3mM Mg^{2+} was added. After 116 ms, EGTA was added to a final concentration of 5 mM in the absence (o) or presence (•) of 1 mM ATP and the reaction was terminated by the addition of perchloric acid to a final concentration of 2.25 % after an additional variable incubation period (4 to 542 ms).

phosphoenzyme accumulation. The present results do not allow such a distinction to be made in smooth muscle microsomes, however , the similarity in the rates and Ca^{2+}-dependency (5) of E~P formation and patterns of E~P decomposition in skeletal and smooth muscle microsomes suggest that they have similar Ca^{2+}-dependent ATPase reaction mechanisms.

The observed lower level of phosphorylation in aorta microsomes (Fig. 1) compared to those of skeletal and cardiac SR might be due to a contamination by other microsomal fractions, or by other proteins of lower molecular weight(5). Consistent with the lower level of phosphorylation, ATP-dependent Ca^{2+} uptake observed in the same aorta microsomal preparation exhibited a low level of activity (5). However, its dependency on free calcium concentration was the same as that of skeletal SR (4,5) indicating that the range of myoplasmic free Ca^{2+} over which the skeletal and smooth muscle systems operate in the contraction-relaxation cycle are similar. The ATPase might be purified further by means of an affinity column chromatography using calmodulin (9), which has been shown to activate the Ca^{2+},Mg^{2+}-ATPase of microsomes obtained from smooth muscle (10) and cardiac SR (11). Characterization of the phosphorylated intermediate of the purified Ca^{2+},Mg^{2+} -ATPase may supply useful information for elucidating the mechanism of Ca^{2+} regulation by the SR Ca^{2+} pump in smooth muscle.

ACKNOWLEDGMENT

We greatly appreciate the expertise of Dr. Arnold M. Katz of the University of Connecticut Health Center for his helpful discussions and comments regarding this manuscript.

REFERENCES

1. Gabella,G. (1980) "Smooth muscle: an assesment of current knowledge" Edward Arnold Ltd.(London) 1-45.
2. Karaki,H.,Kubota,H.,and Orakawa,N. (1979) European J. Pharmacol. 56, 237-245.
3. Grover,A.K.,Kwan,C.Y.,and Daniel,E.E. (1982) Am. J. Physiol. 242, C278-C282.
4. Wuytack,F.,and Casteels,R. (1980) Biochim. Biophys. Acta 595, 257-263.
5. Sumida,M.,Okuda,H.,Hamada,M.,Inui,M.,and.Tada,M. (1982) manuscript in preparation.
6. Froehlich,J.P.,Sullivan,J.V.,and Berger,R.L. (1976) Anal.

Biochem. 73, 331-341.
7.Sumida,M.,Wang,T.,Mandel,F.,Froehlich,J.P.,and Schwartz,A. (1978) J. Biol. Chem. 253, 8772-8777.
8.Sumida,M.,Wang,T.,Schwartz,A.,Younkin,C.,and Froehlich,J.P. (1980) J. Biol. Chem. 255, 1497-1503.
9.Wuytack,F., Schutter,G.D.,and Casteels,R. (1981) FEBS Letters 129, 297-300.
10.Morel,M.,Wibo,M.,and Godfraind,T. (1981) Biochim. Biophys. Acta 644,82-88.
11.Tada,M.,and Katz,A.M. (1982) Ann. Rev. Physiol. 46, 401-423.

THEORETICAL BASIS FOR THE SEQUENCE ANALYSIS OF THE REACTION STEPS INVOLVED IN THE TRANSPHOSPHORYLATION PROCESS OF THE SARCOPLASMIC CA^{++}-TRANSPORT ATPASE

Madoka Makinose
Werner Boll

Department of Physiology
Max-Planck-Institute for Medical Research
Heidelberg, Federal Republic of Germany

I. INTRODUCTION

The kinetic analysis of the sarcoplasmic Ca^{2+}-transport enzyme was started in 1961 by the discrimination between basic and extra ATP-splitting (7). Later the rate of calcium uptake (8,9), ATP-ADP-exchange (8,12), the level of phosphorylated enzyme (12,20), and the rate of backward reaction (1,13,14,15) were analysed in steady and transient phases of the enzyme activity. On the basis of these data a number of discrete reaction steps for the overall transport process was postulated (4,6,19). The proposed reaction schemes, however, are unsatisfactory in several ways: In general they a priori assume a given reaction mechanism, e.g. classical Michaelis-Menten kinetics. Moreover, the activating factors, calcium and magnesium ions, are not accounted for appropriately. Finally the actual molecular species of the substrate ATP (and ADP in the reverse reaction) is not convincingly defined.

ISBN 0-12-260380-X

In order to overcome these short-comings the present authors use the "elimination principle" (16-18) for their analysis: All possible reaction mechanisms of the Ca^{2+}-transporting process which are consistent with the essential activating conditions (substrates, Ca^{++} and Mg^{++}) will be listed first. Then the mechanisms are eliminated one by one from the list using experimentally determined kinetic patterns and theoretical considerations, until only one, presumably the physiologically relevant mechanism remains.

II. Score of the Analysis

The splitting of ATP by the calcium-transporting ATPase of the sarcoplasmic reticulum occurs in two steps: the first one is the phosphorylation of the enzyme by ATP, and the second step is the hydrolysis of the phosphoenzyme. Both steps are fully reversible (13,14).

The actual enzymatic activity of the calcium transporting enzyme on which our analyses are based is the ATP-ADP-exchange reaction in the steady state. As detailed in refs.8 and 12, this activity was measured as the incorporation rate of radioactive ADP into the ATP fraction under the influence of the enzyme.

The mathematical treatment of the experimental results which has been published only in preliminary form (2,3,16,17) is presented here for the first time in full detail. This analysis enables us to identify the nature of partial reactions and their sequence in the calcium-transporting process.

III. Abbreviations

All abbreviations for the ligands, substrates, inhibitors and enzyme forms used in the equations represent both their species and their concentrations.

A	"Activator", general symbol for the activating cations
α, β	Coefficient of specific radioactivity of the enzyme species
ADP	Adenosine diphosphate as free anion
ADP_t	Sum of ADP and MgADP
ATP	Adenosine triphosphate as free anion
ATP_t	Sum of ATP and Mg ATP
$Ca^{++} = Ca_o^{++}$	Extravesicular calcium ions

Ca_i^{++}	Intravesicular calcium ions
D	ADP or MgADP
E	Non activated form of the enzyme
E^*	Mg-activated enzyme
E^o	Ca-activated enzyme
E_t	Sum of all enzyme species
EA	Activated enzyme form(s) in general
E~P	Phosphorylated enzyme
$E^D_{\sim P}$	Phosphoenzyme-ADP complex
ET	Enzyme-ATP complex
F	A factor relating to α and/or β
F_A	Factor F involving the concentration of activator A
k_1-k_{14}	Rate constants
K_A-K_E	Equilibrium constants
$K_t(K_d)$	Dissociation constant of the MgATP complex (MgADP complex)
LBp	Lineweaver-Burk-plot
m	Slope of the plotted lines
Mg^{++}	Extravesicular magnesium ions
MgS	General form of Mg-substrate complex
S	General form of substrate as free anion
SR	Sarcoplasmic reticulum
T	ATP or MgATP
V_A	Rate of ATP-ADP exchange
V_{max}	maximal value of V_A

IV. Role of Magnesium Ions

The assays for the measurement of ATP-ADP exchange contain ATP and ADP as the substrates and magnesium and calcium as divalent cations (2,3,16,17). The binding of two Ca^{++} to the transport enzyme was accepted as the essential activation step by many authors (6, 4, 19). However, the direct effect of Mg^{++} on the enzyme has scarcely been discussed. Moreover the question of the true substrate ATP or MgATP in the forward reaction and ADP or MgADP in the back reaction has been neglected.

In the following section we deal with the binding steps of Mg^{++} and substrates to the transport enzyme. All possible a priori reaction schemes for the ATP-ADP exchange activity will be listed up and the equations representing the exchange rate for each scheme will be derived theoretically. In these equations, the molecular species of the substrates are still open; they are represented by T (ATP or MgATP) or D (ADP or MgADP). A means Mg^{++} in the schemes of this chapter.

The experiments on which this analysis (2, 3, 16) is based were carried out in the presence of $10\,\mu M\ Ca_o^{++}$ in order to saturate the Ca^{++}-binding sites of the enzyme. Calcium precipitating agents such as orthophosphate or oxalate were not used.

Schemc 1

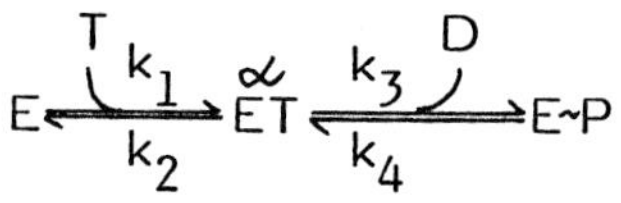

$$V_A^{-1} = \left(\frac{k_2}{k_1\ T} + 1 + \frac{k_3}{k_4\ D}\right) \times F$$

α is the coefficient of the specific radioactivity of the enzyme species (ET) assuming the specific activity of D equals one. In this case

$$\alpha = \frac{k_4 E{\sim}P\ D}{(k_1 E \cdot T) + (k_4 E{\sim}P\ D)} = \frac{k_3 ET}{k_2 ET + k_3 ET} = \frac{k_3}{k_2 + k_3}$$

in the steady state.

$$F = \frac{1}{\alpha k_2 E_t} = \frac{k_2 + k_3}{k_2 k_3} \times \frac{1}{E_t}$$

Scheme 2

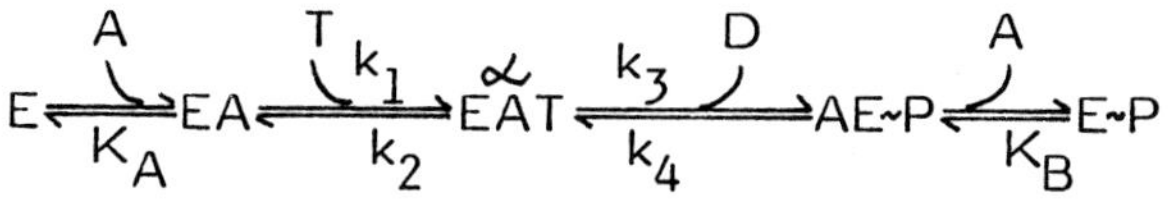

$$V_A^{-1} = \left(\frac{K_A k_2}{k_1 \; T A} + \frac{k_2}{k_1 \; T} + 1 + \frac{k_3}{k_4 \; D} + \frac{k_3 K_B}{k_4 \; D A}\right) \times F$$

$$F = \frac{1}{\alpha k_2 \; E_t} = \frac{k_2 + k_3}{k_2 k_3 \; E_t}$$

Scheme 3

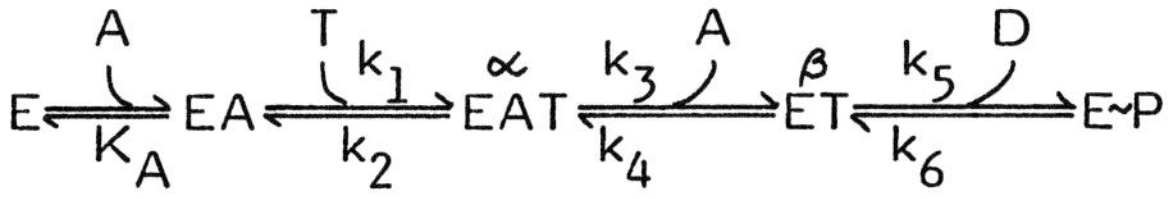

$$V_A^{-1} = \left(\frac{K_A k_2}{k_1 \; T A} + \frac{k_2}{k_1 \; T} + 1 + \frac{k_3}{k_4 \; A} + \frac{k_3 k_5}{k_4 k_6 \; D A}\right) \times F_A$$

$$F_A = \frac{1}{\alpha k_2 \; E_t} = \frac{k_2 k_4 A + k_2 k_5 + k_3 k_5}{k_2 k_3 k_5 \; E_t}$$

Scheme 4

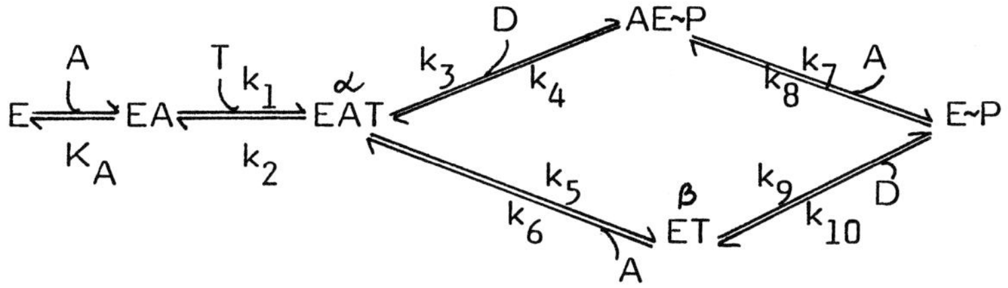

$$V_A^{-1} = \left(\frac{K_A k_2}{k_1 \; T A} + \frac{k_2}{k_1 \; T} + 1 + \frac{k_3}{k_4 \; D} + \frac{k_5}{k_6 \; A} + \frac{k_5 k_9}{k_6 k_{10} D \; A}\right) \times F_A$$

$$F_A = \frac{1}{\alpha k_2 \; E_t} = \frac{k_2 k_6 A + k_2 k_9 + k_3 k_6 \; A + k_3 k_9 + k_5 k_9}{(k_3 k_6 \; A + k_3 k_9 + k_5 k_9) k_2 \; E_t}$$

Scheme 5

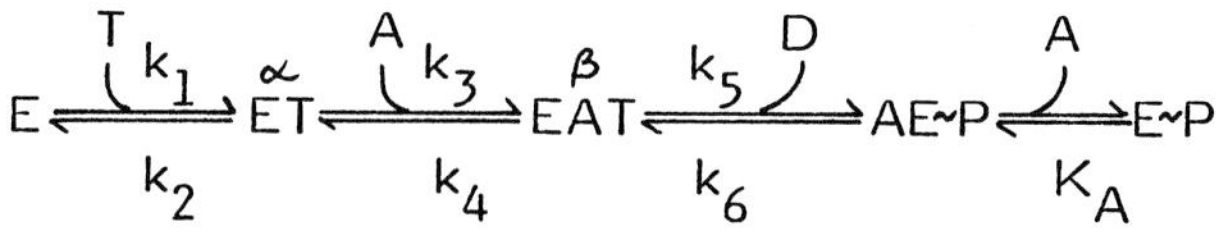

$$V_A^{1} = \left(\frac{k_2}{k_1 T} + 1 + \frac{k_3 A}{k_4} + \frac{k_3 k_5 A}{k_4 k_6 D} + \frac{k_3 k_5 K_A}{k_4 k_6 D}\right) \times F_A$$

$$F_A = \frac{1}{\alpha k_2 E_t} = \frac{k_2 k_4 + k_2 k_5 + k_3 k_5 A}{k_2 k_3 k_5 A E_t}$$

Scheme 6

This scheme is equivalent to

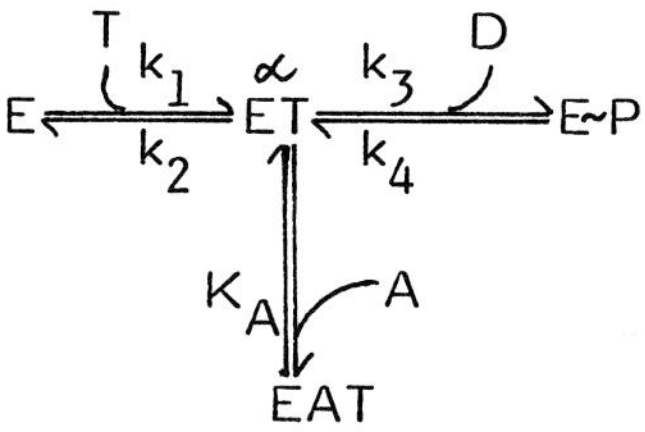

$$V_A^{-1} = \left(\frac{k_2}{k_1 T} + 1 + K_A A + \frac{k_3}{k_4 D}\right) \times F$$

$$F = \frac{1}{\alpha k_2 E_t} = \frac{k_2 + k_3}{k_2 k_3 E_t}$$

Scheme 7

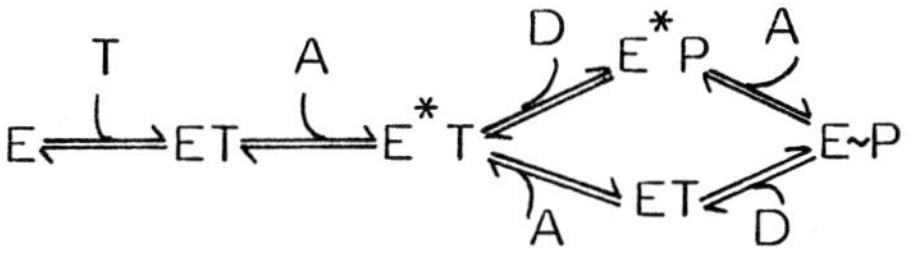

This is equivalent to

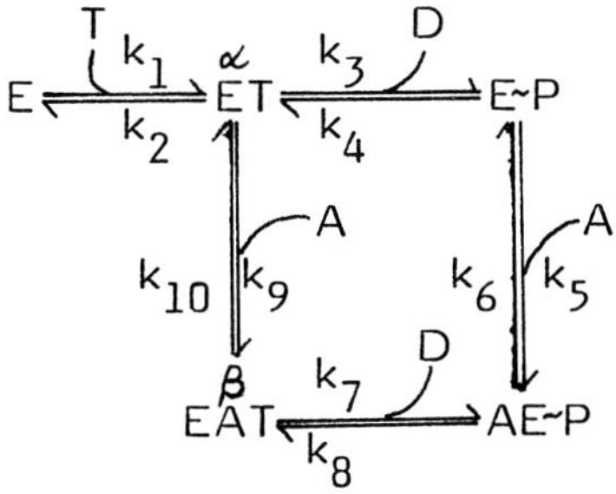

$$V_A^{-1} = \left(\frac{k_2}{k_1\,T} + 1 + \frac{k_9\,A}{k_{10}} + \frac{k_7k_9\,A}{k_8k_{10}\,D} + \frac{k_3}{k_4\,D}\right) \times F_A$$

$$F_A = \frac{1}{\alpha k_2\,E_t} = \frac{k_2k_{10} + k_2k_7 + k_3k_{10} + k_3k_7 + k_7k_9\,A}{(k_3k_{10} + k_3k_7 + k_7k_9\,A)k_2\,E_t}$$

Scheme 8

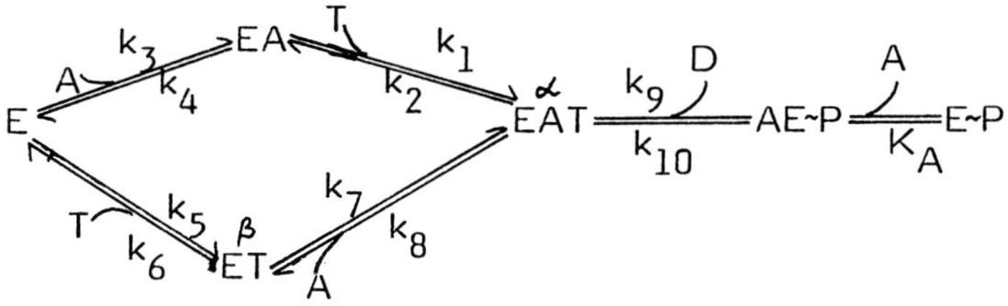

$$V_A^{-1} = \left(\frac{k_2k_4}{k_1k_3\,T\,A} + \frac{k_2}{k_1\,T} + \frac{k_8}{k_7\,A} + 1 + \frac{k_9}{k_{10}\,D} + \frac{k_9K_A}{k_{10}\,D\,A}\right) \times F_A$$

$$F_A = \frac{1}{\alpha\left(k_2 + \frac{k_6 k_8}{k_6 + k_7 A}\right) E_t}$$

where
$$\alpha = \frac{k_6 k_9 + k_7 k_9 A}{k_2 k_6 + k_2 k_7 A + k_6 k_8 + k_6 k_9 + k_7 k_9 A}$$

Scheme 9

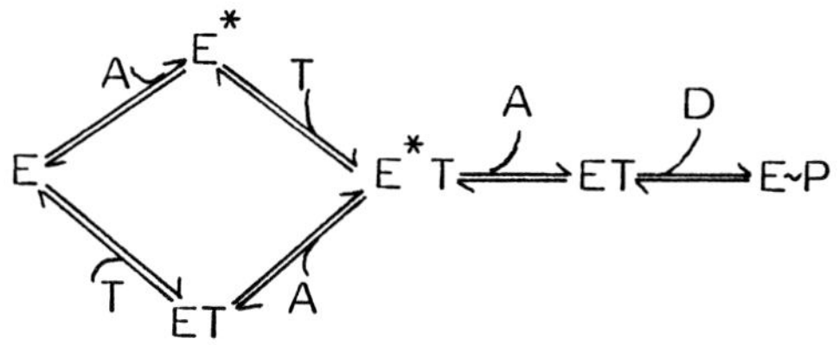

This is equivalent to

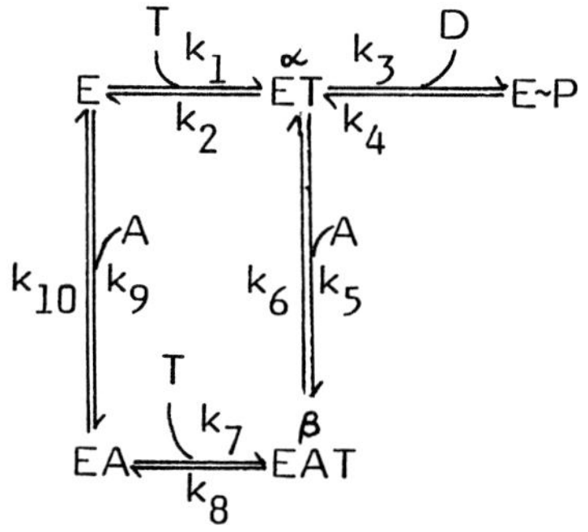

$$V_A^{-1} = \left(\frac{k_2}{k_1 T} + 1 + \frac{k_2 k_9 A}{k_1 k_{10} T} + \frac{k_5 A}{k_6} + \frac{k_3}{k_4 D}\right) \times F_A$$

$$F_A = \frac{1}{\alpha\left(k_2 + \frac{k_5 k_8 A}{k_6 + k_8}\right) E_t}$$

where $$\alpha = \frac{k_3k_6 + k_3k_8}{k_2k_6 + k_3k_6 + k_2k_8 + k_3k_8 + k_5k_8\,A}$$

Scheme 10

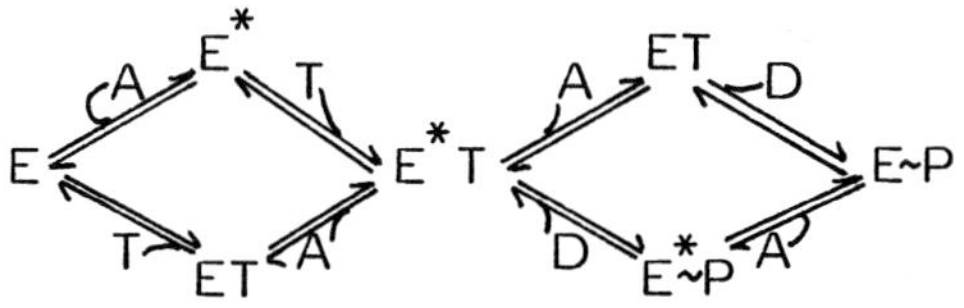

This is equivalent to

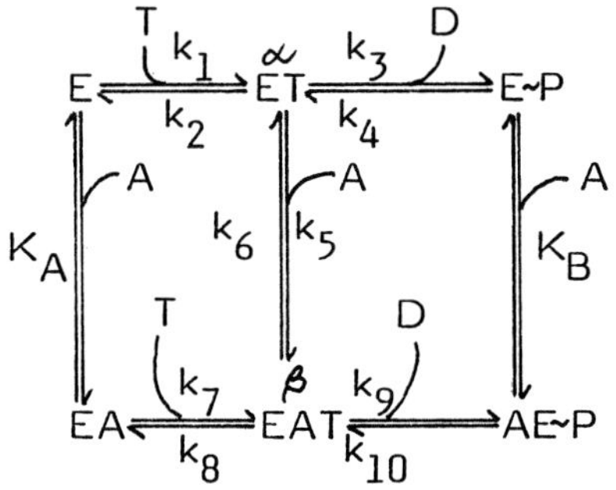

$$V_A^{-1} = \left(\frac{k_2}{k_1\,T} + 1 + \frac{k_5k_8\,A}{k_6k_7\,T} + \frac{k_5\,A}{k_6} + \frac{k_3}{k_4\,D} + \frac{k_5k_9\,A}{k_6k_{10}\,D}\right) \times F_A$$

$$F_A = \left(\alpha k_2 + \frac{k_5k_8\,A\,(k_6\alpha + k_9)}{k_6(k_6 + k_8 + k_9)}\right)^{-1} \times E_t^{-1}$$

where

$$\alpha = \frac{k_3k_6 + k_3k_8 + k_3k_9 + k_5k_9\,A}{k_3k_6 + k_3k_8 + k_3k_9 + k_5k_8\,A + k_5k_9\,A + k_2K_6 + k_2k_8 + k_2k_9}$$

If we take into account that T could mean MgATP or ATP and D could mean MgADP or ADP, the number of alternative reaction schemes rises from 10 to 40.

The prediction of the true reaction scheme is based on the following procedure which involves two steps, namely

a) Lineweaver-Burk plots (V_A^{-1} plotted against S^{-1} and $(MgS)^{-1}$, respectively),

b) plots of the slopes of these Lineweaver-Burk diagrams against Mg^{-1} and Mg, respectively.

In order to illustrate this procedure, scheme 2 (see page 4) may serve as an example. The equation for V_A^{-1} in scheme 2 can be presented as follows

$$V_A^{-1} = \underbrace{\left(\frac{K_A}{A} + 1\right)\frac{k_2F}{k_1}\cdot\frac{1}{T}}_{f(A,T)} + 1 + \underbrace{\left(1 + \frac{K_B}{A}\right)\frac{k_3F}{k_4}\cdot\frac{1}{D}}_{f(A,D)} \qquad \text{(eq. 1)}$$

When the true substrate for the forward reaction is assumed to be free ATP (T = ATP), we obtain

$$V_A^{-1} = \underbrace{\left(\frac{K_A}{Mg} + 1\right)\frac{k_2F}{k_1}}_{m_1}\cdot\frac{1}{ATP} + 1 + f(A,D) =$$

$$\underbrace{(K_A + Mg)\frac{k_2F}{k_1K_t}}_{m_2}\cdot\frac{1}{MgATP} + 1 + f(A,D) \qquad \text{(eq. 2)}$$

where A = Mg, and $K_t = \frac{Mg \bullet ATP}{MgATP}$.

With constant Mg^{++} and ADP concentrations f(A,D) is kept constant. Thus m_1 and m_2 represent the slopes of Lineweaver-Burk-plots in which V_A^{-1} is plotted versus ATP^{-1} and $MgATP^{-1}$, respectively.

In the next step m_1 and m_2 are represented as functions of the magnesium ion concentration:

$$m_1 = \underbrace{\frac{k_2F\,K_A}{k_1}}_{m_{11}}\cdot\frac{1}{Mg} + \frac{k_2F}{k_1} = \underbrace{\frac{k_2F\,K_A}{k_1Mg^2}}_{m_{12}}\cdot Mg + \frac{k_2F}{k_1}$$

$$m_2 = \underbrace{(K_A + Mg)\frac{k_2F\,Mg}{k_1K_t}}_{m_{21}}\cdot\frac{1}{Mg} = \underbrace{\frac{k_2F}{k_1K_t}}_{m_{22}}\cdot Mg + \frac{k_2F\,K_A}{k_1K_2}$$

This means: the m_1 plot versus Mg^{-1} should be linear (m_{11} is Mg independent) and the m_1 plot versus Mg should be nonlinear (m_{12} is Mg dependent). Further, the m_2 plot versus Mg^{-1} (m_{21}) and the m_2 plot versus Mg (m_{22}) should be nonlinear and linear, respectively.

Assuming, that the true substrate is MgATP (T = MgATP), eq.1 has the following form:

$$V_A^{-1} = \underbrace{(\frac{K_A}{Mg} + 1)\frac{k_2F\,K_t}{k_1Mg}}_{m_1}\cdot\frac{1}{ATP} + 1 + f(A,D)$$

$$= \underbrace{(\frac{K_A}{Mg} + 1)\frac{k_2F}{k_1}}_{m_2}\cdot\frac{1}{MgATP} + 1 + f(A,D)$$

Predictions about the slopes of secondary plots can be made in the next step.

$$m_{11} = (\frac{K_A}{Mg} + 1)\frac{k_2F\,K_t}{k_1} \quad ;\ m_1 \text{ plot vs. } Mg^{-1} \text{ is nonlinear}$$

$$m_{12} = (\frac{K_A}{Mg} + 1)\frac{k_2F\,K_t}{k_1Mg^2} \quad ;\ m_1 \text{ plot vs. Mg is nonlinear}$$

$$m_{21} = \frac{k_2K_AF}{k_1} \quad ;\ m_2 \text{ plot vs. } Mg^{-1} \text{ is linear}$$

$$m_{22} = (\frac{K_A}{Mg} + 1)\frac{k_2F}{k_1Mg} \quad ;\ m_2 \text{ plot vs. Mg is nonlinear}$$

In conclusion, from the pattern of the four secondary plots it is possible to deduce whether ATP or MgATP is the true substrate.

For the prediction whether ADP or MgADP is the substrate for the back reaction, the same procedure was carried out with V_A values measured at varied ADP concentrations and constant ATP concentrations.

These calculations illustrated above for scheme 2 were carried out for the 10 reaction schemes. The results are summarised in the Table I.

TABLE I Prediction of 10 alternative schemes and their substrate combinations of the ATP-ADP-exchange reaction. The slope of the plots m_1 and m_2 versus Mg^{-1} and Mg are represented: l = linear, nl = nonlinear and u = linear and parallel to the abscissa.

Alternat. reaction scheme	Assumed as true substrate for T or D	Expected shape of the curve when m_1 is plotted against		m_2 is plotted against	
		Mg^{-1}	Mg	Mg^{-1}	Mg
1	ATP	u	u	nl	l
	MgATP	l	nl	u	u
	ADP	u	u	nl	l
	MgADP	l	nl	u	u
2	ATP	l	nl	nl	l
	MgATP	nl	nl	l	nl
	ADP	l	nl	nl	l
	MgADP	nl	nl	l	nl
3	ATP	nl	nl	nl	nl
	MgATP	nl	nl	nl	nl
	ADP	l	nl	nl	l
	MgADP	nl	nl	l	nl
4	ATP	nl	nl	nl	nl
	MgATP	nl	nl	nl	nl
	ADP	nl	nl	nl	nl
	MgADP	nl	nl	nl	nl
5	ATP	l	nl	nl	l
	MgATP	nl	nl	l	nl
	ADP	nl	nl	nl	nl
	MgADP	nl	nl	nl	nl
6	ATP	u	u	nl	l
	MgATP	l	nl	u	u
	ADP	u	u	nl	l
	MgADP	l	nl	u	u
7	ATP	nl	nl	nl	nl
	MgATP	nl	nl	nl	nl
	ADP	nl	nl	nl	nl
	MgADP	nl	nl	nl	nl
8	ATP	nl	nl	nl	nl
	MgATP	nl	nl	nl	nl
	ADP	nl	nl	nl	nl
	MgADP	nl	nl	nl	nl

9	ATP MgATP ADP MgADP	nl nl nl nl	nl nl nl nl	nl nl nl nl	nl nl nl nl
10	ATP MgATP ADP MgADP	nl nl nl nl	nl nl nl nl	nl nl nl nl	nl nl nl nl
Predicted from		m_{11}	m_{12}	m_{21}	m_{22}

The experimental results (2,16) reveal the following pattern: nl, nl, l, nl for T, and l, nl, nl, l for D. According to Table I this pattern is predicted <u>exclusively</u> by scheme 2 with T=MgATP and D = ADP. Consequently it can be concluded that the true reaction mechanism is

$$E \underset{K_A}{\overset{Mg^{++}}{\rightleftharpoons}} E \underset{k_2}{\overset{MgATP \quad k_1}{\rightleftharpoons}} ET \underset{k_4}{\overset{k_3 \quad ADP}{\rightleftharpoons}} E{\sim}P \underset{K_B}{\overset{Mg^{++}}{\rightleftharpoons}} E{\sim}P$$

and that

$$V_A^{-1} = \left(\frac{K_A K_2}{K_1\ Mg\ MgATP} + \frac{K_2}{K_1\ MgATP} + 1 + \frac{K_3}{K_4\ ADP} + \frac{K_3 K_B}{K_4\ ADP\ Mg}\right) V_{max}^{-1}$$

where

$$V_{max} = \frac{k_2 k_3}{k_2 + k_3} E_t$$

This conclusion is corroborated by deducing V_{max}- from V_A-values using a variety of graphical methods (2).

V. Identification of the Ca_o-binding step

The cleavage or synthesis of one molecule ATP in the sarcoplasmic reticulum is coupled with the translocation of two calcium ions (9,13). This and other findings (4,6,19) suggest that two calcium ions are bound to the transporting enzyme. Starting out from the conclusions of the preceding chapter, the Ca_o-binding step can be identified.

Probably not more than five reaction schemes (Scheme 11 to Scheme 15) are possible:

Scheme 11

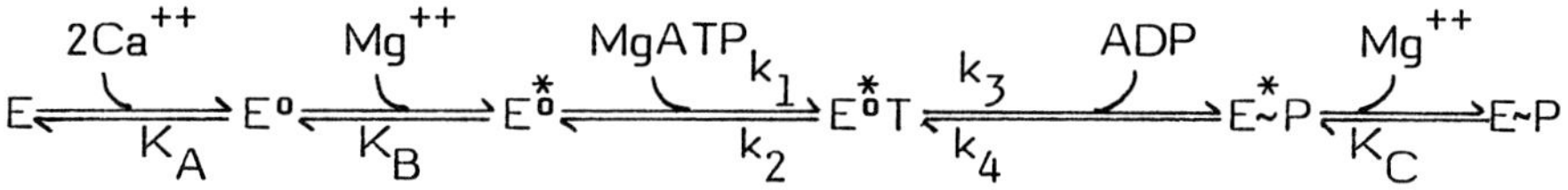

$$V_A^{-1} = \left(\frac{k_2 K_A K_B}{k_1 Ca^2 MgATP\ Mg} + \frac{k_2 K_B}{k_1\ MgATP\ Mg} + \frac{k_2}{k_1\ MgATP} + 1 + \frac{k_3}{k_4 ADP} + \frac{k_3 K_C}{k_4 ADP\ Mg}\right) \times F$$

F is the same as in scheme 2.

Scheme 12

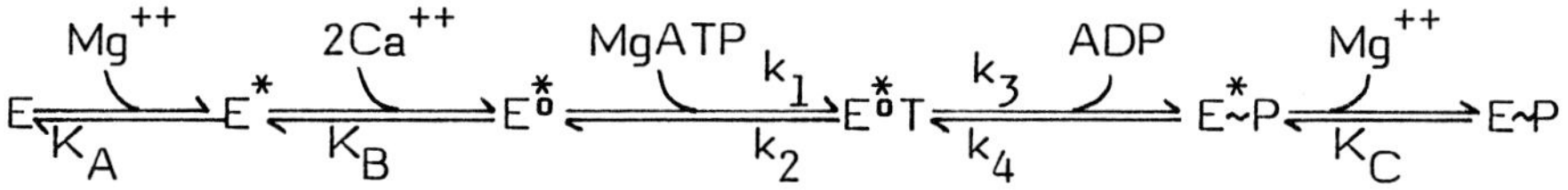

$$V_A^{-1} = \left(\frac{k_2 K_A K_B}{k_1 Ca^2 MgATP\ Mg} + \frac{k_2 K_B}{k_1 Ca^2 MgATP} + \frac{k_2}{k_1 MgATP} + 1 + \frac{k_3}{k_4 ADP} + \frac{k_3 K_C}{k_4 ADP\ Mg}\right) \times F$$

F is the same as in scheme 2.

Scheme 13

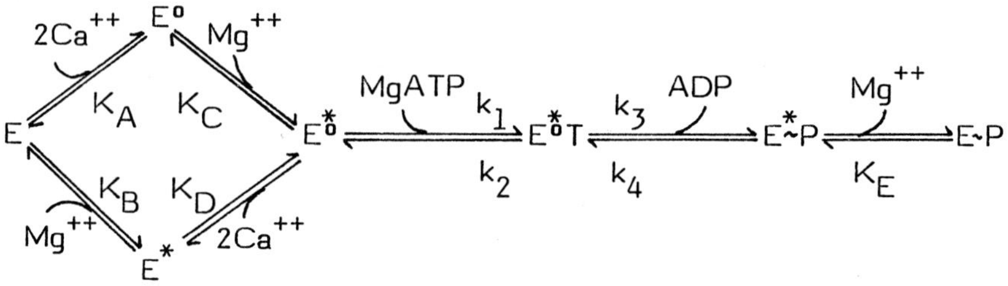

$$V_A^{-1} = \left(\frac{k_2K_BK_D}{k_1Ca^2Mg\,MgATP} + \frac{k_2K_C}{k_1Mg\,MgATP} + \frac{k_2K_B}{k_1Ca^2MgATP} + \frac{k_2}{k_1\,MgATP} + 1 + \frac{k_3}{k_4ADP} + \frac{k_3K_E}{k_4ADP\,Mg} \right) \times F$$

F is the same as in scheme 2.

Scheme 14

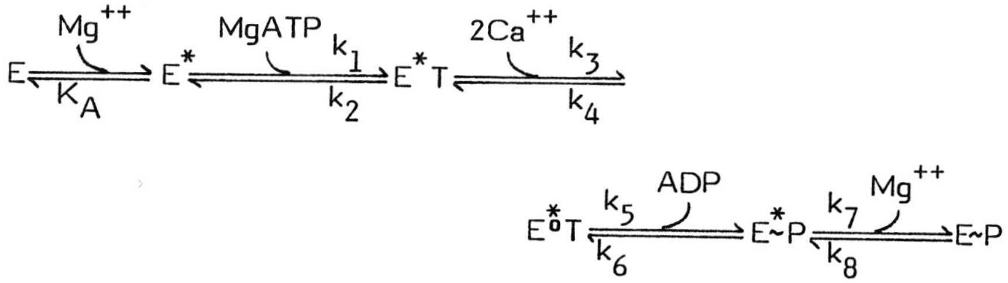

$$V_A^{-1} = \left(\frac{k_2K_A}{k_1Mg\,MgATP} + \frac{k_2}{k_1MgATP} + 1 + \frac{k_3Ca^2}{k_4} + \frac{k_3k_5Ca^2}{k_4k_6ADP} + \frac{k_3k_5k_7Ca^2}{k_4k_6k_8Mg\,ADP} \right) \times F_{Ca}$$

F_{Ca} is the same as F_A in scheme 5 (A = Ca).

Scheme 15

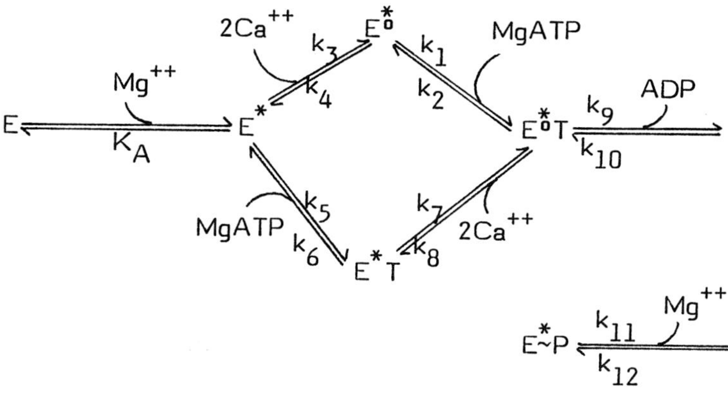

$$V_A^{-1} = \left(\frac{k_2k_4K_A}{k_1k_3Mg\ Ca^2MgATP} + \frac{k_2k_4}{k_1k_3Ca^2MgATP} + \frac{k_2}{k_1MgATP} + \frac{k_8}{k_7Ca^2} + 1 + \frac{k_9}{k_{10}ADP} + \frac{k_9k_{11}}{k_{10}k_{12}ADP\ Mg}\right) \times F_{Ca}$$

F_{Ca} is the same as F_A in scheme 8 (A = Ca).

The screening procedure for these five alternatives (Schemes 11-15) consists also of two steps.

Step 1: V_A values measured with varied ADP concentrations were used for Hanes-Woolf plots (ADP·V_A^{-1} vs. ADP) (10).

The slope (m_3) of this plot for Scheme 11 is for example

$$m_3 = \left(\frac{k_2K_AK_B}{k_1Ca^2MgATP\ Mg} + \frac{k_2K_B}{k_1\ MgATP\ Mg} + \frac{k_2}{k_1\ MgATP} + 1\right) \times F$$

m_3 Values are determined now for several Ca^{++} concentrations.

Step 2: The m_3 values obtained are used for the secondary Hanes-Woolf-plot. ($m_3\ Ca^2$ vs. Ca^2). The slope of this plot is

$$m_4 = \left(\frac{k_2K_B}{k_1\ MgATP\ Mg} + \frac{k_2}{k_1\ MgATP} + 1\right) \times F$$

m_4 Values are determined now for several Mg concentrations.

For the scheme 11 the second Hanes-Woolf plots are expected to be linear (m_4 is Ca^{++} independent) and their slopes m_4 to be negatively dependent on the Mg^{++} concentration.

The m_4-values and the predictions for the other alternatives are:

For scheme 12

$$m_4 = \left(\frac{k_2}{k_1MgATP} + 1\right) \times F$$

m_4 is independent of the Mg^{++}concentration; the second Hanes-Woolf plot is linear.

For scheme 13

$$m_4 = \left(\frac{k_2 K_C}{k_1 \mathrm{MgMgATP}} + \frac{k_2}{k_1 \mathrm{MgMgATP}} + 1\right) \times F$$

m_4 is negatively dependent on the Mg^{++} concentration; the second Hanes-Woolf plot is linear.

For scheme 14

$$m_4 = \left(\frac{K_A}{k_1 \mathrm{MgMgATP}} + \frac{1}{k_1 \mathrm{MgATP}} + \frac{1}{k_2} + \frac{k_3 \mathrm{Ca}^2}{k_2 k_4}\right) \times \frac{1}{E_t}$$

m_4 is negatively dependent on the Mg^{++} concentration; the second Hanes-Woolf plot is nonlinear.

For scheme 15

$$m_4 = \left(\frac{k_2 k_4 K_A}{k_1 k_3 \mathrm{MgCa}^2 \mathrm{MgATP}} + \frac{k_2 k_4}{k_1 k_3 \mathrm{Ca}^2 \mathrm{MgATP}} + \frac{k_2}{k_1 \mathrm{MgATP}} + \frac{k_8}{k_7 \mathrm{Ca}^2} + 1\right) \times F_{Ca}$$

m_4 is negatively dependent on Mg^{++}concentration; the second Hanes-Woolf plot is nonlinear.

The experimental data (2) yield a linear second Hanes-Woolf plot whose slope is independent of the Mg-concentration. This means that scheme 13 represents the real reaction mechanism and that the other possibilities are strictly excluded.

VI. DISCUSSION

The "elimination principle" for elucidating an enzymic reaction mechanism was applied in this paper to the specific case of the calcium-transporting ATPase. The elimination procedures employed here are, of course, not sufficient to identify the true reaction mechanism in all cases. For example, the patterns shown in Table I do not allow to discriminate between schemes 4, 7, 8, 9 and 10. If the true mechanism would belong to this group - which is fortunately not the case - additional elimination methods would have been necessary.

The criteria used here for correlating theoretical predictions with experimental data are a) the dependence of the slopes of the final plots on ligand concentration and b) the linearity of the final plots. To prove that a plotted function is indeed linear, or that two or more lines of a plot are parallel, cannot be achieved theoretically. However, these problems can generally be solved experimentally using statistical methods (2). The possibility of obtaining spurious linearity and/or spurious parallelism in critical plots can be checked experimentally by varying the concentration ranges of substrates and other ligands.

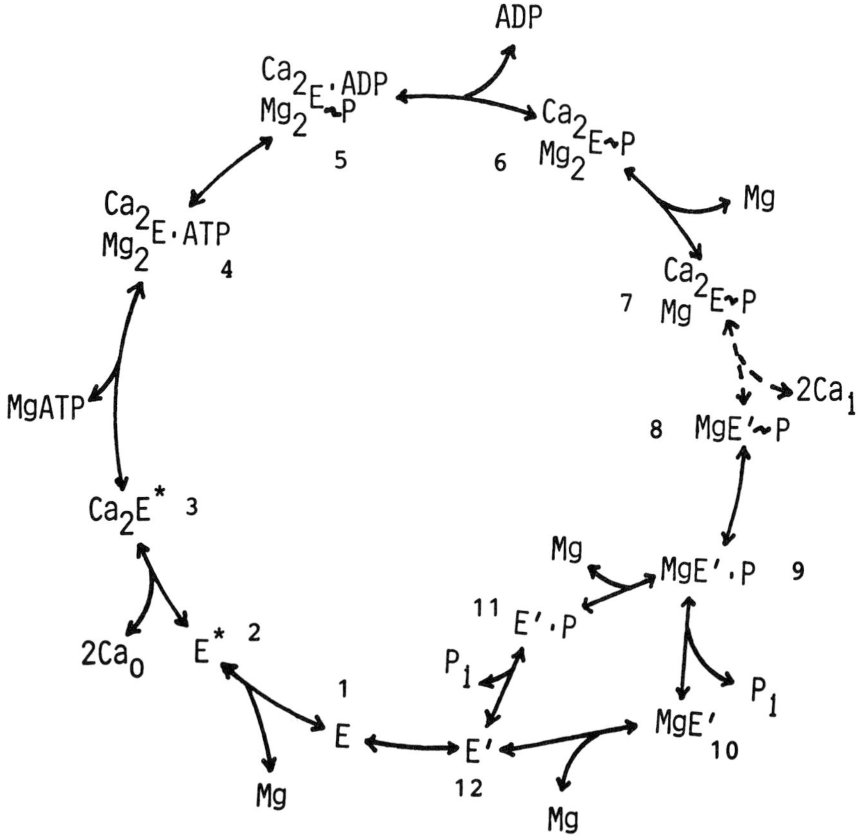

Fig. 1. The reaction scheme of the SR calcium transport ATPase concluded by kinetic analysis with "elimination principle". The partial process from the enzyme species No. 8 to 12 is concluded by the previous work (18).

Fig. 1 shows the reaction cycle of the SR calcium-transportung enzyme as it was determined by the present and another analysis (18) which was also based on the elimination principle. Kinetic experiments using specific inhibitors (10) confirm essential points of this reaction cycle (2,3,17).

In conclusion, Fig. 1 shows the most probable sequence of events when calcium is transported through the membrane of the sarcoplasmic reticulum under (quasi-)physiological conditions.

REFERENCES

1. Barlogie, W., Hasselbach, W. and Makinose, M., FEBS Letters 12: 267 (1971)
2. Boll, W., Dr. med. Thesis, Universität Heidelberg (1982) in preparation for publication
3. Boll, W. and Makinose, M., Z. Naturforsch., (1983) in press
4. DeMeis, L., The Sarcoplasmic Reticulum, John Wiley & Sons, New York (1981)
5. Hanes, C.S., Biochem. J. 26: 1406 (1932)
6. Hasselbach, W., Biochim. Biophys. Acta 515: 23 (1978), Topics in Curr. Chem. 78; 1 (1979)
7. Hasselbach, W. and Makinose, M., Biochem. Z. 333: 518 (1961)
8. Hasselbach, W. and Makinose, M., Biochem. Biophys. Res. Commun. 7: 132 (1962)
9. Hasselbach, W. and Makinose, M., Biochem. Z. 339: 94 (1963)
10. Loer, G., Dr. med. Thesis, Universität Heidelberg (1982) in preparation for publication
11. Makinose, M., Biochem. Z. 345: 80 (1966)
12. Makinose, M., Eur. J. Biochem. 10: 74 (1969)
13. Makinose, M., FEBS Letters 12: 269 (1971)
14. Makinose, M. and Hasselbach, W., FEBS Letters 12: 271 (1971)
15. Makinose, M., FEBS Letters 25: 113 (1972)
16. Makinose, M. and Boll, W. in 'Cation Flux Across Biomembranes' (Y. Mukohata et al., ed.) Academic Press, New York : 89 (1979)
17. Makinose, M. and Boll, W. in 'Function and Molecular Aspects of Biomembrane Transport' (E. Quagliariello et al., ed.) Elsevier/-North Holland Biomedical Press: 115 (1979)
18. Kolassa, N.; Punzengruber, Ch.; Suko, J. and Makinose, M. FEBS Letters 108: 495 (1979)
19. Tada, M., Yamamoto, T. and Tonomura, Y., Physiological Reviews, Vol. 58, No. 1 (1978)
20. Yamamoto, T. and Tonomura, Y., J. Biochem. 64: 137 (1968)

Ca^{2+}-Ca^{2+} EXCHANGE MEDIATED BY THE MEMBRANE-BOUND (Ca^{2+}, Mg^{2+})-ATPase OF SARCOPLASMIC RETICULUM VESICLES

Tohru Kanazawa[1]
Yuichi Takakuwa

Department of Biochemistry
Asahikawa Medical College
Asahikawa, Japan

I. INTRODUCTION

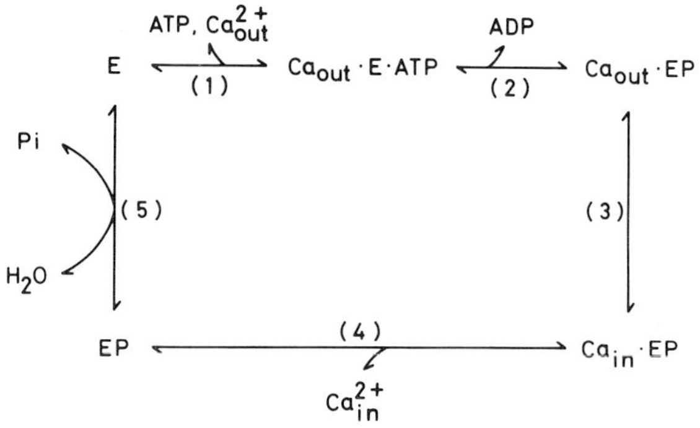

Mg^{2+} plays an important role in the Ca^{2+} transport catalyzed by the membrane-bound (Ca^{2+}, Mg^{2+})-ATPase of sarcoplasmic reticulum vesicles. The cycle of this transport is illustrated by the above diagram (Ca^{2+}_{out} and Ca^{2+}_{in} designate free Ca^{2+} in the external medium and in the internal medium, respectively; Ca_{out}·E[2]·ATP and Ca_{out}·EP are the inter-

[1]This work was supported by a grant from the Ministry of Education, Science and Culture of Japan.

[2]The abbreviations used are; E, dephosphoenzyme; EP, phosphoenzyme; EGTA, ethylene glycol bis(β-aminoethyl ether)-N,N,N',N'-tetraacetic acid; $E^{32}P$, ^{32}P-labeled phosphoenzyme.

ISBN 0-12-260380-X

mediates with bound Ca^{2+} which is exposed to the external medium, whereas $Ca_{in} \cdot EP$ is the intermediate with bound Ca^{2+} which is exposed to the internal medium; the distinction between the form of the enzyme with a high affinity for Ca^{2+} and that of the enzyme with a low affinity for Ca^{2+} is omitted for simplification). This scheme is consistent with earlier findings (1-6).

It has been shown that Mg^{2+} affects several steps in this transport cycle (3-28). However, the role of Mg^{2+} in the step of Ca^{2+} translocation across the vesicular membranes (Step 3) still remains obscure. Makinose (29) demonstrated previously that the ADP-activated Ca^{2+}_{out}-Ca^{2+}_{in} exchange occurred in the early stage before the step of EP hydrolysis. Recently, we suggested that the Ca^{2+}_{out}-Ca^{2+}_{in} exchange was activated by $MgCl_2$ (30). In the present study, in order to reveal the role of Mg^{2+} in the Ca^{2+} translocation more definitely, we made the direct determination of the unidirectional Ca^{2+} efflux in the Ca^{2+}_{out}-Ca^{2+}_{in} exchange through Step 1 to Step 4. The result showed that the Ca^{2+} efflux was markedly activated by Mg^{2+}. It is concluded from the result and other findings that Mg^{2+} is essential for the exposure of the bound Ca^{2+} of EP to the internal medium through Step 3.

II. EXPERIMENTAL PROCEDURES

A. Preparation of Sarcoplasmic Reticulum Vesicles

Sarcoplasmic reticulum vesicles were prepared from rabbit skeletal muscle as described previously (18).

B. Loading of the Vesicles with $^{45}Ca^{2+}$ or Unlabeled Ca^{2+}

The vesicles were incubated at 25°C for 5 min in 1 ml of a mixture containing 4 mg of the vesicles, 0.055 or 0.22 μmol of $^{45}CaCl_2$ (or unlabeled $CaCl_2$), 2 μmol of acetyl phosphate, 3 μmol of $MgCl_2$, 0.1 mmol of KCl, and 20 μmol of Tris/maleate (pH 6.5). About 90% of the $^{45}Ca^{2+}$ (or Ca^{2+}) in the medium was taken up by the vesicles. In order to remove Mg^{2+}, residual $^{45}Ca^{2+}$ (or Ca^{2+}), and substrates in the medium, 0.5 ml of the incubation mixture was passed successively through two columns equilibrated with 100 mM KCl and 20 mM Tris/maleate (pH 6.5) at 6°C, the first being a column (1 × 11 cm) of Sephadex G-50 and the second being a column (1 × 5 cm) of Dowex chelating resin. The turbid fractions (0.4-0.5 mg of the vesicles$\cdot ml^{-1}$) were collected. The amount of $^{45}Ca^{2+}$ in the vesicles thus obtained was 7-33 $nmol \cdot mg^{-1}$.

C. $^{45}Ca^{2+}$ Efflux

The reaction was started at 6°C and pH 6.5 by adding 1 volume of the suspension of the $^{45}Ca^{2+}$-loaded vesicles to 19 or 39 volumes of a buffer containing unlabeled $CaCl_2$, EGTA, and other reagents. At different times, a 0.5-ml portion of the reaction mixture was filtered through a Millipore filter and the radioactivity of the filtrate was measured.

D. $^{45}Ca^{2+}$ Uptake

The reaction was carried out essentially in the same way as described for the $^{45}Ca^{2+}$ efflux determination except that the Ca^{2+} (unlabeled)-loaded vesicles were added to a mixture containing $^{45}CaCl_2$. At different times, a 0.5-ml portion of the reaction mixture was filtered through a Millipore filter. The filter was washed with a buffer containing EGTA and the radioactivity remaining on the filter was measured.

E. Determination of $E^{32}P$ and $^{32}P_i$

Phosphorylation of the Ca^{2+}-loaded vesicles was performed with [γ-^{32}P]ATP. The reaction was quenched with perchloric acid (containing bovine serum albumin in the case of the $E^{32}P$ determination) and the resulting sample was centrifuged. To isolate $E^{32}P$, the acid-denatured vesicles in the pellet were washed with the acid and the radioactivity was measured. Nonspecific adsorption of ^{32}P to the vesicles was measured in the presence of 1 mM EGTA without added $CaCl_2$. The amount of $E^{32}P$ was obtained by subtracting the amount of ^{32}P adsorbed nonspecifically from the total amount of ^{32}P incorporated. To isolate $^{32}P_i$, $^{32}P_i$ in the supernatant was determined as described previously (31).

F. Others

Calculations of the free Ca^{2+}, free Mg^{2+}, and free ADP concentrations were performed using a computer program based on the method of Newton and Raphson. Logarithms of the association constants of metals and H^+ to various ligands at 6°C used in the calculations were as follows. H^+ to ATP^{4-}, 6.55; $HATP^{3-}$, 4.22; ADP^{3-}, 6.50; $HADP^{2-}$, 4.15; $EGTA^{4-}$, 9.92; $HEGTA^{3-}$, 9.18; H_2EGTA^{2-}, 2.65. Ca^{2+} to ATP^{4-}, 4.05; $HATP^{3-}$, 2.28; ADP^{3-}, 2.90; $HADP^{2-}$, 1.61; $EGTA^{4-}$, 11.28; $HEGTA^{3-}$, 5.30. Mg^{2+} to ATP^{4-}, 4.03; $HATP^{3-}$, 2.05; ADP^{3-}, 2.99; $HADP^{2-}$, 1.45; $EGTA^{4-}$, 5.01; $HEGTA^{3-}$, 3.40 (32).

III. RESULTS AND DISCUSSION

In the experiment shown in Fig. 1, the $^{45}Ca^{2+}$ efflux was determined at 6°C and pH 6.5 with the $^{45}Ca^{2+}$-loaded sarcoplasmic reticulum vesicles (33 nmol of $^{45}Ca^{2+}$ per mg of the vesicles) in the presence of 9 μM Ca^{2+}_{out}, 0.1 mM ATP, 0.1 mM ADP, and 1 mM $MgCl_2$. The efflux occurred at a considerably high rate. It was linear for at least 1 minute, and almost all of the $^{45}Ca^{2+}$ in the vesicles flowed out in 10 minutes. The rate of unidirectional Ca^{2+} efflux was determined from this initial linear phase. It was found to be 8.3 $nmol \cdot mg^{-1} \cdot min^{-1}$. When Ca^{2+}_{out} was absent, $^{45}Ca^{2+}$ leaked out much more slowly (0.35 $nmol \cdot mg^{-1} \cdot min^{-1}$). These observations demonstrate that the $^{45}Ca^{2+}$ efflux depends almost entirely on the Ca^{2+}_{out}.

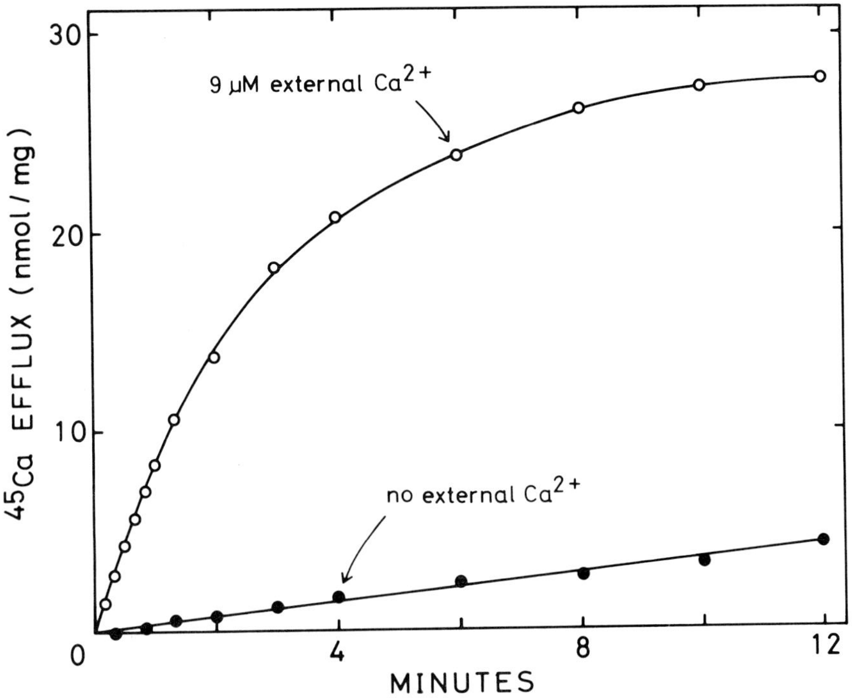

FIGURE 1. $^{45}Ca^{2+}$ efflux from the $^{45}Ca^{2+}$-loaded vesicles. The reaction mixture had a final composition of 20 μg of the vesicles (loaded with 33 nmol of $^{45}Ca^{2+}$ per mg)$\cdot ml^{-1}$, 0.1 mM ATP, 0.1 mM ADP, 1 mM $MgCl_2$, 0.3 mM $CaCl_2$, 0.5 mM EGTA, 100 mM KCl, 20 mM Tris/maleate (pH 6.5). Ca^{2+}_{out} was 9 μM. For the control, the reaction was performed without added $CaCl_2$.

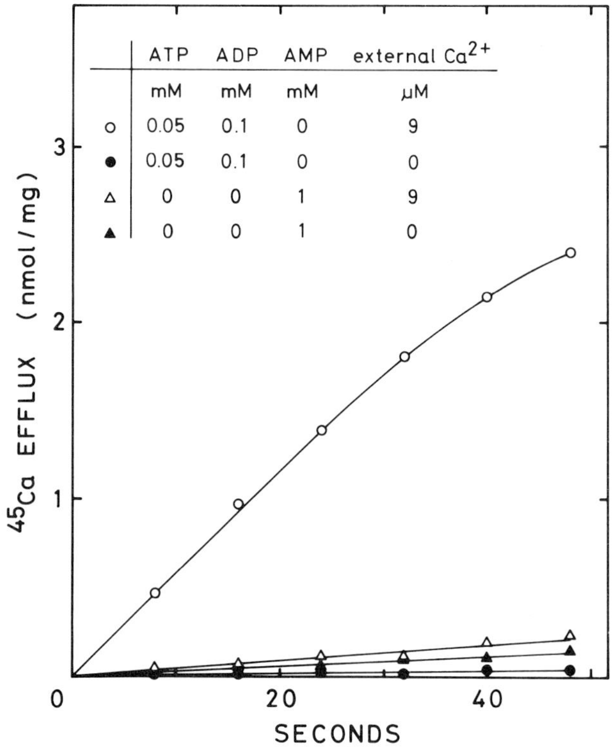

FIGURE 2. Effects of ATP, ADP, AMP, and Ca^{2+}_{out} on $^{45}Ca^{2+}$ efflux. The reaction mixture had a final composition of 13 μg of the vesicles (loaded with 7 nmol of $^{45}Ca^{2+}$ per mg)·ml^{-1}, 0 or 0.05 mM ATP, 0 or 0.1 mM ADP, 0 or 1 mM AMP, 10 mM $MgCl_2$, 0 or 0.3 mM $CaCl_2$, and 0.5 mM EGTA, otherwise as in Fig. 1.

Similar measurements were performed with the $^{45}Ca^{2+}$-loaded vesicles (7 nmol of $^{45}Ca^{2+}$ per mg of the vesicles) in the presence and absence of ATP, ADP, and AMP. The efflux was very slow in the absence of ATP and ADP even when 9 μM Ca^{2+}_{out} was present (data not shown; cf. Fig. 1 B in Ref. 33). Its rate was essentially equivalent to the rate of the slow passive leak in the absence of Ca^{2+}_{out}. It is, therefore, evident that the $^{45}Ca^{2+}$ efflux depends on the presence of ATP and ADP. However, as shown in Fig. 2, 1 mM AMP added in the presence of 9 μM Ca^{2+}_{out} did not induce the $^{45}Ca^{2+}$ efflux significantly. This finding shows that the observed $^{45}Ca^{2+}$ efflux is distinct from the adenine nucleotide-dependent Ca^{2+} induced Ca^{2+} release reported recently (34, 35).

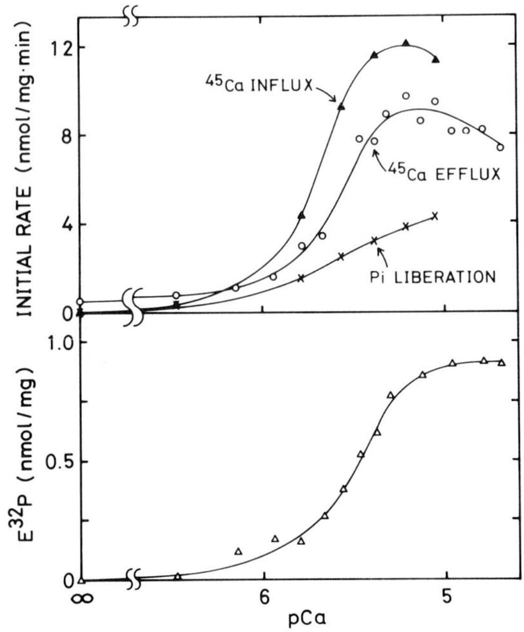

FIGURE 3. Dependences of unidirectional Ca^{2+} efflux, unidirectional Ca^{2+} influx, Ca^{2+}-dependent ATP hydrolysis, and steady level of EP on Ca^{2+}_{out} concentration. $^{45}Ca^{2+}$ efflux from the vesicles (loaded with 33 nmol of $^{45}Ca^{2+}$ per mg) was determined at different $CaCl_2$ concentrations, otherwise as in Fig. 1. The rate of unidirectional Ca^{2+} efflux (O) was obtained from the slope of the initial linear phase. $^{45}Ca^{2+}$ uptake was determined with the Ca^{2+}-loaded vesicles. The rate of unidirectional Ca^{2+} influx (▲) was obtained from the slope of the linear phase immediately following the initial burst of $^{45}Ca^{2+}$ uptake. ATP hydrolysis with the Ca^{2+}-loaded vesicles was performed with 0.1 mM [γ-^{32}P]ATP and the ^{32}Pi liberated was determined. The rate of Ca^{2+}-dependent ATP hydrolysis (X) was obtained by subtracting that of ATP hydrolysis in the presence of 10 mM EGTA without added $CaCl_2$. The Ca^{2+}-loaded vesicles was phosphorylated with 0.1 mM [γ-^{32}P]ATP and the steady level (△) of $E^{32}P$ was determined.

In the experiment given in Fig. 3, the rates of unidirectional Ca^{2+} efflux and influx, the rate of Ca^{2+}-dependent ATP hydrolysis, and the steady level of EP formation were determined at different concentrations of Ca^{2+}_{out} under similar conditions to those given in Fig. 1. All these increased markedly with an increase in Ca^{2+}_{out}. They showed essentially the same concentration dependence on the Ca^{2+}_{out} up to some 10

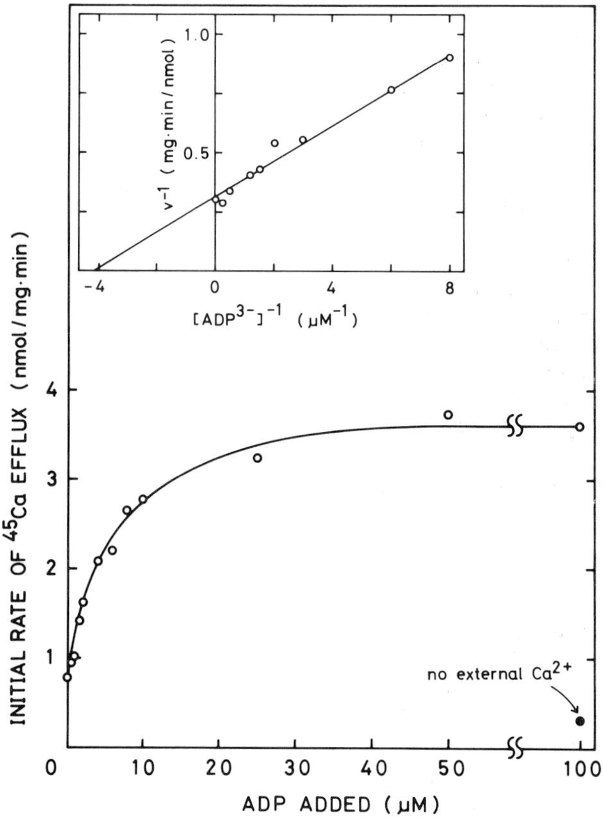

FIGURE 4. Dependence of unidirectional Ca^{2+} efflux on ADP concentration. $^{45}Ca^{2+}$ efflux from the vesicles (loaded with 10 nmol of $^{45}Ca^{2+}$ per mg) was determined at different ADP concentrations in the presence of 0.05 mM ATP and 10 mM $MgCl_2$, otherwise as in Fig. 1. The rate of unidirectional Ca^{2+} efflux was obtained from the slope of the initial linear phase. For the control, the reaction was performed without added $CaCl_2$. The inset shows the double reciprocal plot of the rate (v) of the Ca^{2+}_{out}-dependent part of unidirectional Ca^{2+} efflux against ADP^{3-} concentration.

μM. The Ca^{2+}_{out} concentration to give a half-maximum activation was approximately 2 μM. These results and other findings presented so far strongly suggest an intimate participation of the membrane-bound (Ca^{2+}, Mg^{2+})-ATPase of sarcoplasmic reticulum in this Ca^{2+} efflux. In addition, the data indicate that the rate of the Ca^{2+}_{out}-dependent part of unidirectional Ca^{2+} efflux was nearly equal to the difference between the rate of unidirectional Ca^{2+} influx and the rate of Ca^{2+}-dependent ATP hydrolysis. Since the ATP hydrolysis is coupled to the net Ca^{2+} uptake in the molar ratio of 1 at low temperatures (36-38), this approximate equality shows that the Ca^{2+}_{out}-dependent part of Ca^{2+} efflux represents the Ca^{2+}_{out}-Ca^{2+}_{in} exchange mediated by the enzyme.

Fig. 4 shows the dependence of the rate of unidirectional Ca^{2+} efflux on ADP concentration. In this experiment the $^{45}Ca^{2+}$ efflux from the $^{45}Ca^{2+}$-loaded vesicles (10 nmol of $^{45}Ca^{2+}$ per mg of the vesicles) was determined at different ADP concentrations in the presence of 0.05 mM ATP and 10 mM $MgCl_2$. The rate of the efflux increased markedly with an increase in ADP concentration. The free ADP (ADP^{3-}) concentration to give a half-maximum activation was approximately 0.25 μM. This high affinity for free ADP is in good agreement with the affinity for free ADP in the transphosphorylation from EP to ADP.

All the results presented so far support the conclusion that the observed Ca^{2+} efflux represents the EP-mediated Ca^{2+}_{out}-Ca^{2+}_{in} exchange which is caused by the dynamic reversal of Step 1 to Step 4.

In the experiment shown in Fig. 5, the rate of unidirectional Ca^{2+} efflux and the steady level of EP formation were determined at different concentrations of free Mg^{2+} under similar conditions to those given in Fig. 1. The rate of Ca^{2+} efflux was extremely low in the range of 0.5 to 5 μM free Mg^{2+}, being only slightly higher than the control level without Ca^{2+}_{out}. However, the rate of Ca^{2+}_{out}-dependent part of Ca^{2+} efflux increased remarkably with a further increase in free Mg^{2+}. In contrast, EP formation occurred to a considerable extent at 0.5 μM free Mg^{2+}, and this level was kept constant up to some 30 μM Mg^{2+}. Even when Mg^{2+} concentration was raised to 5 mM, the amount of EP only approximately doubled. Therefore, a major portion of the above Mg^{2+} activation of the Ca^{2+}_{out}-Ca^{2+}_{in} exchange can not be ascribed to the increase in the total amount of EP.

In the following experiments, it was shown that the observed Mg^{2+} activation of the Ca^{2+}_{out}-Ca^{2+}_{in} exchange was not attributable to the Mg^{2+}-induced acceleration (21, 24, 33) of transphosphorylation from EP to ADP (Steps 1 and 2) (data not shown). In these experiments the vesicles were preloaded with unlabeled Ca^{2+}. Then they were incubated with unlabeled ATP and ADP in the presence of 10 μM free Mg^{2+} under otherwise the same conditions as those given in Fig. 1. After EP formation reached a steady level, [γ-^{32}P]ATP was added in trace amounts. The results showed that, even at this low Mg^{2+} concentration, ^{32}P was rapidly and exponentially incorporated into the unlabeled EP without any appreciable lag, and almost all of the EP was labeled in 5 seconds. The first-

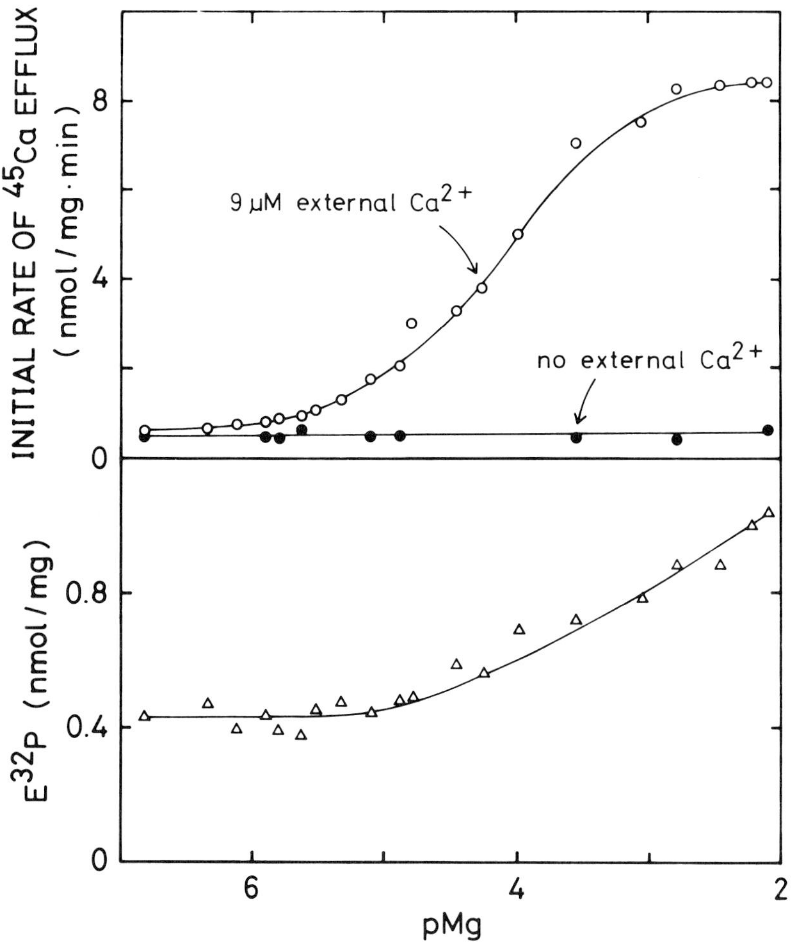

FIGURE 5. Dependences of unidirectional Ca^{2+} efflux and steady level of EP on free Mg^{2+} concentration. $^{45}Ca^{2+}$ efflux from the vesicles (loaded with 33 nmol of $^{45}Ca^{2+}$ per mg) was determined at different concentrations of $MgCl_2$, otherwise as in Fig. 1. The rate of unidirectional Ca^{2+} efflux was obtained from the slope of the initial linear phase. For the control, the reaction was performed in the presence of 1 mM EGTA without added $CaCl_2$. The Ca^{2+}-loaded vesicles was phosphorylated with 0.1 mM [γ-^{32}P]ATP and the steady level of $E^{32}P$ was determined. Contaminant Mg^{2+} in the external medium of the reaction mixture was shown to be less than 1 μM with atomic absorption spectrophotometry.

order rate constant of this incorporation was 35 min^{-1}. On the other hand, the rate of EP hydrolysis under the same conditions was very slow, and its first-order rate constant was only 0.7 min^{-1}. Obviously, this rate is much lower than that of the ^{32}P incorporation. Thus, the observed rapid ^{32}P incorporation should be ascribed to the rapid dynamic reversal of EP formation. The rate of this dynamic reversal of EP formation is much higher than the rate of the Ca^{2+}_{out}-Ca^{2+}_{in} exchange under the given conditions (cf. Fig. 5). It is, therefore, evident that the reversible transphosphorylation from EP to ADP (Steps 1 and 2) does not limit the Ca^{2+}_{out}-Ca^{2+}_{in} exchange even at low Mg^{2+} concentrations.

When a large excess of EGTA was added to the external medium 0.5 second before the addition of [γ-^{32}P]ATP otherwise in the same way as above, the ^{32}P incorporation was entirely prevented. This result shows that Ca^{2+} dissociation from Ca_{out}·E·ATP was so rapid as to be completed within 0.5 second after the EGTA addition. This demonstrates that Ca^{2+} dissociation from Ca_{out}·E·ATP (Step 1) is not rate-limiting in the Ca^{2+}_{out}-Ca^{2+}_{in} exchange.

It is thus clear that the observed Mg^{2+} activation of the Ca^{2+}_{out}-Ca^{2+}_{in} exchange is largely caused by the Mg^{2+}-induced acceleration of an exchange between the bound Ca^{2+} of EP and the Ca^{2+}_{in} through Steps 3 and 4. We showed previously (28) that Ca^{2+} dissociation from Ca_{in}·EP (Step 4) was unaffected by Mg^{2+}. We can conclude, therefore, that Mg^{2+} permits the bound Ca^{2+} of EP to be exposed to the internal medium through Step 3 and as a result activates the EP-mediated Ca^{2+}_{out}-Ca^{2+}_{in} exchange across the vesicular membranes.

The Ca^{2+}_{out}-Ca^{2+}_{in} exchange in the present study is distinct from that reported previously by Beirão and de Meis (39). They demonstrated an ADP-dependent Ca^{2+}_{out}-Ca^{2+}_{in} exchange under conditions in which EP formation was improbable. This exchange did not require Mg^{2+} and its rate increased only twice by addition of 3 mM $MgCl_2$. Katz et al. (40-42) also showed a Ca^{2+} efflux which was estimated indirectly from the determination of Ca^{2+} influx in the presence of ATP without added ADP. This efflux also depended on external Ca^{2+}. However, in contrast to our results, it was markedly inhibited by an elevation of Mg^{2+} in the range 0.1-1.1 mM (42).

REFERENCES

1. Ikemoto, N. (1976) J. Biol. Chem. 251: 7275-7277
2. Verjovski-Almeida, S., Kurzmack, M., and Inesi, G. (1978) Biochemistry 17: 5006-5013

3. Chiesi, M., and Inesi, G. (1979) J. Biol. Chem. 254: 10370-10377
4. de Meis, L., and Vianna, A. L. (1979) Annu. Rev. Biochem. 48: 275-292
5. Yamada, S., and Ikemoto, N. (1980) J. Biol. Chem. 255: 3108-3119
6. Takakuwa, Y., and Kanazawa, T. (1981) J. Biol. Chem. 256: 2691-2695
7. Martonosi, A. (1969) J. Biol. Chem. 244: 613-620
8. Inesi, G., Maring, E., Murphy, A. J., and McFarland, B. H. (1970) Arch. Biochem. Biophys. 138: 285-294
9. Panet, R., Pick, U., and Selinger, Z. (1971) J. Biol. Chem. 246: 7349-7356
10. Barlogie, B., Hasselbach, W., and Makinose, M. (1971) FEBS Lett. 12: 267-268
11. Makinose, M. (1971) FEBS Lett. 12: 269-270
12. Yamada, S., Sumida, M., and Tonomura, Y. (1972) J. Biochem. (Tokyo) 72: 1537-1548
13. Meissner, G. (1973) Biochim. Biophys. Acta 298: 906-926
14. Kanazawa, T., and Boyer, P. D. (1973) J. Biol. Chem. 248: 3163-3172
15. Masuda, H., and de Meis, L. (1973) Biochemistry 12: 4581-4585
16. de Meis, L., and Carvalho, M. G. C. (1974) Biochemistry 13: 5032-5038
17. Vianna, A. L. (1975) Biochim. Biophys. Acta 410: 389-406
18. Kanazawa, T. (1975) J. Biol. Chem. 250: 113-119
19. Kalbitzer, H. R., Stehlik, D., and Hasselbach, W. (1978) Eur. J. Biochem. 82: 245-255
20. Punzengruber, C., Prager, R., Kolassa, N., Winkler, F., and Suko, J. (1978) Eur. J. Biochem. 92: 349-359
21. Makinose, M., and Boll, W. (1979) in Cation Flux across Biomembranes (Mukohata, Y., and Packer, L., eds) pp. 89-100, Academic Press, New York
22. Kolassa, N., Punzengruber, C., Suko, J., and Makinose, M. (1979) FEBS Lett. 108: 495-500
23. Chiesi, M., and Inesi, G. (1980) Biochemistry 19: 2912-2918
24. Hasselbach, W., Fassold, E., Migala, A., and Rauch, B. (1981) Fed. Proc. 40: 2657-2661
25. Lacapère, J. J., Gingold, M. P., Champeil, P., and Guillain, F. (1981) J. Biol. Chem. 256: 2302-2306
26. Suko, J., Plank, B., Preis, P., Kolassa, N., Hellmenn, G., and Conca, W. (1981) Eur. J. Biochem. 119: 225-236
27. Martin, D. W., and Tanford, C. (1981) Biochemistry 20: 4597-4602
28. Takakuwa, Y., and Kanazawa, T. (1982) J. Biol. Chem. 257: 426-431
29. Makinose, M. (1973) FEBS Lett. 37: 140-143

30. Takakuwa, Y., and Kanazawa, T. (1981) J. Biol. Chem. 256: 2696-2700
31. Takakuwa, Y., and Kanazawa, T. (1979) Biochem. Biophys. Res. Commun. 88: 1209-1216
32. Sillén, L. G., and Martell, A. E. (1971) Stability Constants of Metal-ion Complexes, Suppl. 1, The Chemical Society, London
33. Takakuwa, Y., and Kanazawa, T. (1982) J. Biol. Chem. 257: 10770-10775
34. Endo, M. (1981) in the Regulation of Muscle Contraction (Ginnel, A. D., and Brazier, M. A. B., eds) pp. 181-194, Academic Press, New York
35. Morii, H., and Yamamoto, T. (1982) Biophysics 22: S68 (in Japanese)
36. Sumida, M., and Tonomura, Y. (1974) J. Biochem. (Tokyo) 75: 283-297
37. Ikemoto, N. (1975) J. Biol. Chem. 250: 7219-7224
38. Ikemoto, N., Garcia, A. M., Kurobe, Y., and Scott, T. L. (1981) J. Biol. Chem. 256: 8593-8601
39. Beirão, P. S., and de Meis, L. (1976) Biochim. Biophys. Acta 433: 520-530
40. Katz, A. M., Dunnett, J., Repke, D. I., and Hasselbach, W. (1976) FEBS Lett. 67: 207-208
41. Katz, A. M., Repke, D. I., Dunnett, J., and Hasselbach, W. (1977) J. Biol. Chem. 252: 1950-1956
42. Katz, A. M., Repke, D. I., Fudyma, G., and Shigekawa, M. (1977) J. Biol. Chem. 252: 4210-4214

REACTION MECHANISM OF ATP HYDROLYSIS BY SARCOPLASMIC RETICULUM WITH CaATP AS A SUBSTRATE

Munekazu Shigekawa
Shigeo Wakabayashi
Hiroshi Nakamura*

Department of Etiology and Pathophysiology and Department of Molecular Physiology*, National Cardiovascular Center Research Institute, Suita, Osaka, Japan

I. INTRODUCTION

Calcium transport and ATP hydrolysis by SR vesicles are catalyzed by the membrane-bound ATPase (1-3). Ca^{2+} binding to high affinity transport site of the ATPase initiates phosphorylation of the enzyme by ATP (4-7). The phosphoenzyme subsequently undergoes conformational transitions that cause translocation of Ca^{2+} across the SR membranes and a change in Ca^{2+} affinity of the transport site which results in release of Ca^{2+} into the intravesicular lumen (8). Mg^{2+} is required for the optimum activity of the calcium pump (1,2). Mg^{2+} is a component of the MgATP complex which is considered to be the physiological substrate for the ATPase (8,9). Mg^{2+}, in addition, accelerates phosphoenzyme hydrolysis (6,7,10-12), which is presumably responsible for acceleration of ATP hydrolysis and calcium transport by SR vesicles by this ion. These actions of Mg^{2+} can be antagonized by Ca^{2+} because CaATP serves as a substrate for enzyme phosphorylation (13) and because the phosphoenzyme hydrolysis is inhibited by high Ca^{2+} (6,7,10). At present, however, the precise mechanism for the stimulatory effect of Mg^{2+} and the quantitative relationship between the effects of Mg^{2+} and Ca^{2+} are not known. In the present study, we characterized ATP hydrolysis with CaATP as a substrate in order to obtain information about the role of divalent cation in ATP hydrolysis. The existence and properties of this type of ATP hydrolysis have not so far been

ISBN 0-12-260380-X

established. The results indicated that the ATPase can turn over normally with CaATP as a substrate but at a very slow rate. Mg^{2+} was not required for the turnover of the ATPase *per se*.

II. MATERIALS AND METHODS

Sarcoplasmic reticulum vesicles were prepared from rabbit skeletal muscle as described previously (14) with slight modifications. The ATPase protein was purified from the vesicles by solubilizing them with deoxycholate as described by Nakamura and Tonomura (15). The resultant preparations did not accumulate Ca^{2+}. Alkali metal salts were removed from the ATPase preparations by passage through a BIO-RAD AG 50W-x8 (Tris) column. Protein concentration, ATPase activity, phosphoenzyme levels and rapid enzyme phosphorylation were measured as described previously (14). ATPase reaction was started by the addition of ATP to the reaction medium containing Ca^{2+} and ATPase protein.

The amount and rate of release of ^{32}P-label bound to the enzyme were measured by a double-membrane filtration method as described previously (16) with modifications. The experiments were carried out in the presence of 100 mM D-[^{3}H]glucose and 5.8 μM [γ-^{32}P]ATP. At appropriate intervals after the start of reaction, 0.03 ml of an aliquot was taken from the reaction medium and applied to a set of two membrane filters (upper, GS 0.22-μm Millipore filter (Φ = 25 mm); lower, HA 0.45-μm Millipore filter (Φ = 13 mm)) which was then sucked under vacuum at 2°C for less than 1 s. The amount of ^{32}P-label in the filtrate was calculated from the ratio of the radioactivities of ^{32}P and ^{3}H trapped in the lower membrane filter as described previously (16). The amount of the enzyme-bound ^{32}P-label was estimated as the difference between those of ^{32}P-label in the filtrate obtained in the presence or absence of 1.5 mM nonradioactive ATP.

The amount of calcium bound to the ATPase was measured by a membrane filtration method. At appropriate intervals after the addition of EDTA to the reaction mixture as described in the legends to Fig. 4, TABLEs IV and V, 0.1 ml of the reaction mixture was rapidly passed through a membrane filter (HA 0.45-μm Millipore filter) by suction at 2°C. Then, the membrane filter was washed 3 times with 2 ml of a solution containing imidazole/HCl (pH 7.0) and EDTA, the concentrations of which were the same as those in the reaction mixture. Washing usually took approximately 6 s. In the experiment shown in TABLE IV where high KCl was added to induce phosphoenzyme hydrolysis, the first washing could wash KCl away from the ATPase protein within 1 s after the start of filtration. The radioactivity bound to the filter was then measured. The amounts of bound calcium thus obtained were plotted in Fig. 4, TABLEs IV and V against the time at which filtration was initiated. Concentrations of free metals and the metal-ATP complexes were calculated from the total concentrations of calcium, magnesium, potassium and ATP. The following logarithmic association constants were used: H^+ to ATP^{4-}, 6.56 (0.4°C); H^+ to $HATP^{3-}$, 4.29 (0.4°C); Ca^{2+} to ATP^{4-}, 4.10 (0.4°C); Ca^{2+} to $HATP^{3-}$, 2.34 (0.4°C); Mg^{2+} to ATP^{4-}, 3.97 (0.4°C); Mg^{2+} to $HATP^{3-}$, 1.95 (0.4°C); K^+ to ATP^{4-}, 0.9 (25°C) (17). [γ-^{32}P]ATP was synthesized by the method of Glynn and Chappel (18) with slight modifications.

III. RESULTS

A. Enzyme Phosphorylation and its Reversal

When phosphorylation of the ATPase was measured in 5 mM $CaCl_2$, 0.1 M KCl and 20 mM imidazole/HCl (pH 7.0) at 0°C in the presence of 0.25 to 11.0 μM [γ-^{32}P]ATP, its time course exhibited an initial short lag phase followed by a linear increase in the phosphoenzyme level. The initial rate of enzyme phosphorylation (V_f) was estimated from the initial linear portion of the time course. Double reciprocal plot of V_f values versus calculated concentrations of CaATP was found to be linear. The values of the kinetic parameters obtained from the plot were presented in TABLE I. TABLE I also

TABLE I. Kinetic Parameters for Enzyme Phosphorylation

	K_m	V_{max}	
	(μM)	(nmol/mg/s)	(s^{-1})
CaATP	0.332	1.25	(0.26)[a]
MgATP	3.08 3.46[b]	9.71 10.9[b]	(2.01)[a] (2.26)[a]

[a]Values devided by the active site concentration (4.82 nmol/mg) which was estimated in the ATP binding experiment.
[b]Values corrected for the effect of CaATP using equation 2 of "APPENDIX".

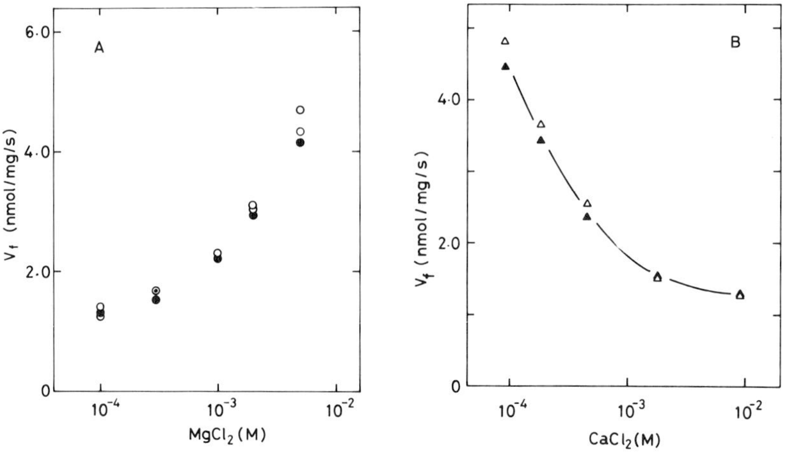

Figure 1. Dependence of V_f values on Ca^{2+} and Mg^{2+} concentrations. V_f values were measured at 0°C with 0.086 mg/ml of ATPase protein in 20 mM imidazole/HCl (pH 7.0) and 0.1 M KCl either in 0.457 mM $CaCl_2$, 5.56 μM [γ-^{32}P]ATP and the concentrations of $MgCl_2$ indicated (o)(A) or in 1.04 mM $MgCl_2$, 8.31 μM [γ-^{32}P]ATP and the concentrations of $CaCl_2$ indicated (△) (B). ● and ▲ indicate the V_f values calculated according to Equation 1 of "APPENDIX" using V_{max} and K_m values of 1.25 nmol/mg/s and 0.332 μM, and 10.9 nmol/mg/s and 3.46 μM for CaATP and MgATP, respectively.

includes values of kinetic parameters for enzyme phosphorylation by MgATP which was measured in 5.2 mM $MgCl_2$, 46 μM $CaCl_2$ and 0.5 to 8.1 μM [γ-^{32}P]ATP under otherwise the same conditions. As shown in the table, the estimated values of the maximum rate of enzyme phosphorylation (V_{max}) and K_m were 8 to 9 times smaller for CaATP than for MgATP. The values of V_{max} and K_m for MgATP may even be underestimated because 46 μM $CaCl_2$ was present in the reaction medium to fully activate the transport site of the ATPase. The CaATP complex formed from this low added $CaCl_2$ and added ATP might have effectively competed with MgATP for the phosphorylation site because of its high affinity although its amount was only 1.2% of the total metal-ATP complex. If Equation 2 of "APPENDIX" was used to obtain the values of K_m and V_{max} for MgATP corrected for the effect of CaATP, 12% greater values were calculated (TABLE I). In the experiment of Fig.1, V_f values were measured in 0.093 to 9.15 mM $CaCl_2$ in the presence of 1.04 mM $MgCl_2$ and 8.3 μM [γ-^{32}P]ATP (B) or in 0.1 to 5.0 mM $MgCl_2$ in the

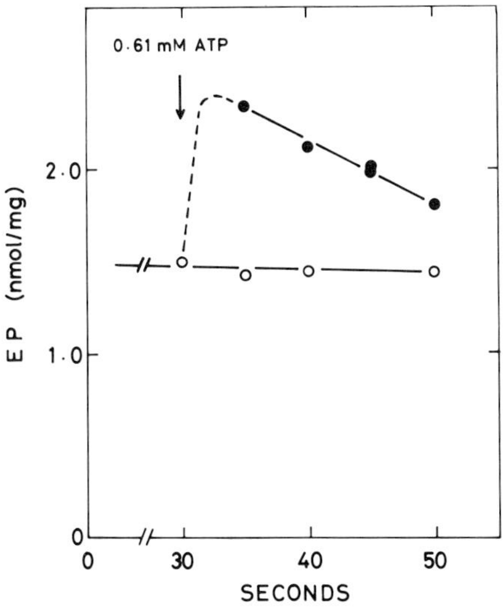

Figure 2. Increase in the ^{32}P-phosphoenzyme level induced by the addition of excess nonradioactive ATP. ATPase protein (0.5 mg/ml) was phosphorylated in 4.3 μM [γ-^{32}P]ATP and no added KCl otherwise under the conditions described in the legend to Table II. At 30 s nonradioactive ATP was added and the time course of the ^{32}P-phosphoenzyme level was followed (●).

presence of 0.457 mM $CaCl_2$ and 5.56 μM [γ-^{32}P]ATP (A). As shown in the figure, the V_f values decreased with increasing concentrations of $CaCl_2$ (B) or decreasing concentrations of $MgCl_2$ (A). The measured values of V_f were found to be in fairly good agreement with the values estimated theoretically using the values of the kinetic parameters listed in TABLE I (see the legend to Fig. 1). The results were consistent with the assumption that CaATP and MgATP react with the phosphorylation site as a competitive inhibitor with respect to each other. The results also appeared to indicate that V_f values were not affected significantly by the concentration ranges of free Ca^{2+} and free Mg^{2+} used.

In the next experiment, reversal of enzyme phosphorylation by added ADP was measured after the ATPase was phosphorylated in 4.4 mM $CaCl_2$ and 0.29 μM [γ-^{32}P]ATP in the presence and absence of 0.1 M KCl. When 1.3 mM ADP and 0.6 mM nonradioactive ATP was added to stop further formation of phosphoenzyme and to induce phosphoenzyme dephosphorylation, dephosphorylation exhibited single exponential time courses at

TABLE II. Existence of Enzyme-ATP Complex during ATP Hydrolysis in High Ca^{2+}

Enzyme phosphorylation was carred out at 0°C with 0.981 mg/ml ATPase protein in 20 mM imidazole/HCl (pH 7.0), 5.0 mM $CaCl_2$ and 5.8 μM [γ-^{32}P]ATP in the presence of KCl and ADP concentrations indicated in the table. At 60 s after the start of reaction, the levels of phosphoenxyme and ^{32}P-label bound to the enzyme were measured as described in "MATERIALS AND METHODS".

KCl	ADP	A:^{32}P bound	B:EP	A-B
(mM)	(μM)		(nmol/mg)	
0	0	3.11 ± 0.02*	1.42 ± 0.03	1.69
	5	3.05 ± 0.02	0.81 ± 0.01	2.24
150	0	2.88 ± 0.03	2.44 ± 0.03	0.44
	50	2.12 ± 0.05	1.07 ± 0.01	1.05

*mean ± S.D.(n=6)

least up to 75 % of its decay after initial short lags (cf. Fig. 5 of ref. 14). The estimated rate constants for dephosphorylation were 1.5 and 2.6 s^{-1} in the presence and absence of 0.1 M KCl, respectively. As less than 1% of the original phosphoenzyme was hydrolyzed to yield P_i during the ADP-induced dephosphorylation, most of the phosphoenzyme transferred its phosphate to ADP to form ATP. In the presence of high Ca^{2+}, therefore, reversal of enzyme phosphorylation was much faster than enzyme phosphorylation.

B. Properties of the Precursor Complex of ATP with Enzyme

In the experiment of TABLE II, total amounts of ^{32}P-label bound to the ATPase and phosphoenzyme levels were measured simultaneously during steady state ATP hydrolysis in 5 mM $CaCl_2$ and 5.8 μM [γ-^{32}P]ATP in the presence and absence of 150 mM KCl. At 150 mM KCl, the fraction of the phosphoenzyme was 85% of the total ^{32}P-label bound to the enzyme while it was less than half of the bound ^{32}P-label in the absence of added KCl. When ADP was included in the reaction medium, the phosphoenzyme levels decreased significantly while the amount of ^{32}P-label bound to the enzyme decreased to a much less extent. Therefore, a significant amount of the enzyme-ATP complex was present during ATP hydrolysis under these conditions especially in the absence of added KCl and/or in the presenc of ADP. Essentially similar results were obtained when a similar type of experiment was carried out in 0.1 mM $CaCl_2$ where the turnover of the phosphoenzyme was observed (cf. TABLE III). The enzyme-ATP complex described above was the precursor of the phosphoenzyme as shown in Fig. 2. In this experiment, the ATPase was first phosphorylated in 5 mM $CaCl_2$ and 4.3 μM [γ-^{32}P]ATP in the absence of added KCl and then excess nonradioactive ATP was added to dilute the radioactivity of ATP. The phosphoenzyme level increased significantly upon addition of nonradioactive ATP, indicating that enzyme phosphorylation from the enzyme-[γ-^{32}P]ATP complex was faster than dissociation of [γ-^{32}P]ATP from the enzyme-[γ-^{32}P]ATP complex. Dissociation of ATP from the enzyme-ATP complex could be measured directly by a membrane filtration method as described in "MATERIALS AND METHODS". In this experiment, the phosphoenzyme was first formed in 5.0 mM $CaCl_2$ and 5.8 μM [γ-^{32}P]ATP in the absence of added KCl (cf. TABLE II). At 30 s after the start of phosphorylation, 1.3 mM ADP and 0.6 mM nonradioactive ATP was added to induce phosphoenzyme dephosphorylation and then the amount of ^{32}P-label liberated in the reaction medium was measured. The ^{32}P-label liberation approximately obeyed first order kinetics and the rate constant of 0.15 s^{-1} was measured in the absence of added KCl. When KCl was added to a final concentration of 0.1 _M_ together with excess ADP and ATP, the rate constant of 0.3 s^{-1} was measured. As the phosphoenzyme dephosphorylation was much faster under the conditions used (see above), the estimated rate constants corresponded to those for ATP liberation from the enzyme-[γ-^{32}P]ATP complex.

C. Turnover of the Phosphoenzyme Formed in the Presence of Ca^{2+} Alone

In the absence of added Mg^{2+} but in the presence of 0.1 mM $CaCl_2$ at 0°C, the ATPase catalyzed slow ATP hydrolysis (TABLE III). The steady state rate of ATP hydrolysis was two-fold less in 0.1 M KCl than in the absence of added KCl, while the steady state phosphoenzyme level was twice greater in 0.1 M KCl. The observed slow ATP hydrolysis was concluded to be due to the turnover of the calcium pump ATPase because no Ca^{2+}-independent, divalent cation-dependent ATPase activity could be detected in the ATPase preparations used in the present study even in the presence of up to 6 mM $MgCl_2$ and under the conditions similar to those of TABLE III.

The phosphoenzyme formed at the steady state in the absence of added KCl could be resolved kinetically into the rapidly decomposing and slowly decomposing fractions in the

TABLE III. Effect of KCl on Partial Reaction of ATP Hydrolysis in 0.1 mM $CaCl_2$

The steady state rate of ATP hydrolysis (V) and the total amount of phosphoenzyme (EP) were measured at 0°C with 0.486 mg/ml of ATPase protein in 20 mM imidazole/HCl (pH 7.0), 0.1 mM $CaCl_2$ and 10.5 μM [γ-^{32}P]ATP in the presence and absence of 0.1 M KCl. The ADP-sensitive (E_1P) and ADP-insensitive (E_2P) phosphoenzyme levels were estimated at 60s after the start of phosphorylation by measuring the time courses of phosphoenzyme decomposition after the addition of 0.80 mM nonradioactive ATP plus 2.14 mM ADP. The amounts of E_2P were estimated by extrapolating the time course of decomposition of the slowly decomposing phosphoenzyme to the time of addition of ATP and ADP (cf. Fig. 3).

	V	EP	E_1P	E_2P	V/E_1P	V/E_2P
	(nmol/mg/s)		(nmol/mg)		(min^{-1})	
no KCl	0.991	1.37	0.83	0.54	1.18	1.84
0.1 M KCl	0.499	2.71	2.65	0.06	0.19	8.33

presence of excess ADP and nonradioactive ATP (cf. Figs. 3 and 4). The rapidly decomposing fraction disappeared without producing P_i, indicating that this fraction represents the ADP-sensitive fraction that reacts with added ADP to form ATP (10,11,19). The slowly decomposing fraction represents the ADP-insensitive phosphoenzyme (cf. 20). In the absence of added KCl, the ADP-insensitive phosphoenzyme accounted for 40% of the phosphoenzyme while it accounted for only less than 5% of the phosphoenzyme in the presence of 0.1 M KCl (TABLE III). The apparent rate constant for conversion of the ADP-sensitive to ADP-insensitive phosphoenzymes estimated as the ratio between the rate of ATP hydrolysis and the level of the ADP-sensitive phosphoenzyme was 1.18 min^{-1} in the absence of added KCl, which decreased to a 6-fold less value in the presence of 0.1 M KCl (TABLE III). In contrast to this observation, the decay constant of the ADP-insensitive phosphoenzyme estimated as the ratio between ATP hydrolysis rate and the level of the ADP-insensitive phosphoenzyme was much greater in the presence of 0.1 M KCl than in the absence of this salt.

The ADP-insensitive phosphoenzyme formed in the presence of low Ca^{2+} alone decomposed at a slow rate. High added KCl accelerated its decomposition markedly (Fig. 3). It should be noted that decomposition of the ADP-insensitive phosphenzyme

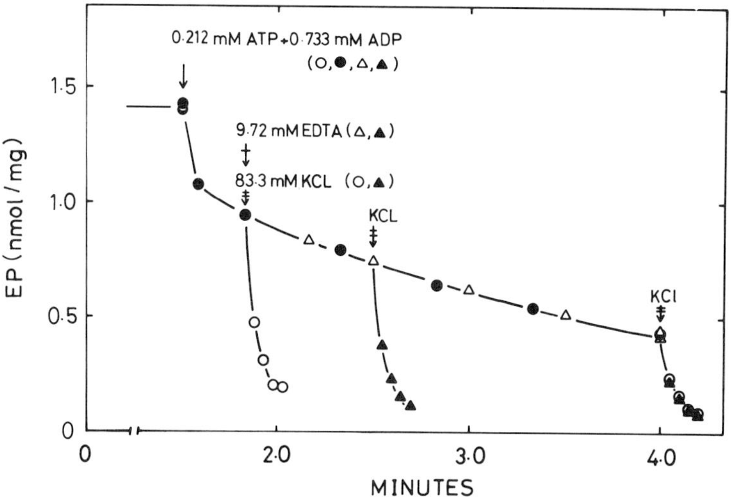

Figure 3. Preservation of reactivity with KCl of ADP-insensitive phosphoenzyme formed in the absence of added Mg^{2+} and added alkali metal salts after a chase with excess EDTA. ATPase protein (0.41 mg/ml) was phosphorylated at 0°C in 20 mM imidazole/HCl (pH 7.0), 30 μM $CaCl_2$ and 1.59 μM [γ-^{32}P] ATP. At appropriate intervals after the start of reaction, ADP and nonradioactive ATP, EDTA, and KCl were added. The time of addition and the final concentration of each reagent are shown in the figure.

occurred at the same rate whether it was measured in the presence of excess ADP and nonradioactive ATP or excess ADP and nonradioactive ATP plus excess EDTA, a chelator of free divalent cations (Fig. 3). The ADP-insensitive phosphoenzyme remained sensitive to added KCl even after the incubation with excess EDTA for longer than 2 min. Qualitatively similar results were obtained when the ADP-insensitive phosphoenzyme was formed in high Mg^{2+} and low Ca^{2+}. The ADP-insensitive phosphoenzyme formed in 2.5 mM $MgCl_2$ and 12 μM $CaCl_2$ under otherwise similar conditions to those used in Fig. 3, decomposed exponentially with a rate constant of 0.45 min^{-1} in the presence of 42 mM EDTA and 1.0 mM ADP. When 83 mM KCl was added to the reaction medium 85 s after the addition of EDTA (cf. Fig. 3), the rate constant of phosphoenzyme decomposition increased by 25-fold. In Fig. 4, it was shown that calcium was tightly bound to the enzyme under the conditions similar to those used in the experiment of Fig. 3. The bound calcium was released from the enzyme in parallel with spontaneous

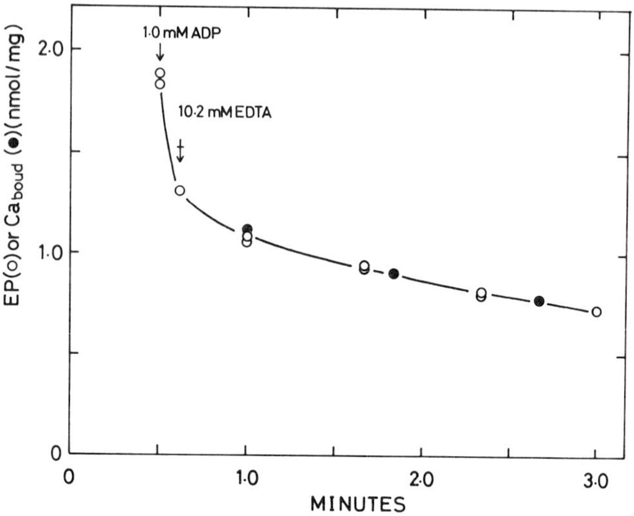

Figure 4. Tight binding of calcium to the enzyme. ATPase protein (1.35 mg/ml) was phosphorylated at 0°C in 20 mM imidazole/HCl (pH 7.0). 50.7 μM $^{45}CaCl_2$ and 20 μM ATP or [γ-^{32}P]ATP. At 30 s and 37 s after the start of the reaction, ADP and EDTA were added to the reaction medium to final concentrations of 1.0 and 10.2 mM, respectively and then the amounts of phosphoenzyme and enzyme-bound calcium were measured as described in "MATERIALS AND METHODS". The amount of enzyme-bound calcium shown in the figure (●) is the average of four determinations. The standard deviation for these determinations is less than ±2% of the average.

decomposition of the ADP-insensitive phosphoenzyme (Fig. 4). When rapid decomposition of the ADP-insensitive phosphoenzyme was induced by the addition of KCl, the amount of calcium released from the enzyme again agreed well with that of the decrease in the ADP-insensitive phosphoenzyme level (TABLE IV). It should be added that inclusion of 5.7 μM Ca^{2+} ionophore A23187 in the reaction medium did not affect the levels of bound calcium measured in these experiments. In contrast to the observations described above, no tightly bound calcium could be detected when enzyme phosphorylation was carried out in the presence of 4 mM $MgCl_2$ under otherwise the same experimental conditions as those used in Fig. 4 (TABLE V).

TABLE IV. Effect of KCl Addition on ADP-insensitive Phosphoenzyme Level and Amount of Calcium Tightly Bound to Enzyme

Reaction was carried out as described in the legend to Fig. 4. At 23 s after the addition of EDTA , KCl was added to a final concentration of 83 mM and phosphoenzyme level and the amount of calcium bound to the enzyme were measured as described in "MATERIALS AND METHODS".

Time after addition of KCl	ADP-insensitive phosphoenzyme	Ca tightly bound to enzyme
(s)	(nmol/mg)	(nmol/mg)
0	0.816, 0.811	0.862, 0.867
10	0.354, 0.376	0.344
12	0.320	0.319, 0.331

IV. DISCUSSION

Yamada and Ikemoto (13) have recently presented evidence that CaATP serves as a substrate for enzyme phosphorylation. In the present study, it was shown that the values of K_m and V_{max} for enzyme phosphorylation at 0°C and pH 7.0 were 8 to 10

TABLE V. Amount of Calcium Tightly Bound to Enzyme Obtained in the Presence of High Mg^{2+}

ATPase protein (1.43 mg/ml) was phosphorylated at 0°C in 20 mM imidazole/HCl (pH 7.0), 4.0 mM $MgCl_2$, 41 μM $^{45}CaCl_2$ or $CaCl_2$, and 20 μM ATP or [γ-^{32}P]ATP. At 15 s after the start of reaction, ADP and EDTA were added to the reaction medium to final concentrations of 1.3 mM and 22 mM, respectively, and the phosphoenzyme level and the amount of enzyme-bound calcium were measured.

Time after addition of ADP and EDTA	ADP-insensitive phosphoenzyme	Ca tightly Bound to enzyme
(s)	(nmol/mg)	(nmol/mg)
25	2.87, 2.86	0.027, 0.026
85	2.05, 2.02	0.018, 0.017

times smaller for CaATP than for MgATP (TABLE I). These substrates appeared to phosphorylate the enzyme as competitive inhibitor with respect to each other (Fig. 1). As Mg^{2+} contaminating the reaction medium used in the present study was found to be less than 1 μM, we concluded that ATP hydrolysis observed in the presence of Ca^{2+} alone was derived from the turnover of the ATPase reaction with CaATP as a substrate.

The phosphoenzyme formed from CaATP turns over normally but at a slow rate as slow ATP hydrolysis was observed (TABLE III). Both the ADP-sensitive and ADP-insensitive phosphoenzymes occurred during ATP hydrolysis. The rate constant for the transition of the ADP-sensitive to ADP-insensitive phosphoenzymes that was estimated as the ratio between ATP hydrolysis rate and the ADP-sensitive phosphoenzyme was 0.19 min^{-1} in 0.1 M KCl, which was more than 10 times slower than that obtained in the presence of high Mg^{2+} and low Ca^{2+} under the comparable experimental conditions (cf. refs 20 and 21). The ADP-insensitive phosphoenzyme formed in low Ca^{2+} alone had kinetic properties similar to those of the ADP-insensitive phosphoenzyme formed in high Mg^{2+} and low Ca^{2+} as both were stable in the presence of excess ADP but their hydrolysis was accelerated by added KCl (Fig. 3 and refs. 20 and 21). It should be noted that the ADP-insensitive phosphoenzyme retained sensitivity to added KCl even in the presence of excess EDTA, indicating that free divalent cations are not necessary for the reactivity of the ADP-insensitive phosphoenzyme. This result is consistent with the observation by Garrahan et al.(12) that hydrolysis of the phosphoenzyme formed in high Mg^{2+} and low Ca^{2+} does not require free Mg^{2+}. It was further shown in the present study that when the enzyme had been phosphorylated in low Ca^{2+}, calcium remained bound to the enzyme even in the presence of excess EDTA. The amount and behavior of the enzyme-bound calcium paralleled with those of the ADP-insensitive phosphoenzyme. These results strongly suggest that the ADP-insensitive phosphoenzyme has tightly bound calcium, its stoichiometry being 1 mole per mole of the phosphenzyme. The source for this enzyme-bound calcium is not known at present. Ca^{2+} which binds to the calcium transport site of the ATPase enzyme is considered to be released from the enzyme upon formation of the ADP-insensitive phosphoenzyme because the ADP-insensitive phosphoenzyme exhibits low affinity for Ca^{2+} (19,22-24). Ca^{2+} also binds to the substrate site of the enzyme as a component of the CaATP complex (13). This calcium appears to remain bound to the enzyme after ADP has been released from the enzyme because free ADP participates in the reversal of enzyme phosphorylation(13). It would be quite possible, therefore, that this calcium would remain bound to all forms of the phosphoenzyme until the ADP-insensitive phosphoenzyme is hydrolyzed, and thus may provide the source for the tightly bound calcium detected in the present study. Such an interpretation is consistent with the present finding that no tightly bound Ca^{2+} could be detected in the enzyme in the presence of high Mg^{2+} where the MgATP complex is the substrate for enzyme phosphorylation.

During ATP hydrolysis with CaATP as a substrate, a significant fraction of the enzyme existed in the form of the enzyme-ATP complex especially in the presence of low KCl and/or in the presence of ADP (TABLE II). This enzyme-ATP complex appeared to be a precursor of phosphoenzyme as the latter was formed from the former and added ADP increased levels of the former while it decreased levels of the latter (TABLE II and Fig. 2). Dissociation of ATP from the enzyme-ATP complex was slow (see "RESULTS"). Existence of the enzyme-ATP complex which was a precursor of phosphoenzyme was detected in our previous study (14) where enzyme phosphorylation was carried out in 1 mM $MgCl_2$ and 5 mM $CaCl_2$ under conditions otherwise similar to those used in the present study. Under these conditions, dissociation of ATP from the precursor enzyme-ATP complex was slow while dissociation of AMP-PCP, an analogue of ATP, from the enzyme-AMP-PCP complex was much faster (14). According to the present result, most of the enzyme was phosphorylated from CaATP even in 1.0 mM $MgCl_2$ and 5 mM $CaCl_2$ (Fig. 1 B). Therefore, the kinetic properties of the enzyme-substrate complex reported previously (14) appears to be primarily those of the enzyme-substrate complex formed from CaATP and the enzyme. Based on the observations described above, the following reaction sequence may be proposed for ATP hydrolysis by the calcium pump ATPase:

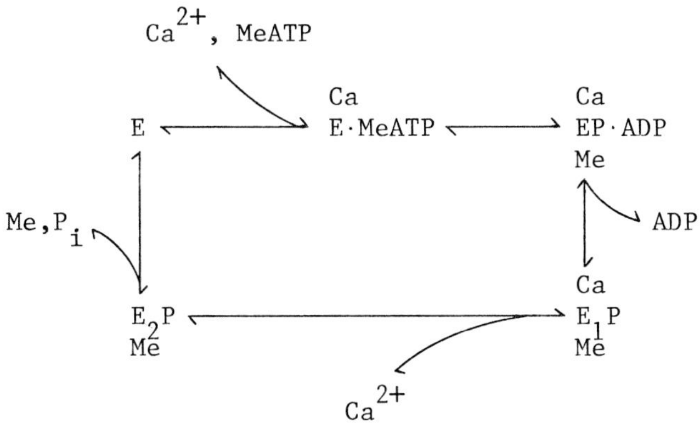

In this scheme, only the fates of the metal-ATP complex and Ca^{2+} which binds to the transport site of the enzyme are shown. Me denotes the metal moiety of the metal-ATP complex.

Dissociation of ATP from E•CaATP was slow. As dissociation of AMP-PCP from the enzyme-AMP-PCP complex was much faster under the same conditions (14), E·CaATP formation is not a simple encounter process. Enzyme phosphorylation by CaATP may include isomerization of the enzyme-substrate complex as proposed previously (14). E_1P and E_2P denote ADP-sensitive and ADP-insensitive, K^+-sensitive phosphoenzymes, respectively. The apparent rate constant for the E_1P to E_2P conversion was very small when CaATP served as a substrate (TABLE III). As the ATPase can complete its catalytic cycle with CaATP as a substrate, Mg^{2+} is not essential for the turnover of the ATPase *per se*.

In the presence of high Mg^{2+}, where MgATP serves as a substrate for ATP hydrolysis, the values of V_{max} and K_m for ATP obtained for enzyme phosphorylation, and the apparent rate constant for the E_1P to E_2P transition were much greater than those obtained in the presence of Ca^{2+} alone (TABLEs I and III, see also data in refs. 20 and 21). At present, however, the precise mechanism for activation of ATP hydrolysis by Mg^{2+} is not known. It is possible that Mg^{2+} derived from the substrate, MgATP, is responsible for faster reaction rate of these partial reactions. Alternatively, accelerated partial reactions may be due to binding of Mg^{2+} to a site which is distinct from the substrate site for MgATP as proposed by Takakuwa and Kanazawa (25). If the former mechanism is the case, it follows that the metal component of the metal-ATP complex determines the apparent affinity of the active site of the ATPase for the substrate and the rate constants of the subsequent reaction steps.

V. APPENDIX

If both CaATP and MgATP interact with the phosphorylation site of the enzyme competitively with respect to each other and if the steady state conditions may be applied to enzyme phosphorylation by each substrate, the initial rate of enzyme phosphorylation could be given by:

$$V_f = \frac{V_{maxA}\cdot(A) / K_{mA} + V_{maxB}\cdot(B) / K_{mB}}{1 + (A) / K_{mA} + (B) / K_{mB}} \quad \text{(Equation 1)}$$

where (A) and (B) denote concentrations of MgATP and CaATP, respectively. K_{mA} and K_{mB} are K_m's for MgATP and CaATP, respectively, while V_{maxA} and V_{maxB} are V_{max}'s for MgATP and

CaATP, respectively. If (B) is much less than (A) and if V_{maxA} / K_{mA} is not much different from V_{maxB} / K_{mB}, $V_{maxA} \cdot (A) / K_{mA} \gg V_{maxB} \cdot (B) / K_{mB}$. Therefore, Equation 1 can be simplified to:

$$V_f = V_{maxA}^* / (1 + K_{mA}^* / (A)) \quad \text{(Equation 2)}$$

where $V_{maxA}^* = V_{maxA} / (1 + \alpha)$, $K_{mA}^* = K_{mA} / (1 + \alpha)$ and $\alpha = (B) \cdot K_{mA} / (A) \cdot K_{mB}$. V_{maxA}^* and K_{mA}^* is the apparent values of V_{max} and K_m for MgATP measured in the experiment of TABLE I. From these values, V_{maxA} and K_{mA} can be calculated.

ACKNOWLEDGMENTS - We are grateful to Drs. Y. Nimura, A. Yamamoto and T. Nakamura of this institute for their encouragements.

REFERENCES

1. Ebashi, S., and Lipmann, F. (1962) J. Cell Biol. 14, 389-400
2. Hasselbach, W., and Makinose, M. (1963) Biochem. Z. 339, 94-111
3. Weber, A., Herz, R., and Reiss, I. (1966) Biochem. Z. 345, 329-369
4. Yamamoto, T., and Tonomura, Y. (1967) J. Biochem. 62, 558-575
5. Makinose, M. (1969) Eur. J. Biochem. 10, 74-82
6. Martonosi, A. (1969) J. Biol. Chem. 244, 613-620
7. Inesi, G., Maring, E., Murphy, A. J., and MacFarland, B. H. (1970) Arch. Biochem. Biophys. 138, 258-294
8. Inesi, G. (1979) in Membrane Transport in Biology (Giebisch, G., Tosteson, D., and Ussing, H., eds.) p. 357, Springer-Verlag, West Berlin and Heidelberg
9. Vianna, A.L. (1975) Biochim. Biophys. Acta 410, 389-406
10. Panet, R., Pic, U., and Selinger, Z. (1971) J. Biol. Chem. 246, 7349-7356
11. Kanazawa, T., Yamada, S., Yamamoto, T., and Tonomura, Y. (1971) J. Biochem.(Tokyo) 70, 95-123
12. Garrahan, P.J., Rega, A.F., and Alonso, G.L. (1976) Biochim. Biophys. Acta 448, 121-132
13. Yamada, S., and Ikemoto, N. (1980) J. Biol. Chem. 255, 3108-3119

14. Shigekawa, M., and Kanazawa, T. (1982) J. Biol. Chem. 257, 7657-7665
15. Nakamura, Y., and Tonomura, Y. (1982) J. Biochem.(Tokyo) 91, 449-461
16. Yamaguchi, M., and Tonomura, Y. (1979) J. Biochem.(Tokyo) 86, 509-523
17. Sillen, L., G., and Martell, A.E. (1971) Stability Constants of Metal-ion Complexes, Suppl. 1, The Chemical Society, London
18. Glynn, I.M., and Chappel, J.B. (1964) Biochem. J. 90, 147-149
19. Shigekawa, M., and Dougherty, J.P. (1978) J. Biol. Chem. 253, 1458-1464
20. Shigekawa, M., and Akowitz, A.A. (1979) J. Biol. Chem. 254, 4726-4730
21. Shigekawa, M., and Dougherty, J.P. (1978) J. Biol. Chem. 253, 1451-1457
22. Knowles, A.F, and Racker, E. (1975) J. Biol. Chem. 250, 1949-1951
23. de Meis, L., and Tume, R.K. (1977) Biochemistry 16, 4455-4463
24. Watanabe, T., Lewis, D., Nakamoto, R., Kurzmack, M., Fronticelli, C., and Inesi, G. (1981) Biochemistry 20, 6617-6625
25. Takakuwa, Y., and Kanazawa, T., (1982) J. Biol. Chem. 257, 426-431

MECHANISM OF ATP SYNTHESIS IN SARCOPLASMIC RETICULUM

Leopoldo de Meis[1]

Instituto Ciencias Biomedicas
Departmento Bioquimica
Universidade Federal Rio de Janeiro
Cidade Univerntaria-Ilha Fundao 21910
Rio de Janeiro - Brasil

Vesicles derived from the sarcoplasmic reticulum retain a membrane bound ATPase which can catalyze both the hydrolysis and the synthesis of ATP. When suspended in a medium containing ATP and Mg^{2+}, these vesicles can reduce the Ca^{2+} concentration of the medium from 10^{-4} to less than 10^{-7} M. In this process the ATPase uses the chemical energy derived from the hydrolysis of ATP to pump Ca^{2+} into the vesicles and to maintain a steep Ca^{2+} concentration gradient that is formed across the vesicle membrane. In 1971 (1-3) it was shown that the entire process of Ca^{2+} transport can be reversed. When a Ca^{2+} concentration gradient is formed across the vesicle membrane, depending on the conditions used, the ATPase can catalyze either the synthesis of ATP or a rapid ATP $\rightleftharpoons$ P_i exchange. During ATP $\rightleftharpoons$ P_i exchange, the ATPase operates simultaneously forward (ATP hydrolysis) and backwards (ATP synthesis from ADP and P_i). When vesicles previously loaded with calcium are incubated in a medium containing EGTA, Ca^{2+} flows out of the vesicles at a slow rate due to the low Ca^{2+} permeability of the membrane. The Ca^{2+}

---

[1]Supported in part by CNPq, FUJB, FINEP and USPH (HL 27867)

ISBN 0-12-260380-X

efflux is sharply increased when ADP, P_i and Mg^{2+} are added to the incubation medium (1). The increment of Ca^{2+} efflux is coupled to the synthesis of ATP (3). For every two calcium ions released from the vesicles, one molecule of ATP is synthesized. The synthesis of ATP is initiated by phosphorylation of the enzyme by Pi (4-6). The acyl phosphate residue thus formed can be measured by incubating vesicles previously loaded with Ca^{2+} in a medium at pH 7 containing EGTA, Mg^{2+}, and (^{32}P) P_i. Under optimal conditions, 2 to 4 μmol of phosphoenzyme is formed per gram of vesicles protein. The phosphorylation of the enzyme by P_i is not accompanied by an increment of Ca^{2+} efflux. The subsequent addition of ADP to the medium leads to a decrease of the steady-state level of phosphoenzyme, synthesis of ATP, and enhancement of the rate of Ca^{2+} efflux (4). Therefore the following minimal reaction sequence was proposed:

$$E + Pi \rightleftharpoons E{\sim}P + HOH \quad (1)$$
$$E{\sim}P + ADP \rightleftharpoons E + ATP \quad (2)$$

Yamada et al., (5,6) reported that at pH 7 and in presence of 5 mM Pi, the enzyme is not phosphorylated by Pi in the absence of a transmembrane Ca^{2+} gradient, and the steady-state level of phosphoenzyme increases progressively as the Ca^{2+} gradient is enhanced.

Another index of the reversal of the Ca^{2+} pump is the ATP$\rightleftharpoons$Pi exchange reaction. In 1971 (2), Makinose observed that when vesicles were incubated in a medium containing 0.2 mM $CaCl_2$, Mg^{2+}, ATP, ADP and ^{32}Pi, Ca^{2+} was accumulated by the vesicles and a steady state was reached in which the Ca^{2+} efflux was balanced by the ATP-driven influx (2,7,8). As soon as net uptake ceased, a steady state of exchange between $^{32}P_i$ and the (γ-^{32}P)ATP began. During the initial phase of Ca^{2+} accumulation, the vesicles catalyze the hydrolysis of ATP but not the incorporation of $^{32}P_i$ into the ATP pool. The ATP $\rightleftharpoons$ P_i exchange was abolished when the membrane of the vesicles was rendered leaky by various procedures that did not affect the transport ATPase (2). ATP$\rightleftharpoons$$P_i$ exchange indicates that the enzyme operates simultaneously forwards (ATP hydrolysis) and backwards (ATP synthesis from ADP and $^{32}P_i$). From these data it was inferred that when the steady state between Ca^{2+} influx and Ca^{2+} efflux was reached, the energy derived from the hydrolysis of ATP was used to maintain the Ca^{2+} gradient, and the energy derived from the gradient was used for the synthesis of ATP (2,4,8).

When the reversal of the Ca^{2+} pump was first described, it was proposed that the energy required for the synthesis of ATP was derived from the osmotic energy of the Ca^{2+} gradient (1-6). This interpretation was particularly attractive in the light of the chemiosmotic hypothesis proposed by Mitchell (9,10) to account for the synthesis of ATP in mitochondria. The osmotic energy of the Ca^{2+} gradient was regarded as driving the first step of ATP synthesis, the phosphorylation of the ATPase by Pi. The finding that the protein phosphate was an energy-rich acyl phosphate contributed to this conclusion. The ΔG^o values calculated from the ratio of Ca^{2+} concentration in the vesicles to that in the medium were compatible with those required for the synthesis of ATP or for the formation of an acyl phosphate residue (2-6).

Beginning in 1972, work in my laboratory was conducted aiming at the understanding of the mechanism by which the sarcoplasmic reticulum ATPase catalyze the synthesis of ATP. Since then it was found that the different intermediary steps of the catalytic cycle of the ATPase can be reversed in the absence of a Ca^{2+} gradient. The ATPase can catalyze both a continuous ATP$\rightleftharpoons$ Pi exchange and the net synthesis of ATP after a single catalytic cycle without the need of using the osmotic energy which might be derived from the Ca^{2+} gradient. The net synthesis of ATP can be promoted by a variety of perturbations, including Ca^{2+}, pH and water activity.

Phosphorylation by Pi in the absence of a Ca^{2+} gradient. The first question we raised when we started to work on the reversal of the Ca^{2+} pump was whether the synthesis of ATP was related to the unidirectional movement of calcium ions forced through the enzyme by the gradient (osmotic energy) or whether it was simply related to the enzyme being exposed to two media of different composition. The enzyme expands through the vesicle membrane. When a Ca^{2+} concentration is formed, the part of the enzyme which faces the outer surface of the vesicle is exposed to a medium without Ca^{2+} (EGTA) while the part of the enzyme which faces the vesicles lumen is exposed to a medium containing a high Ca^{2+} concentration. In order to explore this possibility, leaky vesicles were prepared using different procedures (2, 8,11-14). The membrane of leaky vesicles is disrupted and although the enzyme remains fully active, the vesicles are no longer able to accumulate Ca^{2+}. The first encouraging result was obtained when leaky vesicles were incubated in a medium at pH 7, without Ca^{2+} (excess EGTA), Mg^{2+} and 2 to 4 mM Pi (15). For this condition a small but significant level of phosphoenzyme was attained. This led us to vary the conditions of the assay in the hope to attain a higher level of phosphoenzyme. After different trials it was found that the level of phosphoenzyme increased several fold when the pH of the medium was decreased to 6, reaching values in the same range as those measured in the presence of a Ca^{2+} gradient (15).

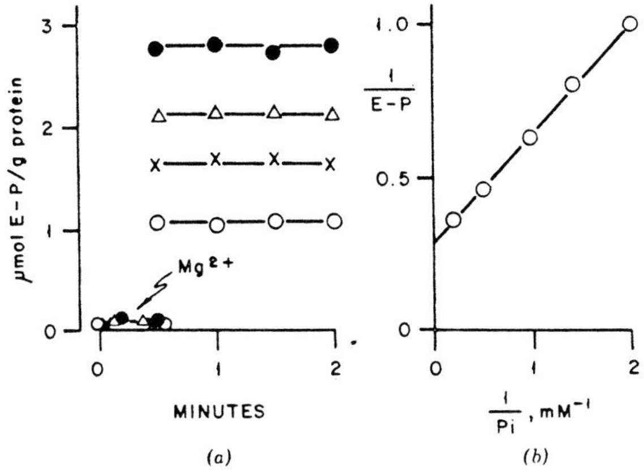

FIGURE 1. Phosphorylation by Pi in absence of a Ca^{2+} gradient. (a) The assay medium composition was 1.0 mM EGTA, 10 mM Tris maleate buffer (pH 6.0), and either (O) 0.5 mM, (X) 1 mM, (△) 2mM, or (●) 4 mM $^{32}P_i$. The reaction was performed at 37°C. It was started by the addition of leaky vesicles and arrested after different incubation intervals by the addition of ice-cold perchloric acid. At the time indicated by the arrow, $MgCl_2$ was added to a final concentration in the assay medium of 10 mM. For further details see Masuda and deMeis, 1973 (15). (b) Double reciprocal plot of the values of the equilibrium values of phosphoenzyme (E-P) as obtained in the left-hand figure.

As in the presence of a Ca^{2+} gradient, the phosphoenzyme formed with leaky vesicles is an acyl phosphate (15,16). This finding clearly shows that the energy derived from the Ca^{2+} gradient is not necessary for the first step of the reversal of the Ca^{2+} pump. A similar pH dependence (Fig 2) is observed when intact vesicles previously loaded with Ca^{2+} are used but in this case the increase of the phosphoenzyme level observed when the pH is lowered from 7 to 6 is less pronounced (17).

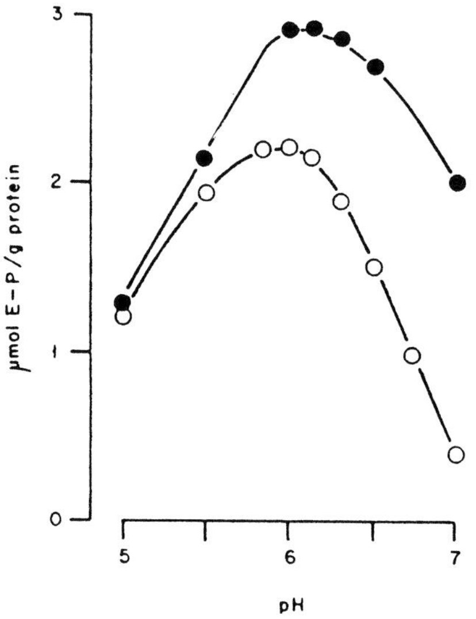

FIGURE 2. Phosphorylation by P_i, in the presence and absence of a transmembrane Ca^{2+} concentration gradient. The assay medium composition was 15 mM EGTA, 25 mM $MgCl_2$, 20 mM Tris maleate buffer at different pH values as shown in the figure 2.5 mM $^{32}P_i$, and 0.7 mg vesicle protein per ml. The reaction was performed at 30°C. It was started with the addition of vesicles and arrested after a 15-sec incubation by the addition of ice-cold perchloric acid. (●) Vesicles loaded with calcium phosphate, (O) leaky vesicles. For further details, see de Meis 1976 (17).

The equilibrium level of phosphoenzyme varies with the P_i concentration in the medium (Fig. 1). Plots of the reciprocal of the phosphoenzyme level versus the reciprocal of the P_i concentration yield a linear relationship. This demonstrates a saturation behavior and suggests involvement of a transient phosphate-enzyme complex $E.P_i$

$$E + P_i \rightleftharpoons E.P_i \overset{Mg^{2+}}{\rightleftharpoons} E\text{-}P + HOH$$

The apparent K_m for P_i varies significantly with the pH of the medium and according to whether leaky vesicles or vesicles previously loaded with Ca^{2+} are used (17,18). Table I shows that the presence of a Ca^{2+} gradient across the membrane promotes an increase of the apparent affinity of the enzyme for P_i. This is more pronounced at pH 7 than at pH 6.

TABLE I. Modification of the apparent K_m for P_i by the Ca^{2+} gradient.

Vesicles	Kp_i for E-P formation (mM)	
	pH 6	pH 7
Leaky	3.5 ± 0.5 (8)	25.3 ± 4.2 (3)
Loaded with Ca^{2+}	0.7 ± 0.1 (7)	2.5 ± 0.4 (4)

The values represent the average ± standard error of the number of experiments shown in parenthesis. For experimental details, see de Meis, (1976) (17) and de Meis and Vianna 1973 (18).

The low affinity for P_i at pH 7 probably accounts for earlier failures to measure phosphorylation by P_i in the absence of a transmembrane Ca^{2+} gradient (2,5,6,19).

The phosphorylation of leaky vesicles by P_i is inhibited by Ca^{2+}. The inhibition is reversible (Fig. 3). The enzyme is rephosphorylated by P_i if the added Ca^{2+} is chelated with an excess of EGTA. Inhibition by Ca^{2+} is observed when either leaky vesicles or vesicles previously loaded with Ca^{2+} are used (15,17). The Ca^{2+} concentration required for half maximal inhibiton of phosphorylation by P_i both in presence and absence of gradient is about 10^{-6} M at pH 7 and it is in the same range that the Ca^{2+} concentration required for half maximal phosphorylation by ATP (15,17). This indicates that inhibition is related to the binding of Ca^{2+} to the enzyme and not to a modification of the Ca^{2+} gradient. This Ca^{2+} binding site is located on a part of the ATPase which faces the outer surface of the vesicles. Were it to face the vesicle's lumen, then in presence of a gradient the enzyme would not be phosphorylated by P_i; the high Ca^{2+} concentration in the vesicle's lumen would have impaired the reaction.

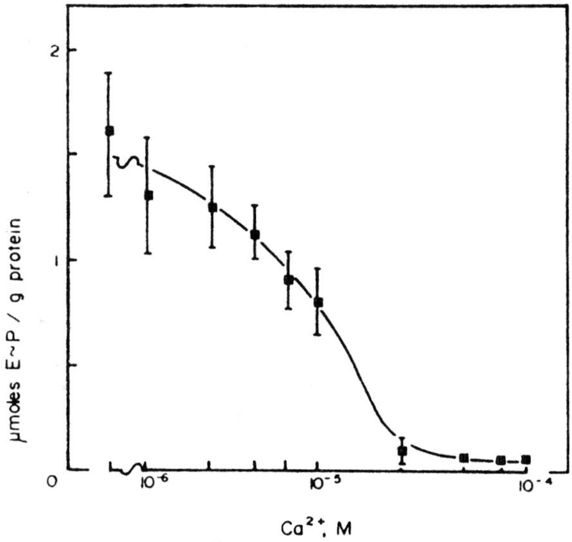

FIGURE 3. Inhibition of E-P formation by Ca^{2+}. The $[^{32}P]$ P_i concentration was 4 mM. For zero Ca^{2+}, no $CaCl_2$ was added and the EGTA concentrations was 1.0 mM. For the calculated free Ca^{2+} concentrations of 1.5, 3.0, 5.0, 7.0, and 10.6 μM, 0.2 mM $CaCl_2$ and 0.70, 0.46, 0.35, 0.30, and 0.26 mM EGTA, respectively, were added to the assay medium. For the Ca^{2+} concentrations of 25, 50, and 100 μM, no EGTA was used. The free Ca^{2+} concentration was calculated using the value of 4×10^{-6} for the dissociation constant of the complex Ca-EGTA. For further details see Masuda and de Meis (15).

These findings led to the conclusion that ATP and P_i interact with a common active site of the enzyme which undergoes a conformational change depending on the binding of Ca^{2+}. In one conformation the ATPase would be phosphorylated only by P_i and in the other conformation it would be phosphorylated by ATP. Thus, the binding of Ca^{2+} would determine the choice of substrate for the phosphorylation reaction (15). It was already known that the enzyme has a Ca^{2+} binding site of high affinity, K_s about 10^{-6} M at pH 7 (20-28). Evidence that the ATPase undergoes a conformational change during the catalytic cycle was already presented by Landgraf and Inesi (1969) and then confirmed by several authors (29-39). The data described above permitted to characterize the properties of the two forms of the enzyme, in one form is phosphorylated by P_i and in the other by ATP and that the simple binding of Ca^{2+} would be sufficient to trigger the conversion of one form into the other (15). The P_i phosphorylation reactions as measured in presence and absence of a Ca^{2+} gradient bear several common features. In both conditions the phosphoryl group of the phosphoenzyme formed is an acyl phosphate, (15,16) both reactions are impaired by Ca^{2+} in the medium (15-17), and exhibit the same temperature dependence profile (40). However, they differ in two important features, namely the apparent affinity for P_i (Table I) and the ability to transfer the phosphate from the phosphoenzyme to ADP. After phosphorylation by P_i, a decrease in the level of phosphoenzyme is observed when ADP is added to media containing either empty vesicles or vesicles loaded with Ca^{2+} (Fig. 4). Net synthesis of ATP, however, is observed only in the system containing Ca^{2+}-loaded vesicles (17).

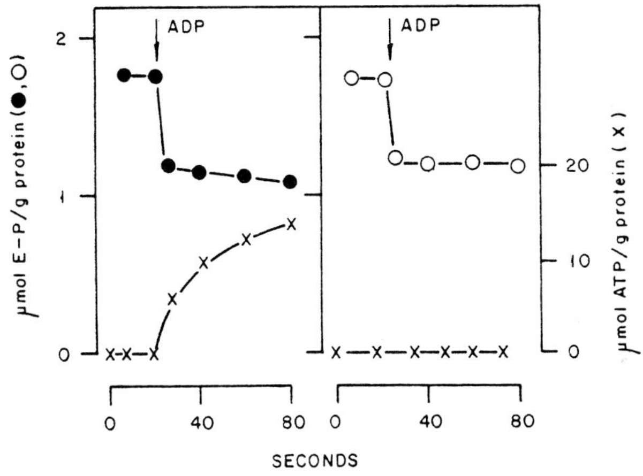

FIGURE 4. Phosphorylation by P_i and ATP synthesis. The assay medium composition was Tris maleate buffer (pH 6.5), 30 mM $MgCl_2$, 15 mM EGTA, and 2 mM (left) or 5 mM (right) $^{32}P_i$. The reaction was started by the addition of vesicles which are loaded with calcium phosphate (left) or empty and leaky

This finding indicates that the energy derived from the Ca^{2+}-concentration gradient is used not for the phosphorylation of the enzyme by P_i, but for the transfer of phosphate to ADP. Therefore, the phosphoenzyme formed in absence of a Ca^{2+} gradient is referred to as "low energy because it is formed spontaneously and is not able to transfer its phosphate to ADP, and that formed in the presence of a Ca^{2+} gradient as "high energy" because it can transfer its phosphate to ADP leading to the synthesis of ATP. In this view the capture of the osmotic energy of the Ca^{2+} gradient would permit the conversion of the "low energy" phosphoenzyme into the "high energy" form. Contrasting with this reasoning is the finding that the phosphoryl group of the phosphoenyzme formed both in the presence and in the absence of a Ca^{2+} gradient is an acyl phosphate (15-18). In water, the ΔG^o of the hydrolysin of both an acyl phosphate residue and ATP is in the range of -8 to -10 Kcal/mol (41-43). The following question then arises: 1) Why and how the ΔG^o of the same acyl phosphate residue vary? 2) How the gradient modify the apparent affinity for P_i? A similar phenomenom was already reported for mitochondria. The elegant $P_i \rightleftharpoons$ HOH exchange performed in Boyer laboratory indicate that for the synthesis of ATP the apparent affinity of the $F_1 - F_o$ complex for P_i varies depending on whether or not the mitochondria is energized (44-47).

Information concerning these two questions will be presented in a subsequent section. The findings described above have been reproduced in several laboratories (48-57).

Simultaneous phosphorylation of the ATPase by Nucleoside Triphosphate and P_i. Phosphorylation of the enzyme by P_i is inhibited by Ca^{2+}. In 1974 we found (16, 58-61) that this inhibition is overcome by the addition to the medium of an energy-yielding substrate such as acetyl phosphate, ITP, GTP or ATP (fig. 5 and table 2).

(right). After different incubation intervals at 30°C, samples were taken to determine phosphoenzyme (E-P; ● and O) and [γ-^{32}P]ATP (X). The arrow indicates the addition of ADP, to a final concentration of 0.05 mM in the assay medium. For further details, see de Meis, 1976 (17).

Table II. Phosphorylation of leaky vesicles by P_i in the presence of Ca^{2+}

Additions to assay medium	μmol $E\text{-}^{32}P$/ g protein
None	0.10 ± 0.05
1 mM acetyl phosphate	1.94 ± 0.18
1 mM GTP	1.76 ± 0.26
1 mM ITP	1.89 ± 0.17
1 mM ATP	0.20 ± 0.08
0.05 mM ATP	1.20 ± 0.13

The assay medium composition was 40 mM Tris-maleate buffer pH 6.7; 20 mM $MgCl_2$, 8 mM $^{32}P_i$, 0.2 mM $CaCl_2$ and 0.2 mg leaky vesicles protein per ml. The values represent the average ± S.E. of four experiments. Data from de Meis and Masuda (16), Carvalhe et. al., (59), de Meis and Boyer (60).

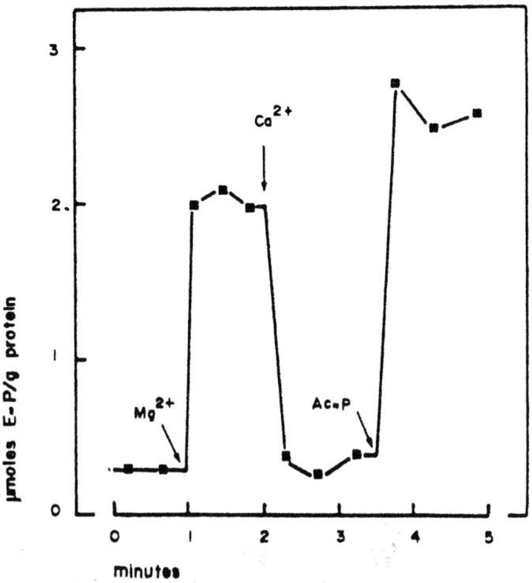

FIGURE 5. Effect of Ca^{2+} and acetyl phosphate on E-P formation. The assay medium composition was 10 mM Tris-maleate buffer (pH 6.5), 4 mM ^{32}P P_i, 0.05 mM EGTA, and 0.6 mg of SRV protein per ml. The reaction was started by the addition of SRV. After different incubation intervals at 37° C, aliquots of 2.5 ml of this medium were removed for E-P

The Ca^{2+} present in the medium activates hydrolysis of the substrate, and phosphorylation by P_i can be detected as long as the substrate is present in the medium. This finding indicates that during the catalytic cycle, an enzyme form is generated which is phosphorylated by P_i but not by nucleoside-5'-triphosphate (NTP). In other words, during the hydrolysis of NTP the enzyme cycles between two forms, one has calcium bound to the site of high affinity and can react with NTP, and the other, in spite of the presence of Ca^{2+} in the medium has no Ca^{2+} bound and is phosphorylated by P_i. Accordingly, table 3 shows that in presence of Ca^{2+} the enzyme is phosphorylated simultaneously by P_i and ITP. This is determined by measuring the amount of radioactive phosphoenzyme formed when the vesicles are incubated with Ca^{2+}, Mg^{2+}, ITP and P_i and only one of the phosphorylating substrates is radioactive.

TABLE III. Simultaneous phosphorylation of leaky vesicles by ITP and P_i

Additions	μmol $E\text{-}^{32}P$/gprot
6 mM $^{32}P_i$	0.10 ± 0.10
0.2 mM ITP plus 6 mM $^{32}P_i$	1.80 ± 0.10
0.2 mM (γ-^{32}P) ITP plus 6 mM P_i	0.80 ± 0.09
0.2 mM (γ-^{32}P) ITP plus 6 mM $^{32}P_i$	2.72 ± 0.15

The values are the average ± S.E. of five experiments. Data from de Meis and Masuda (16), Carvalho et al., (59). Souze and de Meis (62).

During the catalytic cycle of NTP hydrolysis, the proportion of enzyme found which is phosphorylated by P_i depends on the NTP used and its concentration in the medium. Table 2 shows that at concentration of 1 mM, acetyl phosphate, ITP and GTP are equally effective in eliciting phosphorylation by P_i. A small portion of the enzyme is phosphorylated by P_i in the presence of 1 mM ATP. To obtain a high level of phosphorylation by P_i, it is necessary to decrease the ATP

determination. Arrows point to the addition of $MgCl_2$, final concentration in the assay medium of 10 mM; $CaCl_2$, 0.25 mM; AcP, 2 mM. Essentially the same results were observed in three different SRV preparations tested.

concentration to the micromolar range (58-62). In previous reports it was shown that besides being a substrate ATP but not GTP, ITP and acetyl phosphate, can regulate the catalytic cycle of the enzyme. In 1967, Inesi, et. al., (11), measured the rate of ATP hydrolysis of leaky vesicles and found that two different K_m for ATP values can be distinguished on double reciprocal plots, 1 to 3 μM and 50 to 200 μM. Later we observed that in the micromolar concentration range ATP binds only to the catalytic site of the ATPase and at higher concentrations it binds also to a regulatory site which accelerate the velocity of substrate hydrolysis, GTP, ATP and acetyl phosphate being much less effective than ATP (7,63,64). The regulatory role of ATP has now been well documented in different laboratories (59-71). Taken together these data indicate that in the steady state established during the hydrolysis of ITP or GTP, a significant portion of the enzyme is in the form which is phosphorylated by P_i (Table 2). ATP would regulate the rate of conversion of the two enzyme forms. Thus, in presence of 1 mM ATP most of the enzyme would be in the form phosphorylated by the nucleotide and only a small fraction would be in the form phosphorylated by P_i (Table 2).

The discovery of the regulated role of ATP played a key role on the understanding of the mechanism of energy conversation of the system.

<u>$ATP \rightleftharpoons P_i$ Exchange in the absence of Ca^{2+} gradient</u>. When a gradient is formed, the Ca^{2+} concentration required inside the vesicles for the synthesis of ATP is the range of 1 to 10 mM (2-6). The finding that the enzyme is phosphorylated by P_i during the hydrolysis of ATP led us to question whether the $ATP \rightleftharpoons P_i$ exchange is activated only when a Ca^{2+} gradient is formed or whether a high Ca^{2+} concentration, similar to that found inside the vesicles when a gradient is formed would be sufficient by itself to activate this reaction. In other words, whether the energy derived from the gradient or solely a high Ca^{2+} concentration would allow the phospho-enzyme formed from P_i to be converted from "low energy" into "high energy". In order to untangle these possibilities, leaky vesicles or ATPase solubilized with the detergent X-100 were incubated at pH 7 in media containing different Ca^{2+} concentrations, and the rates of ATP hydrolysis and of ATP synthesis from ADP and P_i were measured simultaneously (72). When the Ca^{2+} concentration in the medium is sufficient to saturate only the high-afffinity binding site (10^{-4} M), the enzyme catalyze only the hydrolysis of ATP (Fig. 6). Increasing the Ca^{2+} concentration in the medium to

millimolar range results in a simultaneous inhibition of the ATPase activity and activation of the ATP synthesis. The lack of a requirement for a transmembrane Ca^{2+} gradient for activation of the ATP $\rightleftharpoons$ P_i exchange reaction is documented further in Fig. 7.

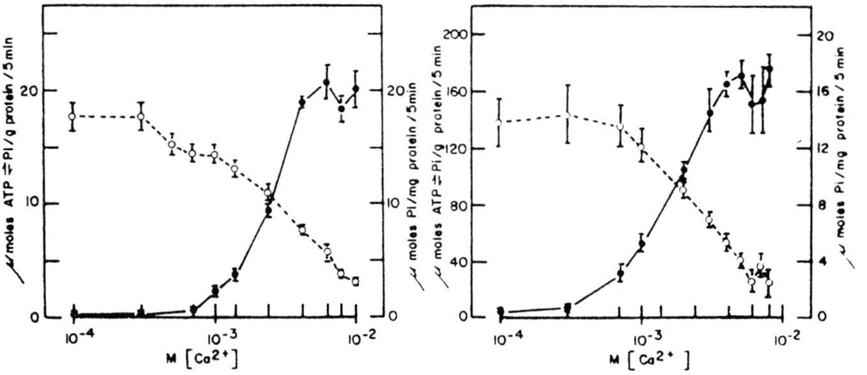

FIGURE 6. Ca^{2+} dependence of ATP $\rightleftharpoons$ P_i exchange. The assay medium composition was 30 mM Tris-maleate buffer (pH 6.8), 10 mM ATP, 6 mM $^{32}P_i$ 20 mM $MgCl_2$, and 0.3 mg/ml of SRV protein. For the ATPase activity (O---O) [γ -^{32}P] ATP was used. For ATP $\rightleftharpoons$ P_i exchange (●-●), $^{32}P_i$ was used. The reactin was performed at 37, (left) solubilized SRV; (right) leaky SRV. The values shown in the figure represent the average ± standard error of five experiments. For details, see de Meis and Carvalho, 1974 (72).

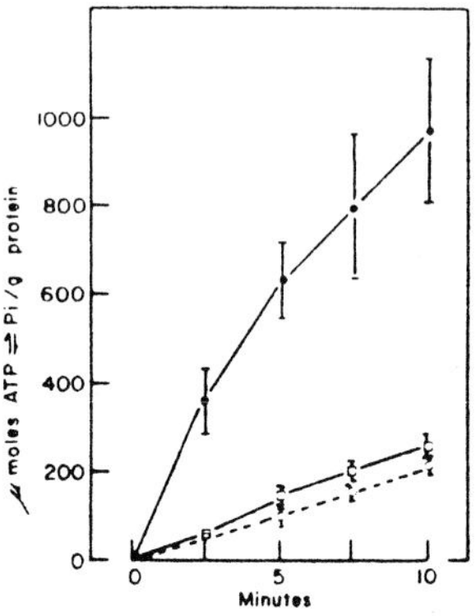

FIGURE 7. The role of the Ca^{2+} concentration gradient. The basic assay medium composition was 30 mM Tris-maleate buffer (pH 6.8), 20 mM $MgCl_2$, 6 mM $^{32}P_i$, and 10 mM ATP. The reaction was performed at 37°C: (●-●) 0.3 mg/ml of SRV protein was incubated in the basic assay medium containing in addition 0.1 mM $CaCl_2$; the rate of ATP $\rightleftharpoons$ P_i exchange was measured after 98% of the Ca^{2+} of the assay medium had been removed by the SRV; (O---O) same as (●-●) but after the vesicles had been loaded, the Ca^{2+} concentration of the medium was raised to 8 mM/ (□-□) 0.3 mg/ml of leaky SRV protein was incubated in the basic assay medium containing in addition 8 mM $CaCl_2$. The values shown in the figure represent the average ± standard error of three experiments. For details see de Meis and Carvalho, 1974 (72).

In this experiment, intact vesicles were incubated in a medium containing 0.1 mM $CaCl_2$ and the rate of ATP synthesis was measured after most of the Ca^{2+} of the medium had been removed by the vesicles. Under these conditions a steep Ca^{2+} concentration gradient is formed across the vesicle membrane. In a parallel experiment, after the vesicles had been loaded, the Ca^{2+} concentration in the medium was raised to 8 mM to approximately equalize the Ca^{2+} concentration on the two sides of the membrane (72). In both cases, ATP $\rightleftharpoons$ P_i exchange was detected, however rate of ATP synthesis is several fold higher in the presence of a Ca^{2+} gradient than in its absence. The rate of ATP $\rightleftharpoons$ P_i exchange varies with the P_i concentration in the medium both in the presence and in the absence of a transmembrane Ca^{2+} gradient. However, the apparent K_m for P_i differs in these two conditions, being in the range of 40 to 60 mM when measured in leaky vesicles and in the range of 2 to 4 mM when a gradient is formed (72). Thus the variation of the enzyme affinity for P_i seems to be a recurring theme. However, at this stage the possibility was raised that the apparent K_m for P_i might vary due to a simple kinetic parameter - the regulatory effect of ATP - and not because of a thermodynamic barrier overcome by the energy derived from the Ca^{2+} gradient. This in fact proved to be the case (Table 4). Notice that the experiments of Figs. 6 and 7 a high ATP concentration was used. Both in the presence and in the absence of a Ca^{2+} gradient, essentially the same apparent K_m for P_i and for a given P_i concentration, essentially the same rate of ATP$\rightleftharpoons P_i$ exchange was attained when (a) ATP was replaced by ITP (59,73), (6) the ATP concentration was decreased to the micromolar range (66) or (c) when in presence of a high ATP concentration, the regulatory effect of ATP was impaired with the use of silver (58). The finding that the ATPase is able to drive an ATP $\rightleftharpoons P_i$ exchange in the absence of a transmembrane Ca^{2+} gradient indicates that the system must be able to conserve some of the energy released from ATP hydrolysis in a form which permits the resynthesis of a new ATP molecule from ADP and P_i. The degree of energy conservation can be estimated from the ratio between the rates of hydrolysis and synthesis. Table 4 shows that depending on the experimental conditions used, the same degree of energy conservation can be attained in presence and absence of a transmembrane Ca^{2+} gradient and that, as in the case of the apparent K_m for P_i, the rates hydrolysis: synthesis of ATP depends on kinetic parameter related to the regulatory effect of ATP and not from a thermodynamic barrier overcome by the energy derived from the gradient.

TABLE IV. Dependence for NTP $\rightleftharpoons$ P_i exchange

Vesicles	Substrate	K_m of P_i (mM)	Ratio NTP hydrolysis / NTP synthesis
Intact vesicles, Ca^{2+} gradient	ATP, 5 to 10 mM	3 to 4	2 to 4
	ITP, 2 to 5 mM	2 to 4	-
Leaky vesicles in 8 mM $CaCl_2$	ATP, 5 to 10 mM	40 to 60	20 to 50
	ATP, 0.05 mM	2 to 4	2 to 4
	ITP, 2 to 5 mM	2 to 4	4 to 6
Leaky vesicles pluse Ag^+ and 8 mM $CaCl_2$	ATP, 5 mM	6 to 8	1 to 2

Data from de Meis and Carvalho, 1974 (72), de Meis and Sorenson, 1975 (58), Carvalho et. al., 1976 (59) and Scofono and de Meis, 1981 (66).

From the experiments of ATP $\rightleftharpoons$ P_i exchange it was concluded that the conversion of the phosphoenzyme from "low energy" into "high energy" is promoted by the binding of Ca^{2+} to a low affinity site of the enzyme (K_s 1 to 2 mM at pH 7) located in the inner surface of the vesicles membrane. It was known that besides the high affinity binding site, the ATPase has a second class of binding site saturated in the presence of a high Ca^{2+} concentration (K_s about 1 mM at pH 7). However, it was thought that this second class of binding site was unspecific because many different proteins bind Ca^{2+} when the concentration of this cation is in the millimolar range. The finding that saturation of this site elicit a simultaneous activation of ATP $\rightleftharpoons$ P_i exchange (72) and inhibition of the ATPase activity (25,72) clearly shows that the low affinity Ca^{2+} binding site has a functional role in the regulation of the catalytic cycle of the enzyme. Shortly after the elegant experiments of Ikemoto (26) and Watanabe et. al., (74) demonstrated that during the catalytic cycle the high affinity Ca^{2+} binding site is converted into a low affinity binding site.

<u>Reaction sequence</u> on the basis of the data described above in 1976 we proposed the following reaction sequence (59).

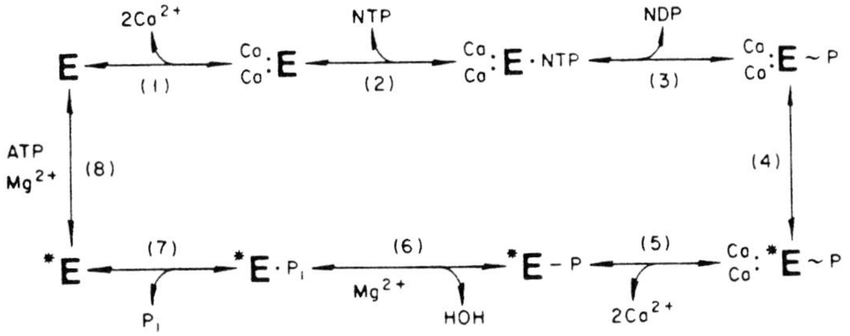

This sequence includes two distinct functional states of the enzyme, E and $^{+}$E. The Ca^{2+} binding site in the E form faces the outer surface of the vesicles and has an apparent K_s for Ca^{2+} in the range of 1 to 2 μM at pH 7 (high affinity). In the *E form, the Ca^{2+} binding site faces the inner surface of the vesicles and has an apparent K_s for Ca^{2+} in the range of 1 to 2 mM at pH 7 (low affinity). After binding of Ca^{2+}, the E form is phosphorylated by different nucleoside triphosphates but not by P_i, whereas the form *E is phosphorylated by P_i but not by nucleoside triphosphate. Reaction 8 is the rate-limiting step regulated by ATP. Conversion of *E into E occurs very slowly in the absence of ATP. Increasing concentration of nucleotide increases the rate of interconversion, ITP and GTP being much less effective than ATP. Among the different phosphoenzyme forms, only $^{Ca}_{Ca}$E∿P is of "high energy", i.e., is able to transfer its phosphate to ADP (reaction 3). In the absence of nucleotide triphosphate, the reactants included in the medium allow the occurrence of a few but not all of the reactions involved in the complete catalytic cycle. Thus a true equilibrium, and not a steady state is established. In the absence of a Ca^{2+} gradient, phosphorylation by P_i involves reactions 8,7 and 6.

$$E \rightleftharpoons {}^{*}E \xrightleftharpoons{P_i} {}^{*}E.P_i \xrightleftharpoons{HOH} {}^{*}E\text{-}P \qquad \text{(equilibrium 1)}$$

A different equilibrium is attained when the vesicles are phosphorylated by P_i in the presence of a Ca^{2+} gradient. In this case, the binding of Ca^{2+} to the low affinity site drives reaction 5 to the right, and the equilibrium includes other forms of phosphoenzyme.

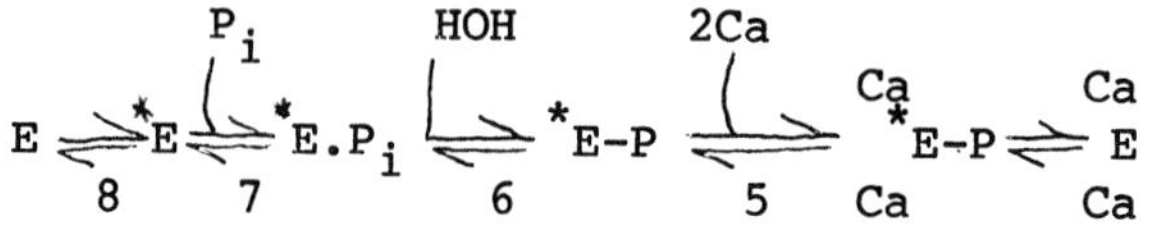 (equilibrium 2)

The apparent K_m for P_i depends on the rate constants (forward and backward) of all the reactions included in equilibria 1 and 2 above. Thus, the lower apparent K_m for P_i measured in presence of a Ca^{2+} gradient can be accounted for by the formation of the phosphoenzyme species $\mathrm{Ca}\overset{\mathrm{Ca}}{\overset{*}{\mathrm{E}}}$-P and $\mathrm{Ca}\overset{\mathrm{Ca}}{\mathrm{E}}\sim$P. Binding of Ca^{2+} to the enzyme's high affinity site (Reaction 1) inhibits phosphorylation by P_i, whether this is measured in the presence or absence of a transmembrane Ca^{2+} gradient. This involves step 1. The equilibrium reached between the enzyme forms Ca_2E and either *E-P (no gradient) or Ca_2 E$\sim$P (with gradient) would be such that in both cases practically all the enzyme units are in the P_i-unreactive form 2Ca.E. In presence of Ca^{2+}, Mg^{2+} and nucleoside triphosphate, all the intermediary reaction are activated and during steady state the enzyme formed prevailing depends on the rate limiting step of the cycle. With the use of GTP or ITP reaction 8 is the rate limiting step. Thus, the enzyme form *E is accumulated and available to be back phosphorylated by P_i (reactions 7 and 6). Because ATP accelerates the conversion of *E into E, in presence of 1 mM ATP the amount of *E available to be phosphorylated by P_i during steady state is small (Tables 2 and 3). ATP $\rightleftharpoons P_i$ or ITP $\rightleftharpoons P_i$ exchange involves reactions 2 to 7. For each P_i molecule released during hydrolysis of NTP (reaction forward 2 to 7), one molecule of *E become available to participate in the reversal reaction, leading to synthesis of one NTP molecule (reaction 7 to 2 backward). The ratio of hydrolysis to synthesis is higher than one to the extent that some of the enzyme units *E are converted into Ca_2E instead of being driven in the reverse direction through phosphorylation by P_i. Therefore, the faster the forward rate of reactions 8 and 1, the faster the rate of hydrolysis and the slower the ratio of hydrolysis to synthesis shown in Table 4. Further evidences supporting the reaction sequence shown above have been obtained in different laboratories. These include transient kinetic studies which were discussed in further details in a previous review (18,75,76).

<u>Net synthesis of ATP - Ca^{2+} jump</u>. In 1975, Knowles and Racker (49,) demonstrated that leaky vesicles can synthesize a small amount of ATP in the absence of a Ca^{2+} gradient. This was achieved by a two step procedure where, initially, the enzyme was phosphorylated by P_i at pH 6.3 in presence of EGTA. Subsequently, upon the addition of ADP and 10 mM $CaCl_2$ (Ca^{2+} jump), it was found that the phosphate of the phosphoenzyme was transferred to ADP, forming ATP. Fig. 8 shows a millisecond mixing and quenching experiment confirming Knowles and Racker observation (77). This experiment shows that after the addition of Ca^{2+} and ADP, the disappearance of phosphoenzyme and synthesis of ATP are synchronous, the time for half maximal ATP synthesis and phosphoenzyme dephosphorylation being in the range of 20 to 40 ms.

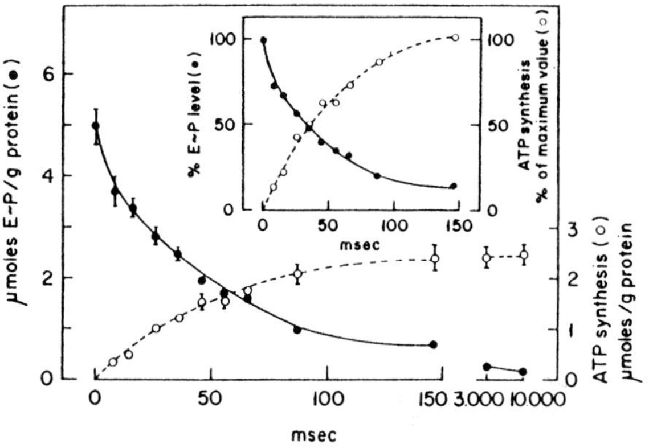

FIGURE 8. Dephosphorylation of E∿P and ATP formation. Syringe A contained 20 mM Tris-maleate, 2 mM EGTA, 20 mM $MgCl_2$, 8 mM $^{32}P_i$ (4×10^7 cpm/µmol) and 1 mg Ca^{2+} + Mg^{2+} dependent ATP. The pH of this phosphorylating solution was 6.0. Syringe B contained 40 mM $CaCl_2$, 4 mM ADP, 10 mM glucose and 4 units of hexokinase. Equal volumes (about 1 ml) of A and B were force mixed together through a capillary tube. After quenching in 250 mM perchloric acid, the content of phosphoenzyme and of (γ-^{32}P)ATP synthesized were analyzed. Each point represents the mean ± standard error of three experiments. From de Meis and Tume, 1977 (77).

The Ca^{2+} concentration added in the second step of the Ca^{2+} jump procedure is sufficient to saturate both Ca^{2+} binding sites of low and high affinity. After binding of Ca^{2+} to the high-affinity site, the phosphoenzyme is hydrolyzed. In transient kinetic experiments, it was found that the time for half maximal conversion of $^{*}E-P$ into Ca_2E through reactions 6,7,8 and 1 shown in the sequence above is in the range of 800 to 1,500 ms (77). Thus, in Fig. 8 net synthesis of ATP was observed because the rate at which the phosphoenzyme transfers its phosphate to ADP (reactions 5,4 and 3 backwards) is much faster than the rate of hydrolysis of the phosphoenzyme promoted by the binding of Ca^{2+} to the high affinity sites (reaction 6,7, 8 and 1 forward). The amount of ATP synthesized by the Ca^{2+}-jump procedure is proportional to and never exceeds the number of enzyme sites phosphorylated by P_i (49,77). In other words, the Ca^{2+} jump permits the enzyme to complete only one catalytic cycle, after which the system reaches equilibrium and no more ATP is synthesized. The ATP synthesized after the Ca^{2+} jump does not remain bound to the enzyme, being found free in solution (49,77).

Proton gradient and synthesis of ATP. The binding of Ca^{2+} to the high-and low-affinity sites of the ATPase is altered by changing the pH of the medium (Table 5). Both sites exhibit an increased affinity for Ca^{2+} with high pH and reduced affinity at low pH values the relative difference of affinities of the two sites remaining unchanged in the range of pH 6 to 8. This we did monitored indirectly by measuring

TABLE V. Effect of pH on the apparent K_s of the high-and low-affinity Ca^{2+} binding sites.

pH	Apparent K_s	
	High Affinity	Low Affinity
	(μM)	(mM)
5.0	2000	-
6.0	10-20	20-30
7.0	1-3	1-3
8.0	0.3-0.4	0.4

Data from Verjovski-Almeida and de Meis, 1977 (78) and de Meis and Tume, 1977 (77).

the Ca^{2+} requirement for activation of NTP $\rightleftharpoons P_i$ exchange, phosphorylation of leaky vesicles by ITP and inhibition by Ca^{2+} of the enzyme phosphorylation by P_i (77,78).

Based on this finding the rationale was developed that no ATP synthesis will be observed when the same Ca^{2+} concentration exists on both sides of the membrane and at a level sufficient to saturate the high-affinity site but not sufficient to allow binding to the low affinity site. However, the system could synthesize ATP if the affinity of the two Ca^{2+} binding sites is modified by varying the pH on the two sides of the membrane, making it acidic in the assay medium and alkaline inside the vesicles. In such a situation the phosphorylation by P_i will no longer be inhibited because the Ca^{2+} concentration will no longer be sufficient so saturate the external Ca^{2+} binding site and the Ca^{2+} concentration inside the vesicles could become sufficient to allow the transfer of the phosphate to ADP because the increased affinity of the internal Ca^{2+} binding site. We tested this hypothesis using leaky vesicles and the technique described for the Ca^{2+} jump except that the Ca^{2+} concentration was maintained constant and the affinity of the two Ca^{2+} binding sites was varied by a sudden change of pH of the medium 77). Fig. 9 shows that in presence of EGTA, the level of phosphoenzyme decreases when the pH value of the medium is decreased from 6 to 5. However this leads also to a marked reduction of the inhibitory effect of Ca^{2+}. At pH 6, 0.6 mM $CaCl_2$ completely inhibited phosphorylation by P_i, whereas at pH 5 the same $CaCl_2$ concentration produced only 20 to 30 % inhibition.

In order to test the effect of pH on the transfer of phosphate to ADP, leaky vesicles were first phosphorylated by P_i at pH 6 and then the medium was diluted in solution containing different Ca^{2+} concentrations adjusted at different pH values. Fig. 10 shows that at pH 6 there is no synthesis of ATP in presence of 0.6 mM $CaCl_2$ whereas at pH 8.1 the same Ca^{2+} concentration trigger the transfer of most of the phosphoenzyme phosphate to ADP.

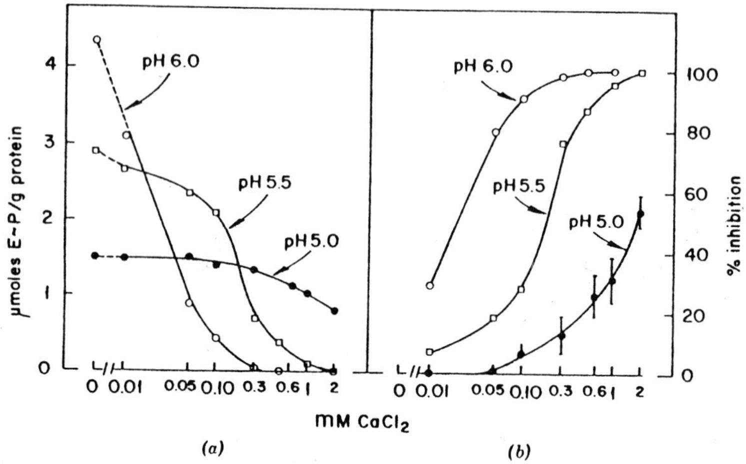

FIGURE 9. Effect of Ca^{2+} concentration and pH on the phosphorylation of the Ca^{2+}-dependent ATPase. The reaction medium contained 20 mM Tris-maleate buffer, 8 mM $^{32}P_i$ (10^7 cpm/μmol), 20 mM $MgCl_2$, 1 mg of Ca^{2+}-dependent ATPase protein, and the $CaCl_2$ concentrations shown in the Figure. For 0 Ca^{2+} concentration $CaCl_2$ was omitted and 2.5 mM EGTA was added. The final volume was 1 ml, and the pH of the mixture was 6.0 (O), 5.5 (□) and 5.0 (O). The reaction was performed at 25°C, it was started by the addition of the enzyme, and it was arrested after 30 sec of incubation by the addition of 3 ml of an ice-cold solution of perchloric acid (125 mM) containing 2 mM orthophosphate. The values represent either the average of two experiments (O, □) or the average ± standard error of five experiments (●). From de Meis, and Tume, 1977 (77).

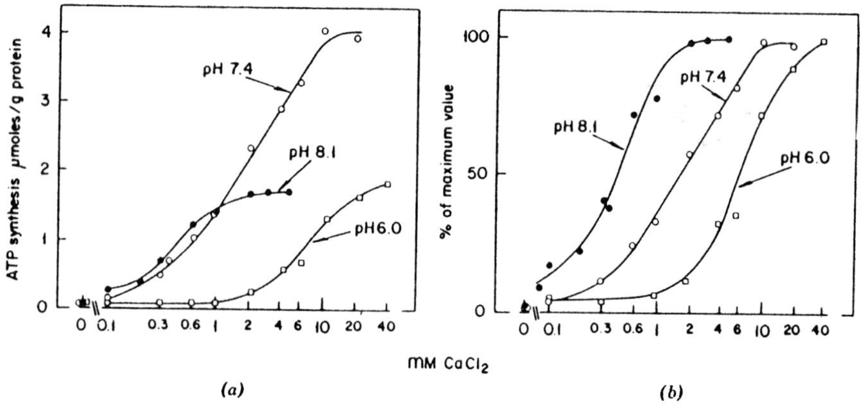

FIGURE 10. Ca^{2+} dependence for ATP synthesis. The reaction media (0.5 ml) contained 20 mM $MgCl_2$, 8 mM $^{32}P_i$, 2 units of hexokinase, and 1 mg of Ca^{2+} + Mg^{2+} -dependent ATPase. The pH of the mixture was 6.0. The reaction was

started by the addition of Ca^{2+} + Mg^{2+} -dependent ATPase and after 30 sec it was added to 0.5 ml of a solution containing 10 mM glucose, 4 mM ADP, various concentrations of $CaCl_2$ to give the final concentrations shown in the figure, and KOH in

---

Fig. 11 shows that after phosphorylation at pH 5 in the presence of 0.6 mM $CaCl_2$, net synthesis of ATP is measured following the addition to the medium of ADP and KOH in an amount sufficient to raise the pH of the mixture from 5 to 8. Synthesis of ATP is not detected when Ca^{2+} is omitted from the assay medium.

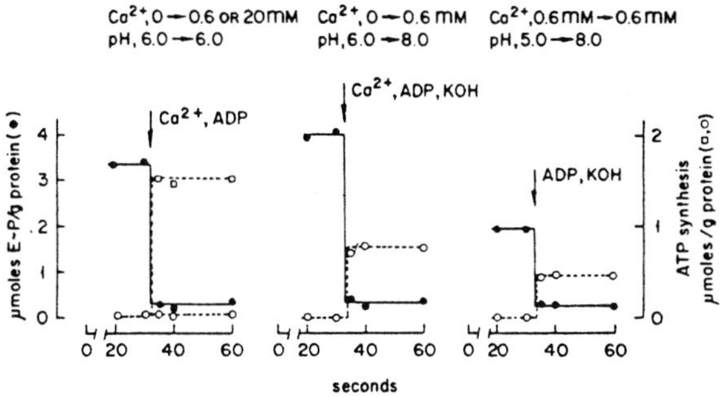

FIGURE 11. pH and Ca^{2+} jump (a) The reaction medium (0.5 ml) contained 20 mM $MgCl_2$, 8 mM $^{32}P_i$ (2.5 X 10^7 cpm/μmol), 2 units of hexokinase and 1 mg of Ca^{2+} + Mg^{2+} dependent ATPase protein. The pH of the medium was 6.0. The reaction was performed at 25°C, it was started by the addition of the enzyme, and after 30 sec of incubation (arrow) 0.5 ml of a solution containing 4 mM ADP, 10 mM glucose, and either 1.2 (O) or 40 mM (□) $CaCl_2$ was added. Note that the concentrations of these reagents were halved after mixing. The pH of the assay medium remained constant (6.0) after this additon. The reaction was terminated at different incubation intervals as shown in the figure by the addition of 0.1 ml of 100% Cl_3AcOH (w/v). Aliquots of the deproteinized supernatant and the protein precipitates were analyzed for (γ-^{32}P)ATP and E $\sim$ ^{32}P formation, respectively. (b) The experimental conditions and assay medium were the same as above, except that the second solution, in addition to 4 mM ADP, 10 mM glucose, and 1.2 mM $CaCl_2$ also contained 23 mM KOH. Thus, the pH of the medium increased from 6.0 to 8.0 after mixing. (c) The reaction medium (0.8 ml) contained 9 mM maleic acid, 20 mM $MgCl_2$, 8 mM $^{32}P_i$, 0.6 mM $CaCl_2$, 2 units of hexokinase, and 1 mg of Ca^{2+} + Mg^{2+}-dependent ATPase protein. The pH was 5.0. The reaction was started by the addition of the enzyme and 30 sec later 0.2 ml of a mixture

a concentration sufficient to increase the pH of the final mixture from 6.0 to either 7.4 (O) or 8.1 (●). KOH was not added to those incubations which were maintained at pH 6.0 (□). The reaction was stopped 1 min later by the addition of 0.1 ml of 100% Cl_3AcOH (w/v). Each point represents the average of two experiments. From de Meis and Tume, 1977 (77).

When extended to sealed vesicles, these data indicate that the sarcoplasmic reticulum ATPase can catalyze the synthesis of ATP when there are equal Ca^{2+} concentrations but different pH values on the two side of the membrane. In such a situation, then, there is a transmembrane H^+ gradient and not a Ca^{2+} gradient, and the synthesis of ATP is not due to the H^+ gradient. The difference in pH is required only to modify the affinity of the two Ca^{2+} binding sites of the enzyme. Recently, evidence obtained in different laboratories indicate that a small H^+ gradient is formed across the vesicles membrane during Ca^{2+} transport. (79-81). Maintenance of a large H^+ gradients seems not possible because of the high H^+ permeability of the sarcoplasmic reticulum membrane. Besides its implication on the mechanism of Ca^{2+} transport, the data described above raise the possibility that in membrane system other than the sarcoplasmic reticulum , when there is a transmembrane H^+ gradient, different transport ATPases can be mobilized for the synthesis of ATP, besides those that specifically transport H^+ (77).

Binding energy. From the data presented it is concluded that each of the individual reactions involved in the hydrolysis and synthesis of ATP is reversible and can flow forward and backward, without using the osmotic energy that might be derived from the concentration gradient of Ca^{2+} across the membrane. For the synthesis of ATP in sealed vesicles, the large difference of Ca^{2+} concentration on the two sides of the membrane is needed only to meet the difference in affinity of the external and internal Ca^{2+} binding sites. In this view, the synthesis of ATP is promoted by the asymmetrical binding of Ca^{2+} on the two sides of the membrane. For the enzyme to be phosphorylated by P_i, it is

---

of ADP, glucose, $CaCl_2$, and KOH was added to give final concentrations of 2 mM ADP, 5 mM glucose, and 40 mM KOH. The $CaCl_2$ concentration remained constant (0.6 mM). After this addition, the pH of the medium increased from 5.0 to 8.0. Other experimental conditions were as described in (a): (O) ATP synthesis; (●) $E{\sim}P$. From de Meis and Tume, 1977 (77).

necessary that Ca^{2+} not be bound to the high-affinity site located on the outer surface of the membrane (Figs. 3 and 9). In order to transfer the phosphate from the phosphoenzyme to ADP, Ca^{2+} must bind to the low-affinity site located on the inner surface of the vesicles membrane Figs. 6,8, 10 and 11). This concept assumes that the energy derived from the binding of Ca^{2+} to the enzyme is used for the synthesis of ATP. The standard free energy, derived from the binding of Ca^{2+} to the low affinity site can be calculated using the equation $\Delta G^o = -RTl_n K_a^{in}$ where K_a^{in} is the association constant for Ca^{2+}. For two Ca^{2+} bound for each enzyme unit at 30°C, this gives a value of $\Delta G^o = - 8.2$ Kcal per mol. After binding of Ca^{2+}, the phosphoenzyme undergoes a conformational change and the binding site is convert into high affinity (reaction 4). For this transition $\Delta G^o = - RTl_n K_a^{out}/K_a^{in}$ where K_a^{out} is the association constant of the high affinity binding site which at pH 7 it is 10^6M. For two calcium bound for each enzyme unit at 30 C this gives a value of $\Delta G^o = - 8.2$ Kcal per mol. Therefore for the overall process, binding of Ca^{2+} to the low affinity site and conversion of the low into the high-affinity site the ΔG^o is - 16.4 Kcal per mol. This is sufficient for the synthesis of one mol of ATP for each mol of enzyme (41-43). In order to initiate a new catalytic cycle it is necessary to dissociate the Ca^{2+} bound to the high-affinity site. During reversal of the Ca^{2+} pump, continuous synthesis of ATP is measured because at the end of each cycle, Ca^{2+} dissociates from this site due to the very low Ca^{2+} concentration found in the medium. Therefore continuous synthesis of ATP is observed because the enzyme cycle between two media of different composition, one in which the activity of Ca^{2+} is high (vesicles lumen) and another where the activity of Ca^{2+} is very low (assay medium). The energy derived from the large difference of Ca^{2+} activity found is these two compartment is converted into binding energy. This is used for the conversion of the phosphoenzyme from low-into high energy (conformational energy) and synthesis of ATP (chemical energy). Thus, during reversal of the Ca^{2+} pump different forms of enery are interconverted.

Role of water on the conversion of the phosphoenzyme from low-into high-energy. During the last decade the concept of high-energy compound has been revised. Until 1969 it was thought that both opposing resonance and electrostatic repulsion within the molecule were the main contribution to the large negative free energies of hydrolysis of ATP, ADP and other high-energy compounds (82-85). In 1970, George et al., (42) analyzing the thermodynamics of hydrolysis of different pH forms of pyrophosphate, acetyl phosphate and other high energy molecules concluded that intramolecular effects such as opposing resonance and electrostatic repulsion were of secondary importance and suggested that difference of solvation energy of reactants and products were the predominant free-energy contributions of the large negative free energy of hydrolysis of these compounds. Hayes et. al., (43) calculated the reaction energies and electronic structures of several high-energy compounds in gas phase and compared these values with those reported in water. The values obtained support the conclusion of George et al., (42) that by far the most important factor in determining the energy of hydrolysis are the relative solvation energies of reactant and products. In Table 6 the energies of hydrolysis of pyrophosphate and acetyl phosphate calculated by Hayes et al., (43) for nonsolvated reactants (gas phase) are compared with the values obtained in water (solvated).

TABLE VI. ΔH of hydrolysis in gas phase and in water

Reaction	ΔH, Kcal/mol	
	Gas phase	Solution in water
$H_3P_2O^-_7 + H_2O \rightarrow H_3PO_4 + H_2PO^-_4$	-0.9	-7.3
$H_2P_2O^{2-}_7 + H_2O \rightarrow 2H_2PO^-_4$	-75.2	-6.8
$CH_3COOPO_3H^- + H_2O \rightarrow CH_3COOH + H_2PO^{2-}_4$	+5.4	-8.6
$CH_3COOPO^{2-}_3 + H_2O \rightarrow CH_3COOH + HPO^{2-}_4$	+32.5	-

Notice that in gas phase, acetyl phosphate is not a high-energy compound. On the contrary, the large positive ΔH of hydrolysis indicates that when reactant and products are not solvated, acetyl phosphate is more stable than its products of hydrolysis. Acetyl phosphate and pyrophosphate have similar energies of hydrolysis in water, but there is a large difference between the corresponding energies of hydrolysis when the reactants are not solvated. Thus, in gas phase an acyl phosphate cannot transfer its phosphate to pyrophosphate or a similar compound such as ADP. Based on the observation of Hayes et al., (43) we raised the possibility that in the absence of a Ca^{2+} gradient, the catalytic site of the enzyme is hydrophobic so that P_i and the aspartic acid would react as if in a gas phase; the ΔH for the formation of an acylphosphate residue would be negative, and the reaction would occur spontaneously. In this case,the major thermodynamic barrier for the phosphorylation of the enzyme by P_i would be the entry of P_i into the catalytic site, i.e., the partition of the P_i from the assay medium into an hydrophobic environment. Factors facilitating this partition should also facilitate the phosphorylation of the enzyme by P_i. In the absence of a transmembrane Ca^{2+} gradient, the phosphoenzyme formed from P_i would not be able to transfer its phosphate to ADP because of the larger difference of ΔH of hydrolysis of the acylphosphate and pyrophosphate residues of ATP (Table 6). The binding of Ca^{2+} to the low-affinity site of the enzyme and its conversion into a site of high affinity would permit the entry of water into the catalytic site, with subsequent solvation of both acylphosphate residues and ADP. As a result, the ΔH of hydrolysis of the acylphosphate and pyrophosphate residues would become equal, and the synthesis of ATP would proceed spontaneously. Experimental conditions which impair the entry of water into the catalytic site should also impair the synthesis of ATP. According to this hypothesis, the concept of "high-energy" and "low-energy" forms of phosphoenzyme would be related solely to the water activity in the catalytic site (86). This hypothesis was tested by measuring the phosphorylation of the enzyme by P_i and the synthesis of ATP in the presence of various water-organic solvent mixtures, namely dimethylsulfoxide (DMSO), glycerol, and N,N-dimethylformamide (DMFA).

In presence of these solvents the phosphorylation of leaky vesicles by P_i is largely facilitated (86-88). The apparent K_m for P_i no longer varies with the pH of the medium. In addition, there is a significant increase of the apparent affinity of the enzyme for P_i which is more pronounced at alkaline pH values. In presence of DMSO, the apparent K_m for P_i decreases three (pH6) to five (pH8) order of magnitude (Table 7).

TABLE VII. Effect of Organic Solvents on the Apparent K_m for P_i[ab]

Organic Solvent Added (v/v)	Apparent K_m fo P_i (M)		
	pH 6.0	pH 7.0	pH 8.0
none	3.5×10^{-3}	25.3×10^{-3}	$> 25.3 \times 10^{-3}$
glycerol, 40%	5×10^{-4}	10^{-3}	10^{-3}
DMFA, 26%	4×10^{-4}	-	7×10^{-4}
DMSO, 20%	7×10^{-5}	2.5×10^{-5}	4×10^{-5}
40%	7×10^{-6}	7×10^{-6}	2×10^{-6}

[a]The assay medium consisted of 50 mM Tris-maleate buffer, 0.5 mM EGTA, 10 mM $MgCl_2$, different $^{32}P_i$ concentrations and the percentages of organic solvents shown in the table. For details see de Meis et. al., 1980 (86), Otero and de Meis, 1982 (87), de Meis et. al., 1982 (88).

The decrease of the apparent K_m for P_i seems to be related to an increase of the hydrophobicity of the medium promoted by the addition of organic solvents. If the catalytic site of the enzyme is hydrophobic, then the partition of P_i from the assay medium into the catalytic site should be facilitated when the difference of hydrophobicity between these compartments is decreased. To test this hypothesis, we measured the partition of P_i between the water phase (phosphorylation medium) and different mixtures of benzene-isobutyl alcohol (organic phase). The hydrophobicity of the organic phase increased with an increase in the benzene/isobutyl alcohol ratio. Isobutyl alcohol is freely soluble in benzene but poorly soluble in water (1:20), and under the conditions described in Table 8, remained in the organic layer with benzene while DMSO and glycerol remained in the aqueous phase. DMFA was not used in this experiment because it distributed into the two layer. In accordance with the hypothesis raised above, it was found that P_i entered in significant amounts into the organic layer only when DMSO or glycerol were included in the medium (Table 8). The amount P_i which entered the organic layer increased as the ratio of benzene to isobutyl alcohol decreased. DMSO was more effective than glycerol in promoting the entry of P_i

into the organic layer.

TABLE VIII. Partition of P_i Between Media of Different Hydrophobicity

Additions to Assay Medium	Organic/Aqueous Phase Partition Coefficient: Benzene	75% Benzene 25% Isobutyl Alcohol	55% Benzene 45% Isobutyl Alcohol
None	n.d.	n.d.	n.d.
DMSO, 40% (v/v)	n.d.	2.0×10^{-5}	1.6×10^{-4}
Glycerol, 40% (v/v)	n.d.	n.d.	10^{-5}

From de Meis et. al., 1980 (86). The assay medium composition was 50 mM Tris-Maleate buffer, 0.5 mM EGTA 10 mM $MgCl_2$, 2 mM $^{32}P_i$ (4×10^8 cpm/μmol) and the DMSO or glycerol concentration shown in the table. To 1 ml of the assay media 1 ml of benzene or benzene-isobutyl alcohol was added. The tube was vigorously stirred for 60 sec. After phase separation an aliquot of the benzene-isobutyl alcohol layer was counted in a liquid scintillation counter. With this method, the minimal concentration of P_i which can be detected in the organic phase is 5×10^{-10} M. n.d., nondetectable. The partition coefficient was calculated by dividing the concentration of P_i in the organic phase by its concentration in the aqueous phase.

Phosphorylation of leaky vesicles by P_i is impaired at low temperatures (40). In presence of organic solvents however, this effect of temperature is abolished and the enzyme is phosphorylated by P_i even at -5^oC. This made it possible to determine the equilibrium constant (K_{eq}) of phosphoenzyme formation in presence of DMSO over a large temperature range and to calculate the ΔH, of formation of the phosphoenzyme. At each given temperature the K_{eq} was calculated by dividing the time constant of phosphorylation by the time constant of phosphoenzyme hydrolysis. It was found that the K_{eq} increases as the temperature is decreased (86,88) that is, the phosphoenzyme is more stable at 8^oC than at 32^oC (ΔS^o<-65. e. μ).

The ΔH, calculated for the formation of the phosphoenzyme is -21.2 kcal/mol, which is equivalent to +21.2 kcal/mol for the phosphoenzyme hydrolysis. When compared with the data of Hayes et. al., (Table 6), this large positive value of ΔH, for the hydrolysis indicates that in the presence of DMSO, P_i and the aspartic acid residue of the catalytic site of the enzyme react as if they were nonsolvated. The synthesis of ATP, measured by the Ca^{2+} jump procedure, is abolished in the presence of 40% DMSO (Fig. 12). After the addition of ADP and Ca^{2+}, the phosphoenzyme is slowly hydrolyzed. This is not accompanied by the synthesis of ATP. The inhibition of the ATP synthesis seems to be specifically due to the presence of DMSO. A rapid cleavage of the phosphoenzyme coupled with a stoichiometric synthesis of ATP was observed if, after the addition of ADP and Ca^{2+} the DMSO concentration was suddenly decreased from 40% to 2%.

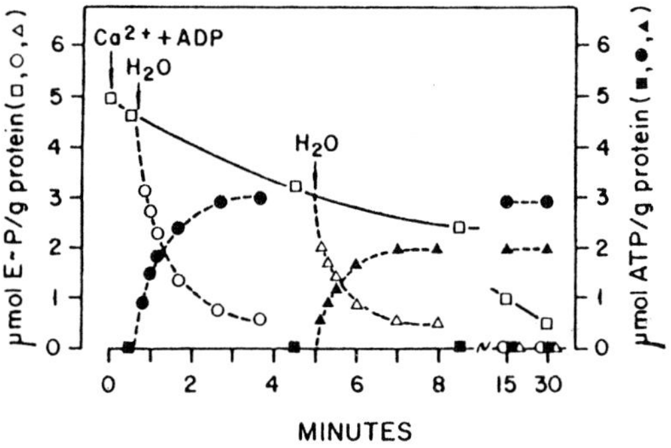

FIGURE 12. Inhibition of ATP synthesis by DMSO. Leaky vesicles reformed from a purified ATPase (7 mg protein/ml) were phosphorylated at 0^oC in 1.7 ml of a medium containing 30mM Tris-maleate buffer (pH 8.0), 0.05 mM EGTA, 5 mM $MgCl_2$, 1 mM $^{32}P_i$, and 40% (v/v) DMSO; 15 min after the addition of enzyme 0.017 ml ADP 20 mM and 0.02 ml $CaCl_2$ 20 mM was simultaneously added. The final concentrations of Ca^{2+} and ADP were 0.15 mM and 0.40 mM respectively. The reaction was arrested at different incubation intervals after the addition of ADP and Ca. In a parallel set of tubes, after the addition of ADP and $CaCl_2$, 0.1 ml of the mixture was diluted in 1.9 ml of a solution of identical composition at 0°, except

The hypothesis proposed states that "high" and "low" energy forms of the phosphoenzyme are correlated with the availability of water at the catalytic site of the enzyme. The experimental results are consistent with this hypothesis. As predicted, phosphorylation of the enzyme by P_i is facilitated when the hydrophobicity of the medium is increased by the addition of organic solvents. Moreover, in the presence of DMSO, the value of the ΔH, of the hydrolysis of the phosphoenzyme is in the range calculated by Hayes et. al., for the hydrolysis of acetylphosphate in the gas phase. Finally, once the phosphoenzyme is formed, synthesis of ATP is observed only if the concentration of DMSO is decreased, that is, when the hydrophobicity of the medium is decreased by the addition of water.

In a previous section of this review it is mentioned that during reversal of the Ca^{2+} pump different forms of energy are interconverted. The data obtained with the use of organic solvents indicate that conformation energy and solvation energy are interconverted. Thus, the following sequence of energy transduction it is proposed.

ΔCa^{2+} activity $\longrightarrow$ binding energy $\longrightarrow$ conformational energy $\longrightarrow$ solvation energy $\longrightarrow$ chemical energy

Where ΔCa^{2+} activity refers to the difference of Ca^{2+} concentration found in the vesicles lumen and in the assay medium when the Ca^{2+} pump is reversed.

Final Remarks. Taken together the data presented rise the possibility that among the different systems available for energy transduction in the cell there is one in which energy is not used directly for the synthesis of high energy molecules but it is used to create and maintain media of different compositions. In one medium the molecule would be of low energy and could be formed spontaneously. The molecule would then be transferred to a second medium where it would become of high energy. In this mechanism synthesis can occur in presence of a large excess of water because it is not dependent on the water concentration in the medium but on the manner by which water is organized around the molecules. Media of different composition can be found in

---

that $^{32}P_i$ and DMSO were omitted. In the figure this is represented as ($\downarrow H_2O$). Open symbols, phosphoenzyme level; close symbols, ATP synthesis; (□,■) 40% DMSO; (O, ●, △, ▲) after dilution of DMSO from 40% to 2% (v/v). From de Meis et. al., 1980 (86).

two compartments of the cell as for instance, the cytosol and sarcoplasmic reticulum lumen or at the molecular level in the catalytic site of an enzyme. Energy would be used by the enzyme to change its conformation and two different microenvironment would be found in the catalytic site, one before and another after the conformational change. (89)

REFERENCES

1. Barlogie, B., Hasselbach, W., and Makinose, M. (1971) FEBS Lett. 12, 267.
2. Makinose, M. (1971) FEBS Lett. 12, 269.
3. Makinose, M. and Hasselbach, W. (1971) FEBS Lett. 12, 271.
4. Makinose, M. (1972) FEBS Lett. 25, 113.
5. Yamada, S. Sumida, M. and Tonomura, Y. (1972) J. Biochem. (Tokyo) 72, 1537.
6. Yamada, S. and Tonomura, Y. (1972) J. Biochem. (Tokyo) 72, 417.
7. de Meis, L. (1971) J. Biol. Chem. 246, 4764.
8. de Meis, L. and Carvalho, M.G. C. (1974) Biochemistry 13, 5032.
9. Mitchell, P. (1966) Chemiosmotic Coupling in Oxidative and Photosynthetic Phosphorylation, Glynn Research, Bodmin, U.K.
10. Mitchell, P. (1979) Eur. J. Biochem. 95, 1.
11. Inesi, G., Goodman, J.J. and Watanabe, S. (1967) J. Biol. Chem. 242, 4637.
12. Duggan, P.F. and Martonosi, A. (1970) J. Gen. Physiol. 56, 147.
13. MacLennan, D.H. (1970) J. Biol. Chem. 245, 4508.
14. MacLennan, D.H., Seeman, P., Iles, G.H. and Yip, C.C. (1971) J. Biol. Chem. 246, 2702.
15. Masuda, H. and de Meis, L. (1973) Biochemistry 12, 4581.
16. de Meis, L. and Masuda, H. (1974) Biochemistry 13, 2057.
17. de Meis, L. (1976) J. Biol. Chem. 251, 2055.
18. de Meis, L. and Vianna, A.L. (1979) Ann. Rev. Biochem. 48, 275.
19. Kanazawa, T. and Boyer, P.D. (1973) J. Biol. Chem. 248, 3163.
20. Carvalho, A.P. (1969) J. Gen. Physiol. 51, 427.
21. Carvalho, A.P. and Leo, B. (1967) J. Gen. Phsyiol. 50, 1327.
22. Chevalier, J. and Butow, R. A. (1971) Biochemistry 10, 2733.

23. Fiehn, W. and Migala, A. (1971) Eur. J. Biochem. 20, 245.
24. Meissner, G., Conner, G.E. and Fleischer, S. (1973) Biochim. Biophys. Acta 298, 246.
25. Ikemoto, N. (1974) J. Biol. Chem. 249, 649.
26. Ikemoto, N. (1975) J. Biol. Chem. 250, 7219.
27. MacLennan, D.H. and Holland, P.C. (1975) Ann. Rev. Biophys. Bioeng. 4, 377.
28. Inesi, G., Kurzmack, M., Coan, C. and Lewis, D.E. (1980) J. Biol.Chem. 255, 3025.
29. Landgraf, W.C. and Inesi, G. (1969) Arch. Biochem. Biophys. 130, 111.
30. Pang, D.C. and Briggs, F.N. (1974) Arch. Biochem. Biophys. 164, 332.
31. Tonomura, Y. and Morales, M.F. (1974) Proc. Natl. Acad. Sci. USA 71, 3687.
32. Champeil, P., Bastide, F., Taupin, C. and Gary-Bobo, C.M. (1976) FEBS Lett. 63, 270.
33. Coan, C.R. and Inesi, G. (1976) Biochem. Biophys. Res. Commun. 71, 1283.
34. Coan, C.R. and Inesi, G. (1977) J. Biol. Chem. 252, 3044.
35. Dupont, Y. (1976) Biochem. Biophys. Res. Commun. 71, 544.
36. Yamamoto, T. and Tonomura, Y. (1977) J. Biochem. (Tokyo) 82, 653.
37. Andersen, J.P. and Moller, J.V. (1977) Biochim. Biophys. Acta 485, 188.
38. Thorley-Lawson, D.A. and Green, N.M. (1977) Biochem. J. 167, 739.
39. Dupont, Y. and Leigh, J.B. (1978) Nature 273, 396.
40. Masuda, H. and de Meis, L. (1977) J. Biol. Chem. 252, 8567.
41. Alberty, R.A. (1968) J. Biol. Chem. 243, 1337.
42. George, P., Witonsky, R.J., Trachtman, M., Wu, C., Dorwost, W., Richman, L., Richaman, W., Shurayh, F. and Lentz, B. (1970) Biochim. Biophys. Acta 223, 1.
43. Hayes, D.M., Kenyon, G.L. and Kollman, P.A. (1978) J. Am. Chem. Soc. 100, 4331.
44. Boyer, P.D., Cross, R.L., and Momsen, W. (1973) Proc. Nat. Acad. Sci. USA 70, 2837.
45. Cross, R.L., and Boyer, P.D. (1975) Biochemistry 14, 392.
46. Boyer, P.D., Chance, B., Ernster, L., Mitchell, P., Racker, E., and Slater, E.C. (1977) Ann. Rev. Biochem. 49, 955.
47. Hutton, R.L., and Boyer, P.D. (1979) J. Biol. Chem 254, 9990.
48. Kanazawa, T. (1975) J. Biol. Chem. 250, 113.
49. Knowles, A.F. and Racker, E. (1975) J. Biol. Chem. 250, 1949.

50. Beil, F.U., Chak, D. and Hasselbach, W. (1977) Eur. J. Biochem. 81, 151.
51. Lacapere, J-J., Gingold, M.P., Champeil, P., and Guillain, F. (1981) J. Biol. Chem. 256, 2302.
52. Boyer, P.D., de Meis, L., Carvalho, M.G. C. and Hackney, D.D. (1977) Biochemistry 16, 136.
53. Rauch, B., Chak, D. and Hasselbach, W. (1977) Z. Naturforsch. 32c, 828.
54. Punzengruber, C., Prager, R., Kolassa, N., Winkler, F. and Suko, J. (1978) Eur. J. Biochem. 92, 349.
55. Kolassa, N., Punzengruber, C., Suko, J. and Makinose, M. (1979) FEBS Lett. 108, 495.
56. Prager, R. Punzengruber, C., Kolassa, N. Winkler, F. and Suko, J. (1979) Eur. J. Biochem. 97, 239.
57. Martin, D., and Tanlord, C. (1981) Biochemistry 20, 4597.
58. de Meis, L. and Sorenson, M.M. (1975) Biochemistry 14, 2739.
59. Carvalho, M.G.C., Souza, D.O. and de Meis, L. (1976) J. Biol. Chem. 251, 3629.
60. de Meis, L. and Boyer, P.D. (1978) J. Biol. Chem. 253, 1556.
61. Vieyra, A., Scofano, H.M., Guimaraes-Motta, H., Tume, R.K. and de Meis, L. (1961) Biochim. Biophys. Acta 568, 437.
62. Souza, D.O., and de Meis, L. (1976) J. Biol. Chem. 251, 6355.
63. de Meis, L. and Hasselbach, W. (1971) J. Biol. Chem. 246, 4759.
64. de Meis, L. and de Mello, M.C.F. (1973) J. Biol. Chem. 248, 3691.
65. Scofano, H.M., Vieyra, A. and de Meis, L. (1979) J. Biol. Chem. 254, 10227.
66. Scofano, H.M., and de Meis, L. (1981) J. Biol. Chem. 256, 4282.
67. Dupont, Y. (1977) Eur. J. Biochem. 72, 185.
68. Froehlich, J.P. and Taylor, E.W. (1975) J. Biol. Chem. 250, 2013.
69. Froehlich, J.P. and Taylor, E.W. (1976) J. Biol. Chem. 251, 2307.
70. Takisawa, H. and Tonomura, Y. (1978) J. Biochem. (Tokyo) 83, 1275.
71. Wang, T., de Gende, A.O.G. and Schwartz, A. (1979) J. Biol. Chem. 254, 10675.
72. de Meis, L., and Carvalho, M.G.C. (1974) Biochemistry 13, 5032.
73. Plank, B., Hellmann, G., Punzengruber, C. and Suko, J. (1979) Biochim. Biophys. Acta 550, 259.

74. Watanabe, T., Lewis, D., Nakamoto, R., Kurzmack, M., Fronticelli, C., and Inesi, G. (1981). Biochemistry 20, 6617.
75. de Meis, L. (1981) The sarcoplasmic reticulum, Transport and Energy Transduction. The Wiley Series: Transport in the Life Sciences, Bittar, E.E. ed. vol. 2. John Wiley and Sons, New York.
76. de Meis, L. and Inesi, G. (1982) The transport of calcium by sarcoplasmic reticulum and various microsomal preparations in membrane transport of calcium (E. Carafoli, ed.) Academic Press , New York, page 141.
77. de Meis, L. and Tume, R.K. (1977) Biochemistry 16, 4455.
78. Verjovski-Almeida, S. and de Meis, L. (1977) Biochemistry 16, 329.
79. Madeira, V.M. (1978) Arch. Biochem. Biophys. 185, 316.
80. Madeira, V.M. (1979) Arch. Biochem. Biophys. 193, 22.
81. Chiesi, M. and Inesi, G. (1980) Biochemistry 19, 2912.
82. Kalckar, H.M. (1941) Chem. Rev. 28, 71.
83. Hill, T.L. and Morales, M.H. (1951) J. Am. Chem. Soc. 73, 1606.
84. Pullman, A. and Pullman, B. (1963) Quantum Biochemistry, Interscience, New York.
85. Boyd, D.B. and Lipscomb, W.N. (1969) J. Theor. Biol. 25, 403.
86. de Meis, L., Martins, O.B. and Alves, E.W. (1980) Biochemistry 19, 4252.
87. Otero, A.S., and de Meis, L. (1982) Z. Naturforsch. 37, 527.
88. de Meis, L., Otero, A.S., Martins, O.B., Alves, E.A., Inesi, G. and Nakamoto, R. (1982) J. Biol. Chem. 257, 4993.
89. de Meis, L. (1983). Annals New York Academy of Sciences, in press.

THE Ca-ATPase OF SARCOPLASMIC RETICULUM CATALYSES A MONOVALENT CATION (Na,K) EXCHANGE

Yves Dupont[1]
Nelly Bennett[1]

Laboratoire de Biologie Moléculaire et Cellulaire
Centre d'Etudes Nucléaires de Grenoble
Grenoble, France

We have previously described a fast EGTA-induced potassium (sodium) efflux from sarcoplasmic reticulum vesicles (1). Further characterization of this mechanism is described in this communication.

I. EGTA Induces an Exchange of Internal Potassium (or Sodium) for the External Monovalent Cation.

The EGTA-induced dissipation of a potassium diffusion potential (Fig.1A) and the EGTA-induced 22Na efflux (Fig.1B) in the absence of a membrane potential are studied with either sodium or Tris ions outside the vesicles: it is clear that the two processes are significantly inhibited in the presence of Tris as the only external monovalent cation. In Fig.1A, the effect of EGTA is also compared to that of the ion channel gramicidin. With sodium outside, the potassium diffusion potential generated upon injection of valinomycin (fluorescence decrease) is rapidly dissipated (fluorescence increase) by gramicidin, which increases the sodium conductance (G.Na) to a value close to that of potassium conductance in the presence of valinomycin (G.K). The same effect is obtained with EGTA, indicating that EGTA increases G.Na (as also shown by Fig.1B).

1- Supported by CNRS grant: ATP: conversion de l'énergie dans les membranes biologiques.

ISBN 0-12-260380-X

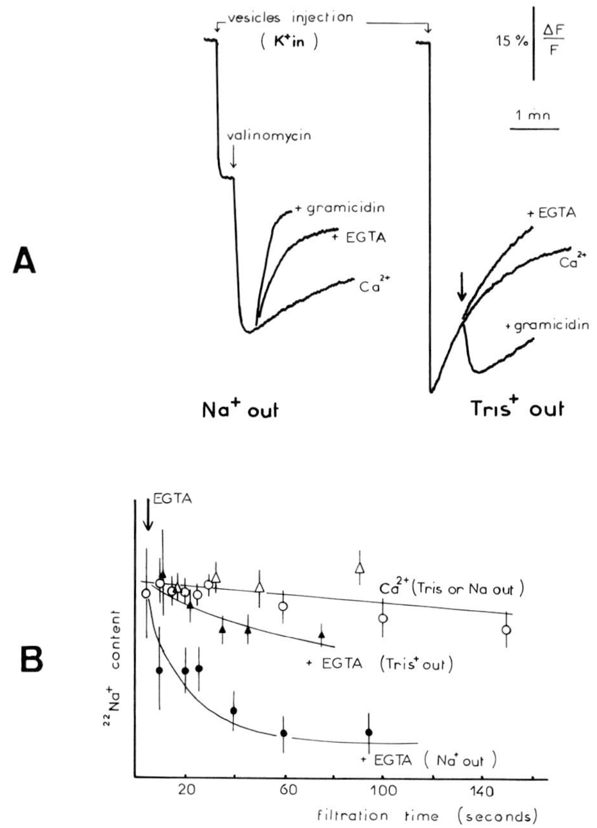

FIGURE 1. EGTA induces an exchange of internal K (Na) for the external monovalent cation.

A. Measurement of membrane potential changes with the potential sensitive fluorescent dye: 3,3'-dipropylthiodicarbocyanine iodide (diS-C3-(5) see ref.3), pH 8.0, 0°C. A potassium diffusion potential is established by suspending the vesicles (internal medium: 100mM K, 220mM Epps, 4mM Mg-gluconate) into either sodium + valinomycin (0.25 μM), or Tris without valinomycin. External medium : 100mM Na or Tris, 220mM Epps, 4mM Mg-gluconate, 50μM calcium. Additions: 0.1μM gramicidin, or 500μM EGTA. The fluorescence of diS-C3-(5) (1μM) is measured at 670 nm (excitation at 622 nm).

B. Measurement of ^{22}Na efflux by Millipore filtration, pH 7.8, 0°C. Vesicles loaded with ^{22}Na (10mM Tris, 20mM Na, Mops) are washed with cold medium containing either sodium: (10mM Tris, 20mM Na, Mops) or Tris as the only monovalent cation (30mM Tris, Mops), in the presence of calcium (50μM) or EGTA (400μM). The amount of ^{22}Na remaining inside the vesicles is measured as a function of the duration of washing.

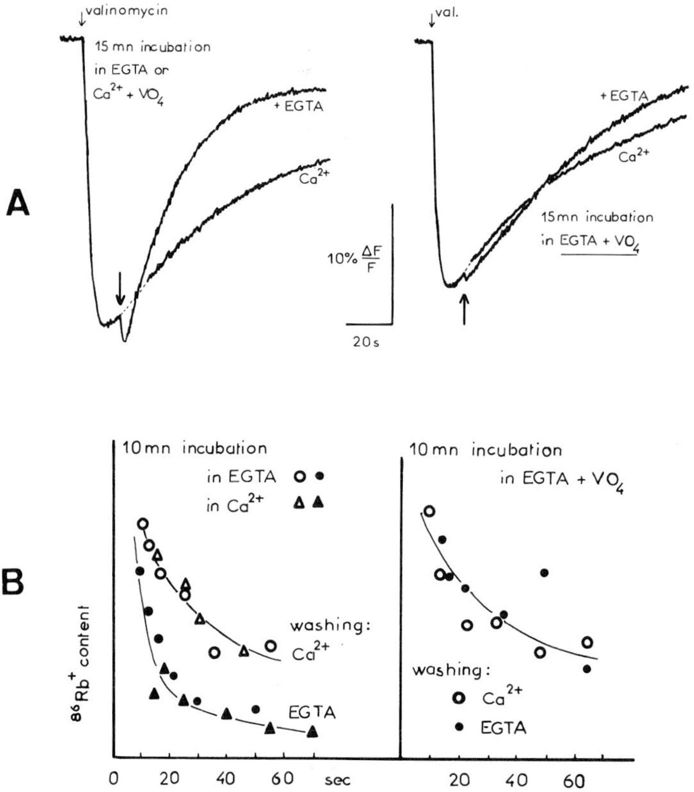

FIGURE 2. Inhibition by vanadate of the monovalent cation efflux induced by EGTA.

A. Measurement of potassium efflux in the presence of a potassium diffusion potential with diS-C3-(5), pH 8.0, 0°C. Internal medium: 100mM K, 220mM Mops, 4mM Mg-gluconate, 0.1mM Ca-gluconate, pH7.0. External medium : 100mM Na, 220mM Epps, 4mM Mg-gluconate, 0.1mM Ca-gluconate, pH8.0, 0.25µM valinomycin. Before suspension into the external medium, membranes are incubated at 20°C in the internal medium containing EGTA (500µM), or vanadate (100µM), or EGTA + vanadate. Incubation is performed at pH 7.0 in order to limit denaturation of the membranes by EGTA at alkaline pH (4).

B. Measurement of ^{86}Rb efflux by Millipore filtration (pH 8.2, 22°C). Internal medium: 100mM K-gluconate (+ 86Rb), 10mM K-Mops, 5mM Mg-gluconate, 20µM calcium or 500µM EGTA. Washing : same media without ^{86}Rb. Before filtration, the vesicles are incubated 10 minutes in the internal medium, with or without calcium (left), or without calcium in the presence of 100µM vanadate (right).

When the potassium loaded vesicles are suspended in Tris, a potassium diffusion potential is created without valinomycin since the permeability of the membrane for Tris is lower than for potassium (2). In that case, gramicidin and EGTA have different effects on the membrane potential: while EGTA very weakly accelerates its dissipation, gramicidin (or valinomycin, not shown) restores the potassium diffusion potential when added after partial dissipation of the membrane potential by passive Tris influx. This indicates that gramicidin (or valinomycin) increases the G.K / G.Tris ratio, whereas EGTA does not. Since we have shown that EGTA increases G.Na (G.K), the rate of the EGTA-induced potassium efflux must be limited by the rate of Tris influx. These experiments therefore indicate that the EGTA-induced monovalent cation efflux is not performed by an electrogenic ion channel such as gramicidin or valinomycin, but most probably by an exchange carrier, the rate of transport being faster for sodium and potassium than for Tris.

II. The EGTA-Induced Potassium (Sodium) Efflux is Inhibited by Vanadate.

Fig. 2 shows that preincubation of the vesicles in the presence of vanadate and EGTA (but not vanadate and calcium) at 20°C inhibits the EGTA-induced cation efflux revealed both by dissipation of the membrane potential (Fig.2A) and by 86Rb efflux measured by Millipore filtration (Fig.2B). The inhibition by vanadate of the EGTA-induced potassium (sodium) efflux requires the same conditions which have been shown to block the Ca-ATPase in its calcium free conformation, i.e. exclusively when both EGTA and vanadate are present in the incubation medium (5). This suggests that the monovalent cation carrier is the ATPase itself.

CONCLUSION

The results described here indicate that the Ca-ATPase is able to catalyse a monovalent cation (Na, K) exchange in the absence of ATP. We propose that this exchange occurs during the E $\rightleftharpoons$ E* transconformation, calcium ions and vanadate inhibiting the exchange by stabilizing the E and E* state respectively:

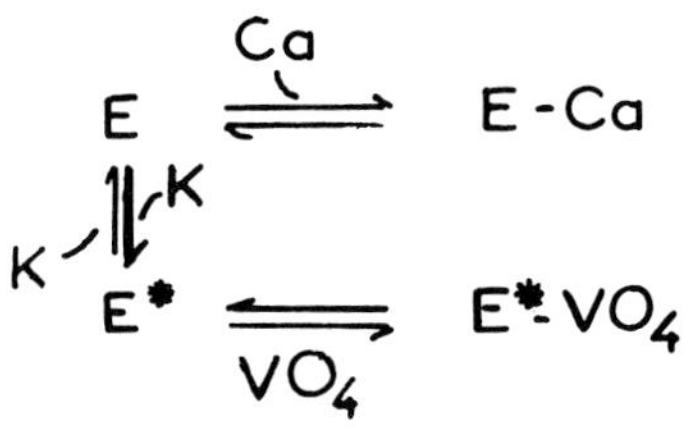

These results suggest that the Ca-ATPase might be able to transport monovalent cations during the active transport of calcium. It is indeed tempting to consider a monovalent cation exchange occuring during the E ⇌ E* reaction as the counterpart for the Ca - Ca exchange mediated by the phosphoenzyme (E-P ⇌ E*-P) (6).

REFERENCES

1. Bennett, N. and Dupont, Y. (1981), FEBS Lett. 128, 269-274
2. Kometani, T. and Kasai, M. (1978), J.Memb.Biol. 41, 295-308
3. Sims, P.J., Waggoner, A.S., Chao-Huei, W. and Hoffman, J.F. (1974) Biochemistry 13, 3315-3330
4. Duggan, P.F. and Martonosi, A. (1970) J.Gen.Phys. 56, 147-167
5. Dupont, Y. and Bennett, N. (1982), FEBS Lett. 139, 237-240
6. Makinose, M. (1973), FEBS Lett. 37, 140-143

Section IV

The Ca^{2+} Release Process

RAPID AND REVERSIBLE CALCIUM RELEASE FROM SARCOPLASMIC RETICULUM IN VITRO

Noriaki Ikemoto[1]
Do Han Kim

Department of Muscle Research
Boston Biomedical Research Institute
and
Department of Neurology
Harvard Medical School
Boston, Massachuesetts

I. INTRODUCTION

Extensive efforts have been made in many laboratories in an attempt to elucidate the molecular mechanism of Ca^{2+} release. Release of Ca^{2+} from sarcoplasmic reticulum (SR) both in vivo and in vitro can be induced by several different methods; for instance, 1) an increase of the extravesicular Ca^{2+} from about 0.1 μM to several μM (so-called Ca^{2+}-induced Ca^{2+} release, refs. 1 -- 4), 2) the addition of drugs such as caffeine (5 -- 8), 3) substitution of permeable anions for impermeable ones, or substitution of impermeable cations for permeable ones, which is expected to produce membrane depolarization (9 -- 12), and 4) the addition of proton ionophores (13).

If an experimentally produced Ca^{2+} release is to be a candidate for a physiological mechanism, it should meet at least two requirements. First, both the rate and the amount of Ca^{2+} release should be sufficiently large to account for the rapid time course of tension development in the intact muscle fiber. Second, the release should be a transient one

[1]Supported by grants from NIH (AM 16922) and MDA.

ISBN 0-12-260380-X

and the released Ca^{2+} should be spontaneously transported back into SR; in other words, Ca^{2+} release should be reversible. So far, these requirements have not been fully satisfied in the Ca^{2+} release experiments in vitro.

Recently we began kinetic studies of Ca^{2+} release in vitro with particular attention to both of the above requirements. We found that the observed V_{max} value (designated as V'_{max}) of the Ca^{2+}-induced Ca^{2+} release in the presence of increasing concentrations of caffeine or quercetin is about the same as that of the depolarization-induced Ca^{2+} release. Furthermore, this is comparable or even higher than the estimated rate of Ca^{2+} release in vivo. We also found that in the presence of ATP and an ATP regenerating system, the released Ca^{2+} is re-accumulated spontaneously. Thus, the in vitro Ca^{2+} release discussed in this study satisfies the criteria for the physiological mechanism at least in terms of rapidity and reversibility.

The results reported here also permit some insights into the classes of the molecular components that are involved in the Ca^{2+} release machinery. One class consists of several different 'receptors' for different triggering signals, which are distinguishable by selective inhibitors (e.g. dantrolene sodium and procaine) and by different activating concentration ranges of extravesicular Ca^{2+} ($[Ca_o^{2+}]$). The other class consists of putative 'channels', whose kinetic properties are essentially identical, independent of the types of triggering signals, as evidenced by the facts that the observed V_{max} values of Ca^{2+} release produced by different triggering methods are about the same and that agents such as higher $[Ca_o^{2+}]$ and ruthenium red block all types of Ca^{2+} release investigated.

II. EXPERIMENTAL PROCEDURES

A. SR Preparation

About 100 g muscle was homogenized in a Waring blender with four volumes of 2.5 mM NaOH solution for 2 min (30 s x 4). During the homogenization the pH was adjusted to 6.8. The suspension was centrifuged at 10,000 x g for 3 min in a GSA rotor (Sorvall). The supernatant fraction was filtered through Whatman filter papers (No. 2) and after re-adjusting the pH to 6.8 the filtrate was centrifuged again at 17,000 x g for 30 min in an SS-34 rotor (Sorvall). The pellets were suspended in a solution containing 150 mM KCl and 20 mM MES, pH 6.8 and re-centrifuged at 17,000 x g for 30 min. The pellets were resuspended in the same buffer and the final protein concentration was adjusted to 20--30 mg/ml and stored at 0-4°. The preparation was used within four days after preparation.

B. Stopped-flow studies of Ca^{2+} release

The SR vesicles were loaded with Ca^{2+} by ATP dependent Ca^{2+} accumulation in a reaction mixture containing 1.6 mg protein per ml, 50 μM $CaCl_2$, 150 mM KCl, 20 mM MES (pH 6.8), 0.5 mM Mg.ATP, with an ATP regenerating system including 2.5 mM phosphoenolpyruvate and 10 units per ml pyruvate kinase at 27° for 1 - 2 min. The reaction solution was loaded in syringe A of a Durrum stopped-flow apparatus (referred to as syringe A solution). At the steady state of Ca^{2+} uptake, when the extravesicular concentration of Ca^{2+} became nearly zero, the contents of syringes A and B (1:1 v/v) were mixed in order to trigger Ca^{2+} release (for details, see figure legends). Changes in $[Ca_o^{2+}]$ were

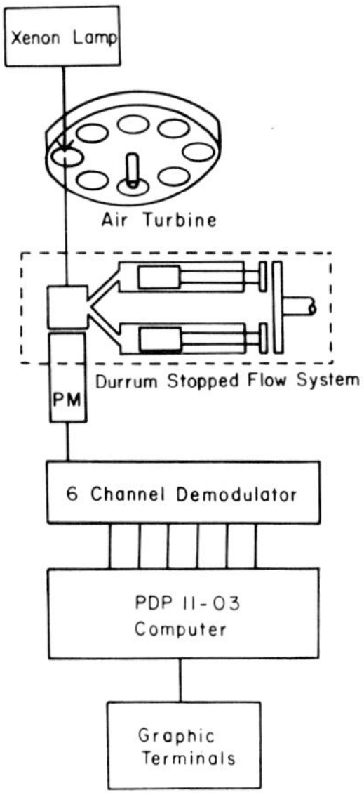

FIGURE 1. Outline of the multiple-channel stopped-flow spectrophotometer system used in this study. For details, see text. Air turbine and demodulator were constructed at the Johnson Research Foundation, University of Pennsylvania.

monitored with the use of a multiple channel stopped-flow spectrophotometer system illustrated in Fig. 1. The system consists of i) a rotating air tubine containing 8 different interference filters, ii) a turbine demodulator containing an I to log I converter (I = transmittance), iii) Durrum stopped flow apparatus model D-110 and iv) a dedicated PDP 11 03 computer, a graphics terminal and a digital plotter. Changes in the difference of absorbance at 700 nm and 650 nm were recorded in the computer and signal-averaged, and various kinetic parameters were calculated by an iterative fitting algorithm.

C. Efflux Assay of Passively Loaded Ca^{2+}

SR vesicles (15 mg per ml) were incubated overnight in a solution containing 150 mM KCl, 20 mM MES (pH 6.8) and 5 mM $CaCl_2$ containing $^{45}CaCl_2$ (10 μCi per ml) at 0°. In order to initiate Ca^{2+} efflux, 20 μl of the loaded SR was diluted 50 fold into a 'dilution' solution containing different concentrations of Ca^{2+} or different reagents, or both (for details, see legend to Fig. 7). At various times after dilution, the reaction was quenched by the addition of an equal volume of a solution containing 10 mM $LaCl_3$ and 20 mM $MgCl_2$. The quenched solution was filtered through a Millipore filter and the radioactivity retained on the filter was determined by scintillation counting.

III. RESULTS

A. Kinetic Characteristics of Various Types of Calcium Release

Spectrophotometric monitoring of changes in the extravesicular Ca^{2+} concentration ($[Ca_o^{2+}]$) with arsenazo III as a Ca^{2+} indicator is one of the most successful methods in reported studies of Ca^{2+} uptake and Ca^{2+} release kinetics (e.g. refs. 4, 8, 12). However, the rate of Ca^{2+} release is too rapid to follow by hand mixing in the conventional dual wavelength spectrophotometric method. Rapid release of Ca^{2+} that had been loaded actively in the SR vesicles in the presence of 0.5 mM Mg.ATP and an ATP-regenerating system could be monitored with the use of the stopped-flow multiple channel spectrophotometer system illustrated in Fig. 1.

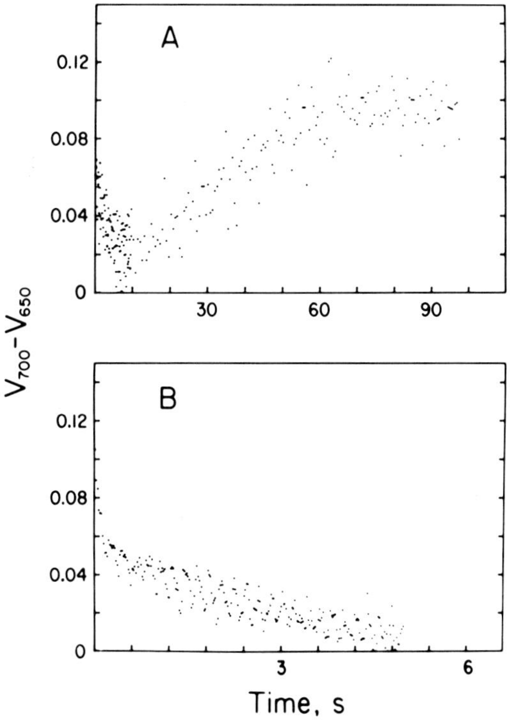

FIGURE 2. Stopped-flow recording of the depolarization-induced Ca^{2+} release. Active Ca^{2+} loading was started by adding 0.5 mM Mg.ATP to a solution containing 150 mM KCl, 1.6 mg/ml SR, 20 mM MES (pH 6.8), 50 µM $CaCl_2$, 2.5 mM PEP, 10 units/ml pyruvate kinase and 9 µM arsenazo III, and immediately loaded in syringe A of a Durrum stopped flow apparatus and incubated at 27°. At the steady state of Ca^{2+} uptake (1 - 2 min) Ca^{2+} release was trigged by mixing one part of the syringe A solution with one part of the syringe B solution containing 150 mM choline chloride, 20 mM MES (pH 6.8) and 9 µM arsenazo III. To determine the time course of Ca^{2+} release, changes in the transmittance of arsenazo III at 700 and 650 nm (I_{700} and I_{650}) were recorded through interference filters built in an air-turbine, converted to log (I_{700}/I_{650}) with a demodulator, and then recorded in a dedicated PDP 11 03 computer. Eight to ten traces were signal-averaged. Downward excursion represents release of Ca^{2+} and upward excursion represents the reuptake of the released Ca^{2+}.

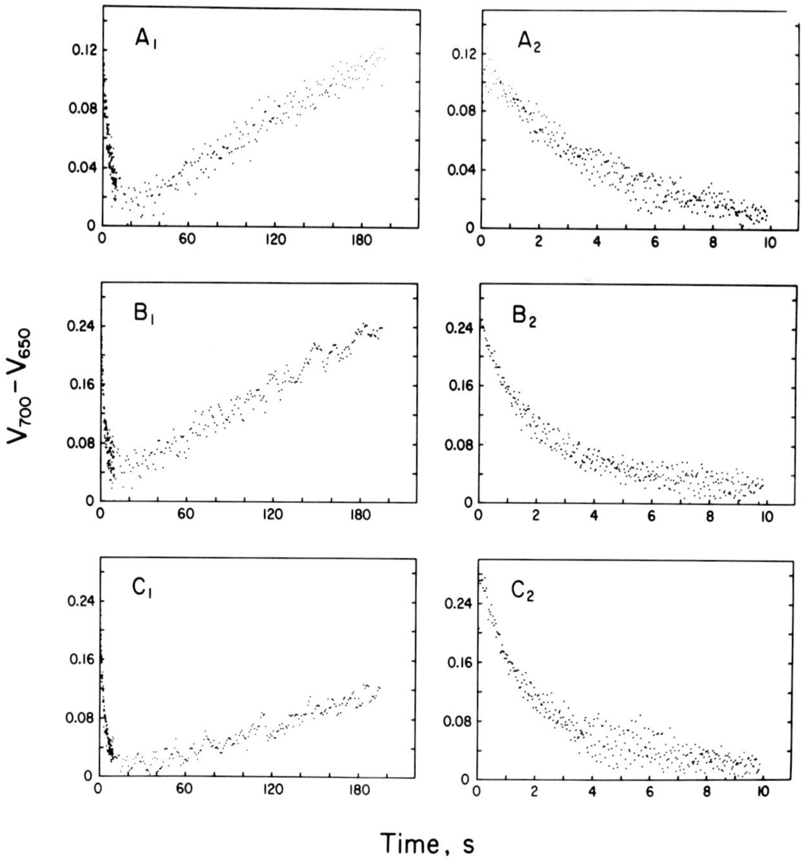

FIGURE 3. Stopped-flow recording of rapid and reversible Ca^{2+} release induced by Ca^{2+} (A), caffeine (B) and quercetin (C). Active Ca^{2+} loading was carried out as described in the legend to Fig. 2. The Ca^{2+} release reaction was triggered by mixing one part of the syringe A solution described in the legend to Fig. 2 with one part of syringe B solution containing 150 mM KCl, 20 mM MES (pH 6.8), 9 µM arsenazo III and various triggering agents: 100 µM added Ca^{2+} (A), 3.0 mM caffeine (B) and 140 µM quercetin (C). The time courses of Ca^{2+} release and re-uptake were monitored as described in the legend to Fig. 2.

1. Depolarization-Induced Calcium Release

Caswell et al (14) carried out stopped-flow studies of a rapid Ca^{2+} release induced by substitution of Cl^- for gluconate$^-$ using chlorotetracycline as an optical probe. However, under the conditions used in the present studies both caffeine and quercetin interfere with the fluorescence properties of chlorotetracycline. Furthermore, the rate of Ca^{2+} re-uptake became much faster when Cl^- was substituted for gluconate$^-$: this creates a problem for the subtraction of the control curve from the experimental curve in order to calculate the time course of the depolarization-induced Ca^{2+} release. Gluconate$^-$ also decreased the sensitivity of arsenazo III to Ca^{2+}. These problems can be circumvented by substituting choline$^+$ for K^+ to induce Ca^{2+} release and using arsenazo III for monitoring of $[Ca_o^{2+}]$. Fig. 2 illustrates such experiments carried out at $[Ca_o^{2+}] \simeq 0$. The time course of Ca^{2+} release consists of two phases, rapid and slow (Fig. 2B). The maximal observed value of the initial rate of Ca^{2+} release is 19.0 nmol Ca^{2+}/mg SR/s. As shown in Fig. 2A, the released Ca^{2+} is spontaneously reaccumulated within about 60 s, indicating that Ca^{2+} release observed here is a reversible process. In order to achieve complete reversibility, the presence of an ATP-regenerating system is essential (cf. ref. 15).

2. (Calcium and Drug)-Induced Calcium Release

A1, B1 and C1 in Fig. 3 illustrate representative Ca^{2+} release and reuptake time courses triggered by Ca^{2+}, caffeine and quercetin, respectively. Details of the time courses of the Ca^{2+} release phase of the corresponding traces are shown in A2, B2 and C2 (Fig. 3). Ca^{2+} release is reversible as in the case of depolarization-induced Ca^{2+} release, although the re-uptake rates are considerably slower, particularly in the presence of quercetin due to the fact that quercetin inhibits the Ca^{2+} pump (16, 17).

Fig. 4 illustrates that all of the Ca^{2+}-, caffeine and quercetin-induced Ca^{2+} releases are regulated by $[Ca_o^{2+}]$ in an essentially identical fashion, although the $[Ca_o^{2+}]$-dependence is less steep for the quercetin-induced Ca^{2+} release in the low $[Ca_o^{2+}]$ range.

Fig. 5 shows the results of stopped-flow studies of the Ca^{2+} release rate with increasing concentrations of caffeine and quercetin in the presence of optimum $[Ca_o^{2+}]$ (12.5 μM). The V'_{max} value for caffeine-induced Ca^{2+} release in nmol/mg SR/s (21.0) and that for quercetin-induced

Ca^{2+} release (25.2) are about the same, although the effective ranges of drug concentration differ considerably (K_m caffeine = 0.2 mM; K_m quercetin = 3.3 μM). The calculated Hill coefficient is very close to 1 in both cases regardless of $[Ca_o^{2+}]$. These results suggest that caffeine and quercetin compete for the same regulatory site, and also that the number of potentiator binding sites is one per Ca^{2+} channel. In further support of this conclusion, the addition of both caffeine and quercetin together produces no further increase of V'_{max}. The fact that the maximal activation of Ca^{2+} release by caffeine or quercetin requires the presence of an optimal $[Ca_o^{2+}]$ suggests that caffeine or quercetin bind to the same regulatory unit as the one which is responsible for the Ca^{2+}-induced Ca^{2+} release.

It is important to note that the V'_{max} values for

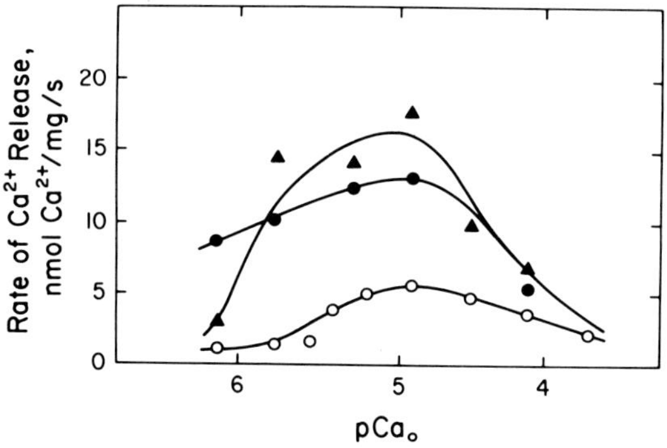

FIGURE 4. The $[Ca_o^{2+}]$-dependence of the (calcium + drug)-induced Ca^{2+} release. Active Ca^{2+} loading was carried out as described in the legend to Fig. 2. At the steady state of Ca^{2+} uptake one part of the syringe A solution described in the legend to Fig. 1 was mixed with one part of the syringe B solution containing various free concentrations of Ca^{2+} (as determined by a Ca^{2+}-selective electrode) with various additions: 4 mM caffeine (▲), 100 μM quercetin (●), and no addition (○). Stopped-flow recording was carried out as described in the legend to Fig. 3. The initial rates of Ca^{2+} release were calculated as the derivation at t = 0 of the computer-fitted Ca^{2+} release curves.

caffeine and quercetin are essentially identical to the initial rate of depolarization-induced Ca^{2+} release observed in the presence of the same concentrations of Mg.ATP and an ATP-regenerating system. This finding suggests the hypothesis that although the triggering mechanism for the (Ca^{2+} + drug)-induced Ca^{2+} release and that for the depolarization-induced Ca^{2+} release are different, as discussed below, the putative channel through which Ca^{2+} is released is kinetically identical regardless of the method of triggering.

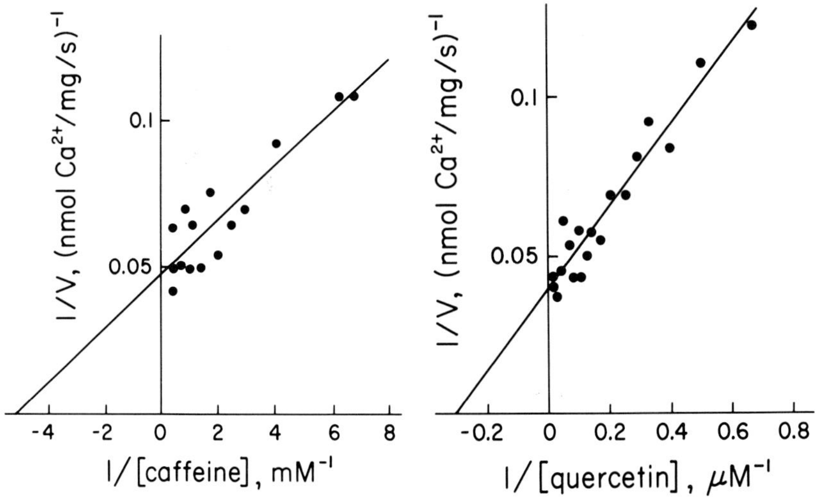

FIGURE 5. The effects of increasing concentrations of caffeine and quercetin on the rate of Ca^{2+} release. Active Ca^{2+} loading was carried out as described in the legend to Fig. 2. At the steady state of Ca^{2+} uptake, Ca^{2+} release was triggered by mixing the syringe A solution described in the legend to Fig. 2 with one part of syringe B solution containing 150 mM KCl, 20 mM MES (pH 6.8), 9 μM arsenazo III, 100 μM added $CaCl_2$, and different concentrations of caffeine (left) or quercetin (right).

B. Selective and Common Inhibitors of Calcium Release

Table I depicts the preliminary results of our studies on the effects of various agents on different types of Ca^{2+} release. In agreement with previous reports (18, 19), dantrolene sodium blocks the (calcium + drug)-induced Ca^{2+} releases but does not inhibit (rather potentiates) the depolarization-induced Ca^{2+} release; hence it belongs to the category of a 'selective' inhibitor. Almost complete inhibition of all types of Ca^{2+} release is produced by the addition of low concentrations of ruthenium red (e.g. 0.25 µM). Thus, ruthenium red may be regarded as a 'common' inhibitor, which probably acts on the channel.

It is interesting to note that the depolarization-induced Ca^{2+} release (Fig. 6) as well as the (calcium + drug)-induced Ca^{2+} release (cf. Fig. 4) are sharply inhibited at $[Ca_o^{2+}]$ higher than 10 µM, although activating $[Ca_o^{2+}]$ ranges are different for different types of Ca^{2+} release (compare Fig. 6 with Fig. 4). Thus, it appears that a high extravesicular Ca^{2+} concentration also works as a common blocker.

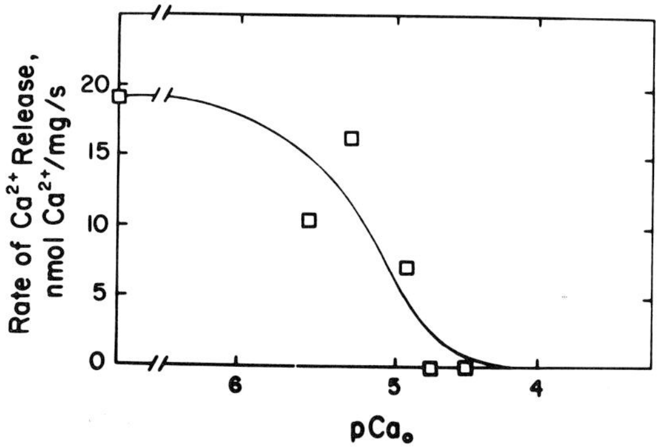

FIGURE 6. The $[Ca_o^{2+}]$-dependence of the depolarization-induced Ca^{2+} release. Active Ca^{2+} loading and triggering were carried out as described in the legend to Fig. 2, except that various amounts of Ca^{2+} were added to the syringe B solution to create various $[Ca_o^{2+}]$ as indicated.

TABLE I. Selective and Common Inhibitors of Calcium Release

Affectors	Methods of triggering	% Control
20 μM dantrolene Na	2 mM caffeine	69
"	12.5 μM Ca^{2+}	136
"	70 μM quercetin	98
"	$K^+ \rightarrow choline^+$ substitution	100
1 μM ruthenium red	2 mM caffeine	0
"	12.5 μM Ca^{2+}	0
"	50 μM quercetin	14
"	$K^+ \rightarrow choline^+$ substitution	0

We have initiated studies of several calcium antagonists. It appears that one of these reagents, nitrendipine, blocks both the (calcium + drug)-induced and the depolarization-induced Ca^{2+} releases.

C. Spontaneous Closing of Channels

Since the released Ca^{2+} is spontaneously re-accumulated into SR vesicles as described above (cf. Figs. 2 and 3), there must be some mechanism by which the direction of Ca^{2+} flux is switched from the release mode to the reuptake mode. As evidenced from the above observation that in the $[Ca_o^{2+}]$ range higher than 10 μM both types of Ca^{2+} release are sharply inhibited, one such mechanism is the elevation of $[Ca_o^{2+}]$ due to the released Ca^{2+}.

In order to investigate the Ca^{2+} permeability properties intrinsic to the membrane, we have studied efflux of passively loaded $^{45}Ca^{2+}$ in the absence of Mg.ATP by the method described previously (20) with the modification that both 10 mM lanthanum and 20 mM Mg^{2+} were used for the chemical quenching of rapid Ca^{2+} efflux (Fig. 7). The Ca^{2+} efflux does not follow the first order kinetics. The most interesting observation in these experiments is that Ca^{2+} release is attenuated at about the same time after the start of efflux independent of the rate and the amount of Ca^{2+} release in the initial phase. Various agents that affect both the rate and the amount of Ca^{2+} release (e.g. $[Ca_o^{2+}]$, caffeine, quercetin and Mg^{2+}) have little or no effect on the $t_{1/2}$ of the attenuation of Ca^{2+} efflux.

These results suggest that opened channels close spontaneously and that the life-time of the opened channel is essentially independent of the extent (or the number) of opened channels, as shown in Model 1.

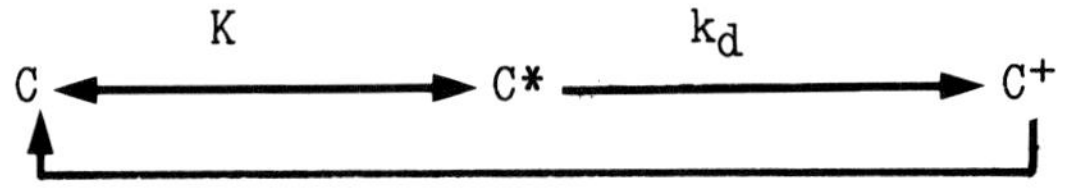

Model 1

The model assumes that the putative channel can take three states: C is a ground state that is closed but capable of opening; C*, opened; and C^+, refractory state. We assume

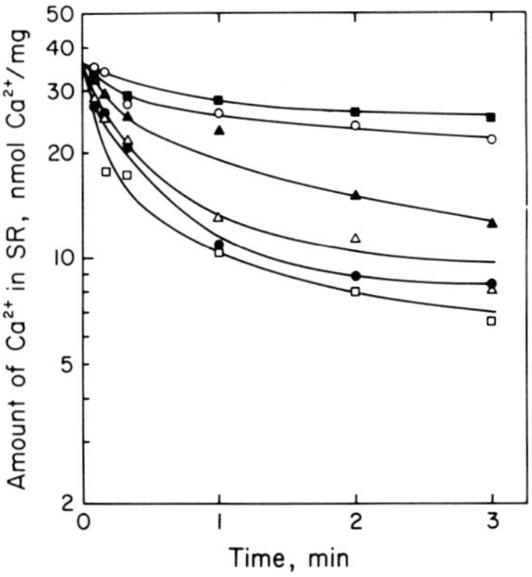

FIGURE 7. The time courses of efflux of the passively loaded $^{45}Ca^{2+}$ at various extravesicular Ca^{2+} concentrations. SR vesicles (15 mg/ml) were incubated overnight in a solution containing 150 mM KCl, 20 mM MES (pH 6.8) and 5 mM $CaCl_2$ including $^{45}CaCl_2$ (10 μCi per ml) at 0-4°C. To initiate the efflux experiments, one part of the above solution was mixed with 50 parts of a 'dilution' solution containing 150 mM KCl, 20 mM MES (pH 6.8), and different extravesicular Ca^{2+} concentrations. Ca^{2+} efflux was quenched by mixing one part of the above solution with one part of the 'quenching' solution containing 10 mM $LaCl_3$ and 20 mM $MgCl_2$. A 0.5 ml portion of the quenched mixture was filtered and washed with a 2 ml solution of 10 mM $LaCl_3$ and 20 mM $MgCl_2$. The filters were air-dried, and the amount of ^{45}Ca retained on the filter was determined by scintillation counting. $[Ca_o^{2+}]$, μM: ○, 0.1; △, 1.0; □, 3.3; ●, 5.0; ▲, 10.0; ■, 100,0.

that the C $\longleftrightarrow$ C* reaction is a rapid equilibrium with the equilibrium constant K, whereas the C* $\longrightarrow$ C^+ reaction proceeds slowly with the rate constant k_d. The C^+ state would eventually return to the C state. The rate of Ca^{2+} release is a function of [C*], which is a function of $\exp(-k_d t)$. Therefore, the ln $[^{45}Ca_i]$ versus t plot shows in fact an exponential time course even in the semi-logarithmic plot, as shown in Fig. 7. These data also explain the fact that the Ca^{2+} release phase is spontaneously terminated even in the continued presence of triggering agents such as caffeine and quercetin.

IV. DISCUSSION

The methods described here permit both the induction and the measurement of rapid Ca^{2+} release from SR in vitro. The observed maximum of the Ca^{2+} release rate is 19.2 nmol Ca^{2+}/mg SR/s in the case of depolarization-induced Ca^{2+} release. The V'_{max} values for caffeine and quercetin (the experimentally achievable maximum of the Ca^{2+} release rate), calculated from 1/V versus 1/[drug], are similar to each other, and are about the same as the V'_{max} value of the depolarization-induced Ca^{2+} release. Since it is assumed that the rate of Ca^{2+} release in vivo is in the range of 5--10 nmol Ca^{2+}/mg/s (e.g. refs. 14, 21), the rate constants of Ca^{2+} release described above are comparable with or even higher than the physiological release rate.

Another important feature described here is the re-accumulation of the released Ca^{2+}, a phenomenon which is equivalent to the rapid decay of the transient tension developed in skinned fibers. However, the rate of re-uptake demonstrated in this study ($t_{1/2}$ = 100s) is considerably slower than that of tension decay ($t_{1/2}$ = 40 ms, ref. 22). It may well be that the conditions used for the in vitro experiments, e.g. arsenazo III, inhibit the Ca^{2+} pump. It is also possible that Ca^{2+} pumping is more labile than Ca^{2+} release during the isolation of SR membranes.

The kinetic characteristics of the two types of Ca^{2+} release; viz. the (calcium + drug)-incuced and the depolarization-induced Ca^{2+} releases, are distinguishable by several criteria: a) the optimum $[Ca_o^{2+}]$ range for the activation of the former is several μM, whereas it is less than 0.1 μM for the latter, b) dantrolene-sodium inhibits the former, but rather potentiates the latter. Several examples of such selective inhibitors, e.g. procaine, are reported in the literature (23). On the other hand, kinetic properties of the putative channels are identical regardless of the methods of triggering as evidenced by the facts that a) the V'_{max} values are roughly identical, and b) some agents such as ruthenium red, high $[Ca_o^{2+}]$ and the Ca^{2+} antagonist nitrendipine block both types of Ca^{2+} release. These results suggest that the Ca^{2+} release machinery consists of at least two classes of molecular components: one class consists of different receptors for different triggering signals which are distinguishable by different $[Ca_o^{2+}]$-dependences of activation and selective inhibitors; the other class consists of kinetically indistinguishable channels. According to this hypothesis, the molecular units that are responsible for the (calcium + drug)-induced Ca^{2+} release and the depolarization-induced Ca^{2+} release can be described as R_1-C and R_2-C, respectively: R_1 represents the receptor to which binding of Ca^{2+}, and caffeine or quercetin, takes place; R_2, another receptor which is sensitive to the changes in membrane potential; and C represents the channel moiety. In view of the rapidity and reversibility of Ca^{2+} release described here, it is very likely that the putative Ca^{2+} channels described here are identical with those that operate in situ. However, different type of receptors other than R_1 and R_2 might operate for triggering of Ca^{2+} release in situ.

Of various types of SR preparations we have investigated, the preparation described in the Experimental Procedures, which is essentially identical to the ones described previously (24), consistently showed the highest Ca^{2+} release function. The preparation is characterized by the presence of several additional protein components in the M_r range of 30,000 -- 70,000, some of which appear to be identical with the transverse (T) tubule membrane components. Furthermore, this preparation has a higher reactivity with several monoclonal antibodies produced against the purified T-tubule membranes (M. Rosemblatt et al, the manuscript in preparation). Thus, it appears that the T-tubule moiety is involved, at least partly, in the Ca^{2+} release functions described here. Studies of the effects of various reagents and antibodies on the receptor and channel functions are currently in progress in order to gain a clearer picture of the molecular components involved in the Ca^{2+} release mechanism.

ACKNOWLEDGMENTS

We wish to thank Dr. T. S. Ohnishi for his collaboration in the initial phase of this work. The effects of quercetin on Ca^{2+} release were found during the course of collaborative work with Dr. D. H. MacLennan on the interaction of quercetin and the Ca^{2+} pump. Thanks are also due to Drs. M. Endo and A. Fabiato for their valuable criticisms.

REFERENCES

1. Ford, L. E., and Podolsky, R. J., Fed. Proc. 27: 375 (1968).
2. Endo, M., Tanaka, M., and Ebashi, S., Proc. Intern. Congr. Physiol. Sci. 7: 126 (1968).
3. Fabiato, A., and Fabiato, R., J. Physiol., London 249: 469 (1975).
4. Ohnishi, S. T., J. Biochem. 86:1147 (1979).
5. Weber, A., and Herz, R., J. Gen. Physiol. 52: 750 (1968).
6. Ogawa, Y., J. Biochem. 67: 667 (1970).
7. Endo, M., Proc. Japan Acad. 51: 479 (1975).
8. Ohnishi, S. T., in "The Mechanism of Gated Calcium Transport Across Biological Membranes" (Ohnishi, S. T., and Endo, M. eds.) pp. 275 (1981).
9. Kasai, M., and Miyamoto, H., J. Biochem. 79: 1053 (1976).
10. Kasai, M., and Miyamoto, H., J. Biochem. 79: 1067 (1976).
11. Endo, M., and Nakajima, Y., Nature New Biol. 246: 216 (1973).
12. Ohnishi, S. T., Biochim. Biophys. Acta. 587: 217 (1979).
13. Shoshan, V., MacLennan, D. H., and Wood, D. S., Proc. Natl. Acad. Sci. 78: 4828 (1981).
14. Caswell, A. H., and Brandt, N. R., J. Mebr. Biol. 58: 21 (1981).
15. Ogawa, Y., and Ebashi, S., J. Biochem. 80: 1149 (1976).
16. Shoshan, V., Campbell, K. P., MacLennan, D. H., Frodis, W., and Britt, B. A., Proc. Natl. Acad. Sci. 77: 4435 (1980).
17. Shoshan, V., and MacLennan, D. H., J. Biol. Chem. 256: 887 (1981).
18. Ellis, K. O., and Carpenter, J. F., Arch. Exptl. Pathol. Parmakol. 275: 83 (1972).
19. Van Winkle, W. B., Science 193: 1130 (1976).
20. Nagasaki, K., and Kasai, M., J. Biochem. 90: 749 (1981).
21. Stephenson, E. W., J. Gen. Physiol. 71: 411 (1978).
22. Fitts, R. H., Courtright, J. B., Kim, D. H., and Witzmann, F. A., Am. J. Physiol. 242: c65 (1982).
23. Thorens, S., and Endo, M., Proc. Japan Acad. 51: 473 (1975).
24. Ogawa, Y., Harigaya, S., Ebashi, S., and Lee, K., in "Methods in Pharmacology" (Schwartz, A., ed.) 1: 327 Appleton-Century-Crofts, New York (1971).

MECHANISM OF CALCIUM RELEASE FROM SARCOPLASMIC RETICULUM: EFFECT OF CAFFEINE ON MAGNESIUM BINDING TO FRAGMENTED SARCOPLASMIC RETICULUM FROM BULLFROG SKELETAL MUSCLE[1), 2)]

Yasuo Ogawa
Hikaru Harafuji
Nagomi Kurebayashi

Department of Pharmacology
Juntendo University School of Medicine
Hongo, Bunkyo-ku, Tokyo

I. INTRODUCTION

While the idea that sarcoplasmic reticulum (SR) plays the central role in excitation-contraction coupling has been established on the basis of findings that the microsomal fraction of muscle homogenate, fragmented sarcoplasmic reticulum (FSR), accumulates Ca in the presence of ATP (1), it is assumed that SR could release Ca under appropriate conditions. There was no demonstration of Ca release of physiological or pharmacological significance until A. Weber and her colleagues (2-4) showed caffeine-induced release from frog and rabbit skeletal muscle FSR (although with much less probability of success in the latter species). They indicated that Ca ions

[1)] This study was supported in part by Grants-in-Aid for Scientific Research from the Ministry of Education, Science and Culture of Japan.

[2)] In this paper, "Ca^{2+}" is used when free calcium ions are referred to; otherwise "Ca" is used. This is also the case with magnesium.

ISBN 0-12-260380-X

stimulated caffeine-induced Ca release, whereas Mg ions competed with the action of Ca^{2+}. Endo et al. (5) showed that repeated contraction with near maximum tension was developed by skinned fiber on application of caffeine under appropriate conditions. They also showed that a minute amount of Ca^{2+} itself caused Ca release from SR with resulting tension development; this was also demonstrated independently by Ford and Podolsky (6). Thus the idea of so called "Ca^{2+}-induced" Ca release has emerged and it appears to be an intrinsic property of SR. It is considered that caffeine labilizes the "Ca^{2+}-induced" Ca release. Although it is controversial at the present time whether the "Ca^{2+}-induced" Ca release is involved in the physiological contraction of skeletal muscle (7), it is certain that it has pharmacological or pathological significance, e.g., caffeine-induced muscle contraction (2-8), malignant hyperthermia (9-11), porcine stress syndrome (11) and so on. It is probably related to caffeine-induced hyperpolarization of sympathetic ganglion cell (12). It is also likely to play an essential role in the physiological contraction of cardiac muscle (13) or hyperpolarization on the fertilization of eggs (14, 15). In view of these observations it is worthwhile to examine the mechanism of Ca release by caffeine.

Ogawa (8) observed re-uptake of Ca^{2+} released from FSR prepared from bullfrog skeletal muscle with the addition of caffeine, confirming the findings by Weber and her collaborators (2-4) and indicating that Ca release cannot be the effect of a reduced capacity of Ca uptake unlike their suggestion. This is consistent with a repeated tension development by skinned fiber in the presence of caffeine. Neither Ca uptake nor the level of EP of FSR was affected by 5 mM caffeine when the activity was supported by carbamylphosphate, whereas Ca uptake driven by ATP decreased (16). Ca uptake activity itself may thus not be reduced by caffeine but the reduction of the net activity may be apparent because of the accompanying Ca release. Our investigations of drugs thought to stimulate "Ca^{2+}-induced" Ca release on Ca^{2+}-ATPase activity and its partial reactions revealed no correlation between the Ca release and inhibition of the ATP hydrolysis process (ref. 17 and see also the paper by Kurebayashi and Ogawa in this volume). On the other hand, a nonhydrolyzable ATP analog, β,γ-methylene adenosine triphosphate (AMPOPCP) induced Ca release similar, but not identical, to caffeine in the mode of the action (16). This would not exclude the possibility that Ca^{2+}-ATPase activity might be affected in some ways during Ca release. Investigations concerning the effect of caffeine on Ca^{2+}-ATPase activity indicate that it is similar to that of lowering Mg^{2+}

concentration (17). It is possible that caffeine may reduce the affinity for Mg^{2+} of FSR and that the effect might be related to the Ca releasing action (17). To clarify this, we examined the properties of Mg binding to FSR from bullfrog skeletal muscle and its modification by caffeine. Here, it is more likely that the dissociation of Mg from the sites responsible for Mg binding may open the channel for Ca^{2+} on Ca release.

II. EXPERIMENTAL PROCEDURES

Fragmented sarcoplasmic reticulum was prepared from bullfrog skeletal muscle as described previously (18). The heavy fraction of FSR (HFSR) was incubated at the final concentration of 0.4 mg protein/ml in the reaction mixture (3 ml) containing various amounts of $MgCl_2$ (0-4 mM $MgCl_2$ added), 100 mM KCl, 20 mM Tris, 11 mM maleate, and others if required, with or without caffeine. The pH was set at pH 6.50 by KOH except as otherwise indicated. The concentration of Ca^{2+} in the medium was adjusted either at about 40 μM by addition of 0.2 mM $CaCl_2$ and 0.2 mM GEDTA (EGTA, ethylene glycol bis(β-aminoethyl ether)-N,N,N',N'-tetraacetic acid) or at 0.02 μM by addition of 6 mM GEDTA alone. After 10 min incubation on ice following the addition of HFSR, the mixture was centrifuged at 100,000 x _g_ for 20 min. Mg and Ca contents of the supernatants and the pellets were determined by a Perkin-Elmer 603 atomic absorption spectrophotometer. The precise concentrations of Mg^{2+} and Ca^{2+} including contaminating metals were calculated on the basis of determinations by atomic absorption spectrophotometry. Corrections were made for [^{14}C]-sucrose space in the pellet, which was 50% of the wet weight of the pellet assuming the density as unity. HFSR was used exclusively since LFSR, the usual microsomal fraction, was less sensitive to caffeine. With LFSR, it was very difficult to obtain the results described here. All of the experiments were carried out in the absence of ATP, because a change in Mg content accompanying ATP-driven Ca uptake (19, 20) would make the interpretation of the results more difficult. The protein concentration of HFSR was determined according to the biuret method.

III. RESULTS

A. "Ca Release" by Caffeine

In contrast to FSR from rabbit skeletal muscle, that from bullfrog skeletal muscle retains some Ca inside the vesicles without prior incubation in a high concentration of Ca^{2+}; this is released on the addition of A23187 (see Fig. 10A of ref. (21)). As shown in Fig. 1A, Ca content in HFSR vesicles remained unchanged in the presence of 0.3 mM Mg^{2+} or more at pH 6.5. However, it decreased as Mg^{2+} concentrations decreased to less than 0.3 mM. Caffeine at 10 mM caused a marked reduction in the Ca content and the effect was reversed to

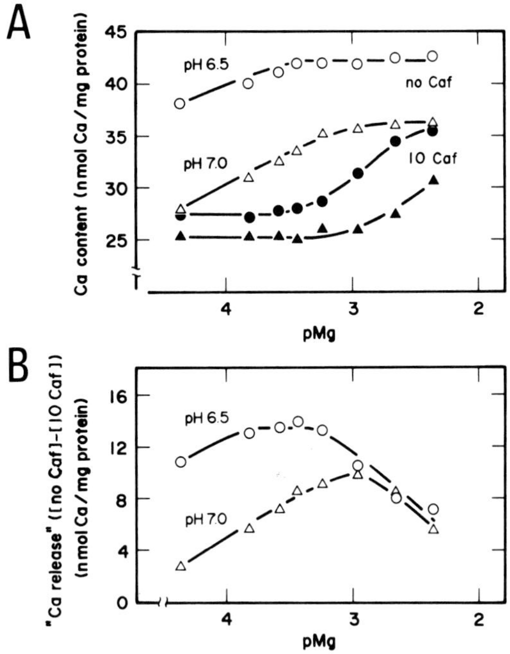

FIGURE 1. Effect of pH on caffeine-induced "Ca release". (A) Mg^{2+}-dependent change in the amounts of Ca in HFSR by caffeine. For details, see EXPERIMENTAL PROCEDURES. ○,●: at pH 6.5; △,▲: at pH 7.0; without (open symbols, ○,△) or with (closed symbols, ●,▲) 10 mM caffeine. (B) Mg^{2+}-dependent "Ca release" by caffeine at different pH. The difference in the amount of Ca in HFSR with and without 10 mM caffeine is referred to as "Ca release". ○, at pH 6.5; △, at pH 7.0.

a greater extent by a higher concentration of Mg^{2+}. At pH 7.0, the amount of Ca in HFSR vesicles was less than that at pH 6.5 even in the presence of 4 mM Mg^{2+}. The decrease in Ca amount with the decrease in Mg^{2+} concentrations was more marked at pH 7.0. Caffeine reduced Ca content in HFSR to lower levels. Furthermore, the reversal by higher concentrations of Mg was less marked. These findings may correspond to the properties of so-called "Ca^{2+}-induced" Ca release: stimulated by caffeine and at an alkaline pH, and inhibited by Mg^{2+}. If the difference in the Ca content of HFSR with and without caffeine is referred to as "Ca release" by caffeine, the Mg^{2+} dependence of the "Ca release" is shown in Fig. 1B. A complicated biphasic dependence on Mg^{2+} concentrations at pH 7.0 is apt to be apparent, because Ca release was already observed in the absence of caffeine. Therefore, most experiments were carried out at pH 6.5.

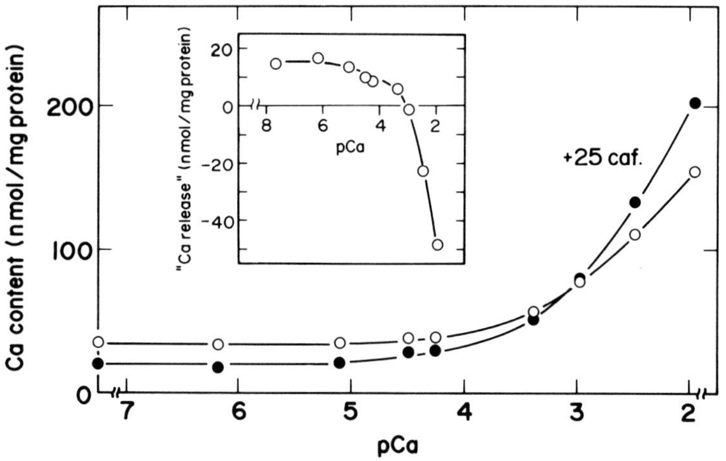

FIGURE 2. Caffeine-induced increase in "permeability to Ca^{2+}". No Mg was added. A concentration of Ca^{2+} between pCa 7 and 4 was adjusted by 0.2 mM $CaCl_2$ and an appropriate amount of GEDTA was calculated using an apparent binding constant for GEDTA according to Harafuji and Ogawa (22). The addition of 6 mM GEDTA without addition of Ca makes pCa 7.7. At Ca^{2+} concentrations higher than 0.1 mM, $CaCl_2$ alone was added as indicated in the abscissa. ○, without caffeine; ●, with 25 mM caffeine. Inset shows Ca^{2+}-dependence of "Ca-release".

The effect of 25 mM caffeine on Ca content in the vesicles was determined in the presence of various concentrations of Ca^{2+} with no added Mg (Fig. 2). In the absence of caffeine, the amount of Ca in HFSR was unchanged up to about 10 μM Ca^{2+} and then it increased as Ca^{2+} concentrations increased. The presence of 25 mM caffeine decreased the amount of Ca in HFSR when Ca^{2+} concentrations were less than 1 mM. At higher than 1 mM Ca^{2+}, in contrast, caffeine increased Ca content. The amount of "Ca release", namely, the difference with and without caffeine, was plotted against Ca^{2+} concentration in the medium in the inset of Fig. 2. This observation indicates that caffeine increases the permeability to Ca^{2+} of the sarcoplasmic reticulum membrane in both directions, and that the final result of Ca loss or gain depends largely on the concentration gradient of Ca^{2+} between outside and inside the vesicles under these experimental conditions. This corresponds to the results by Kitazawa and Endo (23). The effect of Mg^{2+} on "Ca release" by caffeine was examined at fixed concentrations of Ca^{2+} (Fig. 3). With no added Mg, the amount

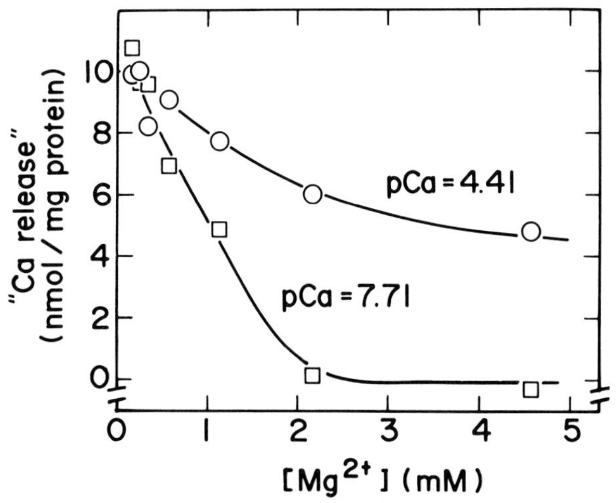

FIGURE 3. Reversal by Mg^{2+} of caffeine-induced "Ca release". The concentration of caffeine was 10 mM. □, in the presence of 6 mM GEDTA which produced pCa 7.71. ○, in the presence of 0.2 mM $CaCl_2$ and GEDTA which produced pCa 4.41 at pH 6.5.

of Ca released by 10 mM caffeine in the virtual absence of Ca^{2+} (pCa 7.71, 0.02 μM Ca^{2+}) was almost the same as that at pCa 4.41 (about 39 μM Ca^{2+}). However, the "Ca release" at pCa 7.71 decreased more sharply as Mg^{2+} concentrations increased. It was abolished at 1-2 mM Mg^{2+}, whereas the "Ca release" at pCa 4.4 was observed even at 4 mM Mg^{2+}, although in lesser amounts (Figs. 3 and 5). This is consistent with Endo's results (24), but at variance with the results by Yamamoto and Kasai (25). In the presence of Ca^{2+} as high as 10 mM where caffeine caused Ca gain instead of Ca release as the results of increased permeability to Ca^{2+}, the caffeine-induced Ca gain was also controlled by Mg^{2+} concentration (Fig. 4). It is likely that the permeability of the HFSR membrane to Ca^{2+} is regulated by Mg^{2+} at the concentration of 10 mM Ca^{2+} even in the absence of caffeine (Fig. 4).

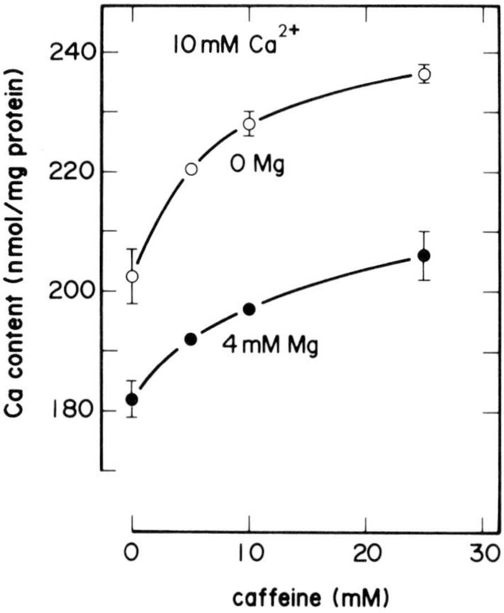

FIGURE 4. Mg^{2+}-dependence of caffeine-induced increase in the "permeability to Ca^{2+}" in the presence of 10 mM Ca^{2+}. ○, with no added Mg. ●, with 4 mM $MgCl_2$ added. Results show the average of 2-4 determinations with standard error or half the range (in the case of duplicate determinations). Results in the presence of 5 mM caffeine with no added Mg ranged within the circles. Other results without bars are of single determination.

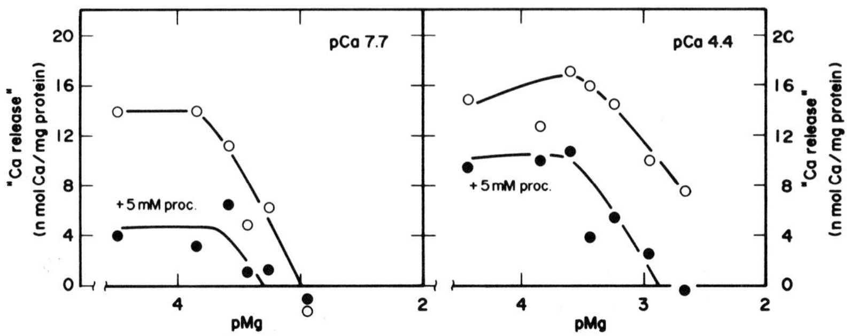

FIGURE 5. The reversal by procaine of caffeine-induced "Ca release". "Ca release" was caused by 5 mM caffeine. ○, without procaine; ●, with 5 mM procaine.

Another important characteristic of the "Ca release" by caffeine is its reversal by procaine (Fig. 5). This was observed not only in the presence of about 40 μM Ca^{2+} but also in the virtual absence of Ca^{2+}; however, it appears that it is more marked in the absence of free Ca^{2+}. It should be noted that a high concentration of Mg^{2+} is required in addition to procaine for complete abolition of "Ca release" by caffeine. Stronger suppression of Ca release by Mg^{2+} or procaine in the virtual absence of Ca^{2+} is at variance with the results reported by Yamamoto and Kasai (25). A possible explanation may be that they examined the effect on choline influx, whereas we watched the change in Ca content.

B. Properties of Mg Binding: Its Modulation by Caffeine

The results presented so far reveal that the characteristics of caffeine-induced change of Ca content in HFSR correspond well to those of Ca release by caffeine reported earlier (7). Mg binding to HFSR and its modification by caffeine were then examined in order to deepen our understanding of the action of caffeine. Experiments were first carried out in the presence of about 40 μM Ca^{2+}, since it was known that the presence of Ca^{2+} favored Ca releasing action of caffeine. As shown in Fig. 6, the Scatchard plot for Mg binding to HFSR in the absence of caffeine was not linear, which was similar to the results shown previously (17). When we analysed it under the assumption that binding sites should consist of two heterogeneous independent classes, the results were best fitted by

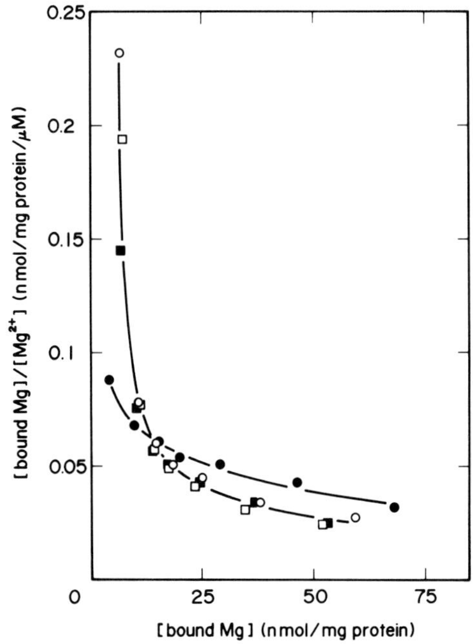

FIGURE 6. Effect of caffeine on Mg binding and the reversal of the effect by procaine. Experiments were carried out as in Fig. 5 in the presence of 0.2 mM $CaCl_2$ and GEDTA. ○, without caffeine or procaine; ●, with 5 mM caffeine, but without procaine; □, with 5 mM procaine, but without caffeine; ■, with 5 mM caffeine and 5 mM procaine. The Scatchard plot for Mg binding data was not linear. For details, see text.

the curve composed of the two classes (Fig. 6), i.e., a high affinity site (K_1 = 11.42 × 10^4 M^{-1}, n_1 = 7.32 nmol/mg protein) and a low affinity site (K_2 = 148 M^{-1}, n_2 = 218 nmol/mg protein). The characteristic parameters determined here differed from the previous results (17) because the earlier ones came from an inappropriate graphical analysis of the Scatchard plot. Here, we determined parameters which should give best fit to the experimental results according to the equation, bound Mg = n_1 K_1 [Mg^{2+}] / (1 + K_1 [Mg^{2+}]) + n_2 K_2 [Mg^{2+}] / (1 + K_2 [Mg^{2+}]). Caffeine at 5 mM caused remarkable change in Mg binding: decreased it for the high affinity site and increased it for the low affinity site. Analysis of the results in Fig. 6 revealed that the parameters were changed to K_1 =

$2.08 \times 10^4\ M^{-1}$, $n_1 = 3.25$ nmol/mg protein, $K_2 = 345\ M^{-1}$, $n_2 = 156$ nmol/mg protein. These effects were reversed by 5 mM procaine as in the case with the Ca release (Figs. 5 and 6). It should be pointed out that procaine alone does not change the

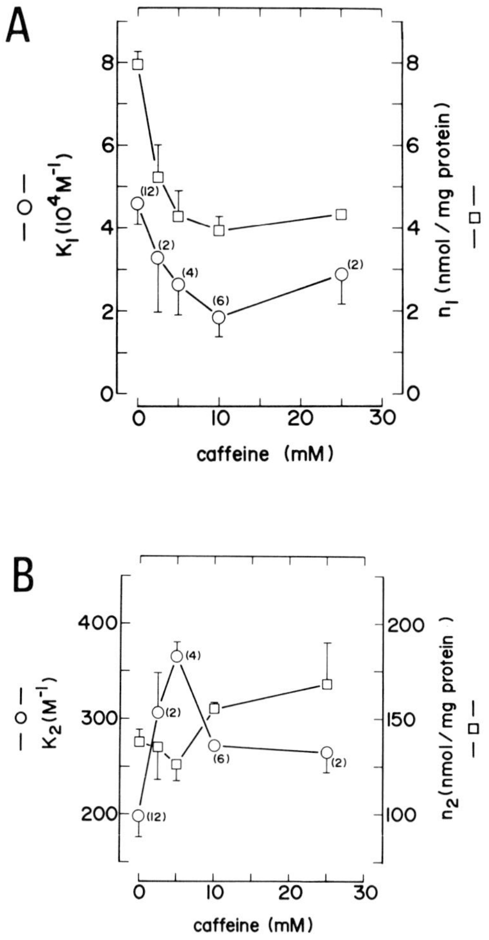

FIGURE 7. Effect of caffeine on Mg binding. (A) Class of high affinity sites. (B) Class of low affinity sites. ○, apparent affinity for Mg^{2+}; □, the number of binding sites. Results are presented as mean values with S.E. or half the range (in the case of duplicate determinations). Figures in parentheses show the number of determinations.

properties of the Mg binding; it prevents only those changes caused by caffeine. The effects of caffeine were determined at its various concentrations in the presence of about 40 μM Ca^{2+} (Fig. 7). Both the apparent affinity (K_1) and the number (n_1) of the high affinity binding sites were decreased dose-dependently by caffeine. The apparent affinity (K_1) was decreased from about $5 \times 10^4\ M^{-1}$ in the absence of caffeine to a minimum value of $2 \times 10^4\ M^{-1}$ in 10 mM of its presence (Fig. 7A). Correspondingly, the number of sites (n_1) also was diminished from 8 to about 4 nmol/mg protein in the presence of 10 mM caffeine. In contrast, the low affinity sites were affected in an erratic dose-effect relationship as shown in Fig. 7B. It may be considered that the changes relevant to Ca release thus occur in the high affinity site and therefore, we refer to only the high affinity site hereafter.

In order to understand how Ca^{2+} is involved in the action of caffeine, we examined Mg binding under a condition free of Ca^{2+}. Its removal increased the apparent affinity for Mg^{2+} without change in the number of binding sites, suggesting that Ca^{2+} compete with Mg^{2+} for these sites (cf. Fig. 8). In the absence of Ca^{2+}, caffeine decreased only the apparent affinity and the extent of decrease in K_1 was more marked (Table I). However, the apparent affinity in the presence of caffeine was similar in value in the two cases. This effect also was reversed by procaine. It did not reduce the

Table I. Ca^{2+} Dependent Effect of Caffeine on Mg Binding

Ca^{2+} (μM)	relative value[a)]	
	K_1	n_1
0.02	0.35 ± 0.07 (2)	0.96 ± 0.07 (2)
~40	0.52 ± 0.10 (8)	0.46 ± 0.04 (8)

a) The relative value represents the ratio of the result with 10 mM caffeine to that without caffeine. Results were presented in mean ± S.E. (number of determinations). For duplicate determinations, half the range was used instead of S.E.

number of binding sites, however, in marked contrast to the case in the presence of Ca^{2+} where both K_1 and n_1 were decreased. These findings are consistent with the results that Ca release by caffeine is more sensitive to Mg^{2+} in the absence of Ca^{2+} than in its presence. The possibility is thus excluded that the decrease in apparent affinity for Mg^{2+} might be ascribed to the increased apparent affinity for Ca^{2+} competing against Mg^{2+}.

As mentioned, it was suggested that Ca^{2+} may compete with Mg^{2+} for the binding sites. In view of the observations that Ca^{2+} and Mg^{2+} compete in giving rise to Ca release, it is important to determine the Mg binding to HFSR in the presence of various concentrations of Ca^{2+}. As shown in Fig. 8, the apparent affinity for Mg^{2+} decreased as Ca^{2+} concentrations increased. The number of Mg binding sites, however, was unchanged, being about 8 nmol/mg protein. Under the assumption that Mg^{2+} and Ca^{2+} bind competitively to the same sites, it could be calculated that the apparent affinity for Mg^{2+} and Ca^{2+} at pH 6.5 is 10^5 M^{-1} and 1-3 × 10^4 M^{-1}, respectively.

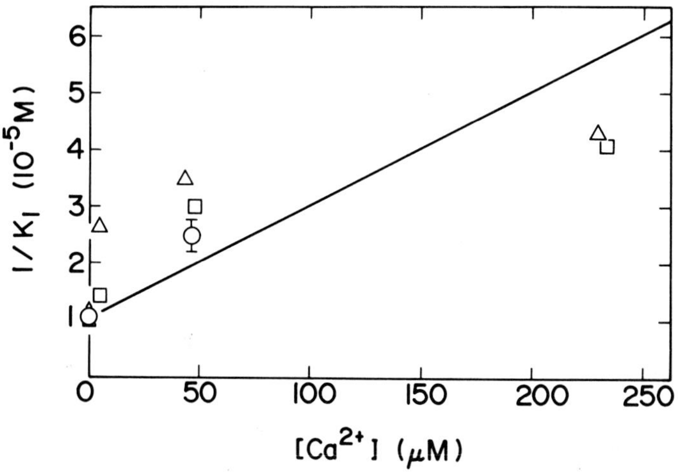

FIGURE 8. Ca^{2+}-dependence of the apparent affinity for Mg^{2+} of the high affinity sites. △, □, a series of determinations with the same batch of HFSR preparations. ○, the average of several determinations. Around pCa 7.7, 4 determinations including △ and □. Around 40 μM Ca^{2+}, 12 determinations including △ and □. n_1 was unchanged, being 8 nmol/mg protein. The straight line represents the case of K_{Mg} = 10 × 10^4 M^{-1}, K_{Ca} = 2 × 10^4 M^{-1}.

IV. DISCUSSION

In brief, our observations mentioned here may be summarized as follows. HFSR from bullfrog skeletal muscle has a class of Mg binding sites ($K_1 = 10^5\ M^{-1}$ at pH 6.5, n_1 = 8 nmol/mg protein). These sites compete with Ca^{2+}, for which they have an apparent affinity of 1-3 × $10^4\ M^{-1}$. In the virtual absence of Ca^{2+}, caffeine reduces K_1 and is accompanied by Ca release, but it causes no change in n_1. These effects are reversed by procaine. The reduction in K_1 can be considered a primary action of caffeine but not a secondary effect of an increased apparent affinity for Ca^{2+}, because of the virtual absence of Ca^{2+}. In contrast, caffeine decreases both K_1 and n_1 in the presence of about 40 μM Ca^{2+}, being accompanied by Ca release. All of the effects are prevented by procaine. Calcium ions not only compete with Mg^{2+} for the binding sites but also decrease the number of Mg binding sites in combination with caffeine. In the presence of caffeine such ions might affect sites other than Mg binding sites thereby decreasing the number of the Mg sites.

The effect intrinsic to caffeine reduces the apparent affinity for Mg^{2+}, which in turn probably causes Ca release. It may be said that the apparent affinity for Mg^{2+} corresponds to the "gate for the Ca channel" and that the number of Mg sites could refer to the size of this "Ca channel". Thus caffeine in the absence of Ca^{2+} might affect only the "gate". In the presence of Ca^{2+}, in contrast, it might make also the "Ca channel" too wide to be covered by the gate molecules. This interpretation is not incompatible with the findings that Ca release by caffeine is still observable at 4 mM Mg^{2+} in the presence of 40 μM Ca^{2+}, whereas it is completely abolished by 1-2 mM Mg^{2+} in the absence of Ca^{2+} (cf. Figs. 3 and 5). Thus it may be concluded that the decrease of Mg bound to the high affinity sites is the first step not only for caffeine-induced Ca release but also for so-called "Ca^{2+}-induced" Ca release.

Endo (40) estimated on the basis of the reversal by Mg^{2+} of "Ca^{2+}-induced" Ca release that the apparent affinity for Mg^{2+} would be approximately 3 × $10^3\ M^{-1}$ in the presence of about 20 μM Ca^{2+}. Although he did not describe the pH of the solution, it may be reasonably assumed that it would be 7.1 or so, where most of his experiments were carried out. Surprisingly, our preliminary results showed that the apparent affinity for Mg^{2+} at pH 7.0 in the presence of Ca^{2+} decreased to about one-tenth that at pH 6.5; thus our results correspond well to Endo's.

To get some ideas about the molecules responsible for Mg binding which may be involved in the Ca release, the possible

Table II. Summary of Mg Binding and Possible Roles of Mg^{2+} in Sarcoplasmic Reticulum

1. formation of an appropriate substrate (ref. 26)

2. *E + Pi ⟶ *E-P (refs. 27-32)
   $K_{Mg}^{app} \simeq 200\ M^{-1}$

3. stimulation of ATP hydrolysis
   i) direct activation of the enzyme
      $K_{Mg}^{app} \simeq 7\text{-}9 \times 10^{3}\ M^{-1}$ (refs. 33, 34)
   ii) $Ca_{out}\cdot E\sim P \longrightarrow Ca_{in}\cdot E\sim P$ (ref. 35)
   iii) E~P ⟶ *E-P (ref. 36)
   iv) *E-P ⟶ E + Pi
      $K_{Mg}^{app} = 3.3 \times 10^{3}\ M^{-1}$ (ref. 37)

4. competition with Ca^{2+} for "Ca^{2+}-activation" of ATP hydrolysis (refs. 38, 39)
   $K_{Mg}^{app} = 200\ M^{-1}$ (ref. 39)

5. modulation of "Ca^{2+}-induced" Ca release (refs. 4-7)
   $K_{Mg}^{app} = 3 \times 10^{3}\ M^{-1}$ (in the presence of about 20 μM Ca^{2+}) (ref. 40)

This work : Mg binding relevant to the action of caffeine (at pH 6.5)
   $K_{Mg}^{app} = 10^{5}\ M^{-1}$, $K_{Ca}^{app} = 1\text{-}3 \times 10^{4}\ M^{-1}$,
   n = 8 nmol/mg protein

roles of Mg^{2+} in the sarcoplasmic reticulum are summarized (Table II). On the basis of the number of binding sites, 8 nmol/mg protein, ATPase protein could be a candidate (41-44). However, the apparent affinity for Ca^{2+} would be too low in view of reported values (42-44) for ATPase protein from rabbit FSR, except for Ikemoto's results (41). It should be mentioned that we have not succeeded in demonstrating as high an affinity for a Ca binding site in bullfrog FSR as results reported for rabbit FSR. Furthermore, the reported apparent affinities for Mg^{2+} of Ca^{2+}-ATPase protein from rabbit FSR appear to be grouped into two categories although determined under various conditions: 10^3-10^4 M^{-1} and about 200 M^{-1} (see Table II). More experiments, however, are required to conclude whether these sites are related to the modulation of Ca^{2+}-ATPase activity by caffeine (17). What molecules may be concerned in the Mg binding responsible for Ca release also remains to be determined.

REFERENCES

1. Ebashi, S. & Endo, M., Prog. Biophys. Mol. Biol. 18: 123 (1968)
2. Herz, R. & Weber, A., Fed. Proc. 24: 208 (1965)
3. Weber, A. & Herz, R., J. Gen. Physiol. 52: 750 (1968)
4. Weber, A., J. Gen. Physiol. 52: 760 (1968)
5. Endo, M., Tanaka, M. & Ogawa, Y., Nature 228: 34 (1970)
6. Ford, L. E. & Podolsky, R. J., Science 167: 58 (1970)
7. Endo, M., Physiol. Rev. 57: 71 (1977)
8. Ogawa, Y., J. Biochem. 67: 667 (1970)
9. Takagi, A., Sugita, H., Toyokura, Y. & Endo, M., Proc. Japan Acad. 52: 603 (1976)
10. Ogawa, Y. & Kurebayashi, N., J. Biochem. 92: 899 (1982)
11. Aldrete, J. A. & Britt, B. A., ed., "Second International Symposium on Malignant Hyperthermia", Grune & Stratton, New York, 1978.
12. Kuba, K., J. Physiol. 298: 251 (1980)
13. Fabiato, A., Fed. Proc. 41: 2238 (1982)
14. Ridgway, E. B., Gilkey, J. C. & Jaffe, L. F., Proc. Natl. Acad. Sci. USA 74: 623 (1977)
15. Miyazaki, S. & Igusa, Y., in "The Mechanism of Gated Calcium Transport across Biological Membranes" (S. T. Ohnishi & M. Endo. ed.). p. 305., Academic Press, New York, 1981.
16. Ogawa, Y. & Ebashi, S., J. Biochem. 80: 1149 (1976)

17. Ogawa, Y., Kurebayashi, N. & Harafuji, H., in "The Mechanism of Gated Calcium Transport across Biological Membranes" (S. T. Ohnishi & M. Endo. ed.), p.265. Academic Press, New York, 1981.
18. Kurebayashi, N., Kodama, T. & Ogawa, Y., J. Biochem. 88: 871 (1980)
19. Carvalho, A. P. & Leo, B., J. Gen. Physiol. 50: 1327 (1967)
20. Ogawa, Y. & Ebashi, S., in "Organization of Energy-Transducing Membranes" (Nakao, M. & Packer, L., ed.) p.127. University of Tokyo Press, Tokyo, 1973.
21. Ogawa, Y. Harafuji, H. & Kurebayashi, N., J. Biochem. 87: 1293 (1980)
22. Harafuji, H. & Ogawa, Y., J. Biochem. 87: 1305 (1980)
23. Kitazawa, T. & Endo, M., Proc. Japan Acad. 52: 599 (1976)
24. Endo, M., in "The Mechanism of Gated Calcium Transport across Biological Membranes" (S. T. Ohnishi & M. Endo, ed.). p.257. Academic Press, New York, 1981
25. Yamamoto, N. & Kasai, M., J. Biochem. 92: 465, 477 and 485 (1982)
26. Weber, A., Herz, R. & Reiss, I., Biochem. Z. 345: 329 (1966)
27. Kanazawa, T., J. Biol. Chem. 250: 113 (1975)
28. Punzengruber, C., Prager, R., Kolassa, N., Winkler, F. & Suko, J., Eur. J. Biochem. 92: 349 (1978)
29. Epstein, M., Kuriki, Y., Biltonen, R. & Racker, E., Biochemistry 19: 5564 (1980)
30. Martin, D. W. & Tanford, C., Biochemistry 20: 4597 (1981)
31. Lacapère, J. J., Gingold, M. P., Champeil, P. & Guillain, F., J. Biol. Chem. 256: 2302 (1981)
32. de Meis, L., de Souza Otero, A., Martins, O. B., Alves, E. W., Inesi, G. & Nakamoto, R., J. Biol. Chem. 257: 4993 (1982)
33. Garrahan, P. J., Rega, A. F. & Alonso, G. L., Biochim. Biophys. Acta 448: 121 (1976)
34. Makinose, M. & Boll, W., in "Cation Flux across Biomembranes" (Y. Mukohata & L. Packer ed.), p.89. Academic Press, New York, 1979.
35. Takakuwa, Y. & Kanazawa, T., J. Biol. Chem. 257: 10770 (1982)
36. Nakamura, Y. & Tonomura, Y., J. Biochem. 91: 449 (1982)
37. Yamada, S. & Ikemoto, N., J. Biol. Chem. 255: 3108 (1980)
38. Ogawa, Y., Kurebayashi, N., Irimajiri, A. & Hanai, T., in "Molecular and Cellular Aspects of Muscle Function" (E. Varga, A. Köver, T. Kovács & L. Kovács ed.), p.417. Pergamon Press, London and Akadémiai Kiadó, Budapest, 1981.

39. Guillain, F., Gingold, M. P. & Champeil, P., J. Biol. Chem. 257: 7366 (1982)
40. Endo, M., Abstr. of 8th Intern. Congr. Pharmacol. #1478 (1981)
41. Ikemoto, N., J. Biol. Chem. 250: 7219 (1975)
42. Tada, M., Yamamoto, T. & Tonomura, Y., Physiol. Rev. 58: 1 (1978)
43. Inesi, G., in "Membrane Transport in Biology" Vol. II. (G. Giebisch, D. C. Tosteson & H. H. Ussing ed.), p.357. Springer-Verlag, Berlin, Heidelberg and New York, 1979.
44. Hasselbach, W., in "Membrane Transport" (S. L. Bonting & J. J. H. H. M. de Pont, ed.), p.183. Elsevier/North-Holland Biomedical Press, Amsterdam, 1981.

Ca RELEASING ACTION OF QUERCETIN ON THE SARCOPLASMIC RETICULUM OF FROG SKELETAL MUSCLE[1),2)]

Nagomi Kurebayashi
Yasuo Ogawa

Department of Pharmacology
Juntendo University School of Medicine
Hongo, Bunkyo-ku, Tokyo

I. INTRODUCTION

Quercetin, a flavonoid, has been reported to inhibit the activity of several enzymes involved in energy conversion reactions: Na^+, K^+-ATPase (1, 2); F_1, the ATPase of bovine heart mitochondria (1, 3); CF_1, the chloroplast ATPase (4) and the ATPase of *Escherichia coli* (5). The Ca^{2+}-ATPase of sarcoplasmic reticulum (SR) (6-9) is no exception. Shoshan *et al*. (8) claimed that quercetin inhibited Ca^{2+}-ATPase activity and Ca uptake activity of SR, but that it had no effect on Ca release from SR. If so, it should be a good tool to use in investigating the relation between Ca uptake and Ca release as they claimed (8). Recently Kirino and Shimizu (10) reported that quercetin stimulated Ca release from fragmented sarcoplasmic reticulum (FSR) in the virtual absence of Ca^{2+}. However, they did not comment on whether this was a caffeine-like Ca release or Ca release due to the non-specific increase

1)This study was supported in part by Grants-in-Aid for Scientific Research from the Ministry of Education, Science and Culture of Japan.

2)In this paper, "Ca^{2+}" is used when free calcium ions are referred to; otherwise "Ca" is used. This is also the case with magnesium.

ISBN 0-12-260380-X

in permeability of FSR membrane (referred to hereafter as Ca leakage). Shoshan *et al.* (8) observed a slow increase in tension of a skinned fiber after an initial lag with the addition of 100-200 μM quercetin alone, and this was ascribed to the passive leakage of Ca from SR when the Ca uptake system was inhibited by quercetin. Watras *et al.* (11) suggested the possibility of Ca^{2+}-induced Ca release under conditions other than physiological. While we carried out experiments to further our understanding of its action on SR, we have found that quercetin causes caffeine-like Ca release not only from FSR from bullfrog skeletal muscle but also SR in frog skeletal skinned fiber under physiological conditions, and that this release cannot be ascribed to the inhibition of Ca uptake or Ca^{2+}-ATPase activity.

II. CHEMICALS

Quercetin, obtained from Wako Pure Chemicals, was dissolved in dimethyl sulfoxide (DMSO). The final concentration of DMSO was at most 1%, which has no effect on FSR activity. However, the reversible inhibition of Ca^{2+}-induced tension development by skinned fiber was observed in the presence of 1% DMSO at low concentrations of Ca^{2+}. In some cases, the final DMSO concentration was reduced to 0.5% or less to avoid the inhibitory effect on the contractile system when precipitation of quercetin was observed. Alcohol was not used for the solvent because it exerted an effect not only on FSR but also on skinned fiber (both SR and contractile system) at a concentration as low as 1%, contrary to the claim by Shoshan *et al.* (8).

III. EXPERIMENTS WITH FRAGMENTED SARCOPLASMIC RETICULUM

Fragmented sarcoplasmic reticulum was obtained from bullfrog and rabbit skeletal muscle according to the method already reported (12, 13). The heavy fraction of FSR from bullfrog skeletal muscle (HFSR) was used in most experiments since this was more sensitive to the Ca releasing action of quercetin than the light fraction (LFSR), as is true of other caffeine-like Ca releasing drugs (12, 14). Ca^{2+}-ATPase activity, the level of EP and the ATP-ADP exchange rate were determined as reported previously (15, 16). Ca release and Ca uptake were examined in the presence of murexide with an Aminco dual-wavelength spectrophotometer, also as described previously (12, 14). Correction was made for the change of the absorbance difference due to addition of quercetin in the presence of 0.33 mM ethylene glycol bis(β-aminoethylether)-N,N',N'-tetraacetic acid (GEDTA, EGTA).

1. Inhibition of Ca^{2+}-ATPase Activity and Ca Uptake. The extent of inhibition of Ca^{2+}-ATPase activity corresponded well to the rate of Ca uptake of bullfrog FSR in various concentrations of quercetin, both of which were determined on the basis of the values at 30 s and 20 s, respectively, after the addition of ATP at 15°C. In the presence of 30 μM quercetin, the activities were reduced to about half of the control, which was similar to the results with rabbit FSR reported by Shoshan and MacLennan (9). The decrease in Ca^{2+}-ATPase activity by 30 μM quercetin was independent of free Ca^{2+} concentrations between 5×10^{-8} and 10^{-4} M. The simultaneous determination using [γ-^{32}P]ATP revealed unchanged levels of acid stable EP as well as reduced rates of ATP hydrolysis, thus indicating that the steps for EP hydrolysis were affected. ATP-ADP exchange rates were reduced in the presence of various concentrations of Ca^{2+}. This suggests two possibilities: an increased fraction of ADP-insensitive EP or a decreased transphosphorylation rate to produce ATP from ADP-sensitive EP and ADP. The latter may be more likely because the ATP-ADP exchange rate in the presence of 0.1 mM Ca^{2+} or more was somewhat increased by quercetin (cf. references 15, 16). The unchanged level of EP and reduced rates of EP hydrolysis as well as the transphosphorylation of EP suggest that the rate of EP formation by ATP must also decrease. These are consistent with the conclusions of Shoshan and MacLennan (9).

2. Ca Releasing Action. Quercetin caused Ca release from loaded bullfrog FSR when it was added immediately after ATP-driven Ca uptake reached a steady state (Fig. 1A). The amount of Ca release at 5 s after addition of quercetin in the case in Fig. 1A was 30 nmol Ca/mg protein, which was followed by an additional Ca leakage of about 8 nmol Ca/mg protein, *i.e.*, further increase in Ca^{2+} concentration in the medium at a much slower time course. When Ca uptake was supported by carbamylphosphate, however, the amount of Ca release was as little as 3 nmol Ca/mg protein at 5 s after the addition of quercetin. The following Ca leakage, in contrast, reached about 30 nmol Ca/mg protein. Quercetin decreased the rate of Ca uptake driven by ATP to 67% of the control (Fig. 1A), whereas it went to 30% when the uptake was supported by carbamylphosphate (Fig. 1B). Ca release of a rapid time course therefore cannot be due to a non-specific increase in Ca permeability of FSR membrane, nor to the decrease of the rate of Ca uptake by

quercetin. Ca leakage of a much slower time course may be related to the inhibition of Ca uptake by quercetin. It might also be due to a non-specific increase in Ca permeability of FSR. Referring exclusively to the amount of Ca release at 5 s after addition of quercetin, this may be ascribed to a specific increase in Ca permeability of FSR membrane. The amount of Ca released by 30 μM quercetin from HFSR was an average of 32.2 ± 1.5 (S.E.) nmol Ca/mg protein (25 determinations), ranging from 19 to 46 nmol Ca/mg protein (cf. Fig. 2). The release was reversed by 5-10 mM procaine. In contrast, the effect of procaine on Ca^{2+}-ATPase activity was synergistic with quercetin, providing further evidence that the Ca releasing action cannot be the result of inhibition of Ca^{2+}-ATPase activity or of Ca uptake. Higher concentrations of Mg^{2+} reduced the amount of Ca released by quercetin: 40.9 ± 0.8 nmol Ca/mg protein (7 determinations) in the presence of 1 mM Mg^{2+}, but 27.2 ± 0.6 nmol Ca/mg protein (4 determinations) at 10 mM Mg^{2+}. Higher concentrations of ATP stimulated Ca release: 13.7 ± 0.6 nmol Ca/mg protein in the presence of 3 μM ATP with the ATP regenerating system of phosphoenolpyruvate and pyruvate kinase, and 35.9 ± 1.8 nmol Ca/mg protein in the

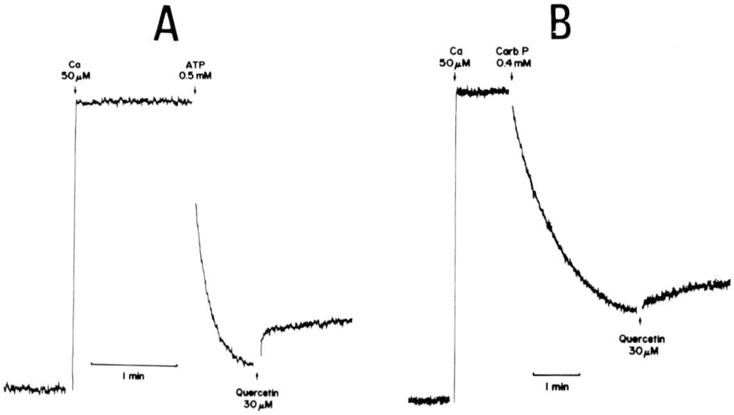

FIGURE 1. Quercetin-induced Ca release in the presence of ATP (A) or carbamylphosphate (B). The reaction mixture (3 ml) contained 100 mM KCl, 20 mM Tris-maleate pH 6.80, 1 mM $MgCl_2$, 0.15 mM murexide and 0.2 mg protein/ml of HFSR. After addition of 50 μM $CaCl_2$ (as a final concentration), Ca uptake was driven by ATP or carbamylphosphate (CarbP). Immediately after Ca uptake reached the steady state, quercetin was added. Temperature, 15°C.

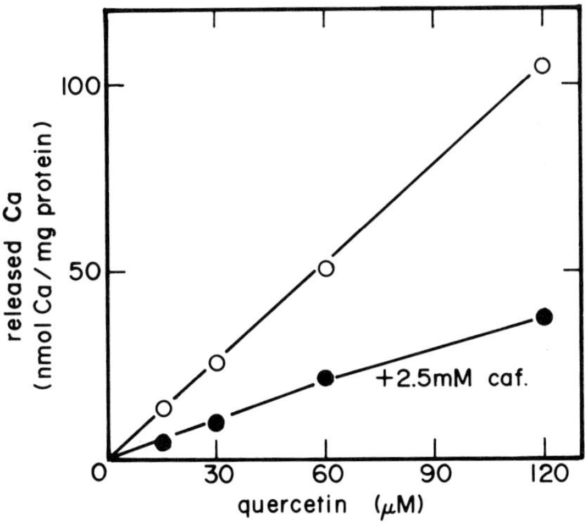

FIGURE 2. Effect of pretreatment of caffeine on the Ca releasing action of quercetin. HFSR was first loaded with Ca in the presence (•) or absence (o) of 2.5 mM caffeine, and then quercetin was added as in Fig. 1A. Each value is an average of 5 or 6 determinations.

presence of 0.5 mM ATP. When Ca uptake was driven by carbamylphosphate, Ca release was minimal as mentioned earlier (Fig. 1B). However, the presence of β,γ-methylene adenosine triphosphate stimulated Ca release as in the case with caffeine (19). These observations are very similar to those with caffeine.

HFSR from bullfrog skeletal muscle released more Ca at a lower temperature, whereas a higher temperature resulted in greater release by FSR from rabbit skeletal muscle; this was also observed with halothane (14).

To clarify whether the mode of action is similar to that of caffeine, we examined the amounts of Ca released by quercetin from FSR loaded in the presence of caffeine and compared them with those from FSR loaded in its absence (Fig. 2). As shown in Fig. 2, the amount of Ca released from the former was less than that from the latter. Similar results were obtained when the quercetin concentration was fixed and when the concentration of caffeine was varied, suggesting that the mode of action of quercetin may be the same as or quite similar to that of caffeine. Quercetin may, however, be different in its action from halothane.

IV. EXPERIMENTS WITH FROG SKINNED FIBER PREPARATION

Since quercetin solution is yellow, some correction is required in experiments using murexide. To confirm the findings with FSR and to deepen our understanding of the action of quercetin, we carried out experiments with frog single skinned fiber preparation.

A skinned fiber was prepared mainly from m. semitendinosus or m. iliofibularis of R. japonica, nigromaculata or catesbiana according to the methods of Endo et al. (17). The tension was determined isometrically with a tension transducer in an experimental set-up similar to that of Endo et al. (17). Quercetin did not induce contracture of frog skeletal

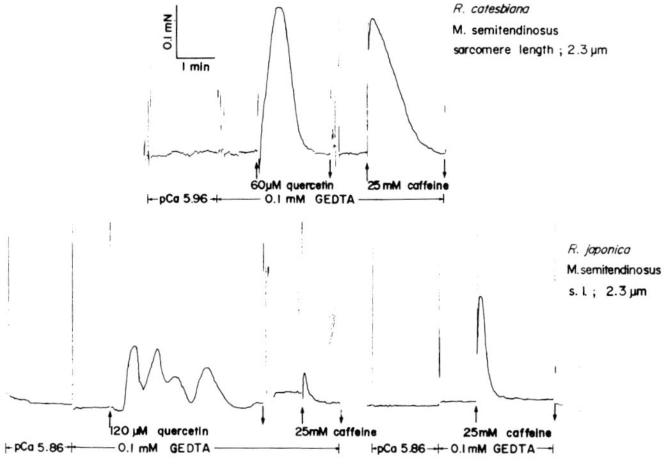

FIGURE 3. Quercetin-induced contraction of skinned skeletal muscle fibers. A single fiber from frog skeletal muscle was skinned in relaxing solution containing 115 mM KCl, 20 mM Tris-maleate pH 6.8, 4.5 mM $MgCl_2$ total, 4.0 mM ATP total, 2 mM GEDTA (EGTA) at 4°C as in Endo et al. (17). After discharge of Ca from SR by 25 mM caffeine, the skinned fiber was immersed for 2 min at 4°C in a Ca-loading solution containing 1 mM GEDTA instead of 2 mM, adjusted at an indicated pCa which was calculated according to the results of Harafuji and Ogawa (20). After rinsing with a washing solution containing 0.1 mM GEDTA without addition of Ca, the skinned fiber was challenged with quercetin. Note the repeated contraction during its application. The remaining Ca in SR was estimated by exposure to 25 mM caffeine.

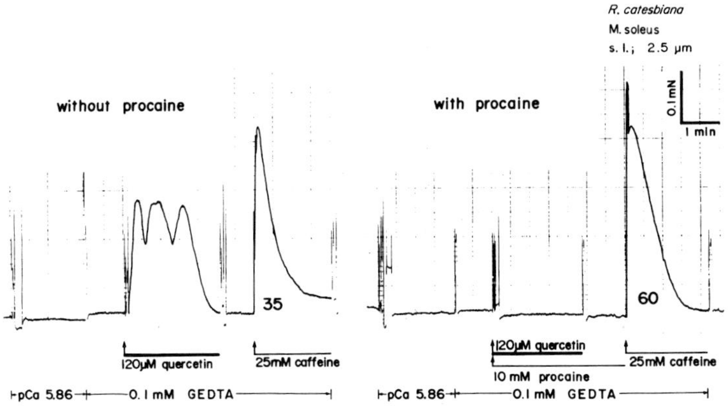

FIGURE 4. The reversal by procaine of quercetin-induced contraction of skinned fiber. Experiments were carried out as in Fig. 3. The amount of Ca remaining in SR of the skinned fiber after treatment was estimated by the area covered by the contracture tension curve induced by 25 mM caffeine (shown in figures), following Endo (18).

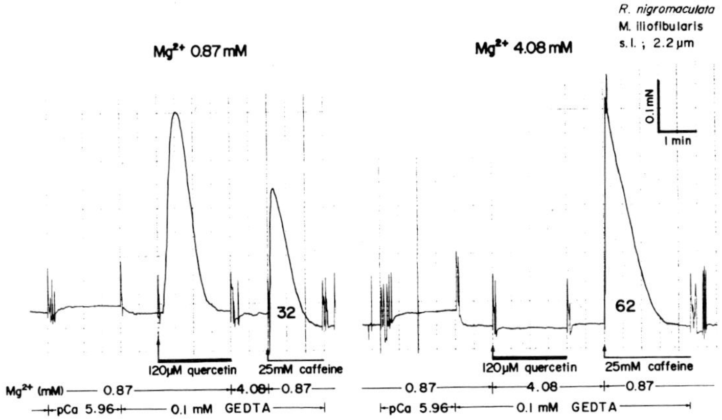

FIGURE 5. Prevention by a higher concentration of Mg^{2+} of quercetin-induced contraction. Free Mg^{2+} concentration was calculated from total concentrations of Mg and ATP using the results of Khan and Martell (21). See also legend to Fig. 4.

muscles, fiber bundles or single fibers. With a mechanically skinned fiber, however, it caused contracture on its addition with a delay in some cases and without delay in others (Fig. 3). When skinned fibers were treated with detergents such as Triton X-100, no contracture was observed on the addition of quercetin. These observations indicate that the sarcolemma is hardly permeable to quercetin and that the drug causes contracture by releasing Ca from SR in skinned fibers. Sometimes repeated contraction was observed during quercetin application as shown in Figs. 3 and 4. The Ca release thus cannot be the effect of Ca uptake inhibition and is very similar to caffeine (17). This is also supported by the following observations. Quercetin-induced contracture was completely prevented by procaine (Fig. 4) or a higher concentration of Mg^{2+} (Fig. 5). The amount of Ca remaining in SR in either case was estimated by the area of caffeine-induced contracture according to Endo's procedure (18). It can be concluded that more Ca remained in SR when contracture was prevented on the application of quercetin, either by procaine or higher concentrations of Mg^{2+}. Then we can conclude that quercetin causes caffeine-like Ca release from SR.

When the results with skinned fiber were compared with those with FSR, higher concentrations of quercetin appeared required to induce Ca release from SR in a skinned fiber. One explanation may be that there is a difference in the extent of loading with Ca: it is a full extent with FSR but a limited saturation in SR of skinned fibers. In support of this, quercetin contracture was observed at 30 μM when incubation in the Ca loading solution was prolonged to 5 min.

ACKNOWLEDGMENTS

We thank Dr. A. Takagi for his advice in setting up the apparatus for tension measurement in the skinned fiber experiments. We are indebted to Dr. N. Ikemoto for advising us of the paper by Watras *et al*. (11).

REFERENCES

1. Carpenedo, F., Bortignon, C., Bruni, A. & Santi, R., Biochem. Pharmacol. 18:1495 (1969).
2. Kuriki, Y. & Racker, E., Biochemistry 15:4951 (1976).
3. Lang, D. & Racker, E., Biochim. Biophys. Acta 333:180 (1974).

4. Deters, D. W., Racker, E., Nelson, N. & Nelson, H., J. Biol. Chem. 250:1041 (1975).
5. Futai, M., Sternweis, P. C. & Heppel, L. A., Proc. Natl. Acad. Sci. USA 71:2725 (1974).
6. Fewtrell, C. M. S. & Gomperts, B. D., Nature 265:635 (1977).
7. Suolinna, E-M., Buchsbaum, R. N. & Racker, E., Cancer Res. 35:1865 (1975).
8. Shoshan, V., Campbell, K. P., MacLennan, D. H., Frodis, W. & Britt, B. A., Proc. Natl. Acad. Sci. USA 77:4435 (1980).
9. Shoshan, V. and MacLennan, D., J. Biol. Chem. 256:887 (1981).
10. Kirino, Y. & Shimizu, H., J. Biochem. 92:1287 (1982).
11. Watras, J., Glezen, S., Seifert, S. & Katz, A. M., Biophys. J. 37:113a (1982).
12. Ogawa, Y., J. Biochem. 67:667 (1970).
13. Kurebayashi, N., Kodama, T. & Ogawa, Y., J. Biochem. 88:871 (1980).
14. Ogawa, Y. & Kurebayashi, N., J. Biochem. 92:899 (1982).
15. Ogawa, Y. & Kurebayashi, N., J. Muscle Res. Cell Motil. 3:39 (1982).
16. Kurebayashi, N. & Ogawa, Y., J. Biochem. 92:907 (1982).
17. Endo, M., Tanaka, M. & Ogawa, Y., Nature 228:34 (1970).
18. Endo, M., Physiol. Rev. 57:71 (1977).
19. Ogawa, Y. & Ebashi, S., J. Biochem. 80:1149 (1976).
20. Harafuji, H. & Ogawa, Y., J. Biochem. 87:1305 (1980).
21. Khan, M. M. T. & Martell, A. E., J. Am. Chem. Soc., 88:668 (1966).

THE EFFECTS OF ADENINE NUCLEOTIDE AND EXTERNAL Ca^{2+} ON Ca^{2+} RELEASE FROM FRAGMENTED SARCOPLASMIC RETICULUM IN THE PRESENCE OF Mg^{2+} [1]

Hiroshi Morii

Department of Biology, Faculty of Science
Osaka Univeristy
Toyonaka, Osaka, Japan

I. INTRODUCTION

Muscle contraction and relaxation are regulated by the intracellular Ca^{2+} concentration (1-3). When a muscle fiber is excited, Ca^{2+} is rapidly released from the sarcoplasmic reticulum (SR) inducing muscle contraction. When Ca^{2+} is reaccumulated into the SR, the muscle relaxes. Ca^{2+} accumulation is tightly coupled to ATP hydrolysis by the Ca^{2+} ATPase and the mechanism of this reaction has been extensively investigated (4). By contrast, the mechanism of Ca^{2+} release from SR remains unsolved (5,6). Ca^{2+} release is known to be induced from skinned muscle fibers or fragmented SR (FSR) by the following stimuli: the addition of a small amount of external Ca^{2+} (5,6), depolarization (5-18), increasing pH (19,20), Mg^{2+} reduction (21,22), rapid cooling (23-25), exposure to hypotonic solutions (13,14,26-28), the addition

[1] *This study was supported by Grants-in-Aid to Professor Y. Tonomura from the Ministry of Education, Science and Culture of Japan, the Ministry of Health and Welfare of Japan, and the Muscular Dystrophy Association, Inc.*
Abbreviation: SR, sarcoplasmic reticulum; FSR, fragmented sarcoplasmic reticulum; ATPase, adenosine triphosphatase; EGTA, ethyleneglycol-bis (β-aminoethylether)-N,N,N',N',-tetraacetic acid; AMPPNP, adenyl-5'-yl-imidodiphosphate

ISBN 0-12-260380-X

of drugs (5,6), and reversal of the Ca^{2+}-pump (4,5).

Endo *et al.* (29) and Ford and Podolsky (30) found that Ca^{2+} in the cytoplasmic space can induce Ca^{2+} release from the SR in skinned muscle fibers. Since then, the properties of Ca^{2+}-induced Ca^{2+} release have been investigated in both skeletal and cardiac muscle fibers. Ca^{2+}-induced Ca^{2+} release is believed to play an important role in excitation-contraction coupling (EC coupling) in cardiac muscle fibers (5,6). However, there are several lines of evidence against its physiological role in EC coupling in skeletal muscle fibers: (i) Ca^{2+}-induced Ca^{2+} release is inhibited by Mg^{2+} (5,6). (ii) Contraction of skinned muscle fibers is induced by excitation even when Ca^{2+}-induced Ca^{2+} release is inhibited by procaine (31). (iii) Contraction is unaffected by adenine inhibition of Ca^{2+}-induced Ca^{2+} release (32).

The properties of Ca^{2+}-induced Ca^{2+} release must be investigated in more detail in order to make its physiological significance clear. Furthermore, Ca^{2+}-induced Ca^{2+} release is interesting from the point of view of regulation of ion permeability of biological membranes in general. For these reasons, we have investigated the mechanism of Ca^{2+}-induced Ca^{2+} release from FSR under physiological conditions. In this paper, we review the general properties of Ca^{2+}-induced Ca^{2+} release from FSR of skeletal muscle, and then we describe our recent work on the action of adenine nucleotides and Ca^{2+} on Ca^{2+} release from FSR (33).

II. GENERAL ASPECTS OF Ca^{2+}-INDUCED Ca^{2+} RELEASE FROM FSR

FSR is a favorable material for the study of the mechanism of Ca^{2+}-induced Ca^{2+} release at the molecular level, since a large amount of FSR vesicles can be easily prepared, and since the protein and lipid composition of the SR membrane has been studied extensively (4).

Weber *et al.* (34,35) first reported that Ca^{2+} permeability of the FSR is increased by external Ca^{2+}. Inesi and Malan (36) pointed out the similarity between the enhancement by external Ca^{2+} of Ca^{2+} efflux from FSR and Ca^{2+}-induced Ca^{2+} release in skinned muscle fibers. Katz *et al.* (37) subsequently demonstrated that Ca^{2+} efflux from FSR increases with increasing external Ca^{2+} concentration. The properties of Ca^{2+}-induced Ca^{2+} release from FSR have been extensively studied, using $^{45}Ca^{2+}$ (38) or metallochromic indicators (39) to measure Ca^{2+} release. The properties have been found to be very similar to those of SR in skinned muscle fibers and include inhibition by a high concentration of Mg^{2+} or procaine and stimulation by ATP, caffeine, and halothane (38,39). Furthermore, in the absence of Mg^{2+}, Ca^{2+} release from FSR and from skinned muscle fibers is enhanced by μM external Ca^{2+} and inhibited by a high concentration of external Ca^{2+}.

FSR is composed of heavy and light vesicles, which are derived from the terminal cisternae and the longitudinal portion of the SR, respectively (40). Ca^{2+}-induced Ca^{2+} release is observed only with heavy SR (18,41), indicating that Ca^{2+}-induced Ca^{2+} release occurs at the terminal cisternae, as suggested by electron-probe analysis (42).

Specific inhibitors and stimulators have been used as tools to characterize the channel for Ca^{2+}-induced Ca^{2+} release. Caffeine and halothane at high concentrations enhance Ca^{2+}-induced Ca^{2+} release, in frog FSR and less reliably in rabbit FSR (38-39,43-45). The stimulatory effect of caffeine is inhibited by decreasing external Ca^{2+} concentration, increasing Mg^{2+} concentration, or addition of procaine. Several local anesthetics, such as procaine or tetracaine, inhibit Ca^{2+}-induced Ca^{2+} release (38,39,45) while others, such as cocaine or lidocaine, have no effect on Ca^{2+} release (46,47). Ca^{2+}-induced Ca^{2+} release is inhibited by ruthenium red at a concentration of 10-20 μM, the lowest inhibitory concentration among inhibitors reported so far (18,39,41,45). Ca^{2+} antagonists, such as verapamil and diltiazem, do not affect Ca^{2+} release (45). Recently Shoshan *et al.* (48,49), Kirino and Shimizu (45), and Yamada *et al.* (50) reported that quercetin, a flavonoid compound, inhibits Ca^{2+}-pump activity without affecting Ca^{2+} release.

Previous studies on Ca^{2+}-induced Ca^{2+} release have usually been performed in the absence of Mg^{2+}, since Mg^{2+} is a potent inhibitor of this reaction (5,6,37-39). However, even in the presence of a high concentration of Mg^{2+}, Ca^{2+}-induced Ca^{2+} release has been found to be accelerated by β,γ-methylene adenosine triphosphate, an unhydrolyzable ATP analog (51). Beirao and de Meis (52) also reported an ADP activation of Ca^{2+} efflux in the presence of a high concentration of Mg^{2+}. In the following sections, we will discuss the interaction of Ca^{2+} and nucleotide in stimulating Ca^{2+} release under physiological conditions, *i.e.*, in the presence of 5 mM Mg^{2+} and 0.1 M KCl at pH 7.0.

III. ACCELERATION BY ADENINE NUCLEOTIDE OF Ca^{2+}-INDUCED Ca^{2+} RELEASE IN THE PRESENCE OF A HIGH CONCENTRATION OF Mg^{2+}

The FSR of rabbit skeletal muscle prepared as described previously (53) was passively preloaded with $^{45}Ca^{2+}$, and its Ca^{2+}-induced Ca^{2+} release was measured by a Millipore filtration method (54,55). FSR was preloaded with $^{45}Ca^{2+}$ by incubation overnight at 0°C with a buffer solution composed of 10 mM $^{45}CaCl_2$, 5 mM $MgCl_2$, 0.1 M KCl, and 50 mM Tris-maleate (pH 7.0). $^{45}Ca^{2+}$ release was started by diluting $^{45}Ca^{2+}$-preloaded FSR 200 fold with a buffer solution containing 1 mM $CaCl_2$, various concentrations of EGTA, various concentrations of nucleotide, 5 mM $MgCl_2$, 0.1 M KCl, and 50 mM Tris-maleate (pH 7.0) at 0°C. As shown in Fig. 1, the Ca^{2+}-induced Ca^{2+} release was inhibited by Mg^{2+} (●), as previously reported by other workers (5,6,37-39). In the presence of 5 mM Mg^{2+} and 15 μM external free Ca^{2+}, Ca^{2+} was released at a slow linear rate: In the presence of 5 mM Mg^{2+}, 15 μM external free Ca^{2+}, and 10 mM AMP (■), Ca^{2+} was released with a rapid initial phase followed by a slow phase which was almost equal to the rate in the absence of AMP. Since the rapid initial phase was observed only in the presence of AMP and external Ca^{2+}, we conclude that the rapid initial phase corresponds to Ca^{2+}-induced Ca^{2+} release.

The accelerating effect of AMP was reversible. When the AMP concentration was diluted from 1 mM to 10 μM in the assay buffer, the Ca^{2+} release rate in the rapid initial phase decreased to a rate similar to that observed in the absence of AMP. AMP at 10 μM had almost no effect on the rate of Ca^{2+}-induced Ca^{2+} release (data not shown).

Ca^{2+}-induced Ca^{2+} release in the presence of 5 mM Mg^{2+} was enhanced not only by AMP but also by other adenine nucleotides, such as ATP, ADP, or AMPPNP, but not by cAMP, adenosine, GTP, ITP, CTP, UTP, or acetyl phosphate. These findings were similar to earlier observation with skinned muscle fibers (56).

The Ca^{2+}-ATPase does not seem to play a major role in Ca^{2+}-induced Ca^{2+} release, as previously reported by Endo (56). In this study, we found that the nucleotide specificity for the acceleration of Ca^{2+}-induced Ca^{2+} release differed from the nucleotide specificity for binding to the Ca^{2+}-ATPase, as determined by measurement of changes in the EPR spectrum of spin-labeled FSR upon the addition of various phosphorylated compounds (data not shown). Furthermore, no Ca^{2+} release was observed from LSR (Sec. II), although LSR is greatly enriched in Ca^{2+}-ATPase (40).

Fig. 1 shows that the rate of Ca^{2+}-induced Ca^{2+} release

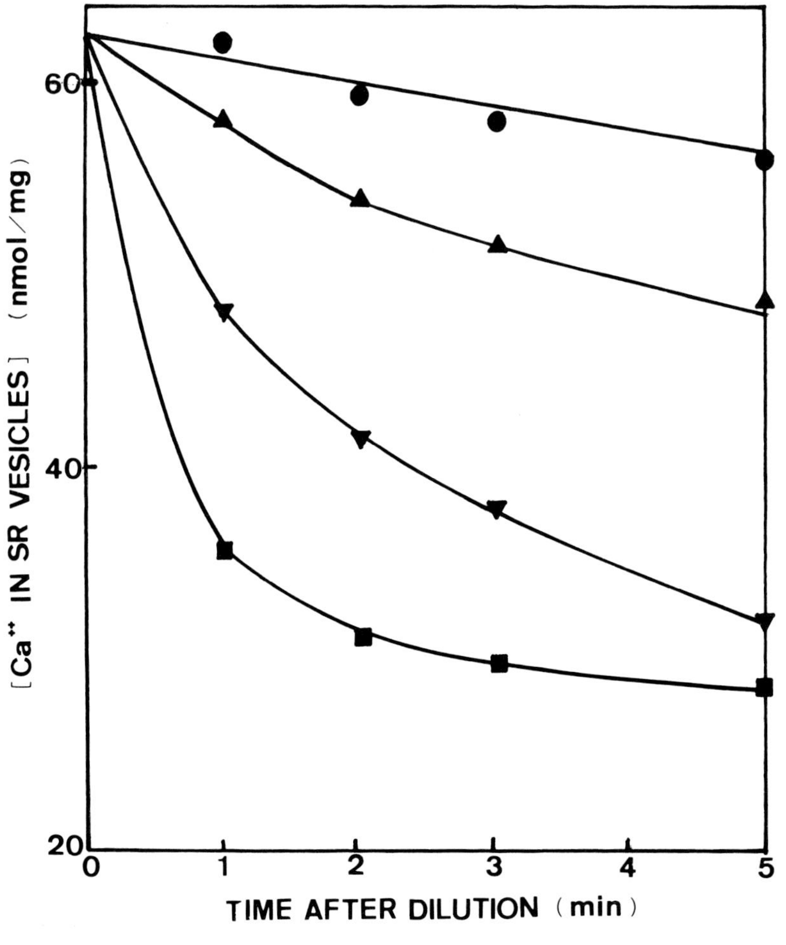

FIGURE 1. Acceleration by AMP of Ca^{2+}-induced Ca^{2+} release from FSR. FSR was preloaded with $^{45}Ca^{2+}$ by incubation overnight at 0°C with a buffer solution (5 mM $MgCl_2$, 0.1 M KCl, and 50 mM Tris-maleate, pH 7.0) containing 10 mM $^{45}CaCl_2$. $^{45}Ca^{2+}$ release was initiated by diluting $^{45}Ca^{2+}$-preloaded FSR 200 fold into the same buffer solution containing 1 mM $CaCl_2$ and 1 mM EGTA (15 μM free Ca^{2+}) in the presence (▲,▼,■) and absence (●) of AMP. Concentrations of AMP :●, 0 ; ▲, 0.1 ;▼, 1 ;■, 10 mM.

increased with increasing AMP concentration, while the final amount of Ca^{2+}-induced Ca^{2+} release was almost independent of AMP concentration. The dependence of the apparent first-order rate constant on AMP concentration followed a Michaelis-Menten type equation with an "apparent" dissociation constant of about 2 mM.

The findings on the accelerating effect of adenine nucleotide can be explained by the following reaction scheme:

$$\text{channel} + N \rightleftharpoons \text{channel}^*\text{-}N$$

where N represents adenine nucleotide and asterisk represents Ca^{2+}-activatable conformation. The equilibrium between them shifts to an activatable conformation when 1 mol of adenine nucleotide binds to 1 mol of the repressed channel. In the above scheme, external Ca^{2+} is assumed not to affect the equilibrium.

IV. ALL-OR-NONE RESPONSE OF THE CHANNEL TO EXTERNAL Ca^{2+}

Fig. 2 shows that the channel for Ca^{2+}-induced Ca^{2+} release opens and closes reversibly with changes in external Ca^{2+} concentration. External Ca^{2+} at 50 µM induced Ca^{2+} release in the presence of 1 mM AMP and 5 mM Mg^{2+} (■). When the external Ca^{2+} concentration was decreased from 50 µM to 50 nM by the addition of 2 mM EGTA (↓) during the rapid initial phase, the rate of Ca^{2+} release immediately decreased (●) to a rate approximately equal to that in the absence of AMP (□). When 2 mM $CaCl_2$ was added to the same reaction mixture at various time intervals (↓), the rate of Ca^{2+} release immediately recovered to the high level (▲).

The effect of external Ca^{2+} on Ca^{2+} release is quite different from the effect of adenine nucleotides. Adenine nucleotides increased the rate without changing the final amount of Ca^{2+}-induced Ca^{2+} release. By contrast, increasing external Ca^{2+} concentration increased the final amount of Ca^{2+}-induced Ca^{2+} release in a graded fashion, as shown in Fig. 3. In the experiments shown in Fig. 3, it was difficult to determine the precise rate of Ca^{2+} release, since the rate of release was so high. With FSR preparation, in which the Ca^{2+} release rate was slow even in the presence of adequate external Ca^{2+}, the rate of Ca^{2+} release was measured at various concentrations of external Ca^{2+}. The time for half maximal Ca^{2+} release was found to be virtually independent of external Ca^{2+} concentration (data not shown).

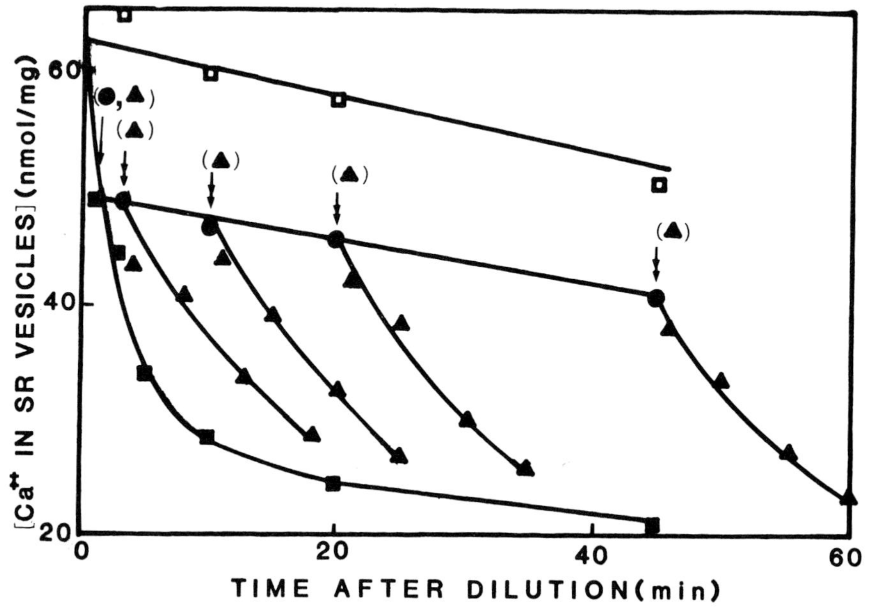

FIGURE 2. Reversible opening and closing of the Ca^{2+} channel by changes in the concentration of external Ca^{2+}. FSR preloaded with $^{45}Ca^{2+}$ by incubation overnight at 0°C with a buffer solution (5 mM $MgCl_2$, 0.1 M KCl, and 50 mM Tris-maleate, pH 6.5) containing 10 mM $^{45}CaCl_2$ was diluted into the same buffer solution containing $CaCl_2$ at a final concentration of 50 µM in the presence (■) and absence (□) of 1 mM AMP at pH 6.5. One minute later, 2 mM EGTA (●) was added (↓), and then at various times (3, 10, 20, and 45 min), 2 mM $CaCl_2$ (▲) was added (↓).

We assume that our FSR vesicles consisted of vesicles with and without the channel for Ca^{2+}-induced Ca^{2+} release, since we were unable to release more than about one half of the amount of preloaded Ca^{2+}. Thus, the finding that the amount of Ca^{2+}-induced Ca^{2+} release is increased by external Ca^{2+} in a graded fashion suggests that the SR contains only a few channels so that most FSR vesicles have no or only a few channels and that the Ca^{2+} gate opens in an all-or-none fashion at external Ca^{2+} concentration above the threshold. Since the rate of Ca^{2+} release was independent of external Ca^{2+}, the effect of external Ca^{2+} cannot be explained by the

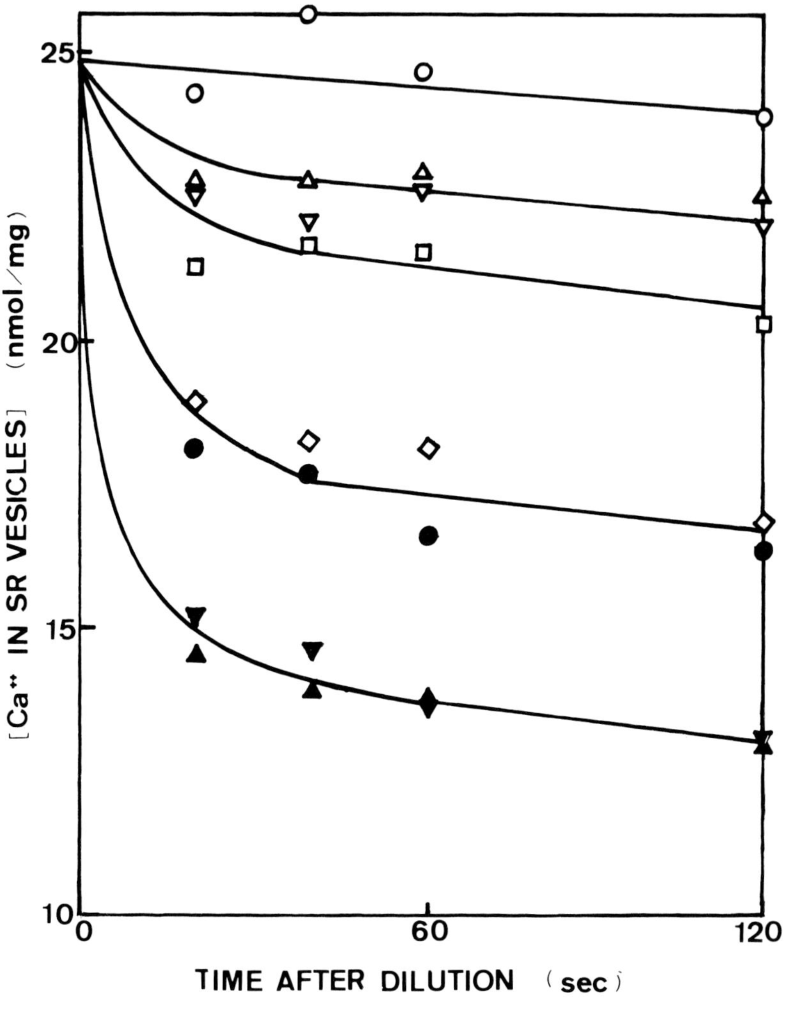

FIGURE 3. The response of the Ca^{2+} channel to varied external Ca^{2+} concentrations. $^{45}Ca^{2+}$ release was measured at pH 7.0 in the presence (△, ▽, □, ◇, ●, ▲, ▼) and the absence (○) of 1 mM AMP at various concentrations of external free Ca^{2+}. External free Ca^{2+} concentrations : △, 0.13 ; ▽, 0.15 ; □, 0.24 ; ◇, 0.34 ; ●, 0.56 ; ▲, 4.2; ▼, 14.7 μM.

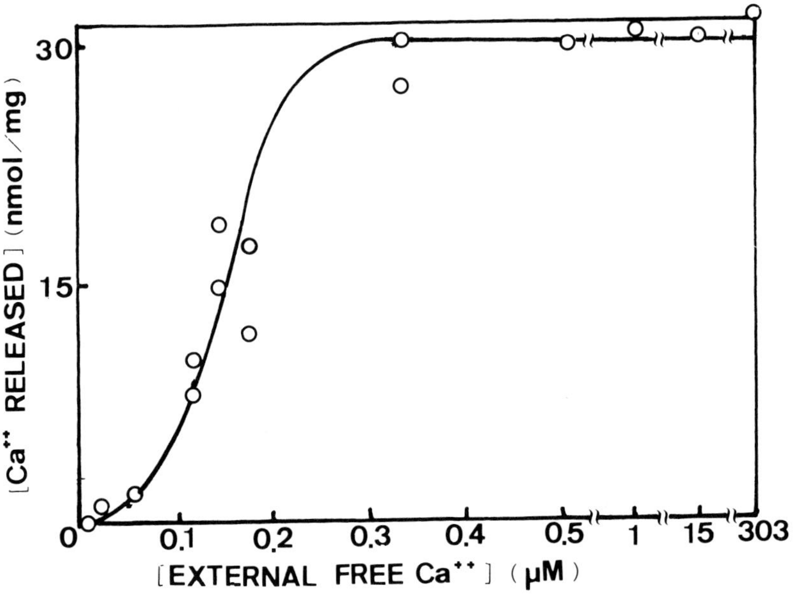

FIGURE 4. Ca^{2+} threshold for activation of the Ca^{2+} channel. The amount of Ca^{2+} released is shown as the difference between the amount of Ca^{2+} remaining in FSR in the presence and absence of 1 mM AMP 2 min after initiation of the Ca^{2+}-induced Ca^{2+} release assay (see FIGURE 3).

cooperative binding of external Ca^{2+}. Recently, Lew *et al.* (57) and Garcia-Sancho *et al.* (58) reported independently that the amount of K^+ released from vesicles of red cell membranes increases in a graded fashion with increasing Ca^{2+} concentration, and concluded that the Ca^{2+}-activated K^+ channel also responds in an all-or-none fashion to Ca^{2+}.

Fig. 4 shows that the channels for Ca^{2+}-induced Ca^{2+} release have heterogeneous Ca^{2+} thresholds. The number of open channels began to increase at about 50 nM Ca^{2+} and reached a maximal number at about 0.3 μM external Ca^{2+}. About 70% of the total channels had a Ca^{2+} threshold in the range of 0.1 to 0.2 μM Ca^{2+}.

The effect of external Ca^{2+} in the absence of Mg^{2+} is different from that in the presence of Mg^{2+}. In the absence of Mg^{2+}, external Ca^{2+} changed the rate of Ca^{2+} release but not the amount of Ca^{2+}-induced Ca^{2+} release (data not shown; cf. ref. 38). Furthermore, the dependence of the rate of

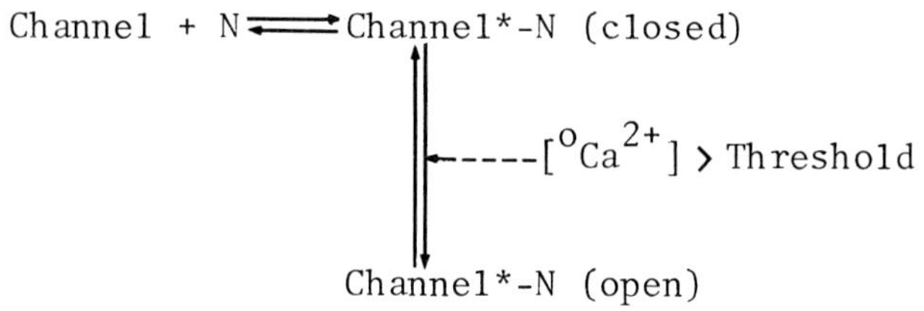

FIGURE 5. A reaction mechanism for the gating behavior of the channel for Ca^{2+}-induced Ca^{2+} release. N represents adenine nucleotide and asterisk represents Ca^{2+}-activatable conformation.

Ca^{2+} release on external Ca^{2+} concentration, in the absence of Mg^{2+}, exhibited a bell-shaped curve as previously reported by Nagasaki and Kasai (38). It remains to be determined whether the Ca^{2+} channel in the presence of Mg^{2+} is different from the channel in the absence of Mg^{2+}.

Fig. 5 shows a reaction scheme for the function of external Ca^{2+} in addition to that of adenine nucleotide. Adenine nucleotide converts repressed channel into the activatable channel by stoichiometric binding to the channel. The gate of the activatable channel opens in all-or-none fashion at external Ca^{2+} concentrations above its threshold, and closes at external Ca^{2+} concentrations below its threshold.

V. NO EFFECT OF INTERNAL Ca^{2+} ON THE CHANNEL

As shown in Fig. 6, the amount of preloaded Ca^{2+} and the amount of Ca^{2+}-induced Ca^{2+} release increased with increasing concentrations of Ca^{2+} in the preloading mixture (2-20 mM). However, the ratio between the amount of Ca^{2+}-induced Ca^{2+} release and the amount of preloading Ca^{2+} was independent of the Ca^{2+} concentration used for preloading. The time for half maximal Ca^{2+}-induced Ca^{2+} release was also independent of the amount of preloaded Ca^{2+}. Furthermore, the distribution of the Ca^{2+} threshold for external Ca^{2+} was independent of the concentration of Ca^{2+} used for preloading (5-20 mM) (data not shown). Thus, internal Ca^{2+} does not affect the gating behavior of the channel for Ca^{2+}-induced Ca^{2+} release and the amount of Ca^{2+} released through the open

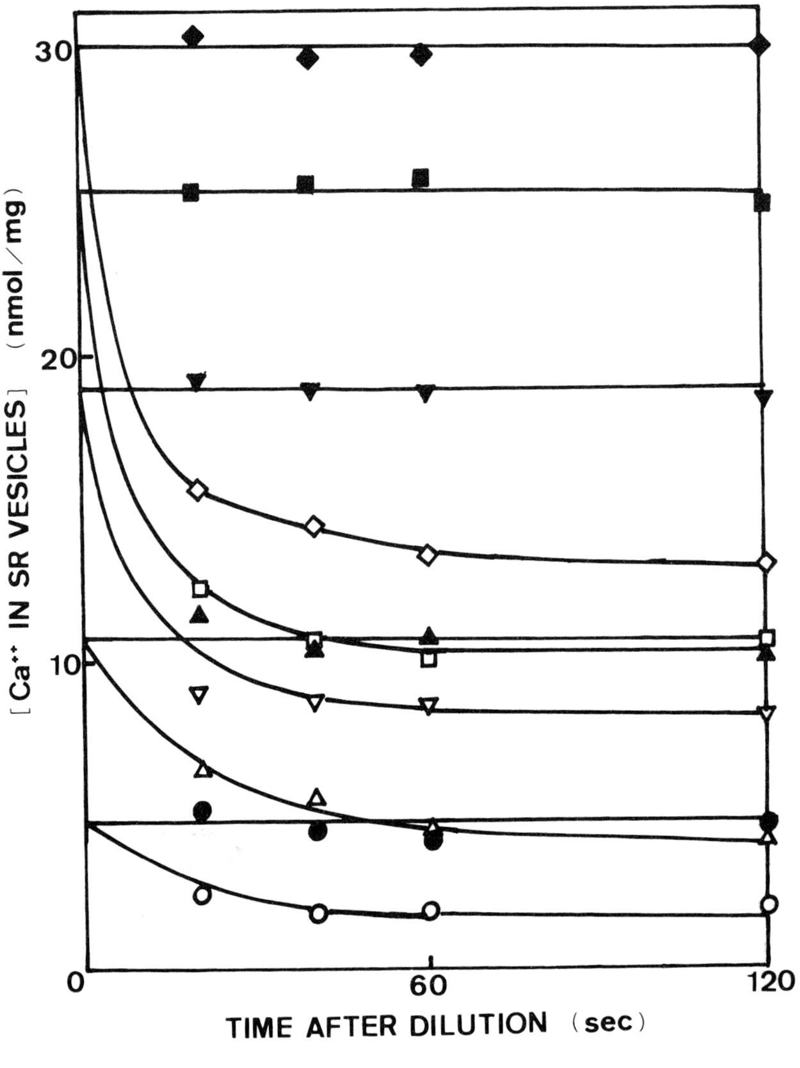

FIGURE 6. Lack of effect of internal Ca^{2+} concentration on Ca^{2+} release. FSR was preloaded at pH 7.0 with 2 (●, ○), 5 (▲, △), 10 (▼, ▽), 15 (■, □), and 20 mM $^{45}CaCl_2$ (◆, ◇). $^{45}Ca^{2+}$ release was measured in the presence (○, △, ▽, □, ◇) and the absence (●, ▲, ▼, ■, ◆) of 1 mM ATP, 10 mM creatine phosphate, and 0.1 mg/ml creatine kinase.

gated channel is proportional to the concentration of internal Ca^{2+}.

ACKNOWLEDGMENTS

The author sincerely thanks to the late Professor Y. Tonomura for his valuable suggestions, useful discussions and constant encouragement. The author is also grateful to Dr. T. Yamamoto, Dr. H. Takisawa, and Dr. Y. Nakamura for helpful advice and discussions and to Professor D.H. MacLennan for many helpful comments on the manuscript.

REFERENCES

1. Hasselbach, W., *Progr. Biophys. Mol. Biol. 14:*167 (1964).
2. Weber, A., in *"Current Topics in Bioenergetics"* (D. R. Sanadi, ed.), p. 203. Academic Press, New York, 1966.
3. Ebashi, S. and Endo, M., *Progr. Biophys. Mol. Biol. 18:* 123 (1968).
4. Tada, M., Yamamoto, T., and Tonomura, Y., *Physiol. Rev. 58:*1 (1978).
5. Endo, M., *Physiol. Rev. 57:*71 (1977).
6. Fabiato, A. and Fabiato, F., *Circ. Res. 40:*119 (1977).
7. Costantin, L. L. and Podolsky, R. J., *J. Gen. Physiol. 50:*1101 (1967).
8. Ford, L. E. and Podolsky, R. J., *Science 167:*58 (1970).
9. Ford, L. E., Spotnitz, A. J., and Sonnenblick, E. H., *J. Gen. Physiol. 55:*138 (1970).
10. Endo, M. and Nakajima, Y., *Nature New Biol. 246:*216 (1973).
11. Endo, M. and Kitazawa, T., *Proc. Japan Acad. 52:*595 (1976).
12. Stephenson, E. W., *J. Gen. Physiol. 71:*411 (1978).
13. Kasai, M. and Miyamoto, H., *FEBS Lett. 34:*299 (1973).
14. Kasai, M. and Miyamoto, H., *J. Biochem. 79:*1053 (1976).
15. Kasai, M. and Miyamoto, H., *J. Biochem. 79:*1067 (1976).
16. Campbell, K. P. and Shamoo, A. E., *J. Membrane Biol. 54:* 73 (1980).
17. Caswell, A. H. and Brandt, N. R., *J. Membrane Biol. 58:* 21 (1981).
18. Miyamoto, H. and Racker, E., *J. Membrane Biol. 66:*193 (1982).
19. Nakamaru, Y. and Schwarts, A., *J. Gen. Physiol. 59:*22 (1972).

20. Shoshan, V., MacLennan, D. H., and Wood, D. S., *Proc. Natl. Acad. Sci. USA 78:*4828 (1981).
21. Thorens, S. and Endo, M., *Folia Pharmacol. Japan 69:*312P (1973).
22. Stephenson, E. W., *J. Gen. Physiol. 77:*419 (1981).
23. Conway, D. and Sakai, T., *Proc. Natl. Acad. Sci. USA 46:* 897 (1960).
24. Sakai, T., *Jikeikai Med. J. 12:*88 (1965).
25. Sakai, T. and Kurihara, S., *Jikeikai Med. J. 21:*47 (1974)
26. Huxley, H. E., Page, S., and Wilkie, D. R., *J. Physiol. (London) 169:*312 (1963).
27. Lannergren, J. and Noth, J., *J. Gen. Physiol. 61:*158 (1973).
28. Endo. M. and Thorens, S., *J. Physiol. Soc. Japan 37:*422 (1975).
29. Endo. M., Tanaka, M., and Ogawa, Y., *Nature 228:*34 (1970)
30. Ford, L. E. and Podolsky, R. J., *Science 167:*58 (1970).
31. Thorens, S. and Endo, M., *Proc. Japan Acad. 51:*473 (1975)
32. See M. Endo's contribution in this book.
33. Morii, H. and Tonomura, Y., *J. Biochem.,* In press (1983).
34. Weber, A., Herz, R., and Reiss, I., *Biochem. Z. 345:*329 (1966).
35. Weber, A., *J. Gen. Physiol. 57:*64 (1971).
36. Inesi, G. and Malan, N., *Life Sci. 18:*773 (1976).
37. Katz, A. M., Repke, D. I., Fudyma, G., and Shigekawa, M., *J. Biol. Chem. 252:*4210 (1977).
38. Nagasaki, K. and Kasai, M., *J. Biochem. 90:*749 (1981).
39. Ohnishi, S. T., *J. Biochem. 86:*1147 (1979).
40. Meissner, G., *Biochem. Biophys. Acta 389:*51 (1975).
41. Miyamoto, H. and Racker, E., *FEBS Lett. 133:*235 (1981).
42. Somlyo, A. V., Gonzalez-Serratos, H., Shuman, H., McClellan, G., and Somlyo, A. P., *J. Cell Biol. 90:*577 (1981).
43. Weber, A. and Herz, R., *J. Gen. Physiol. 52:*750 (1968).
44. Weber. A., *J. Gen. Physiol. 52:*760 (1968).
45. Kirino, Y. and Shimizu, H., *J. Biochem. 92:*1287 (1982).
46. Sakai, T., *Jikeikai Med. J. 10:*113 (1963).
47. Bianchi, C. P. and Bolton, T. C., *J. Pharmacol. Exptl. Therap. 157:*388 (1967).
48. Shoshan, V., Campbell, K. P., MacLennan, D. H., Frodis, W., and Britt, B. A., *Proc. Natl. Acad. Sci. USA 77:*4435 (1980).
49. Shoshan, V. and MacLennan, D. H., *J. Biol. Chem. 256:*887 (1981).
50. Yamada, S., Endo, S., Iwadare, T., and Katayama, H., *7th Annual Meeting of the Japan Bioenergetics Group,* Abstract (in Japanese) p. 104, 1981.
51. Ogawa, Y. and Ebashi, S., *J. Biochem. 80:*1149 (1976).

52. Beirao, P. S. and de Meis, L., *Biochem. Biophys. Acta 433:*520 (1976).
53. Yamada, S., Yamamoto, T., and Tonomura, Y., *J. Biochem. 67:*789 (1970).
54. Sumida, M. and Tonomura, Y., *J. Biochem. 75:*283 (1974).
55. Meissner, G. and McKinley, D., *J. Membrane Biol. 30:*79 (1976).
56. Endo, M., Kakuta, Y., and Kitazawa, T., in *"The Regulation of Muscle Contraction"* (A. D. Ginnel and M. A. B. Brazier, eds) , p. 181. Academic Press, New York, 1981.
57. Lew, V. L., Muallem , S., and Seymour, C. A., *Nature 296:* 742 (1982).
58. Garcia-Sancho, J., Sonchez, A., and Herreros, B., *Nature 296:*744 (1982).

CALCIUM-INDUCED CALCIUM RELEASE FROM FRAGMENTED SARCOPLASMIC RETICULUM[1]

Yutaka Kirino
Masanori Osakabe
Hiroshi Shimizu

Faculty of Pharmaceutical Sciences
University of Tokyo
Bunkyo-ku, Tokyo

I. INTRODUCTION

Since its discovery in 1970 (1,2), Ca-induced Ca release has been primarily studied using skinned muscle fibers (3,4). Experiments with fragmented sarcoplasmic reticulum (FSR) have been rare until recently. Although there were some reports of Ca-induced Ca release from FSR (5,6) and related phenomena (7-9), it is well known that Ca release is more difficult to induce in FSR than in SR of skinned fibers (3,10). Caffeine, which is believed to enhance Ca-induced Ca release (3), is not very effective in its induction from rabbit skeletal FSR, a preparation widely used by many investigators to study the Ca pump of SR.

Recently, Miyamoto and Racker (11,12) and we (13) have reported Ca- and caffeine-induced release of Ca from the heavy fraction of FSR which is known to be derived from terminal cisternae of SR. It was demonstrated that almost all of the characteristics of Ca-induced Ca release observed in skinned fibers under conditions believed to be physiological are reproduced in heavy FSR vesicles under the same conditions (13). Under physiological conditions where ATP, Mg and Ca are present, only net uptake or net release of Ca can be determined since the Ca pump is also in operation.

---

[1] This study was supported in part by a grant from Ho'ansha Foundation awarded to Y. K.

ISBN 0-12-260380-X

Care must also be taken to distinguish Ca-induced Ca release from Ca exchange through the Ca pump (14,15).

In order to elucidate the mechanism of Ca-induced Ca release, the measurement of unidirectional, passive Ca efflux is desirable. This was first carried out by Nagasaki and Kasai with FSR (16) and by Endo with skinned fibers (17). Yamamoto and Kasai (18,19,20) measured Ca-dependence of the permeability of FSR membranes to choline by the light scattering technique and suggested that the Ca-gated cation channel and the Ca-gated Ca channel, through which Ca-induced Ca release is thought to occur, might be the same entity.

In the present study we investigated in detail passive Ca efflux from heavy and light fractions of rabbit skeletal FSR and attempted to characterize the putative Ca-gated Ca channel.

II. MATERIALS AND METHODS

The heavy (H) and light (L) fractions of FSR were isolated from rabbit white skeletal muscle as described in a previous report (13). The L fraction was placed on top of a continuous linear sucrose density gradient (20 - 60 w/v %) and centrifuged at 30,000 rev./min for 6 hr in a Hitachi RPS40T rotor. Fractions obtained from the 40 - 45 % sucrose layer and the 50 - 52 % layer were collected and denoted as LS and HS fractions, respectively. Preparations were used in experiments within three days after isolation.

FSR preparations were incubated overnight at the concentration of 10 mg protein/ml in a solution containing 3 mM $^{45}CaCl_2$, 0.1 M KCl, 20 mM MOPS (pH 7.0) at 4 $^{\circ}$C. One hundred microliters of the suspension was admixed at 4 $^{\circ}$C with 9.9 ml of a dilution solution containing 0.1 M KCl, 20 mM MOPS (pH 7.0), 1 mM EGTA and appropriate additives such as various concentrations of calcium chloride, drugs and others. Free Ca ion concentration was calculated using the value of $10^{6.4}$ for the apparent binding constant of Ca ion and EGTA at pH 7.0 (24). At an appropriate time after the dilution, a 1 ml aliquot of the diluted suspension was filtered by suction through a Millipore PHWP02500 membrane filter (0.3 µm pore size). The Millipore filter was immediately washed twice with 2 ml of a washing solution containing 0.1 M KCl, 20 mM MOPS (pH 7.0) and 4 mM magnesium chloride and then dried in a scintillation vial (13). The radioactivity was measured after the addition of a toluene-based scintillation mixture and the amount of calcium remaining in FSR vesicles adsorbed in the filter was determined.

III. RESULTS AND DISCUSSION

A. Time Course of Ca^{2+} Release

The calcium efflux from FSR, which had been loaded passively with calcium-45, was followed as a function of time after dilution 100 times with a medium containing 0.1 M KCl, 20 mM MOPS (pH 7.0), 1 mM EGTA and various concentrations of calcium chloride. The semilogarithmic plot of Fig. 1 shows some of the time courses of calcium efflux from H fraction and LS fraction. Little difference was observed between H and HS fractions. In both of the H and LS fractions, the curves are composed of two components, fast and slow. Calcium efflux rate in the fast component depended strongly on the free calcium concentration in the dilution medium while that in the slow component did not. The size of the fast component was much larger in H fraction than in LS fraction, comprising about 85 % in the former but only 40 % in the latter. The size of each component was independent of experimental conditions. Considering these facts, we may assume the occurrence of two kinds of SR vesicles: those with and without Ca-gated Ca channels. The former gives the fast component of the Ca efflux curve and comprises most of the

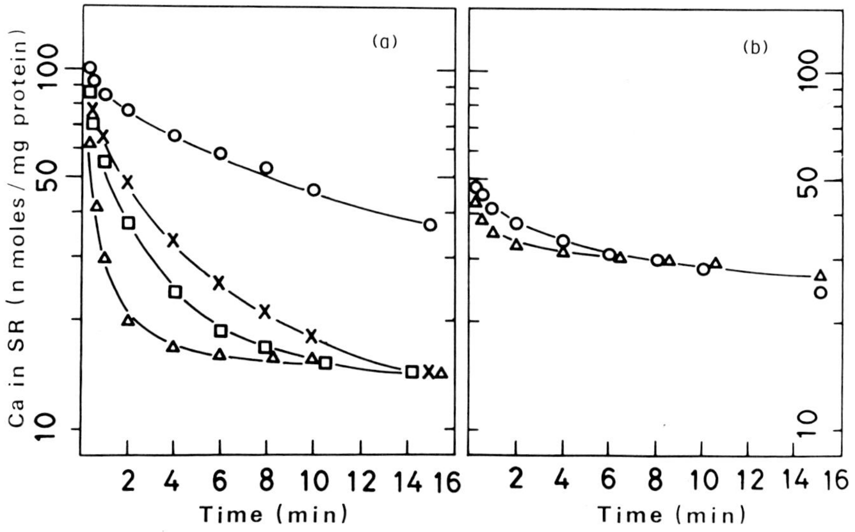

FIGURE 1. Semilogarithmic plot of Ca efflux from (a) heavy and (b) light fractions of FSR. Extravesicular free Ca concentration (μM): 0.12 (o), 0.30 (x), 2.0 (Δ), 50.0 (□).

vesicles in the heavy fraction of FSR, while the latter comprises the major portion of the light vesicles giving the slow component of the Ca efflux curve.

The amount of calcium in FSR at time zero, which was determined by extrapolating a very slow efflux curve, was found to be much larger than that estimated on the basis of the vesicle inner volume and the equilibration of intravesicular Ca with extravesicular Ca (3 mM). The estimated value is 13.5 nmol Ca/mg protein if we adopt the value of 4.5 μl/mg protein for the intravesicular water space (22). Contrary to this estimation, the observed amount of Ca in FSR at time zero was actually about 100 nmol/mg protein for H fraction and about 50 nmol/mg protein for LS fraction. This fact clearly indicates that most of the Ca ions inside FSR are not in the form of free ions but in the bound form. The most probable binding site of calcium ions is calsequestrin, a peripheral calcium-binding protein inside SR (23). Calsequestrin is known to be abundant in the FSR heavy fraction but is scarce in the light fraction (24-26). In fact, the difference in the amount of Ca passively loaded in H (or HS) and LS fractions of FSR could be accounted for by the difference in the relative amount of calsequestrin in the

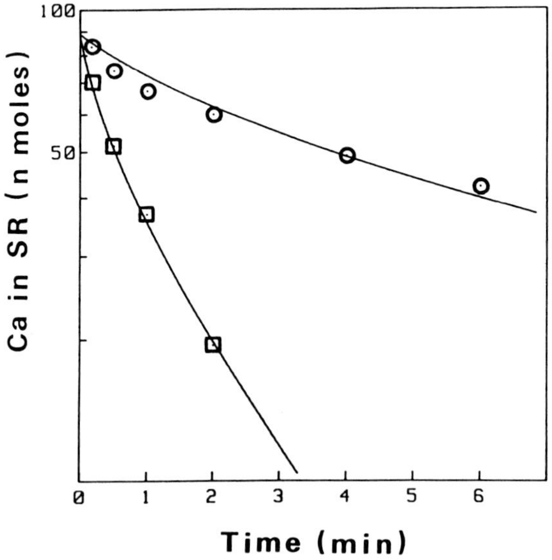

FIGURE 2. Fast component of Ca efflux in H fraction. Points designate experimental values and lines refer to curves calculated on the assumption described in III-B. Extravesicular free Ca is at 0.12 μM (o) or 50 μM (□). (27).

preparations, as estimated from the area of the 65 k dalton band on an SDS-polyacrylamide gel electrophoretic densitogram according to Laemmli (28).

After the subtraction of the slow component, which was approximated to be linear, from the curve in Fig. 1a the fast component was replotted as shown in Fig. 2. Clearly this semilogarithmic plot is not straight, which seems to indicate that the calcium efflux is not a single first-order process. This was explained by a linear combination of two or three exponentials (16). It is also possible to explain by the assumption of a time-dependent inactivation of the calcium-gated calcium channel. In contrast to these explanations, we will show in the next section that the consideration of calcium binding inside FSR as described above leads to a non-straight semilogarithmic plot for the calcium efflux in terms of a single rate process without any further assumption.

B. Analysis of Time Course of Ca^{2+} Release

We consider three compartments for Ca ions: Ca binding sites (calsequestrin) and free water space, both inside FSR vesicles, and a bulk solution where free Ca concentration is kept very low by EGTA. When the concentration of binding sites occupied by Ca and the concentration of free Ca inside FSR are B and F, respectively, at time t, the rate equation is

$$\frac{d(F + B)}{dt} = - kF \tag{1}$$

where k is the rate constant for Ca efflux. If rapid equilibrium holds between F and B,

$$\frac{F(E - B)}{B} = K \tag{2}$$

where E is the concentration of total Ca binding sites and K is the dissociation constant of calsequestrin for Ca and set at 0.8 mM (29). Using eq. (2), eq. (1) can be integrated to

$$-(1+\frac{E}{K})\ln(\frac{F}{F_o}) + \frac{E}{K}\ln(\frac{K+F}{K+F_o}) - E(\frac{1}{K+F} - \frac{1}{K+F_o}) = kt \tag{3}$$

where F_o refers to F at t = 0, which is 3 mM throughout this study. The sum of B + F is known experimentally if the Ca amount in the FSR measured is converted to a concentration using the value of 4.5 µl/mg protein for the intravesicular space (22). The value of E is evaluated from eq. (2) with the value of $F_o + B_o$ at t = 0 and F_o = 3 mM.

Therefore, using eq. (3) a value of k can be calculated for each experimental point on a single time course. The k values thus derived were found to be close to one another and the average was adopted as the k value for the time course. The agreement of the time course simulated using the model described above and experimental points was good as shown in Fig. 2.

C. Dependence of Calcium Efflux Rate Constant on Extravesicular Calcium Concentration and Its Pharmacological Modification

The rate constant for Ca efflux as determined by the method described above is plotted as a function of extravesicular free Ca concentration in Fig. 3 (o). It exhibits a maximum when the extravesicular Ca is at about 3 μM. This bell-shaped dependence has been shown in SR of skinned fiber (17) although the maximum is at about 20 μM Ca thererin. It should be noted that the plot is similar to the Ca-dependence of choline permeability of FSR (18).

In the range of low Ca concentration, Ca enhances the Ca efflux, that is, it opens the Ca channel. If a double reciprocal plot is made, a straight line is obtained only

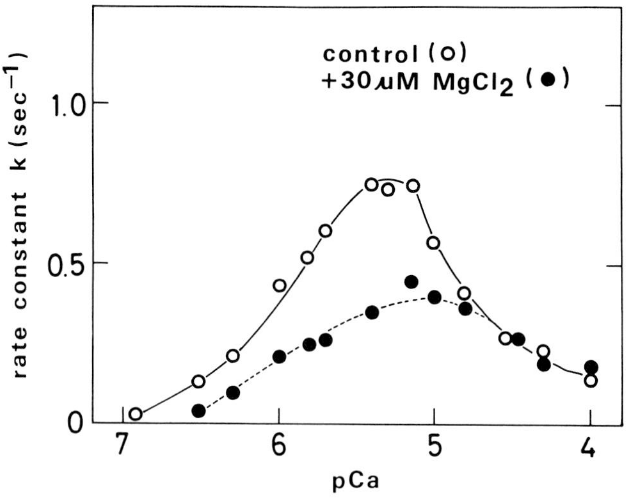

FIGURE 3. Dependence of Ca efflux rate constant on extravesicular free Ca concentration in the absence of or presence of added Mg.

when 1/k is plotted against the reciprocal of the square of Ca concentration. This may indicate that two molecules of Ca are required to open the channel. This is in accordance with the result on skinned fiber SR (17) but at variance with that reported on the Ca-gated cation channel of FSR (19).

Inhibitory effect on Ca efflux of ruthenium red, magnesium ion and procaine was examined. Half inhibition concentration was 0.04 μM for ruthenium red, 20 μM for Mg and 1 mM for procaine when extravesicular Ca was at 2.0 μM. The inhibitory effect of Mg ions was observed to vary somewhat from experiment to experiment. This is probably due to the various magnitudes of Mg contamination in the preparations.

The effects of Mg ions, procaine and caffeine on Ca dependence of Ca efflux were examined. The Ca-dependence profile with 30 μM magnesium chloride is shown in Fig. 3 (o). Note that Mg has no effect in the higher Ca concentration range.

Caffeine greatly enhanced Ca efflux in the range of lower extravesicular Ca concentration but slightly inhibited it in the higher Ca concentration range (Fig. 4). This inhibitory effect was not observed in the skinned fiber study (17). Also noted was that caffeine increased the maximum value of the rate constant at most 50 % in this study while a much larger increase was observed in the skinned fiber study (17).

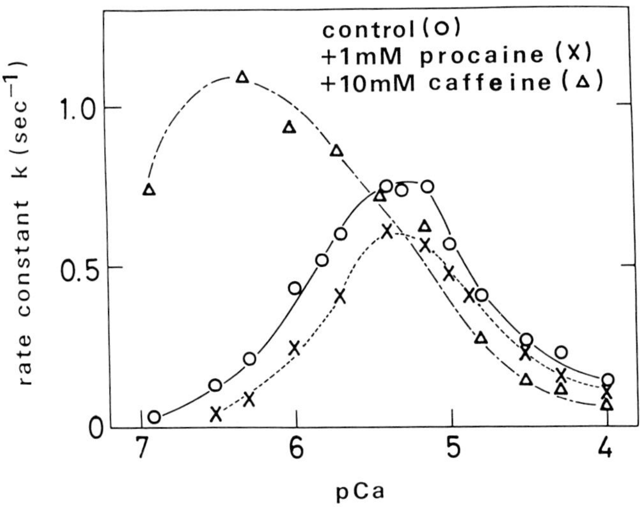

FIGURE 4. Effect of caffeine or procaine on the dependence of Ca efflux rate constant on extravesicular Ca concentration.

Figure 4 also includes the result with procaine. Procaine inhibited Ca efflux more or less evenly throughout the range of extravesicular Ca concentration. This is consistent with the result of skinned fiber study (17).

REFERENCES

1. Endo, M., Tanaka, M. and Ogawa, Y., *Nature 228*:34 (1970).
2. Ford, L. E. and Podolsky, R. J., *Science 167*:58 (1970).
3. Endo, M., *Physiol. Rev. 57*:71 (1977).
4. Fabiato, A. and Fabiato, F., *Ann. Rev. Physiol. 41*:473 (1979).
5. Inesi, G. and Malan, N., *Life Sci. 18*:773 (1976).
6. Ohnishi, S. T., *J. Biochem. 86*:1147 (1979).
7. Ogawa, Y. and Ebashi, S., *J. Biochem. 80*:1149 (1976).
8. Katz, A. M., Repke, D. I., Dunnett, J. and Hasselbach, W., *J. Biol. Chem. 252*:1950 (1977).
9. Katz, A. M., Repke, D. I., Fudyma, G. and Shigekawa, M., *J. Biol. Chem. 252*:4210 (1977).
10. Hasselbach, W., *Basic Res. Cardiol. 75*:2 (1980).
11. Miyamoto, H. and Racker, E., *FEBS Letters 133*:235 (1981).
12. Miyamoto, H. and Racker, E., *J. Membrane Biol. 66*:193 (1982).
13. Kirino, Y. and Shimizu, H., *J. Biochem. 92*:1287 (1982).
14. Makinose, M., *FEBS Letters 37*:140 (1973).
15. Takakuwa, Y. and Kanazawa, T., *J. Biol. Chem. 257*:10770 (1982).
16. Nagasaki, K. and Kasai, M., *J. Biochem. 90*:749 (1981).
17. Endo, M., in "*The Mechanisms of Gated Calcium Transport across Biological Membranes*" (S. T. Ohnishi and M. Endo, eds.), p.257. Academic Press, New York, 1982.
18. Yamamoto, N. and Kasai, M., *J. Biochem. 92*:465 (1982).
19. Yamamoto, N. and Kasai, M., *J. Biochem. 92*:477 (1982).
20. Yamamoto, N. and Kasai, M., *J. Biochem. 92*:485 (1982).
21. Harafuji, H. and Ogawa, Y., *J. Biochem. 87*:1305 (1980).
22. Miyamoto, H. and Kasai, M., *J. Biochem. 85*:765 (1979).
23. MacLennan, D. H. and Wong, P. T. S., *Proc. Nat. Acad. Sci. USA 68*:1231 (1971).
24. Meissner, G., *Biochim. Biophys. Acta 389*:51 (1975).
25. Caswell, A. H., Lau, Y. H. and Brunschwig, J. P., *Arch. Biochem. Biophys. 602*:417 (1976).
26. Campbell, K. P., Franzini-Armstrong, C. and Shamoo, A. E., *Biochim. Biophys. Acta 602*:97 (1980).
27. Kirino, Y., Osakabe, M. and Shimizu, H., *J. Biochem.* submitted.
28. Laemmli, U. K., *Nature 227*:680 (1970).
29. Ostwald, T. J. and MacLennan, D. H., *J. Biol. Chem. 249*:974 (1974).

CALCIUM IMPULSE

Y. Yamaguchi
Y. Komatsu
H. Shimizu

Faculty of Pharmaceutical Sciences
University of Tokyo
Hongo, Bunkyo-ku, Tokyo 113

I. INTRODUCTION

The presence of calcium impulses propagating in intracellular space has been reported. In chicken skeletal muscle fibers developing in culture Coleman and others (1) observed a series of self-propagating waves that appeared bright in a phase contrast microscope when a few mM theophyllin was applied. These waves traveled at speeds of 60 microns per second with intervals of 10 microns. However, a culture kept in Ca^{2+}-free saline for 1 hr no longer gave such waves. If calcium was added, propagating waves appeared immediately.

Essentially the same phenomena associating with contractions were observed by Natori and his collaborators (2) when a few mM caffeine was applied to intact or skinned skeletal fibers of toad. Observed contractions were conveniently classified as follows. (i) Fluctuating ripple-like contractions of one or two sarcomere length which do not propagate in fibers. (ii) Contracting waves: a) when waves are of several sarcomere lengths, they decay during propagation and finally disappear when they stop, and b) when waves are 10-12 sarcomeres in length, they propagate without decay and are abolished when they collide with one another. (iii) Periodic contractions propagating as waves from a portion where caffeine contracture is induced by a tiny amount of more than 10 mM caffeine solution.

Propagating waves with a constant velocity at 50 - 100 μm/sec were observed by Fabiato and Fabiato in a single cardiac cell with disrupted sarcolemma (3,4). The wave propa-

ISBN 0-12-260380-X

gation as well as the cyclic contraction was observed when the pCa of the medium was set as 7-6. Rieser et al. (5) observed propagating waves of rhythmic contractions in isolated cardiac cells. The velocity of propagation as well as the beat frequency appeared to increase with elevated Ca^{2+} in the bathing solution.

The underlying mechanisms of the above phenomena might be regenerative calcium release and uptake. At least, in most of the above experiments, the presence of a slow diffusion of calcium between the myoplasmic and the extracellular spaces was necessary for the emergence of propagating waves. However, it has not yet been clarified how the propagating waves are induced and under what condition what kind of contraction appears. These are the main problems to be discussed in the present paper.

II. MODELLING OF CALCIUM-INDUCED CALCIUM RELEASE

A. Basic Model

As shown in Fig. 1, our model is composed of three spaces: sarcoplasmic reticulum (SR), myoplasm and extracellular space filled with a bathing solution. In the SR membrane, the Ca^{2+}-ATPase and the Ca^{2+}-channel are present. The former actively transports calcium ions from the myoplasm into the SR by utilizing the free energy of ATP-hydrolysis while the latter has an ability of autocatalytic Ca^{2+} release from the SR due to Ca^{2+}-induced Ca^{2+} release. The myoplasmic

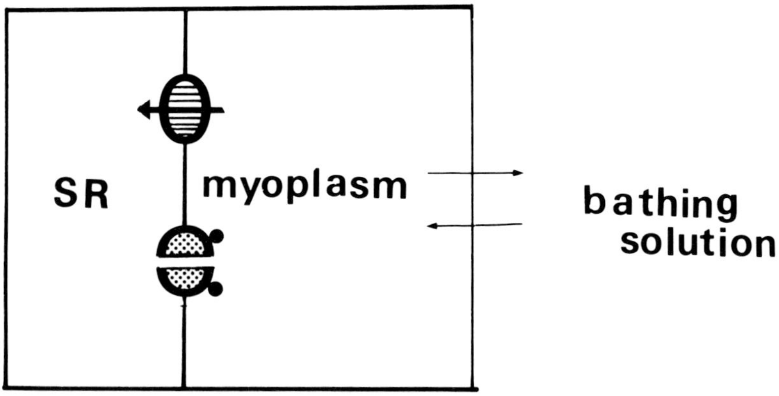

Figure 1. An illustration of the basic model.

and extracellular spaces are partitioned from each other by a "membrane" working as a diffusion barrier for calcium ions.

Calcium concentration in the SR and myoplasmic spaces, x and y, respectively, varies with time t according to the following equations,

$$dx/dt = k_1 R(y - x) + U, \quad (1a)$$

$$dy/dt = - v_f(dx/dt) + k_D(z - y), \quad (1b)$$

where k_1, v_f, and k_D respectively denote the diffusion rate of calcium through the channel, the volume ratio of the SR space to the myoplasmic space, and the diffusion rate of calcium between the myoplasm and the extracellular space. The term z stands for the calcium concentration in the bathing solution and is assumed constant, R the fraction of the channels in the open state, and U the Ca^{2+}-uptake activity of the Ca^{2+}-ATPase.

The detailed molecular mechanism of the Ca^{2+}-channel is still unknown. We will describe the change between the open and closed states of the Ca^{2+}-channel in terms of a Monod-Wyman-Changeux model of an allosteric protein assuming the presence of two calcium binding sites on the outer side of the SR membrane. Calcium binding to the sites leads to a conformational change from the closed to the open form. The kinetic scheme we use for the Ca^{2+}-ATPase activity is determined so as to satisfy the experimental values obtained by Weber (6) in various conditions. Illustrations of state chan-

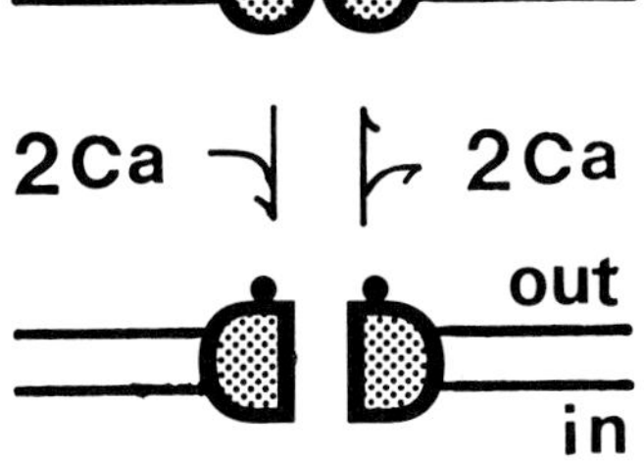

Figure 2a. The conformation change of Ca^{2+}-channel.

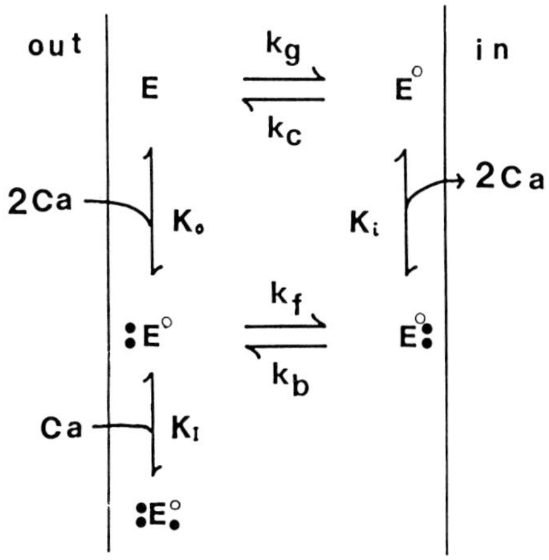

Figure 2b. The kinetic scheme of Ca^{2+}-ATPase.

ges in the channel and the pump are given in Fig. 2, from which we obtain

$$R = [1 + L(1 + \beta K_R y)^2/(1 + K_R y)^2]^{-1}, \tag{2a}$$

$$U = k_a(k_c k_f K_o y^2 - k_g k_b K_i x^2)/[(k_c + k_b K_i x^2)(K_o y^2 + K_I K_o y^3 + 1) + (k_g + k_f K_o y^2)(K_i x^2 + 1)]. \tag{2b}$$

In Eq. (2a), K_R stands for the binding constant of Ca^{2+} of the Ca^{2+}-channel in the open form, and L and β are an allosteric constant and the ratio of the binding constants between the open and closed forms, respectively. The notation of the constants in Eq. (2b) are in agreement with those in Fig. 2. In our model caffeine is assumed to increase K_R by e_f times. We may regard the parameter e_f as a measure of the amount of caffeine applied to the fiber. Our study elucidated that the factor e_f was in proportion to the amount of caffeine in a way such that 3 mM caffeine corresponded to e_f=30. The values of the above parameters which are listed in Table I give reasonable results either qualitatively or semi-quantitatively.

Numerically solving Eq. (1) with (2) using the Runge-Kutta-Gill method, we can obtain the time change of calcium concentration in a spatially homogeneous fiber.

TABLE I. The Values of Parameters in the Basic Model

$k_1 = 1\ sec^{-1}$,	$v_f = 0.1$,	$L = 100$,	$\beta = 0.01$,
$K_R = 10^5$ (M/l fiber)$^{-1}$,	$k_a = 0.02$ M/l SR,		
$k_c = 1\ sec^{-1}$,	$k_b = 0.01\ sec^{-1}$,	$k_f = 110\ sec^{-1}$,	
$k_g = 10^4\ sec^{-1}$,	$K_o = 10^{14}$ (M/l fiber)$^{-2}$,		
$K_i = 10^4$ (M/l SR)$^{-2}$,	$K_I = 1.11\times10^5$ (M/l fiber)$^{-1}$.		

B. One-Dimensional Model

To study the propagation of calcium impulse in muscle fibers, we extend our basic model as is schematically shown in Fig. 3. This extended model can be considered as a one-dimensional array of the basic models; the above basic model may be regarded as a special case of our one-dimensional model. The one-dimensional model is reduced to the basic model when the diffusion of calcium in the longitudinal

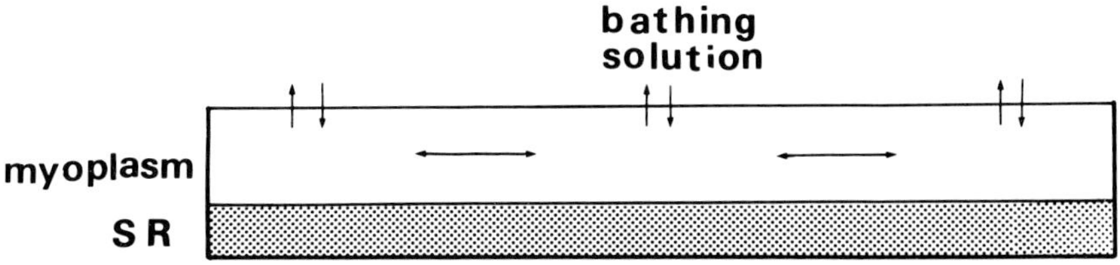

Figure 3. An illustration of one-dimensional model.

direction of the fiber is large enough. In the following discussion, we shall use the basic model when the Ca^{2+} concentration in the myoplasmic space of the fiber is spatially homogeneous or when the diffusion rate is sufficiently large. On the other hand, spatial and temporal changes of the calcium concentration will be studied using the one-dimensional model.

In general, there exists observable diffusion in the myoplasmic space along the longitudinal direction. On the contrary, the SR space is partitioned by the membrane of sequentially distributed SR, where no diffusion occurs along the longitudinal direction. The time-dependence of x and y is respectively described by

$$\partial x/\partial t = f_x(x,y), \quad (3a)$$

$$\partial y/\partial t = f_y(x,y) + D\partial^2 y/\partial r^2, \quad (3b)$$

where f_x and f_y denote the right-hand side (rhs) of Eq. (1a) and (1b), respectively. The last term in the rhs of (3b) represents the diffusion of Ca^{2+} in the myoplasmic space with Fick's diffusion constant D with respect to the space coordinate r in the longitudinal direction. The value of 350 μm^2/sec is used for D, which is one-half of the value obtained in the aqueous solution at 20°C (7). The full length of the fiber is assumed to be 1000 µm and the boundary of the fiber to have no flux of calcium. The Crank-Nicolson method was used in the following computer experiments.

In our model we will disregard the following fiber properties: the suppression of Ca^{2+} release in bathing solution with a high concentration of calcium, calcium binding of calsequestrin and contractile proteins, and Ca transport of mitochondria. These factors make only minor differences in our results mentioned below.

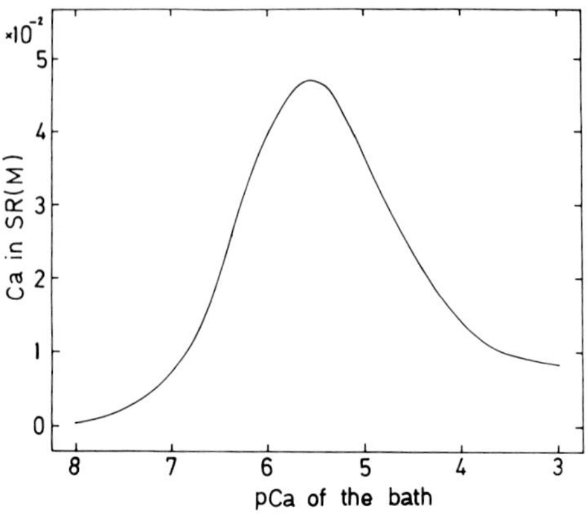

Figure 4. Amounts of Ca^{2+} in SR in the steady state.

III. COMPUTER EXPERIMENTS

A. The Fundamental Nature of Ca^{2+}-Induced Ca^{2+} Release

Let us show the fundamental features of our basic model. Figure 4 represents the steady state value of x, the concentration of Ca^{2+} in SR, against pCa of the bathing solution. In the region of large pCa (small y) the steady state value of x is small due to weak activity of the Ca^{2+}-ATPase. It should be noted however that any initial value of Ca^{2+} concentration in this region given in the SR space is kept without a significant change as if it were in the steady state. As both the rates of release and uptake of Ca^{2+} are small, the rate of relaxation to the true steady state is so slow that any initial value of x appears to be a quasi steady one. Decrease in the concentration of Ca^{2+} in SR occurs at pCa 4-5, which is in a good agreement with experimental data (8).

B. Self-Organization in Ca^{2+}-Induced Ca^{2+} Release

1. Excitability - Calcium Spike Figure 5 shows examples of "calcium spike" of our model found by our computer experiments. The values of parameters used were pCa 6 of the bathing solution, e_f=8 and k_D=15. The initial value of x is taken as 30 mM which is the steady-state value in the present condition. The abrupt increment of calcium in the myoplasmic space, y, from the value in the steady state causes a further increase in y due to the net release of Ca^{2+} stored in SR. The letters given on the curves denotes the value of y at t=0, i.e., the magnitude of stimulation or the pulse-like addition of calcium to the myoplasm. When the initial value

of y is 1.80 μM, no response appears, i.e., no further increase in y. On the other hand, stimulation by 1.81 μM Ca^{2+} induces remarkable calcium release. Thus, a clear threshold

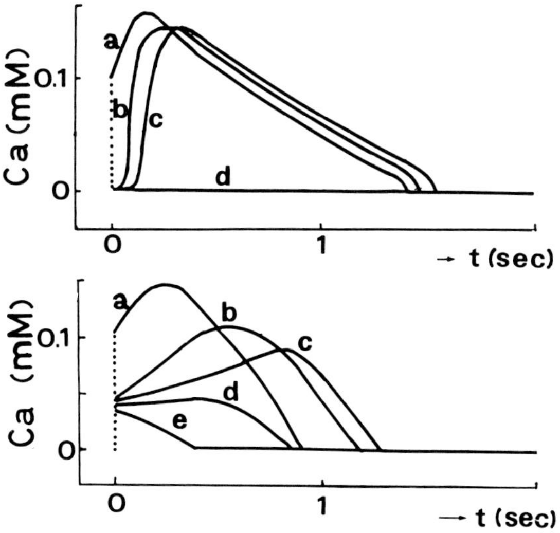

Figure 5. The excitation of SR induced by a pulse-like addition of various amounts of Ca^{2+} in myoplasm under pCa 6 of the bath and k_D=15. Upper: e_f=8, magnitude of stimulation (y_0)=100, 2, 1.81, 1.80 μM in a-d, respectively. Lower: e_f=1, y_0=100, 40, 37, 35, 30 μM in a-e, respectively. (18)

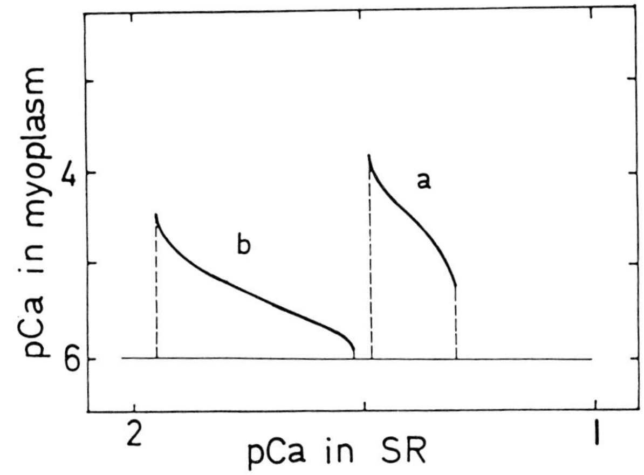

Figure 6. The excitation region with respect to x and the threshold of y with pCa 6 of the bath and k_D=15. a: e_f=1. b: e_f=8.

is found in this figure between 1.80 and 1.81 μM. However, in the absence of caffeine, the time-course of the responses does not as clearly differ between below and above the threshold as is observed in the lower part of Fig. 5.

In Fig. 6 we give in the abscissa a region a of x where a calcium spike appears in the absence of caffeine and in the ordinate the threshold, the minimum value of y, that produces the calcium-spike. In the region of x lower than this excitable one, no increment of y can cause a calcium impulse. In the region of x higher than the right broken vertical line of the region a, the threshold for the calcium impulse vanishes, which means calcium release from SR occurs spontaneously even when no calcium is added in the myoplasmic space. The excitability is found within a moderate region of x. (SR will be bought to a kind of refractory state when calcium is released by excitation.) The threshold value changes according to the amount of added caffeine as shown in the same figure. The increase of caffeine shifts the excitable region of x to the lower side, b, and also diminishes the threshold.

2. Propagation of Calcium Impulse As shown above, SR has

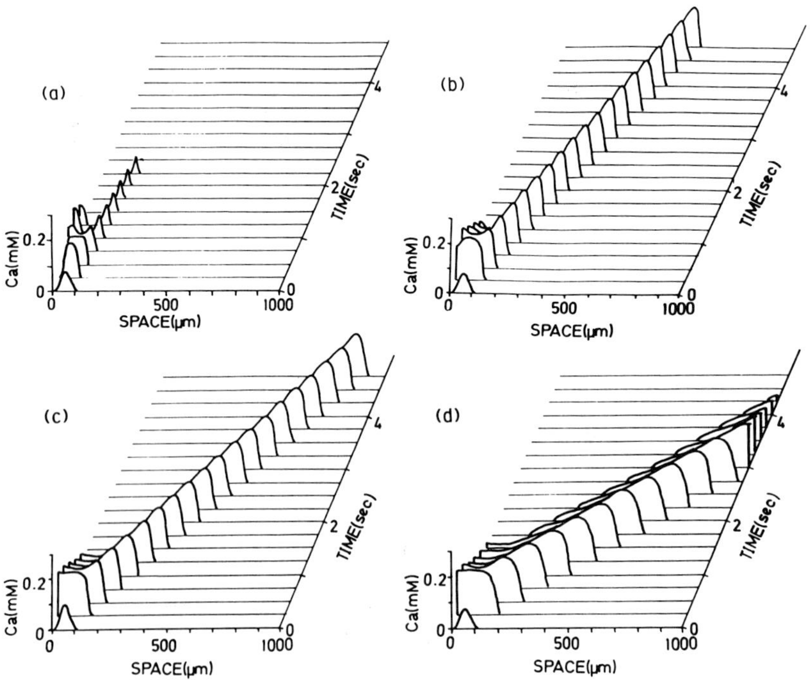

Figure 7a. Propagation of waves for various amounts of caffeine; e_f=1.5, 3, 5, 9 in (a)-(d), respectively.

the ability to produce the excitation, which is analogous to the excitation of the nerve membrane. The excitation produces an impulse under appropriate conditions. Let us show how the calcium impulse propagates. Figure 7a represents results calculated with various amounts of caffeine, e_f, under k_D=12, pCa 9 of the bathing solution and the initial value, 30 mM, of x. The ordinate denotes the calcium concentration of the myoplasm, y, the abscissa the space coordinate in the longitudinal direction of the fiber, and the oblique axis the time. At the initial time, calcium is added in the myoplasmic space, as is seen in the figure, within 100 μm from the left terminal portion and y of pCa 9 is placed in the other portion. The figure shows that an increase in the caffeine concentration leads to an increase in the velocity of the propagation. In the case of e_f=1.5, the calcium impulse gradually decays both in the velocity and in the strength, and disappears. This is consistent with the description of

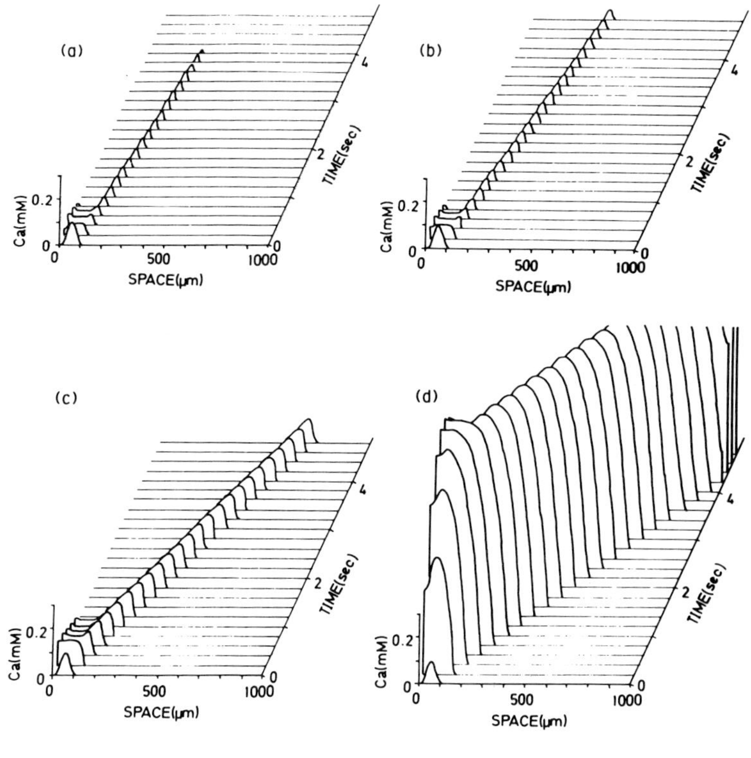

Figure 7b. Propagation of waves for various magnitude of diffusion rates; k_D=37, 35, 20, 1 in (a)-(d), respectively.

Natori et al. that contraction waves decay during their propagation and finally disappear when they stop (2). This could be explained as follows. The calcium impulse goes ahead only when calcium released in a local portion is transferred to the neighboring portion with a level higher than the threshold. A unidirectional propagation is assured by the existence of a refractory state just after the excitation. The strength of the calcium impulse in the local portion and the threshold level of the neighboring portion are the main factors governing wave propagation. The reason why the addition of caffeine increases the propagation velocity is that it decreases the threshold level of y, as shown in Fig. 6. The relation among the amount of caffeine, the velocity and the size of the wave is summarized in Fig. 8a. The propagation velocity linearly increases as the amount of caffeine increases. The width of the wave also increases according to the increase in e_f. The height of the wave remains constant irrespective of the amount of caffeine which might, however, depend on the channel model. Only within the region of e_f plotted by the solid line in the figure is the calcium impulse observed.

The wave propagation under various values of k_D is shown in Fig. 7b and 8b. Increase in the diffusion rate k_D decreases the height of the calcium impulse as is seen in Fig. 8 a. Figure 8b summarizes how the propagation velocity, the height and width of calcium impulses depend on the diffusion rate k_D. The decrease in the size of k_D causes increase in either

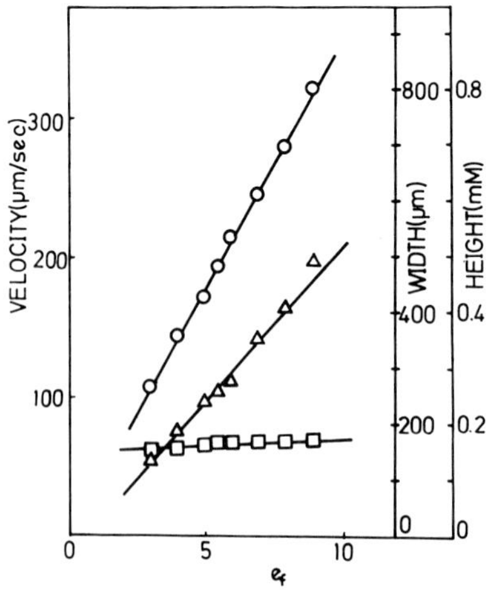

Figure 8a. The dependence of the velocity(O), the width (Δ) and the height(□) of calcium impulse on e_f.

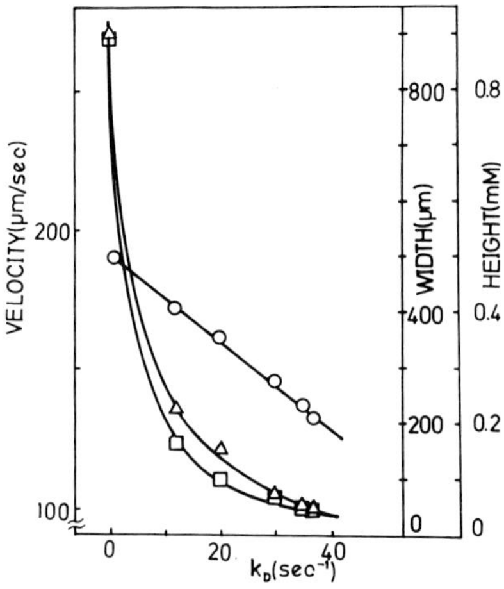

Figure 8b. The dependence of the velocity(O), the width(Δ) and the height(□) of calcium impulse on k_D.

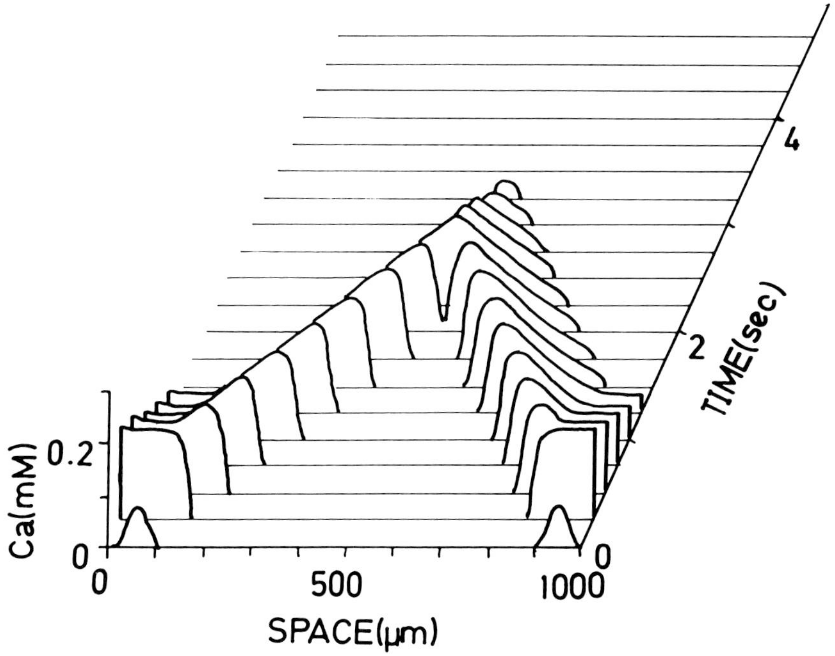

Figure 9. Collision of two calcium impulses with pCa 9 of the bath, k_D=12, e_f=5 and x=30 mM at t=0. (18)

the velocity, wave height or wave width. The latter two show a remarkably good correlation.

Figure 9 shows the collision of two impulses; they disappear when they collide. This phenomenon is also described by Natori et al. (2) in skinned fibers and by Gilkey et al. (9) in medaka eggs. Because SR is brought to a refractory state after excitation, an impulse is not able to propagate on the positions which the other one has passed.

3. Repetitively Generated Calcium Impulses Under appropriate conditions our model shows sustained oscillations with various periods and wave forms. In Fig. 10, a large amount of caffeine is applied to the left terminal portion of the fiber under the condition of the oscillation. In the absence of the diffusion in the longitudinal direction, the left terminal portion is in the steady state where most of the channels are in the open state and, hence, there is little calcium in SR. On the other hand, the portion other than the left terminal is in the state of sustained oscillation when the diffusion in the longitudinal direction is absent. We will tentatively call such a uniform and nonpropagating oscillation bulk osci-

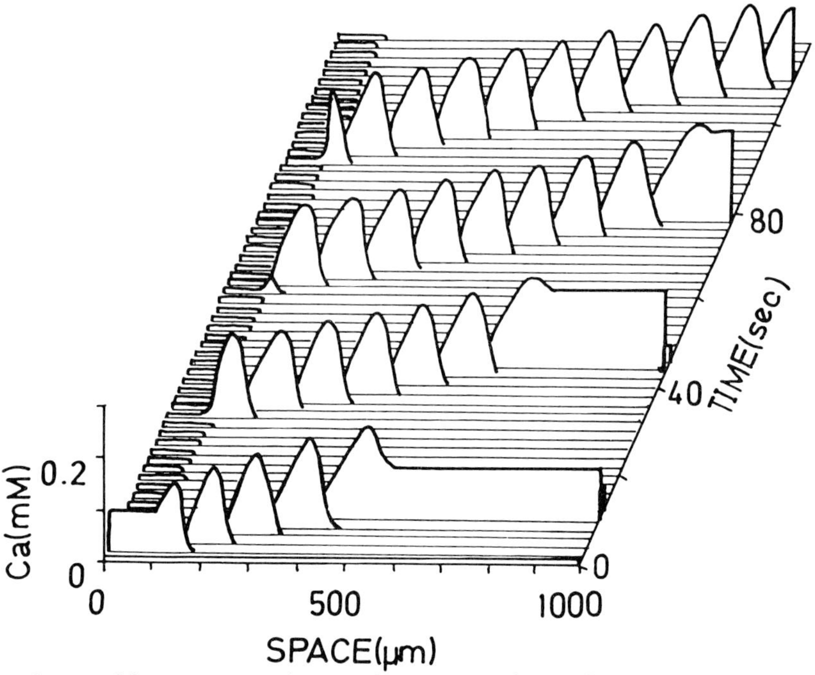

Figure 10. Generation and propagation of calcium impulses; pCa 5 of the bath, k_D=12, e_f=1.5 and 150 in the bulk and terminal portions, respectively. (18)

llation. Results obtained from our one-dimensional model are illustrated in Fig. 10, where we can see two kinds of oscillations with different frequencies; one is caused at the boundary of the left terminal portion and the other at the other portion giving a bulk oscillation. The former oscillation always has a higher frequency than the bulk oscillation and the reason for this can be explained as follows. The marginal portion where the former oscillation emerges contacts the domain of the left terminal portion where y is fixed at a value of the Ca^{2+} concentration of the bathing solution. As a result, it can be considered that the marginal portion has an effectively larger value of k_D. The larger k_D gives higher-frequency oscillation. We will use the term "pace-maker oscillation" for the oscillation repetitively generated at the boundary. As Fig. 10 shows, calcium impulses generated at the pace-maker propagate along the fiber and collide with the spikes of the bulk oscillation. The collapse of such propagating waves concomitant with the bulk oscillation is essentially the same as that in Fig. 9. Imagine that a collision takes place between a calcium impulse and a spike

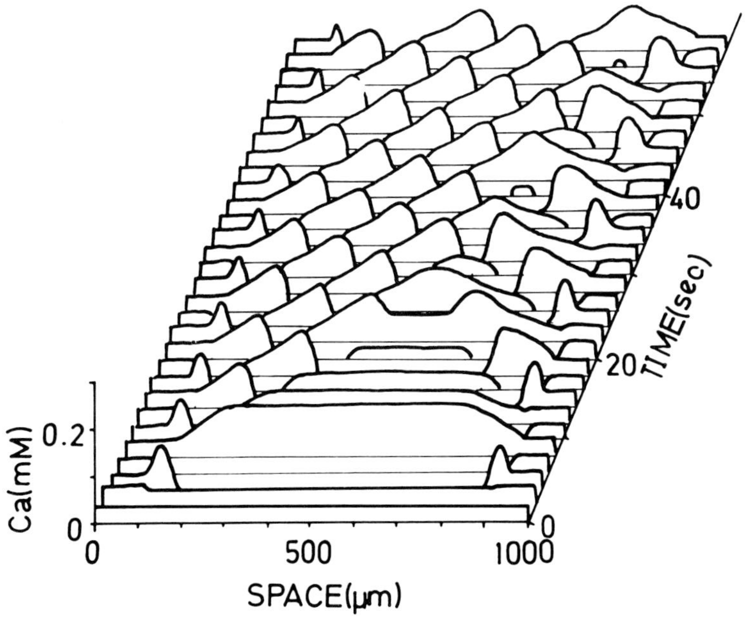

Figure 11. Generation and propagation of two kinds of calcium impulses with pCa 4.52 of the bath and k_D=12. (18)

of the bulk oscillation at place A. Then, the next impulse reaches the same place after one period of the pace-maker oscillation. At that time, the bulk oscillation has not reached the state giving a calcium spike, as the period of the bulk oscillation is longer than that of the pace-maker oscillation. The calcium impulse originated by the pace-maker therefore does not meet the bulk oscillation at place A, and goes ahead till the spike of the bulk oscillation appears. Thus, the calcium impulse with the higher frequency elongates the propagation distance and finally covers the whole space.

In the next figure, Fig. 11, we observe how waves from a pace-maker with a higher frequency at the left win against waves from a similar pace-maker with a lower frequency at the right terminal and govern the entire space. The amounts of caffeine are altered in the right and left terminals as e_f=8 and 500, respectively, while e_f=5 in the middle portion. From the both ends calcium impulses proceed toward the middle part. In the beginning both kinds of impulses elongate the propagation distances through repetitive collisions with the spike of the bulk oscillation. After a while, two kinds of calcium impulses meet each other in the middle part. Then only the calcium impulse of the higher frequency oscillation progresses for a reason similar to the case of the collision with the bulk oscillation spike. In the final stage the calcium impulse with the higher frequency reaches the opposite end dominating the whole space. The dominance of the higher frequency oscillation in propagating waves is commonly seen in the well-known chemical reaction, Belousov-Zhabotinsky reaction (10).

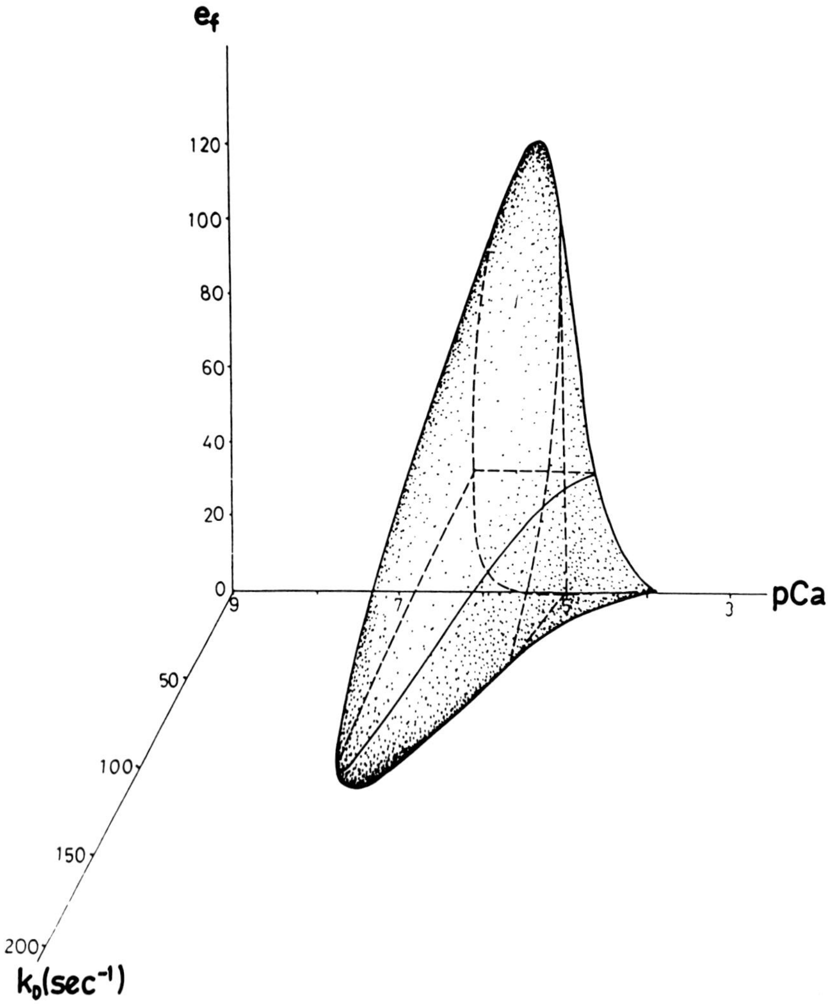

Figure 12. The oscillation region in the parameter space of pCa of the bath, k_D and e_f.

IV. GLOBAL BEHAVIORS

In Fig. 12 we present the results of linear stability analyses of the steady state of the basic model in the parameter space of three kinds of external parameters. The ordinate denotes the effect of caffeine, e_f, the abscissa the pCa of the bathing solution, and the oblique axis the magnitude of the diffusion rate k_D through the cell surface. The inner space surrounded by the dotted surface denotes the conditions where the sustained oscillations appear because the steady state is unstable. Our analysis further shows that damping oscillations are observed in the neighborhood of the region representing the above conditions. The region around pCa 5 where oscillations appear is surrounded along the pCa axis by two kinds of steady state conditions. In one with smaller pCa, Ca^{2+} is almost depleted in SR, while SR is full of Ca^{2+} and ready for excitation in the other. The oscillations can be regarded as a repetitive transition between these two kinds of steady states.

For oscillations to appear, the diffusion rate k_D must be smaller than a certain critical value as is seen in Fig. 12. This is because, in such an oscillation region, small k_D results in a so large amount of Ca^{2+} around SR as to induce the regenerative release of calcium. The condition $k_D=0$, however, gives a special case where the total amount of calcium in the fiber is always conserved. The relation $x+y$ = constant reduces the number of independent variable from two to one. No system with one independent variable yields sustained oscillation.

The addition of a small amount of caffeine enlarges the region where the oscillation appears in the pCa-k_D plane because of the potentiation in the calcium sensitivity of Ca^{2+}-channels. A large amount of caffeine stabilizes the steady state where most of the channels are in the open state and Ca^{2+} in SR is depleted.

Under suitable conditions the steady calcium impulse propagates, though typical examples in computer experiments are not shown here. This occurs in a region of pCa larger than, (the lhs of), that of the oscillation region as well as in the oscillation region. Two conditions are necessary to generate the pace-maker oscillation leading to propagating calcium impulses: the presence of inhomogeneity in the structure of the fiber giving rise to a local portion must satisfy the oscillation condition and the other portion must satisfy the excitation condition. It should be noted, however, that the introduction of the diffusion effect in the longitudinal direction modifies the conditions for the emergence of the oscillation and the excitation.

V. DISCUSSION

Our simple model of Ca^{2+}-induced Ca^{2+} release was found capable of showing various self-organizing behaviors. The essential character of this model is based on the following properties: (i) the autocatalytic property of the Ca^{2+} channel for release from SR, (ii) the active transport of Ca^{2+} into SR by the Ca^{2+}-pump, and (iii) the existence of the slow diffusion of Ca^{2+} between the myoplasm and the extracellular space. It might not be a complete model but it does satisfy the fundamental conditions necessary to exhibit regenerative Ca^{2+} uptake and release. (The presence of these three properties is also assumed in Kuba and Takeshita's model on the oscillation of the membrane potential in sympathetic ganglion cells, which is supposed to be caused by Ca^{2+}-induced Ca^{2+} release (11).) The third property might not be evident but is necessary for giving sustained oscillations. In our model the diffusion rate of Ca^{2+} of the order of 10 sec^{-1} was found to be necessary for a continuous oscillation. This diffusion rate value is far smaller than the corresponding rate of Fick's diffusion in an aqueous solution. The slow diffusion of Ca^{2+} must take place through the cell membrane of intact cells or through the surface area of skinned fibers, at least in most of the experiments cited above. This is not always certain, however, for the experiment by Natori and his colleagues. The frequency of the observed sustained oscillation varies with the change in value of pCa of the bathing solution, which is in agreement with our simulation (data not shown).

Generally speaking, a positive correlation is observed between the velocity and the width in our theoretical results. The propagation velocity in our computer simulation is of the order of 10 - 100 μm/sec, which is the same as experimental values. Though our model is of one-dimension, its simple extension to a two-dimensional model will result in the propagation of the calcium impulse as a concentric wave from an initiating source which was observed by Natori et al. (2) and by Coleman et al. (1).

The source of repetitive propagation of calicum impulse seems to be caused for various reasons. Coleman et al. suggested that the source might be originated from an acetylcholine receptor. Addition of caffeine to a localized portion causes the source in the observation by Natori et al. (2), which is consistently simulated in our model. In the experiment by Rieser et al. repetitive waves were originated from the cell-end (5). When calcium flux is allowed at the boundary of the fiber, the the cell-end has an effectively large k_D in comparison with the other fiber portion, giving a higher frequency oscillation. As was explained before, it is essentially the same as the case with local inhomogeneity of the fiber shown in Fig. 10.

It is shown that Ca^{2+}-induced Ca^{2+} release occurs in eggs after fertilization (9). The local increment of Ca^{2+} concentration in the cytoplasm was observed to propagate at the velocity of 10 μm/sec and width of 10 μm from the animal pole to the antipode. These values are comparable in size with those observed in muscle fibers. The existence of the diffusion of Ca^{2+} between the cytoplasm and the extracellular space was assured in the egg. Our model could be easily extended to that of calcium impulses in fertilization of eggs, where the calcium impulse is known as the trigger of the fusion of cortical vesicles, followed by the formation of the fertilization membrane.

Characteristic behaviors of our SR model are a) excitability for stimulation by a pulse-like increase of calcium in the myoplasmic space, b) calcium impulse propagating in one direction, c) sustained oscillation of calcium concentration, d) periodic emergence and propagation of calcium impulse, and e) collapse of propagating waves when they collide with each other. Essentially similar properties have been found in an electrical impulse in axon (12). As the calcium impulse could be regarded as a signal carrying cytoplasmic information particularly in the development stage after fertilization, a comparison of its nature as an information carrier with that of the nerve impulse is interesting. Table II summarizes the space- and time-scales of these two kinds of information

TABLE II. Comparison between Electrical and Calcium Impulses

	Propagation Velocity	Period of Oscillation
Electrical		
Axon Giant Nerve	10 m/sec (13)	1-10 msec (14,15)
Calcium		
Skeletal Muscle	10-100 μm/sec (1,2)	1 sec - 1 min (1,2,16)
Cardiac Cell	100 μm/sec (3,5)	1-30 sec (4,5)
Medaka Egg	10 μm/sec (9)	—
Sympathetic Ganglion	—	10 sec - 10 min (17)

transmission. Although the Ca^{2+} impulse has essentially the same signal transmission nature as the nerve impulse, their space- and time-scales are quite different. However, the slow velocity of the calcium impulse as well as its large period of repetitive oscillation would be sufficient to transmit cytoplasmic signals to distant portions within the same cell. The calcium impulse would have its advantage in the simplicity of its molecular mechanism in comparison with the electrical excitation in nerve cells. Therefore, it is capable of working commonly in the early stage of development. The significance and existence of signals carried by repetitive calcium impulses are still of interest and under investigation.

REFERENCES

1. Coleman. A. W., Coleman, J. R., Griffin, J. D., Weltman, J. K. and Chapman, K. M., Proc. Nat. Acad. Sci. USA 69:613 (1972).
2. Natori, R., Isojima, C. and Natori, Rb., Jikeikai Med. J. 23:255 (1976).
3. Fabiato, A. and Fabiato, F., Circ. Res. XXXI:293 (1972).
4. Fabiato, A. and Fabiato, F., Eur. J. Cardiol., 1/2:143 (1973).
5. Rieser, G., Sabbadini, R., Paolini, P., Fry, M. and Inesi, G., Am. J. Physiol. 236:70 (1979).
6. Weber A., J. Gen. Physiol. 57:50 (1971).
7. Kushmeric, M. J. and Podolsky, R. J., Science 166:1297 (1969).
8. Endo, M., Proc. Jap. Acad. 51:479 (1975).
9. Gilkey, J., Jaffe, L. F., Ridgway, E. B. and Reynolds, G. T., J. Cell. Biol. 76:448 (1978).
10. Zaikin, A. N. and Zhabotinsky, A. M., Nature 225:537 (1970).
11. Kuba, K. and Takeshita, S., J. Theor. Biol. 93:1009 (1981).
12. Tasaki, I., Biochim. Biophys. Acta. 3:493 (1949).
13. Hodgkin, A. L. and Huxley, A. F., J. Physiol. 117:500 (1952).
14. Guttman, R., Biopyhs. J. 9:269 (1969).
15. Matsumoto, G., in "Nerve Membrane", (G. Matsumoto and M. Kotani, ed.), P. 203, Univ. Tokyo Press, Tokyo, 1981.
16. Endo, M., Tanaka, M and Ogawa, Y., Nature London 228:34 (1970).
17. Kuba, K. and Nishi, S., J. Neurophysiol. 39:493 (1969).
18. Yamaguchi, Y., Komatsu, Y. and Shimizu, H., Biophys. J., submitted.

APPRAISAL OF THE HYPOTHESIS OF THE "DEPOLARIZATION-INDUCED" RELEASE OF CALCIUM FROM THE SARCOPLASMIC RETICULUM IN SKINNED CARDIAC CELLS FROM THE RAT OR PIGEON VENTRICLE[1]

Alexandre Fabiato

Department of Physiology
Medical College of Virginia
Richmond, Virginia

1. This article reports previously unpublished experiments done between 1970 and 1982 in skinned cardiac cells with a variety of methods to test the hypothesis of a "depolarization-induced" release of Ca^{2+} from the sarcoplasmic reticulum (SR).

2. Recordings with microelectrodes impaled in skinned cardiac cells from the rat ventricle did not show any change of potential during Ca^{2+} release induced by Ca^{2+} or caffeine. Anodal electrical stimulation induced either a shrinkage or an artifactual contraction resulting from the high density of the current at the tip of the microelectrode. Electrical polarization using a voltage clamp double sucrose gap did not induce contraction in skinned skeletal muscle fibers.

3. The replacement of potassium propionate (K^+P^-) with potassium chloride (K^+Cl^-) did not induce Ca^{2+} release in single skinned cardiac cells from the rat ventricle. In "chemically-skinned" multicellular preparations of rat ventricular cardiac muscle the replacement of K^+P^- with K^+Cl^- induced a repriming of the low resting potential followed by a slow depolarization and an action potential accompanied by a contraction.

4. For the study of the effect of the replacement of K^+P^- with choline chloride (Ch^+Cl^-), skinned single cardiac cells from the pigeon ventricle were used because this tissue is devoid of transverse tubules. This ionic substitution did

[1] *This study was supported by grant #R01 HL19138, and several previous grants, from the National Heart, Long and Blood Institute, as well as by several grants from the American Heart Association.*

ISBN 0-12-260380-X

not induce Ca^{2+} release in the presence of a low [total EGTA], but did induce a release of strontium when strontium was substituted for Ca^{2+}.

5. Experiments using the "potential-sensitive" dye merocyanine oxazolone (NK2367) were done in skinned single cardiac cells from the pigeon ventricle in which the myosin had been extracted by perfusion with a high ionic strength solution. The replacement of K^+P^- with Ch^+Cl^- induced a light absorption signal in the direction of an increase of net positive charge at the outer surface of the SR membrane even though this ionic substitution did not produce any Ca^{2+} release detectable through the monitoring of aequorin bioluminescence.

6. When a high [total EGTA] was used during the replacement of K^+P^- with Ch^+Cl^- in skinned single cardiac cells from the pigeon ventricle, estimation of the amount of Ca^{2+} remaining in the SR by the application of caffeine in the presence of a low [total EGTA] suggested that the ionic switch could have induced a Ca^{2+} release from the SR. Yet, the observation that replacement of K^+P^- with Ch^+Cl^- produced a long-lasting decrease of the Ca^{2+} sensitivity and maximum tension of the myofilaments sheds some doubt about the actuality of the Ca^{2+} release.

7. In small bundles of about 6 rat ventricular fibers with disrupted but not skinned sarcolemma, the replacement of K^+P^- with Ch^+Cl^- induced a contraction that was attributed to a release of Ca^{2+} bound to the sarcolemma because (a) the loading of the Ca^{2+} store involved in this release required a [free Ca^{2+}] higher than 10^{-5} M and (b) the amount of Ca^{2+} bound was decreased by increasing [total K] or [free Mg^{2+}] as is observed for the Ca^{2+} binding to the isolated sarcolemma. Experiments using neuraminidase gave results that could be consistent with Langer's hypothesis that the sarcolemmal Ca^{2+} was bound to the sialic acid radicals of the glycocalyx, but alternative interpretations of these results are possible.

8. None of the results presented in this article provide any evidence for a "depolarization-induced" release of Ca^{2+} from the cardiac SR, nor do they provide compelling evidence against this hypothesis. Yet, they suggest that the ionic switches can have an effect on the sarcolemma and on the myofilaments which makes demonstration of a Ca^{2+} release from the SR very difficult with this method. Nonetheless the experiments using strontium and those using "potential-sensitive" dyes indicate that the cardiac SR membrane is not totally insensitive to the ionic switches.

INTRODUCTION

As soon as skinned fibers (single fibers with sarcolemma removed) from skeletal muscle were used to study the mechanism of Ca^{2+} release from the sarcoplasmic reticulum (SR), two major hypotheses were proposed which have been termed Ca^{2+}-induced release of Ca^{2+} (1, 2) and "depolarization-induced" release of Ca^{2+} (2-4). All of our reports on the mechanism of Ca^{2+} release from the SR in skinned cardiac cells have been focused on the Ca^{2+}-induced release of Ca^{2+}. Negative results on the "depolarization-induced" release of Ca^{2+} from the cardiac SR have been only briefly mentioned in several articles (5-7). The purposes of the present article are to report these negative results in more detail and to describe more recent experiments thoroughly testing the effect of the replacement of potassium propionate (K^+P^-) with choline chloride (Ch^+Cl^-).

In 1970 and 1971 we reported in two ciné films at the Rennes and Homburg meetings of the Association des Physiologistes de Langue Française and in several abstracts (e.g. ref. 8) that externally-applied anodal polarization induced contraction in cardiac fibers with disrupted but not skinned sarcolemma. Subsequently we found evidence that this result was caused by an artifact, as kindly quoted by Trube (9).

In 1972 (5) we mentioned that the replacement of K^+P^- with K^+Cl^- did not induce a Ca^{2+} release from the cardiac SR. In 1974 Kerrick and Best (10) reported a strong release induced by the replacement of K^+P^- with K^+Cl^- in a preparation that was described as similar to the one we used in 1972 (5). However, we provided evidence (6) that the results of Kerrick and Best (10) were not obtained in a preparation similar to the one we used (5) inasmuch as it was made of about 1,000 cells which were likely to retain most of their sarcolemma. We confirmed (6) our negative results relative to the absence of effect of replacement of K^+P^- with K^+Cl^- in single skinned cardiac cells and suggested that the positive results of Kerrick and Best (10) could be explained by the presence of intact sarcolemma in their preparation inasmuch as action potentials could be recorded in preparations of similar size obtained by microdissection and submitted to "chemical skinning" with ethyleneglycol-bis(β-aminoethylether)-*N*,*N*'-tetraacetic acid (EGTA). We did not exclude, however, the possibility of some effect of this type of ionic substitution on the cardiac SR and reported that the replacement of K^+P^- with K^+Cl^- induced a release of strontium when this ion was substituted for Ca^{2+} (7).

In 1977 we reported preliminary results obtained with a "potential-sensitive" dye and demonstrating an optical signal induced by the replacement of K^+P^- with tris chloride, even though this ionic substitution did not induce a Ca^{2+} release (11).

In 1978 Trube studied in more detail the effect of electrical stimulation on a preparation similar to the one we used in 1972 with no conclusion, however, on the mechanism of the contraction induced by this electrical stimulation (9).

In 1978 Endo and Kitazawa demonstrated that the replacement of K^+ methanesulfonate with K^+Cl^- did not induce a significant release of Ca^{2+} from the SR of a multicellular preparation of cardiac muscle that had been "chemically skinned" with saponin (12). This result was in agreement with our previous negative results with respect to the absence of effect of the replacement of K^+P^- with K^+Cl^- inasmuch as their "anionic switch" is supposed to be more efficient than ours. Endo and Kitazawa (12) also reported that the replacement of K^+ methanesulfonate with Ch^+ methanesulfonate produced a Ca^{2+} release when applied in the presence of a high [total EGTA] with detection of the Ca^{2+} release by the subsequent application of caffeine in the presence of a low [total EGTA].

In 1979 Trube reported that the replacement of K^+Cl^- with K^+ isobutyrate accelerated the frequency of the cyclic contractions attributed to "Ca^{2+}-induced release of Ca^{2+}" in skinned cardiac cells (13). This result could have many explanations, and Trube did not interpret it as a demonstration of a "depolarization" of the SR membrane induced by this anionic substitution.

In 1981 and 1982 I reported the results obtained with three "potential-sensitive" dyes in detail (14, 15). These results demonstrated that the replacement of K^+P^- with K^+Cl^-, or the replacement of K^+P^- with Ch^+P^-, or the replacement of K^+P^- with Ch^+Cl^- induced an optical signal of either fluorescence or differential light absorption even though these ionic substitutions did not induce a Ca^{2+} release from the SR. Yet it was not possible to demonstrate that this signal was related to a change of potential across the SR membrane. Instead it was deemed that this signal was most likely to correspond to a simple change of surface potential related to a change of charge distribution on the SR membrane. This conclusion was prompted by the demonstration by Somlyo et al. (16) that electron-probe microanalysis provided no evidence for a potential across the SR membrane.

Recently Best (17) published an excellent review in which he thoroughly described the different cardiac preparations that are termed "skinned" with conclusions similar in many respects to ours (6, 7, 11). Best agreed that the preparation of Kerrick and Best (10) contained abundant remaining pieces of sarcolemma, and he did not exclude the possibility that this preparation was largely "chemically skinned" by EGTA. However, Best provided evidence that his preparation presented osmotic properties of a "skinned" preparation, at least in the presence of a high [total EGTA]. He also suggested that the mechanical homogenization procedure that he used (a Potter homogenizer) may disrupt tissues in a manner different from that obtained with the blade homogenizer that we have used (5). I agree with his statements and emphasize that I did not use his technique for the experiments (6, 7) attempting to explain the results of Kerrick and Best (10). Section B of the present article should not be interpreted as the only possible explanation of the results of Kerrick and Best (10). This clarification provided by the recent review by Best (17) permitted writing of the present article.

The results reported in this article were obtained with various types of preparations from the rat ventricle. These included skinned cardiac cells obtained by microdissection (18), groups of 2-6 cells with sarcolemma disrupted by homogenization without skinning by microdissection (7) and multicellular preparations of about 50 μm diameter which had been "chemically skinned" by EGTA. In addition, some experiments were done in frog skeletal muscle fibers, either skinned with Natori's technique (19) or with sarcolemma disrupted by homogenization.

Various stability constants were used to compute the free ion concentrations in these early studies. In the earliest experiments no computations at all had been done. Yet, in all experiments all information, including the total calcium contamination of the solution, was available. Accordingly, all the data for the free ion concentrations included in the present article have been either calculated or recalculated with the most recently updated stability constants (20, 21).

The major thrust of the present article is a detailed study of the effects of the replacement of K^+P^- with Ch^+Cl^- which was done in 1981 and 1982. These experiments used the most recently updated stability constants (20, 21) and were done in skinned cardiac cells from the pigeon ventricle. This preparation was selected because the intact cell from the pigeon ventricle is devoid of transverse tubules which eliminated one problem in interpreting the results.

Because the experiments were done over so many years during which the methodology and rationale changed considerably, the article is presented in four parts, each of which includes Rationale, Methods, Results and Discussion.

RESULTS

A. Electrical Stimulation of Rat Ventricular Fibers with Disrupted Sarcolemma

1. Rationale. Since it is not possible to place microelectrodes across the membrane of the SR, the simplest approach for providing evidence for a potential across the SR membrane consisted of applying an electrical polarization by microelectrodes or double sucrose gap. Electrical stimulation through microelectrodes had already been applied to skinned fibers of skeletal muscle (3, 22).

2. Methods. These experiments were done in 1970 and 1971 (e.g., ref. 8). At this time the technique for skinned cardiac cells had not been developed. Thus, cardiac fibers with disrupted sarcolemma were used. These preparations were obtained by a homogenization which disrupted the sarcolemma of a fragment of a single cardiac cell or of groups of 2-6 cells.

The methods for tension recording had not yet been developed. Accordingly, only the shortening of the cell was recorded. This was done with a closed-circuit television camera (Figure 1). A high-resolution television (GPL)

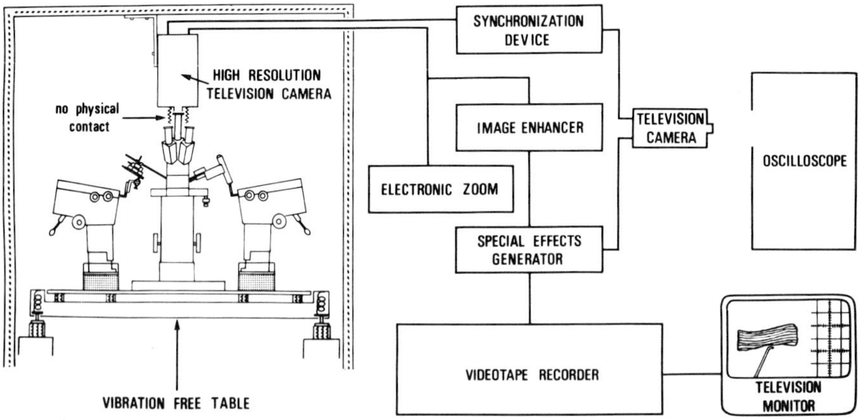

FIGURE 1. *Closed-circuit television system for simultaneous recording of the microscopic image and of the oscilloscope display (setup used in 1970).*

camera recorded the microscopic image. It was attached to the top of the Faraday cage in which the experimental setup was enclosed. There was no physical contact between this camera and the microscope, micromanipulators or vibration-free table to avoid the introduction of any mechanical interference. This camera had a silicon tube which was selected for its high sensitivity and short persistence of the image. In addition, the contrast was increased electronically by an image enhancer (Dynascience 832). The magnification of this image was adjusted with a specially-developed electronic zoom. An ordinary television camera with vidicon tube was used to view the oscilloscope screen on which the intensity of the current was simultaneously displayed. A special effects generator (Shintron, Boston, Mass.) mixed the images of the two television cameras. This composite image was recorded on a television monitor (Panasonic) and stored on a one inch videotape (Panasonic Videomaster). Subsequently the most interesting parts of the videotape image were transferred to ciné film with a special camera synchronized with the television. Figures 2 and 3 show individual frames from this ciné film.

The electrical stimulation was applied either between two microelectrodes (bipolar stimulation) or between a large indifferent electrode in the bath and a microelectrode close to the preparation (unipolar stimulation). The microelectrodes were made of different types of metals, including platinum, tungsten and stainless steel. In all cases the same metal was used for the wire constituting the indifferent electrode. In some experiments electrical stimulation was applied through two glass microelectrodes filled with 3 M KCl, making contact with a chlorided silver electrode. In other experiments the stimulation was applied between a 3 M KCl-filled glass microelectrode and a chlorided indifferent electrode. The results were identical with all types of metal or glass microelectrodes.

The current was applied from a constant current source isolated from the ground. Since metal electrodes or metal wires in glass electrodes were used, a polarization of the metal certainly caused a decrease in the electrical field with time despite the application of a constant current. Two types of electrical stimulation were used: short pulses (10-20 ms) of high intensity (20-50 μA), and long pulses (1 s) with low intensity (1-3 μA).

In some experiments voltage clamp was applied through a double sucrose gap. The chamber was identical to that used by Rougier et al. (23) and was built in the machine shop of the Labatoire de Physiologie Comparée Faculté des Sciences, Orsay, France. These experiments were done only in fibers from the frog semitendinosus that were skinned according to Natori's method (19). This chamber was made up of five compartments separated by about 50 μm wide Vaseline seals. The preparation extended across all the compartments. The two extreme compartments were reference compartments. The central compartment was the test compartment and was 50 μm wide. In the two intermediary compartments about isotonic sucrose was flowing. The reference compartments and the test compartment contained a solution with 0.050 mM total EGTA, pCa 7.50, pMg 3.38, pMgATP 2.45, pH 7.00 buffered with 18 mM imidazole, at 22°C.

3. Failure to Record any Transient Change of Membrane Potential with Microelectrodes during Ca^{2+} Release. Natori has reported "internal action potentials" during recordings with microelectrodes inserted into skinned skeletal muscle fibers (22). Natori suggested that these "internal action potentials" were generated by the SR membrane (22). Inserting a microelectrode into cardiac fibers with disrupted sarcolemma permitted the recording of a negative polarization with respect to the bathing fluid of about -10 mV (5). But, no transient change of this polarization was observed during Ca^{2+} release induced by Ca^{2+} or by caffeine. In all the cases where transient signals were recorded with the microelectrode, it was possible to demonstrate that they were mechanical artifacts caused by the bending of the high-impedance microelectrode tip during contraction.

4. Long, Low-Intensity Unipolar Stimulation. A 1-3 μA unipolar anodal stimulation of 1 s duration produced shortening of the cardiac fiber with disrupted sarcolemma (Figure 2, left column). This shortening became more pronounced as the intensity of the stimulation increased. Eventually the preparation reached a length which was about half of the original length. The effect was more pronounced in the area of the fiber close to the microelectrode. Thus, during anodal stimulation the shortening caused the preparation to bend around the tip of the microelectrode (Figure 2, left column). The shortening of the fiber was not, in fact, caused by a contraction but probably related to the interaction between the negatively charged myofilaments and the electric field created by the anode. The phenomenon was completely reversible provided that the stimulation intensity was not higher than 3 μA and its duration not longer than 1 s. The reversibility was accelerated by the reversal of the electrical polarization which caused a repulsion of the preparation (Figure 2, lowest frame of left column).

A long (1 s) cathodal stimulation of 1-3 μA caused a lengthening and a swelling of the preparation much beyond the original dimensions (data not shown). Simultaneously the preparation moved away from the cathodal microelectrode. It

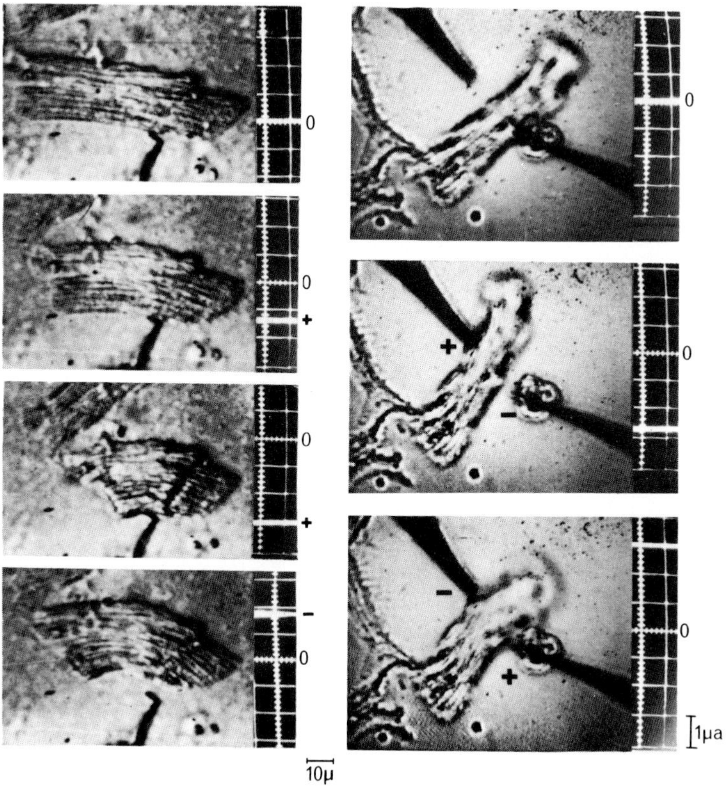

FIGURE 2. *Effects of unipolar (left column) and bipolar (right column) stimulation of cardiac fibers with disrupted sarcolemma from the rat ventricle. The duration of the polarization pulse was 1 s in both experiments. All the electrodes were made of platinum.* *Left column:* *unipolar stimulation applied between a microelectrode which can be seen close to the cell and a large indifferent electrode in the bath. The microelectrode was either the anode (+) or the cathode (-).* *Right column:* *bipolar stimulation applied between two platinum microelectrodes of about 3 μm diameter tip. The signs close to the microelectrode tips indicate which electrode was the anode (+) or the cathode (-). The right sides of the composite television pictures show the oscilloscope recording of the current intensity. The solutions contained 0.050 mM total EGTA, with pCa 7.50, pMg 3.38, pMgATP 2.45, pH 7.00 buffered with 18 mM imidazole, at 22°C.*

was deemed that the electric field created by the cathode caused the negative charges borne by the myofilaments to move away from the cathode. This would explain both the repulsion of the whole preparation and the swelling of the preparation in all directions. The opposite occurred during anodal stimulation (Figure 2, left column).

5. Long, Low-Intensity Bipolar Stimulation. With a 1-3 μA bipolar stimulation of 1 s duration the shortening always started close to the anode. In addition, the anode always attracted the disrupted fiber. Thus, changing the respective polarity of the two microelectrodes resulted in a displacement of the cardiac fiber with disrupted sarcolemma and in a change of its direction of bending (Figure 2, right column). The shortening was slow and progressive for all the time during which the stimulation lasted. Again, this shortening was not caused by a contraction, but was attributed to the interaction between the electric field created by the anode and the negative charges borne by the myofilaments.

6. Brief, High-Intensity Unipolar Stimulation. A 20-40 μA unipolar anodal stimulation of 20 ms duration induced a localized contraction that propagated along the fiber with a velocity of 50-100 μm/s. The contractile wave appeared as a change of optical density of the preparation and was very well visible on the television recording and on the ciné films that were made from the videotape recording. Yet, printing individual frames of this ciné film did not produce pictures of the contractile wave with clarity. For this reason this phenomenon is illustrated in Figure 3 for an experiment done in a fiber from the frog semitendinosus with disrupted sarcolemma. The stimulation was applied at one disrupted end of the preparation and produced a contraction (second frame of Figure 3). After 20 ms the stimulation was interrupted, and the microelectrode was moved away from the preparation. This was done to minimize the likeliness of a stimulation of the preparation by some reaction product at the tip of the microelectrode. The initial contractile wave propagated as in the cardiac preparation but about two times faster: 100-200 μm/s as previously observed by Natori (24). In addition, new waves started repetitively, some time for up to half an hour, always from the end of the preparation from which the wave had been initiated (Figure 3, right panel).

7. Electron Microscopy of the Stimulated Preparations. Electron microscopy was done several years later in cardiac fibers with disrupted sarcolemma and in completely skinned cardiac cells (18) that had been submitted to the same types of electrical stimulations. Electron micrographs of

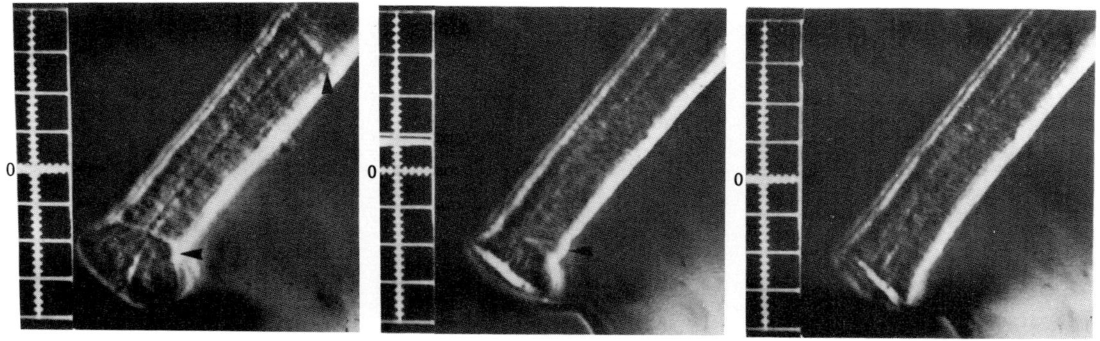

FIGURE 3. *Effect of unipolar electrical stimulation in a skeletal muscle fiber from the frog semitendinosus with disrupted but not completely skinned sarcolemma. The current was applied between a large platinum indifferent electrode in the bath and a platinum microelectrode close to one end of the fiber. As opposed to the convention used for Figure 2, an anodal stimulation corresponded to a downward deflection of the oscilloscope recording. One large graduation of the oscilloscope corresponded to 25 μA. The width of the fibers was about 100 μm (no record of measurement of this width in this particular experiment). The end of the preparation was completely disrupted, while large pieces of sarcolemma remained present laterally. In the second panel, a current of about 20 μA was applied for 20 ms. Then the microelectrode was moved away from the end of the fiber. This current pulse resulted in the induction of a contractile wave which is visible at the upper end of the preparation. This wave propagated at a speed of 175 μm/s as measured on slow playback of the videotape recording. The right panel shows this wave as it has propagated toward the lower end of the picture while a new wave is starting at the upper end. The upper and lower arrows point respectively to the new and previous contractile waves.*

preparations submitted to long, low-intensity stimulation showed a major disorganization of the myofilament lattice and a swelling of the SR all throughout the preparation.

Electron micrographs of preparations submitted to a brief, high-intensity stimulation showed much more dramatic damage which was localized to the 4-8 μm close to the tip of the microelectrode. The myofilaments were coagulated which rendered the striations undiscernable. The SR was considerably swollen in some areas while it appeared like a coagulated magma hardly discernable from the myofilaments in other areas. Finally, the mitochondria were disrupted.

8. Electrical Polarization with Double Sucrose Gap. Electrical polarization with double sucrose gap was possible only with skeletal muscle skinned fibers prepared according to Natori's method (19) because they were long enough to go across the five compartments of the double sucrose gap. The application of current at any intensity and in either direction of polarization did not result in the induction of any detectable contraction. During this electrical polarization the fiber was observed under a stereomicroscope and did not show any sarcomeric movement.

9. Discussion. The shortening induced by the long, low-intensity anodal stimulation was not caused by a contraction resulting from a rapid release of Ca^{2+} from the SR since it was slow and progressed for as long as the current pulse continued. In addition, the lengthening and swelling observed during cathodal stimulation could not be accounted for by a Ca^{2+} reaccumulation in the SR. Thus, the shortening and shrinkage induced by anodal polarization and the lengthening and swelling induced by cathodal polarization were attributed to the interaction between the electric field around the tip of the microelectrode and the negative charges on the myofilaments. The existence of these negative charges is supported by the observation of an attraction of the preparation by an anode and its repulsion by a cathode as clearly demonstrated during bipolar stimulation. Negative charges on the myofilament lattice are also suggested by the recording of an about -10 mV potential with microelectrodes in cardiac fibers with disrupted sarcolemma (5). More careful studies of these negative charges in the myofilament lattice have been done in skinned fibers from skeletal muscle (25).

The electron microscopic findings suggest that the repetitive propagated contraction induced by brief, high-intensity stimulation was caused by a permanent damage resulting in a localized increase of [free Ca^{2+}]. This in turn caused a "Ca^{2+}-induced release of Ca^{2+} from the SR" (26), or rather, cycles of Ca^{2+} release and reaccumulation caused by a mere overload of the SR (27). The mechanism of the propagation of the contractile wave has been elegantly modeled by Prof. H. Shimizu (article in this book). One possibility is that the Ca^{2+} released from one area of the SR would induce an overload of the adjacent areas causing them to release Ca^{2+} which would in turn overload other areas (5). Alternatively, the propagation of the contractile wave could be related to the propagation of a slow action potential along the interconnected network of transverse tubules in skinned fibers from skeletal muscle. In cardiac fibers with disrupted, but not skinned sarcolemma, the remaining sarcolemma may play a role in this propagation since it maintains a resting potential that can be reprimed (see next section). The propagated wave is never observed in a skinned single cardiac cell (18, 6). This is explained by a homogeneous activation of this small preparation by externally applied solutions that diffuse rapidly and homogeneously into all areas of the preparation. Yet, a localized application of Ca^{2+} by microiontophoresis at one end of a skinned single cardiac cell from the dog Purkinje tissue (a tissue devoid of transverse tubules) caused repeated localized contractions by Ca^{2+} release from the SR but no propagated contractile wave (A. Fabiato, unpublished experiment done after a recent discussion with Mark B. Cannel). This observation supports the hypothesis of a role of the interconnected transverse tubular system in the propagation phenomenon observed in skinned fibers from skeletal muscle.

Trube studied in detail the effects of electrical stimulation of cardiac fibers with disrupted sarcolemma using currents of variable intensity (9). Although Trube did not exclude the possibility of a depolarization of the SR membrane, he did not claim that his results supported this possibility as anything more than a mere suggestion (9).

B. Effect of the Replacement of K^+P^- with K^+Cl^- in Skinned Single Cardiac Cells and "Chemically-Skinned" Multicellular Preparations from the Rat Ventricle

1. Rationale. These experiments were done in 1974 after Kerrick and Best (10) provided evidence that the replacement of K^+P^- with K^+Cl^- induced a transient contraction in their preparation, which they attributed to a Ca^{2+} release from the SR. The rationale used by Kerrick and Best was the same as that used for skinned fibers of skeletal muscle (2-4). The preparation was soaked for a long time in a solution containing K^+P^- as the major ionic species. This presumably permitted the diffusion of P^- inside the SR despite the low permeability of the SR membrane to this anion. During its exposure to K^+P^- the preparation was also exposed to a very high

[free Ca^{2+}] of about pCa 4.50 (actual value would be about 0.2 pCa units lower using the stability constants employed in the present article) in the presence of 7 mM total EGTA. Then the preparation was washed with a solution at pCa 6.40 in the presence of 0.050 mM total EGTA for about 1 min. Replacing P^- with Cl^- in the same solution at pCa 6.40 with 0.050 mM total EGTA caused the induction of a large tension transient lasting about 30 s (10). The experiments were done in the presence of a low [free Mg^{2+}] of about pMg 4.00 and of a variable but always very high concentration of imidazole of about 100 mM. Imidazole was used not only as a pH buffer but to adjust the ionic strength while keeping the [total K] constant (28, 29).

2. Methods. To explain the discrepancy between these results and our previous negative results (5), we redid this experiment under similar conditions in skinned single cardiac cells and in "chemically-skinned" multicellular preparations, obtained by dissection under a stereomicroscope, which were of about the same dimensions as those used by Kerrick and Best (10).

Most of the results have been reported in detail, but with no figures, in a previous article (6).

3. Experiments in Skinned Single Cardiac Cells. The experiments (Figure 4) were done in the presence of a low [free Mg^{2+}] (pMg 3.42). The skinned cardiac cell was first exposed to a relaxing solution in the presence of K^+P^-, and then loaded in the presence of a pCa 5.25. Thereafter it was washed with a solution at pCa 7.11 in the presence of a low [total EGTA] with K^+P^- still as the major ionic species. Spontaneous cyclic Ca^{2+} releases and reaccumulations by the SR were observed in the form of cyclic contractions. This contrasts with the absence of cyclic contractions in the records of Kerrick and Best (10) despite the fact that they used a higher [free Ca^{2+}] for washing. The frequency of the cyclic contractions decreased progressively. When the solution was replaced by a solution at the same pCa and the same [total EGTA] but with K^+Cl^- as major ionic species, the frequency of the cyclic contractions was not significantly modified (Figure 4). Their amplitude was often somewhat increased although it was not possible to decide whether this slight difference was significant. This may be related to the consistent but statistically insignificant leftward shift of the force-pCa curve at low [free Ca^{2+}] caused by the replacement of P^- with Cl^- (see Figure 12).

Similar experiments were done with different values of "loading [free Ca^{2+}]" of up to pCa about 3.50 or with varying the [free Mg^{2+}] between pMg about 2.50 and pMg about 4.00.

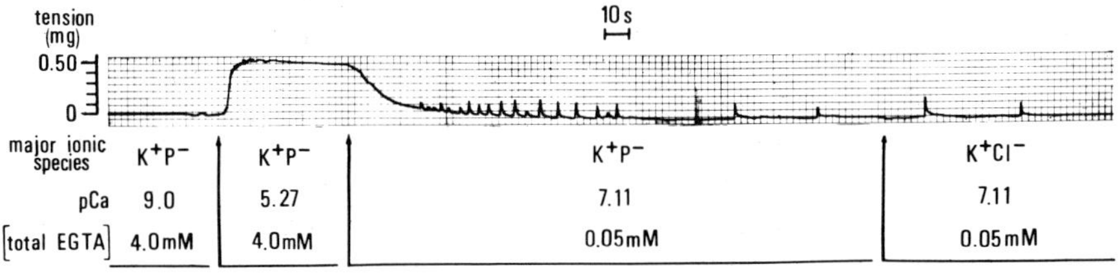

FIGURE 4. *Experiment done in a 12 μm wide, 48 μm long*[2] *skinned single cardiac cell from the rat ventricle under conditions resembling those used by Kerrick and Best (10) to demonstrate the "depolarization-induced" release of* Ca^{2+} *in their multicellular preparations. The pMg was 3.42. pMgATP 2.51, pH 7.00 buffered with 18 mM tris(hydroxymethyl)-aminomethane) (tris). Arrows indicate the time of solution changes.* K^+P^- *stands for potassium propionate.*

In all cases qualitatively similar results were obtained. Thus, these experiments provided no evidence for a release of Ca^{2+} induced by the replacement of P^- with Cl^-. Other experiments demonstrated that the [free Ca^{2+}] used for loading the SR was much too high for inducing such a loading and instead induced Ca^{2+} release from the SR (6).

4. Experiments in "Chemically-Skinned" Multicellular Preparations. These experiments were done with dissected fine trabeculae from the rat ventricle which were about 50 μm in width, i.e. about the diameter of the preparation used by Kerrick and Best (10). The preparation was "chemically-skinned" by an exposure for 1-2 hours to a solution containing 4 mM EGTA, pMg around 4.00, and pMgATP of about 2.50. It was then loaded in the presence of a pCa of about 4.75 with 4 mM EGTA and K^+P^- as the major ionic species. Thereafter the preparation was washed with a solution containing 0.050 mM total EGTA with K^+P^- still as the major ionic species and a pCa of about 6.80. The replacement of P^- with Cl^- induced a large and long phasic contraction on which small amplitude cyclic tension oscillations were superimposed.

Standard microelectrodes, with 1 μm tip diameter and

[2] *In all figures where a length is given for the preparation, the length was that used for tension recording. Recording was done at a sarcomere length giving optimum tension (i.e. about 2.3 μm), although sarcomere length has not been measured in all the experiments.*

filled with 3 mM KCl, were inserted into these preparations. A potential of about -20 mV was recorded when K^+P^- were the major ionic species in the solution containing 0.050 mM total EGTA. The replacement of P^- with Cl^- resulted in an increase of this potential to -40 mV. This repriming was followed by the development of a slow depolarization and a small action potential (Figure 5). The small action potential was associated with a contraction of large amplitude and of 10-30 s duration on which cyclic small and rapid (about 0.5 Hz)

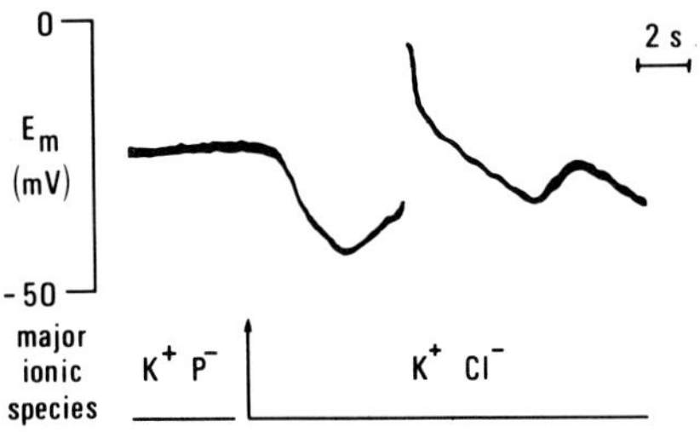

FIGURE 5. *Effect of replacing K^+P^- with K^+Cl^- on the potential recorded by inserting a microelectrode into an about 50 μm wide, and 1 mm long trabecula from the rat ventricle obtained by dissection under a stereomicroscope. The preparation had been "chemically-skinned" in the presence of 4.00 mM total EGTA and a pCa of 9.00 during more than one hour. Then the preparation had been preloaded in the presence of a pCa of 4.74 with 4.00 mM total EGTA during 2 min. This was followed by a washing with a solution containing 0.050 mM total EGTA and a pCa of 6.78. The recording starts at this experimental step. At the time indicated by the arrow, K^+P^- was replaced with K^+Cl^- with no change in the [total EGTA] or the pCa. In all solutions pMg was 4.08, pMgATP 2.51 and pH 7.00 buffered with 18 mM tris. The [total K] was 104 mM at pCa 6.78 and the temperature 22°C.*

tension oscillations were superimposed (data not shown, but see Figure 13). The large tension transient was deemed to be triggered by the action potential while the small and rapid tension oscillations were attributed to cycles of Ca^{2+} accumulation in the SR followed by Ca^{2+} release.

Kerrick and Best (10) did not observe any of the small tension oscillations that we attributed to the SR. It could have been assumed that the SR-generated tension oscillations were so asynchronous in their multicellular preparations that they had become undetectable. Yet, we did observe them in our recordings done in 1974 in 50 μm wide "chemically-skinned" multicellular preparations (not shown). Thereafter other investigators reported well visible SR-generated tension oscillations in "chemically-skinned" cardiac preparations (12, 30) of even larger dimensions than those used by Kerrick and Best (10).

Kerrick and Best (10) used, however, high concentrations of imidazole (28, 29). It has been demonstrated that imidazole, even at much lower concentrations, partially inhibits the Ca^{2+} accumulation in and Ca^{2+} release from the SR (18, 31). Accordingly, experiments were redone in about 50 μm wide "chemically-skinned" multicellular preparations from the rat ventricle but in the presence of 100 mM imidazole. Under this condition it was possible to obtain a large and long transient of tension upon replacement of K^+P^- with K^+Cl^- without superimposition of the cyclic tension oscillations attributed to the SR (Figure 6). It was assumed that this large

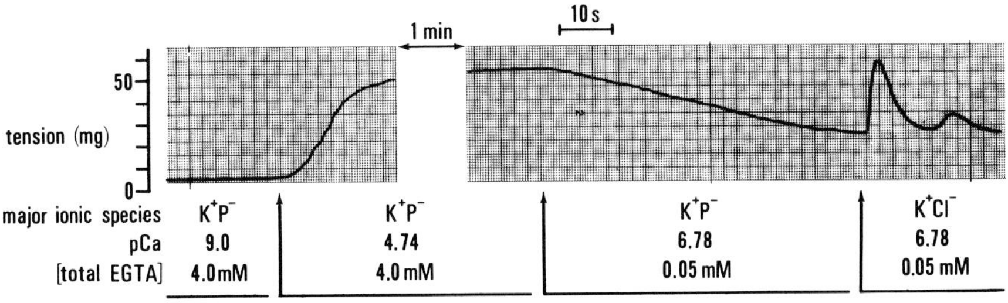

FIGURE 6. *Effect of replacing K^+P^- with K^+Cl^- in an about 50 μm wide and 1 mm long multicellular trabecula from the rat ventricle obtained by dissection under a stereomicroscope. The preparation had been initially "chemically-skinned" in the presence of 4.00 mM total EGTA and a pCa of 9.00 during more than one hour. All the solutions contained 70 mM KCl and 100 mM imidazole at pH 7.00, pMg 4.08, pMgATP 2.51 and 22°C.*

tension transient was caused by the previously described action potential without any participation of the SR.

5. Discussion. These results provide no evidence for a Ca^{2+} release from the SR of skinned single cardiac cells induced by the replacement of K^+P^- with K^+Cl^-. However, the replacement of K^+P^- with K^+Cl^- induced a release of strontium (Sr^{2+}) from the SR of skinned single cardiac cells from the rat ventricle when Sr^{2+} was substituted for Ca^{2+} (7). Similar

results will be presented for the effect of the replacement of K^+P^- with Ch^+Cl^- in the presence of Sr^{2+} in skinned single cells from the pigeon ventricle (see Figure 8). Endo and Kitazawa confirmed the failure of the anion substitution to induce a Ca^{2+} release from the SR of multicellular preparations of cardiac muscle that had been "chemically-skinned" by saponin (12).

Our interpretation (6) of the long Ca^{2+} transient of large amplitude induced by the replacement of K^+P^- with K^+Cl^- in "chemically-skinned" multicellular preparations was the following. These multicellular preparations were indeed rendered highly sensitive to extracellular Ca^{2+} after the 1-2 hour exposure to the solution containing 4 mM total EGTA at pCa 9.00. This high sensitivity to Ca^{2+} did not correspond, however, to a physical removal of any portion of the sarcolemma (30, 32, 33). The loading with a low [free Ca^{2+}] and the subsequent washing with a low [total EGTA] of 0.050 mM reversed the "chemical skinning." This possibility had been clearly pointed out by Winegrad (34) in the initial description of this method of "chemical skinning." In addition, the fact that the exposure to high [free Ca^{2+}] and low [total EGTA] reverses the "chemical skinning" is demonstrated by the successful use of a similar method for loading multicellular cardiac preparations with aequorin (35, 36). These preparations appear to have an intact surface membrane after this procedure (35, 36).

In the present experiments, the cells remained partially depolarized after the reversal of the "chemical skinning." The replacement of K^+P^- with Ch^+Cl^- resulted in a net anionic influx into the cells. This caused the repriming of the membrane potential followed by the development of a slow depolarization and an action potential. In the view of the low resting potential, this action potential probably corresponded to a "slow action potential" in which the initial fast Na^+ inward current was absent but the slow inward current, predominantly carried by Ca^{2+}, was present. When reviewing the present manuscript, Clive Marc Baumgarten pointed out to me somewhat similar findings reported by Carmeliet in his thesis (37). In depolarized intact multicellular preparations from the cardiac Purkinje tissue the replacement of acetylglycinate by Cl^- produced a hyperpolarization followed by a slow depolarization and an action potential (fig. 20, p. 72, in ref. 37).

In the present study the action potential permitted by the repriming of the membrane potential induced the long tension transient of large amplitude. The presence of a very high concentration of imidazole probably prevented (31, 18) any participation of the Ca^{2+} accumulation and release by the SR in this phenomenon. Accordingly, recordings similar to those of Kerrick and Best (10) were obtained.

As emphasized by Best (17), our multicellular preparations were not obtained by the homogenization procedure used by Kerrick and Best (10). Thus, I agree with Philip Best (17) that the application of the preceding interpretation to their results (10) is unwarranted. Perhaps no reversal of the "skinning" (whatever the mechanism of this "skinning" may be) occurred in their preparation.

The most difficult feature of the positive results reported by Kerrick and Best (10) is that the Ca^{2+} loading was applied at a very high level of [free Ca^{2+}] and under low [free Mg^{2+}]. These conditions are appropriate for "loading" the binding sites of the sarcolemma (27) and just the opposite to the conditions necessary to load the SR: a pCa around 4.50 should empty the SR by Ca^{2+}-induced release of Ca^{2+} in any cardiac preparation (5-7, 12, 38). Lowering the [free Mg^{2+}] should decrease the Ca^{2+} accumulation in the SR (39, 20). Then one must hypothesize that the SR in the preparation of Kerrick and Best (10) presents properties very different from those observed in either single skinned cardiac cells (6, 7), or cardiac fibers with disrupted sarcolemma (5, 13), or in "chemically-skinned" multicellular preparations prepared with saponin treatment (12) or with EGTA (6). This suggests that the interpretation of the results obtained with one type of "skinned" preparation may not be applicable to the data from another preparation. This is the major message carried by the recent excellent review by Philip Best (17).

C. Effects of the Replacement of K^+P^- with Ch^+Cl^- in Skinned Single Cardiac Cells from the Pigeon Ventricle

1. Rationale. The rationale for these experiments consisted of attempting to produce a change of polarization of the SR membrane by changing the concentration of anions and cations across it. The skinned single cardiac cell was bathed for 15-30 min in a solution containing K^+P^- as major ionic species. Despite the low permeability of the SR membrane to P^-, it was assumed that this time was long enough to permit the penetration of the anion into the SR. Then the solution was switched to a solution containing Ch^+Cl^- as major ionic species. At the time of this ionic substitution, Cl^- diffused more rapidly into the SR than P^- diffused out. This should have resulted in an increase of negative charge inside the SR membrane. Simultaneously, Ch^+ diffused more slowly into the SR than K^+ diffused out. This should have resulted in an increase of net positive charge outside the SR membrane. Thus, both anionic and cationic substitutions should have resulted in an increase of net positive charge outside the SR

membrane or an increase of net negative charge inside the SR membrane. This should have caused a depolarization of the SR membrane if this membrane was polarized negatively outside and positively inside.

If it had been applied alone, the replacement of P^- with Cl^- would have induced a swelling of the SR because the net influx of anion into the lumen of the SR would have been accompanied by an influx of water in response to the osmotic pressure across the SR membrane (40). To minimize this effect the $[K^+]\cdot[Cl^-]$ product has been maintained constant according to the rationale used for the intact muscle cells (41). There is, however, no evidence that this should prevent the swelling of the SR since the SR membrane has ionic permeabilities that are very different from those of the sarcolemma (40). On the other hand, if it had been applied alone, the replacement of K^+ with Ch^+ should have produced a shrinkage of the SR. Therefore, the combination of the anionic and cationic substitutions together with the keeping of a constant $[K^+]\cdot[Cl^-]$ product theoretically minimized the chances that the ionic substitution produced major osmotic changes. Experiments of each of the types to be described in this section have been redone with the addition of 100 mM sucrose to the solution which should have produced a shrinkage of the SR (42, 43). The results did not differ qualitatively from those obtained in the absence of sucrose.

2. Methods. The experiments reported in this section were done with skinned cardiac cells from the pigeon ventricle because this tissue does not contain transverse tubules. Accordingly, no possible positive results could be attributed to a "depolarization" of the transverse tubules. This eliminated one of the problems encountered in similar experiments done in skinned skeletal muscle fibers (3, 4, 44-46). Single cardiac cells were obtained by homogenization of the tissue into fragments of a length about two-thirds of the original length of the intact cardiac cell. The sarcolemma remained present laterally. It was removed by microdissection together with superficial myofibrils and mitochondria (6, 18, 38). Tension was recorded with a photodiode tension transducer (6, 18, 38).

For the experiments using Sr^{2+} instead of Ca^{2+}, the following absolute stability constants were used in the computer program (20, 21): for the Sr-EGTA complexes $\log K_1 = 8.50$, $\log K_2 = 4.37$; for the Sr-ATP complexes $\log K_1 = 3.60$, $\log K_2 = 2.05$; for the Sr-creatine phosphate complex $\log K_1 = 1.08$. It must be stressed that the apparent stability constant at pH 7.10 for the Sr-EGTA complex is $1.425 \times 10^4\ M^{-1}$, which is much lower than that for the Ca-EGTA complex, $3.976 \times 10^6\ M^{-1}$. Consequently, the experiments done in the presence of a low EGTA buffering used 50 μM total EGTA when Ca^{2+} was used but 500 μM total EGTA when Sr^{2+} was used (see Figure 8). This concentration of 500 μM total EGTA still produced a very weak buffering of the [free Sr^{2+}]. The computed concentration of Sr-EGTA complex was still five times less than the concentration of Ca-EGTA complex, despite the use of a ten times lower [total EGTA] with Ca^{2+}.

Experiments using three "potential-sensitive" dyes in skinned cardiac cells from the rat ventricle have been reported previously (14, 15). In the skinned single cardiac cells from the pigeon ventricle experiments were done only with the dye merocyanine oxazolone, NK2367. It was necessary to eliminate contraction which has been shown to cause an artifact much larger than the light absorption signal even with differential recording (14, 15). This was done by treating the skinned single cardiac cell for 15 min with a solution at an ionic strength of 0.510 M. Simultaneous recordings of tension and aequorin bioluminescence demonstrated that this treatment resulted in a disappearance of the contraction elicited by the Ca^{2+}-induced release of Ca^{2+} mechanism while the amplitude of the aequorin bioluminescence transient decreased to about half of its original value. Return to an ionic strength of 0.160 M resulted in a return of the aequorin bioluminescence signal to its original value while the contraction remained completely abolished. The skinned cell was also treated with 10 mM azide, which was maintained in all solutions in order to dissipate any potential across the membrane of the mitochondria (14, 15). Then the myosin-extracted skinned cell was stained with 0.3 mg/ml of the dye NK2367. Differential absorption measurements were done with the time-sharing, rotating-reel spectrophotometer system that was previously described (47, 14, 15). The recording was done between the optimum wavelength of 670 nm and a wavelength of 700 nm. Since the myosin-extracted skinned cell did not contract, additional experiments were done under the same conditions with aequorin to detect any Ca^{2+} release from the SR. The method for detection of aequorin bioluminescence was the same as recently described (20).

The effect of replacing K^+P^- with Ch^+Cl^- in skinned cardiac cells of the pigeon ventricle was also studied in the presence of a high [total EGTA]. The amount of Ca^{2+} release was inferred from the amplitude of the contraction induced by caffeine in the presence of a low [total EGTA]. This method was similar to that used by Endo and co-workers (26, 4, 12) except that (a) a much stronger pH buffer was used, 30 mM *N*,*N*-bis(2-hydroxyethyl)-2-aminoethane sulfonic acid (BES) which has a pK_a of 7.10 at 22°C; and (b) the amplitude of the caffeine-induced contraction was used rather than the area under the curve for the estimation of the amount of Ca^{2+}

remaining in the SR.

Finally, experiments were done to see if the replacement of K^+P^- with Ch^+Cl^- changed the sensitivity of the myofilaments to Ca^{2+}. This was done by obtaining a force-pCa curve in the presence of a high [total EGTA] with either K^+Cl^-, K^+P^- or Ch^+Cl^- as major ionic species.

3. Absence of a Ca^{2+} Release after the Replacement of K^+P^- with Ch^+Cl^- in the Presence of a Low Total EGTA Concentration. A skinned single cardiac cell from the pigeon ventricle was loaded in the presence of a pCa of 7.40, 0.050 mM total EGTA, pMg 2.50, pMgATP 2.50, pH 7.10 buffered with 30

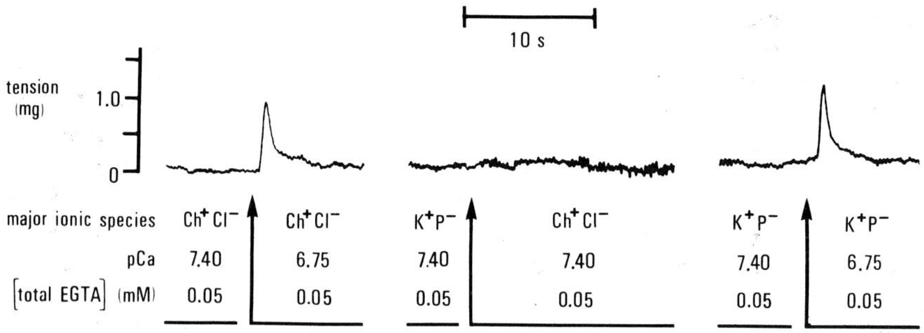

FIGURE 7. *Absence of "depolarization-induced" release of Ca^{2+} by the replacement of K^+P^- with choline chloride (Ch^+Cl^-) and effect of this ionic substitution on the Ca^{2+}-induced release of Ca^{2+}. Arrows indicate the time of solution change. In all cases, the pMg was 2.50, pMgATP 2.50, pH 7.10 buffered with 30 mM N,N-bis(2-hydroxyethyl)-2-aminoethane sulfonic acid (BES), ionic strength 0.160 M and temperature 22°C. The solutions contained 12 mM creatine phosphate and 15 U/ml creatine phosphokinase. The experiment was done in a 9 µm wide, 48 µm long skinned single cardiac cell from the pigeon ventricle. During the ionic substitution the $[K^+]\cdot[Cl^-]$ product was kept constant at 117 mM^2. The middle panel shows that the anionic and cationic substitutions did not induce a tension transient in the presence of a pCa insuring near optimum loading of the SR. Comparison between the first and the last panels shows that the amplitude of the tension transient elicited by Ca^{2+}-induced release of Ca^{2+} was smaller in the presence of Ch^+Cl^- than in the presence of K^+P^-. This may be accounted for by the effect of Ch^+ ion on the tension developed by the myofilaments and their sensitivity to Ca^{2+} (see Figure 12).*

mM BES and 0.160 M ionic strength at 22°C. A pCa of 7.40 in the bulk solution permitted a near optimum loading of the SR with Ca^{2+} (7, 20, 38). At this pCa the skinned cell did not undergo cyclic contractions. Replacement of K^+P^- with Ch^+Cl^- in the presence of pCa 7.40 did not induce any phasic contraction (Figure 7, middle panel). A phasic contraction was induced by an increase of [free Ca^{2+}] from pCa 7.40 to pCa 6.75. The amplitude of this phasic contraction was significantly lower when the major ions were Ch^+Cl^- instead of K^+P^- (Figure 7, right and left panels). This difference in amplitude can be explained by the effect of Ch^+ on the tension developed by the myofilaments and on their sensitivity to Ca^{2+} (see Figure 12).

4. Induction of a Sr^{2+} Release from the SR by the Replacement of K^+P^- with Ch^+Cl^- in the Presence of a Low Total EGTA Concentration. Sr^{2+}-induced release of Sr^{2+} from the SR requires a ten times higher divalent cation concentration than Ca^{2+}-induced release of Ca^{2+} (A. Fabiato, unpublished observations quoted in ref. 48).

Accordingly, it is possible to load the SR with a much higher free divalent cation concentration when Sr^{2+} is used. A skinned cardiac cell from the pigeon ventricle was loaded for more than 2 min in the presence of a pSr 6.75 and 40 mM total EGTA with K^+P^- as major ionic species (Figure 8A). This was followed by a 3 s washing at the same pCa with K^+P^- still as the major ionic species but in the presence of 0.050 mM total EGTA. The replacement of K^+P^- with Ch^+Cl^- without change of pSr and of [total calcium] induced a tension transient.

This tension transient can be attributed to an effect of change of charge distribution on the SR membrane related to the substitution of the major anion and cation. The effect could not be on the transverse tubules since the pigeon ventricular cell is devoid of transverse tubules. The tension transient was certainly caused by a Sr^{2+} release from the SR. Sr^{2+} competes with Ca^{2+} for EGTA and ATP. Thus, the addition of Sr^{2+} resulted in an increase of the [free Ca^{2+}] caused by the interaction between the total calcium contamination of the solution (3 μM) and EGTA and ATP. Yet, computations show that the level of [free Ca^{2+}] remained much below that corresponding to a pCa 8.00 under all conditions. This value of [free Ca^{2+}] is too low to permit any Ca^{2+} accumulation in, or release from the SR.

When the same experiment was repeated using a higher [free Sr^{2+}] for loading the SR (a pSr of 5.25) (Figure 8B), a larger tension transient, corresponding to a larger Sr^{2+} release, was induced by the replacement of K^+P^- with Ch^+Cl^-. This observation is consistent with a higher preloading of

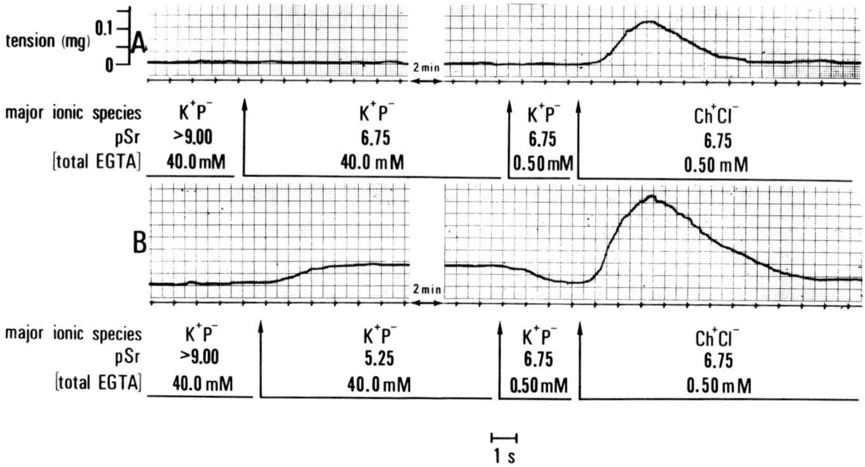

FIGURE 8. *Effect of the replacement of K^+P^- with Ch^+Cl^- in the presence of a low [total EGTA] buffering of the [free Ca^{2+}] and of a Sr^{2+} substitution for Ca^{2+}. The experiment was done in a 9 μm wide, 52 μm long skinned single cell from the pigeon ventricle. Arrows indicate the time of solution change. The $[K^+]\cdot[Cl^-]$ product was maintained constant at 117 mM^2 in the solutions containing 0.050 mM total EGTA. In all cases the pMg was 2.50, pMgATP 2.50, pH 7.10 buffered with 30 mM BES, ionic strength 0.160 M and temperature 22°C. The solutions contained 12 mM creatine phosphate and 15 U/ml creatine phosphokinase.*

the SR by a pSr 5.25 than by a pSr 6.75. It is to be noted that the pSr 5.25 produced some tonic tension because this [free Sr^{2+}] was sufficient for partially activating the myofilaments directly.

5. Effect of the Replacement of K^+P^- with Ch^+Cl^- on the Signal Recorded with a "Potential-Sensitive" Dye and with Aequorin in Myosin-Extracted Skinned Single Cardiac Cells. For the experiments using the "potential-sensitive" dye the skinned single cardiac cells from the pigeon ventricle had their myosin extracted in a 0.510 M ionic strength medium. After this treatment the myosin-extracted skinned single cardiac cell was stained with 0.3 mg/ml of the "potential-sensitive" dye NK2367 during 15 min in a solution containing 1 mM total EGTA, pCa greater than 9.00, pMg 2.50, pMgATP 2.50, pH 7.10 buffered with 10 mM BES, with 0.160 M ionic strength and K^+P^- as the major ionic species, at 22°C. The solution also contained 10 mM azide to dissipate the potential across

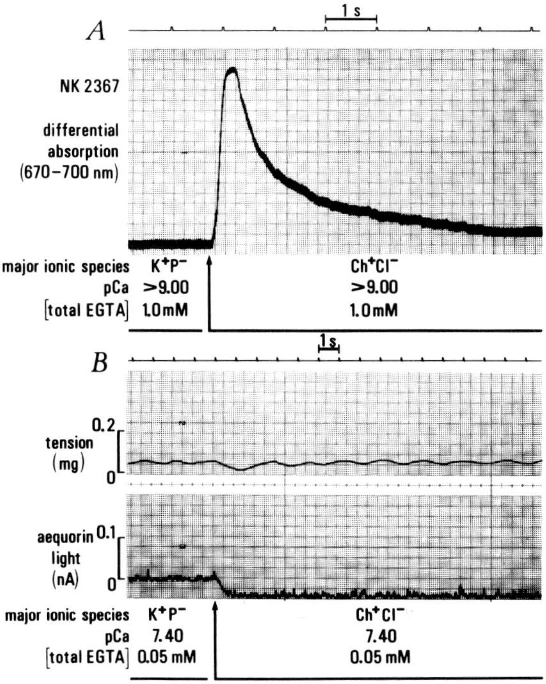

FIGURE 9. *Effect of the replacement of K^+P^- with Ch^+Cl^- on the signals recorded with the "potential-sensitive" dye NK2367 (A) and aequorin (B). The experiments were done in two different skinned single cardiac cells from the pigeon ventricle which were, respectively, 9 μm wide and 46 μm long (A) and 8 μm wide and 38 μm long (B). Both skinned cardiac cells had been treated for 15 min with a solution at high ionic strength (0.510 M) to eliminate the contraction by myosin extraction. The solution used for the experiment itself had in both cases an ionic strength of 0.160 M, pMg 2.50, pMgATP 2.50, pH 7.10 buffered with 30 mM BES. The solution also contained 12 mM creatine phosphate, 15 U/ml creatine phosphokinase and 10 mM azide to dissipate the potential across the membrane of the mitochondria. Arrows indicate the time of solution change. The pronounced downward deflection (direction opposite to that for a Ca^{2+} release) of the aequorin light might correspond to a more depressive effect of 0.160 M ionic strength on the Ca^{2+} binding to aequorin when Ch^+ is the major cation than when K^+ is the major cation. This has not been verified however. See ref. 20 for definition of the zero for the aequorin light recording. The slight downward tension movement caused by the solution change is related to the considerable increase of the compliance of the preparation resulting from the myosin extraction.*

the membrane of the mitochondria. Light absorption was recorded differentially between 670 and 700 nm. The replacement of K^+P^- with Ch^+Cl^- resulted in a signal in the direction of an increase of light absorption (Figure 9 *A*).

Previous experiments in skinned cardiac cells from the rat ventricle had eliminated a number of possible artifacts from this signal permitting the conclusion that it was related to change of charge distribution on the SR membrane (14, 15), although not necessarily to a change of potential across this membrane (49).

In the present experiment (Figure 9 *A*) the SR had been emptied by the perfusion at pCa >9.00 so that the signal could not be attributed to a Ca^{2+} translocation across the SR membrane. Other experiments were done after preloading of the SR in the presence of pCa 7.40, which should have produced a near optimum loading as previously indicated. The amplitude and time course of the signal were not significantly different from those observed in the presence of pCa >9.00. Since Ca^{2+} release from the SR produces a change of optical signal (14, 15), this result already suggests that the replacement of K^+P^- with Ch^+Cl^- did not induce Ca^{2+} release from the SR.

To verify the absence of Ca^{2+} release from the SR, experiments were done in myosin-extracted skinned single cardiac cells from the pigeon ventricle with the detection of aequorin bioluminescence. The SR of the skinned cell was preloaded for 15 min in a solution at the optimum pCa of 7.40 in the presence of 0.050 mM total EGTA, pMg 2.50, pH 7.10 buffered with 10 mM BES, with 0.160 M ionic strength and K^+P^- as the major ionic species, at 22°C. The solutions also contained 10 mM azide. The replacement of K^+P^- with Ch^+Cl^- produced no Ca^{2+} release, as shown on the aequorin light recording (see discussion in the legend of Figure 9 *B*). No tension signal was expected since the myosin had been extracted. Thus, these results confirm that the replacement of K^+P^- with Ch^+Cl^- in the presence of a low [total EGTA] does not induce release of Ca^{2+} from the SR despite the fact that this ionic substitution produced a change of charge distribution on the SR membrane recorded with the "potential-sensitive" dye.

6. Replacement of K^+P^- with Ch^+Cl^- in the Presence of Ca^{2+} and of a High Total EGTA Concentration in Skinned Single Cardiac Cells. The experiments were done according to the method of Endo and co-workers (26, 4, 12). A [total EGTA] of 10 mM was used to load the SR with Ca^{2+} with the optimum pCa of 7.40 as well as to apply the ionic change capable of inducing Ca^{2+} release (i.e., the replacement of K^+P^- with Ch^+Cl^-). This was followed by a 1.5 s washing of the preparation with a relaxing solution containing 10 mM EGTA, a pCa of 9.00 and K^+P^- in all cases. The Ca^{2+} content of the SR

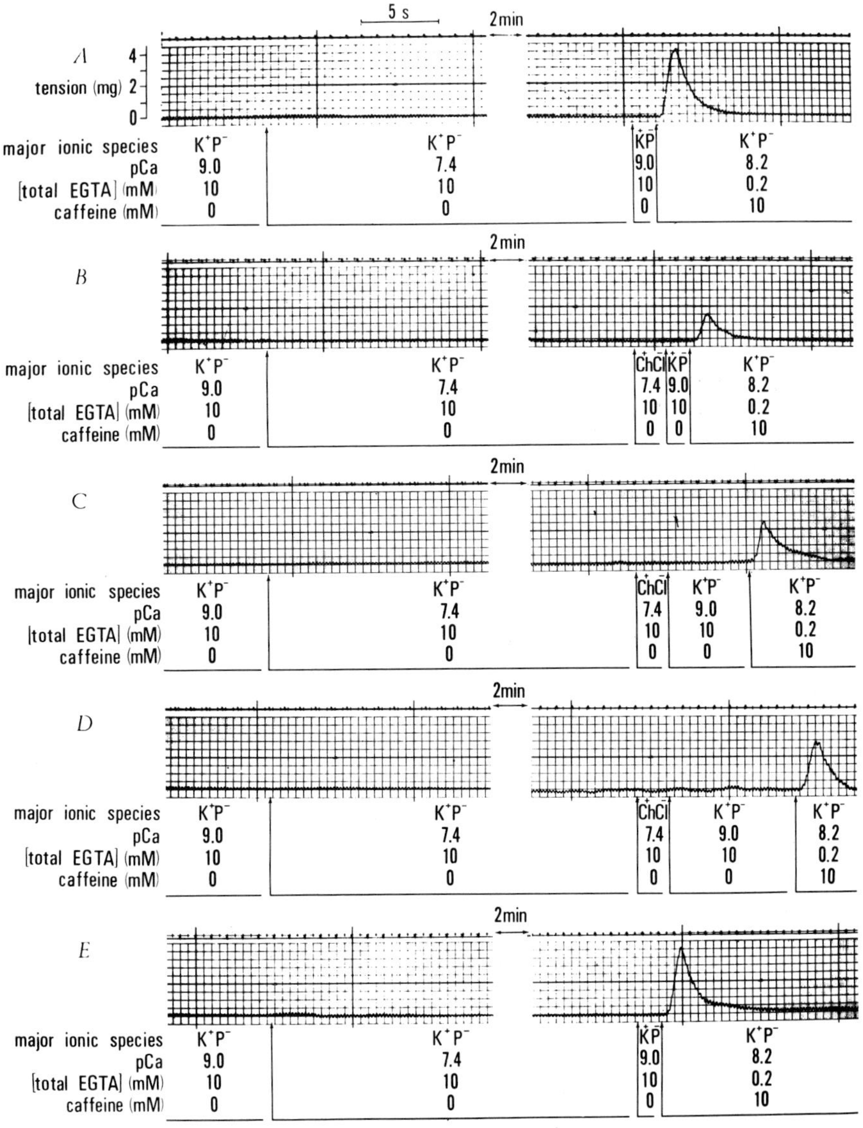

FIGURE 10. *Effect of the replacement of K^+P^- with Ch^+Cl^- in the presence of a high [total EGTA] on the amount of Ca^{2+} remaining in the SR, detected by the subsequent application of caffeine in the presence of a low [total EGTA]. The experiment was done in a 9 μm wide, 53 μm long skinned cardiac cell from the pigeon ventricle. Same pH buffered with BES, pMg, pMgATP, temperature and ATP regenerating system as in Figure 7. Arrows indicate the times of solution changes.*

was estimated by measuring the amplitude of the contraction induced by caffeine in the presence of 0.2 mM total EGTA and K^+P^-.

In the test experiments (Figure 10, *B*, *C*, and *D*) the washing was preceded by a 2 s exposure to a solution containing Ch^+Cl^- as the major ionic species but with the same pCa and [total EGTA] as in the loading solution which contained K^+P^- as major ionic species.

This exposure to Ch^+Cl^- resulted in a clear decrease of the amplitude of the caffeine-induced contraction. From this result it could be inferred that the ionic switch induced a Ca^{2+} release from the SR. However, caution must be applied in deriving this conclusion because it was observed that the magnitude of the decrease of the amplitude of the caffeine contraction decreased when the duration of the washing period with K^+P^-, in the presence of a pCa 9.00, was increased (Figure 10, *B*, *C*, and *D*). Such an observation has never been made when studying Ca^{2+}-induced release of Ca^{2+} with the same method (50, 51). The SR cannot be reloaded by a perfusion at pCa 9.00. Thus, it is possible that the effect of Ch^+Cl^- was, at least in part, a transient effect on the myofilaments rather than on the SR. This is further suggested by the subsequent experiments (Sections 7 and 8).

It must also be emphasized that the control done at the end of the experiment (Figure 10 *E*) showed a complete recovery of the amplitude of the caffeine contraction reaching the value obtained in the initial control (Figure 10 *A*) only in about 40% of the cases. In the other cases the terminal control showed a caffeine-induced contraction of lower amplitude than during the initial control.

7. Effect of the Replacement of K^+P^- with Ch^+Cl^- on the Caffeine-Induced Contraction in Skinned Single Cardiac Cells from the Pigeon Ventricle. A skinned single cardiac cell was loaded with a pCa of 7.40 in the presence of 10 mM EGTA, and after 2.5 min of loading was washed, as previously, in a relaxing solution at pCa 9.00 in the presence of K^+P^- (Figure 11). The Ca^{2+} release was induced by 10 mM caffeine in the presence of 0.2 mM total EGTA. The amplitude of the caffeine-induced contraction was significantly reduced when Ch^+Cl^- was substituted for K^+P^- in the caffeine-containing solution.

The magnitude of the decrease of the amplitude of the caffeine-induced contraction caused by the replacement of K^+P^- with Ch^+Cl^- (Figure 11) was less than that observed in Figure 10, *B*, *C*, and *D*. Thus, it is not claimed that the effect shown in Figure 11 entirely accounts for the observations shown in Figure 10. Accordingly, the induction of a Ca^{2+} release from the SR by the replacement of K^+P^- with

Ch^+Cl^- under this condition of strong EGTA buffering of the [free Ca^{2+}] is not excluded by these experiments.

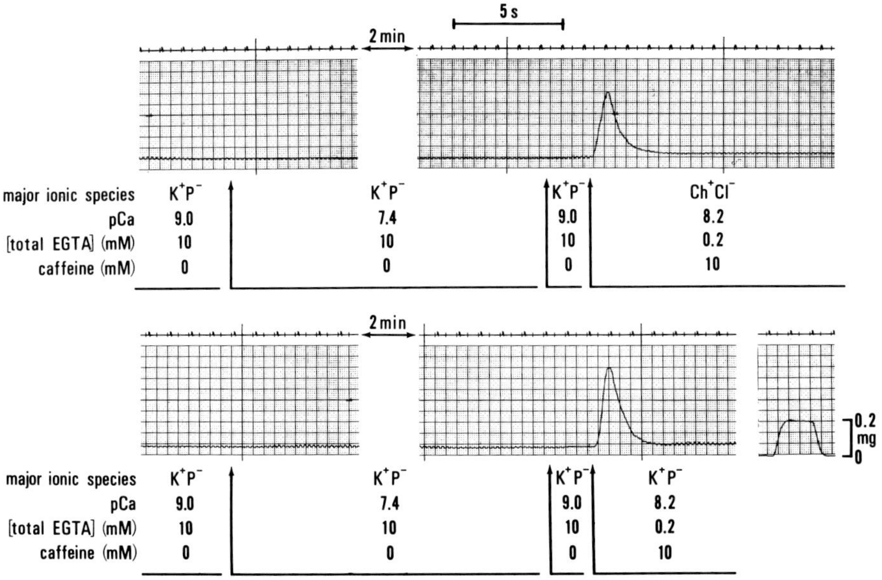

FIGURE 11. *Effect of the replacement of* K^+P^- *with* Ch^+Cl^- *on the contraction induced by 10 mM caffeine. The experiment was done in a 9 µm wide, 45 µm long skinned single cardiac cell from the pigeon ventricle. The total EGTA was 0.2 mM in all steps of the experiment. The pMg was 2.50, pMgATP 2.50, pH 7.10 buffered with 30 mM BES, ionic strength 0.160 M and temperature 22°C. The solutions also contained 12 mM creatine phosphate and 15 U/ml creatine phosphokinase.*

8. Effect of Anionic and Cationic Substitution on the Tension Developed by the Myofilaments. The force-pCa curve was not significantly modified for any point when K^+Cl^- was replaced with K^+P^- (Figure 12). However, there was a consistent shift of the curve to the right for the lower value of [free Ca^{2+}] (pCa>5.50) as previously noted (6). The replacement of K^+Cl^- with Ch^+Cl^- resulted in a significant shift of the force-pCa curve to the right suggesting that Ch^+ ion decreased the sensitivity of the myofilaments to Ca^{2+}. This, perhaps, corresponded to an effect of Ch^+ ion on Troponin C. There was also a significant decrease of the maximum tension developed by the myofilaments. This, perhaps, corresponded

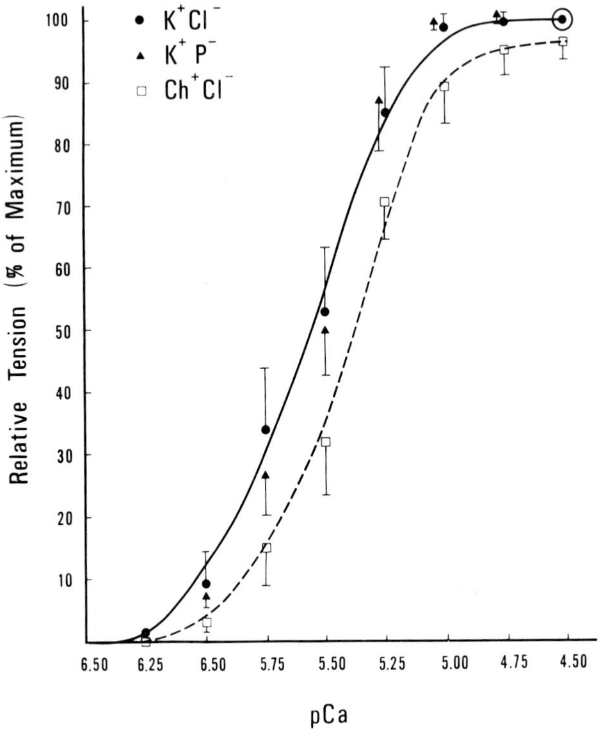

FIGURE 12. *Effect of anionic or cationic substitution on the tension developed by the myofilaments in 7-12 μm wide skinned cardiac cells from the pigeon ventricle. The [total EGTA] was 10 mM, pMg 2.50, pMgATP 2.50, pH 7.10 buffered with 30 mM BES, ionic strength 0.160 M and the temperature 22°C. The solution contained 12 mM creatine phosphate and 15 U/ml creatine phosphokinase. The concentration of K^+Cl^-, K^+P^- or Ch^+Cl^- was the same at each pCa value. The tension was expressed as a percentage of the maximum tension induced in each single skinned cardiac cell by a pCa of 4.50 in the presence of K^+Cl^- and 10 mM EGTA. Each point represents the mean of 5 to 9 determinations, and the vertical bar corresponds to the standard deviation (SD) shown in one direction only for clarity. The solid curve corresponds to the results obtained in the presence of K^+Cl^-. Although the data points obtained in the presence of K^+P^- are consistently lower than those obtained in the presence of K^+Cl^- for pCa values greater than 5.50, the differences were not significant. The dotted curve corresponds to the data obtained in the presence of Ch^+Cl^-. The tension at each pCa is significantly different from that observed in the presence of K^+Cl^-.*

to a direct effect of Ch^+ on actin, myosin or the acto-myosin interaction.

This depressant effect of Ch^+ ion was a long-lasting one. At least 10 s of washing in a solution containing K^+Cl^- was needed for recovery of the tension induced by a given pCa. In addition, in more than 50% of the fibers the recovery was incomplete. This result explains the depression of the caffeine-induced contraction shown in Figure 11. In addition, the long-lasting nature of this depressant effect and the frequent absence of complete recovery from this effect may partly explain the results shown in Figure 10.

9. Discussion. The results shown in Figure 10 are consistent with the data of Endo and Kitazawa (12) who showed that the replacement of K^+ with Ch^+ in the presence of a high [total EGTA] caused a decrease of the amplitude and duration of the caffeine-induced contraction used to measure the amount of Ca^{2+} remaining in the SR (see fig. 14 in ref. 12). This, together with the negative result of Endo and Kitazawa (12) on the effect of the replacement of methanesulfonate by Cl^-, shows that the SR's of the saponin-treated multicellular preparations and of the skinned single cardiac cells respond in the same way, at least to this type of interventions.

Yet, the present results indicate that the long-lasting depressant effect of Ch^+ ion on both the tension developed by the myofilaments and their Ca^{2+} sensitivity may partly explain the decrease of the amplitude and duration of the caffeine-induced contraction observed by Endo and Kitazawa (12). On the other hand, the results obtained with (a) Sr^{2+} substitution for Ca^{2+} and (b) "potential-sensitive" dyes demonstrate that the replacement of K^+ with Ch^+ does indeed affect the SR membrane.

D. Effect of the Replacement of K^+P^- with Ch^+Cl^- on Bundles of Cardiac Cells with Disrupted but Not Skinned Sarcolemma from the Rat Ventricle

1. Rationale. In order to investigate whether the replacement of K^+P^- with Ch^+Cl^- could mobilize Ca^{2+} bound to the sarcolemma, experiments were done in small bundles of about six cells from the rat ventricle with disrupted but not skinned sarcolemma. These preparations underwent the same ionic substitutions as used with the single skinned cell from the pigeon ventricle, but the [free Ca^{2+}] in the solution used for "loading" was higher because the Ca^{2+} binding sites of the sarcolemma have a very low affinity (52-55). In addition, the [free Mg^{2+}] and [total K] were varied since these two cations are known to compete with Ca^{2+} for the

binding sites on isolated sarcolemma (52-55).

In other experiments, done several years earlier and briefly mentioned in another article (51), neuraminidase was used to digest the sialic acid residues that are likely to bind Ca^{2+} in the glycocalyx. These experiments were done following the rationale proposed by Langer et al. (56) and with the kind advice of Prof. Glenn Langer.

2. Methods. The experiments were done on small bundles of about six cells from the rat ventricle in which the sarcolemma had been disrupted by homogenization but not skinned (5). The preparation was exposed for 1 min to a solution of various pCa values with various [free Mg^{2+}] and [total K] levels in the presence of a high [total EGTA] and K^+P^- as the major ionic species. Then K^+P^- was replaced with Ch^+Cl^-.

For the experiments with highly purified neuraminidase (Worthington) the same type of preparation was used but the temperature was 35°C to permit a close to optimum activation of the enzyme, with adjustment of the Ca-EGTA stability constant using unpublished data of Allen and Blinks reported in ref. 21.

3. Induction by the Replacement of K^+P^- with Ch^+Cl^- in Cardiac Fibers with Disrupted Sarcolemma of a Tension Transient Different from that Obtained in Skinned Single Cardiac Cells. The pluricellular preparation with disrupted sarcolemma was exposed during 1 min to pCa varying between 2.50 and 6.50 in the presence of 10 mM total EGTA and 10 mM azide to prevent any Ca^{2+} loading of the mitochondria. Then the preparation was washed in a solution containing 0.050 M total EGTA at pCa 5.00 with K^+P^- as major ionic species. The pMg was 3.50, the pMgATP 2.50, pH 7.10 buffered with 30 mM BES. Replacing this solution with a solution of the same composition but containing Ch^+Cl^- as major ionic species resulted in a large tension transient with superimposed cyclic partial relaxations (Figure 13). The tension transient returned to baseline in approximately 30 s. The cyclic relaxations were attributed to cycles of Ca^{2+} accumulation in the SR followed by Ca^{2+} release from the overloaded SR. These tension oscillations are referred to as cyclic relaxations rather than cyclic contractions because the baseline tension already corresponded to about 50% of the maximum tension that the myofilaments could develop. The large and long tension transient was deemed related to the release of Ca^{2+} bound to the sarcolemma inasmuch as this transient was absent in completely skinned cardiac cells. This was confirmed by the following experiments (Section 4).

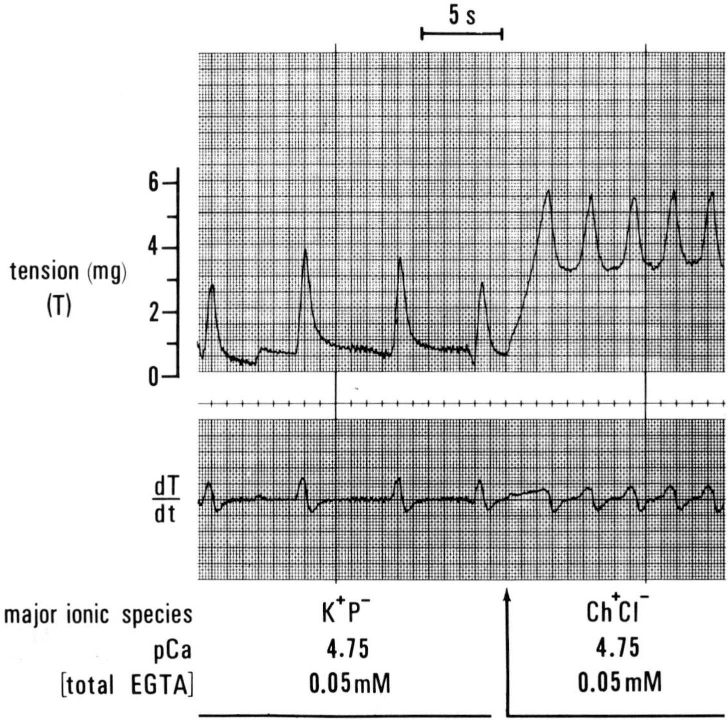

FIGURE 13. *Effect of the replacement of K^+P^- with Ch^+Cl^- in a bundle of about 6 rat ventricular cells with disrupted sarcolemma prepared by homogenization. The preparation was about 25 μm wide and 100 μm long. The preparation was previously exposed for 1 min to a pCa of 5.25 in the presence of 10 mM azide and 10 mM total EGTA with K^+P^- as major ionic species and then washed in a solution containing 0.050 mM total EGTA and pCa of 4.75 with still K^+P^- as major ionic species. At the time indicated by the arrow, the washing solution was replaced by a solution of the same composition but with Ch^+Cl^- as major ionic species. The upward shift of resting tension returned to baseline within 30 s. The pMg was 3.50, pMgATP 2.50, pH 7.10 buffered with 30 mM BES. The solution contained 12 mM creatine phosphate and 15 U/ml creatine phosphokinase.*

4. Demonstration that the Large Tension Transient was Related to the Release of Ca^{2+} Bound to the Sarcolemma by a Study of the Effect of Varying Free Mg^{2+} Concentration and Total K Concentration on the Amplitude of this Transient. Experiments similar to that shown in Figure 13 were repeated with variation of the [total K] and [free Mg^{2+}] in the Ca^{2+} loading solution containing 10 mM EGTA and 10 mM azide which

was applied during 1 min. The relative amplitude of the large and long tension transient decreased when the [total K] or [free Mg^{2+}] were increased (Figure 14). Thus, in the presence of a presumably physiological [free Mg^{2+}] (pMg 2.50) and [total K] (104 mM), the tension transient was 50% of maximum after Ca^{2+} loading with pCa 4.00.

For comparison between different preparations the amplitude of the tension transient, measured from baseline to the peak of the low amplitude tension oscillations (Figure 14 inset), was expressed as a percentage of the submaximum tension released in the same bundle of cells by the application of pCa 5.50 in the presence of 10 mM EGTA, 10 mM azide, pMg 2.50 and pMgATP 2.50. A submaximum tension was used for this normalization to avoid deterioration of the preparation. Unlike skinned single cardiac cells, these pluricellular preparations contract very slowly upon full activation with pCa 4.50. This slow contraction deteriorates the preparation. Reproducible contractions are obtained, however, when a submaximum activation is induced with pCa 5.50. It must be stressed that the pMg and [total K] were varied in the loading solution applied during 1 min. In contrast, pMg and [total K] were constant, at pMg 2.50 and 104 mM total K, in both the washing solution containing 0.050 mM total EGTA and K^+P^- as the major ionic species and the transient-triggering solution which had the same composition but with Ch^+Cl^- as the major ionic species. Since the effects of changing [free Mg^{2+}] or [total K] are not long-lasting, the results shown in Figure 14 cannot be attributed to an effect of these changes in [total K] and pMg on the tension developed by the myofilaments.

These results are in qualitative agreement with those obtained for the Ca^{2+} binding to the isolated sarcolemma (52-55). I am most grateful to Donald Bers for providing me with the following information relative to the Ca^{2+} binding to the isolated sarcolemma. The association constant for Ca^{2+} to the relatively high affinity site of the isolated sarcolemma is 46,000 M^{-1} at low ionic strength and [free Mg^{2+}] (54). Increasing the [total K] to 140 mM decreases this association constant by about 32% (55). Sodium, which was present at about 30 mM in the solution used in the present experiments, depresses the Ca^{2+} binding to the high affinity site of the isolated sarcolemma even more than potassium does (52, 55). Thus, an apparent affinity constant for Ca^{2+} for this high affinity site of about 25,000 M^{-1} should be expected in the presence of 104 mM total K. Finally, [free Mg^{2+}] is known to compete with Ca^{2+} and the monovalent cations for the high affinity binding site of the sarcolemma (52). There are no quantitative data, however, for the effect of Mg^{2+} on the Ca^{2+} binding of this relatively high affinity site in isolated sarcolemmal preparations. Accordingly, the value of

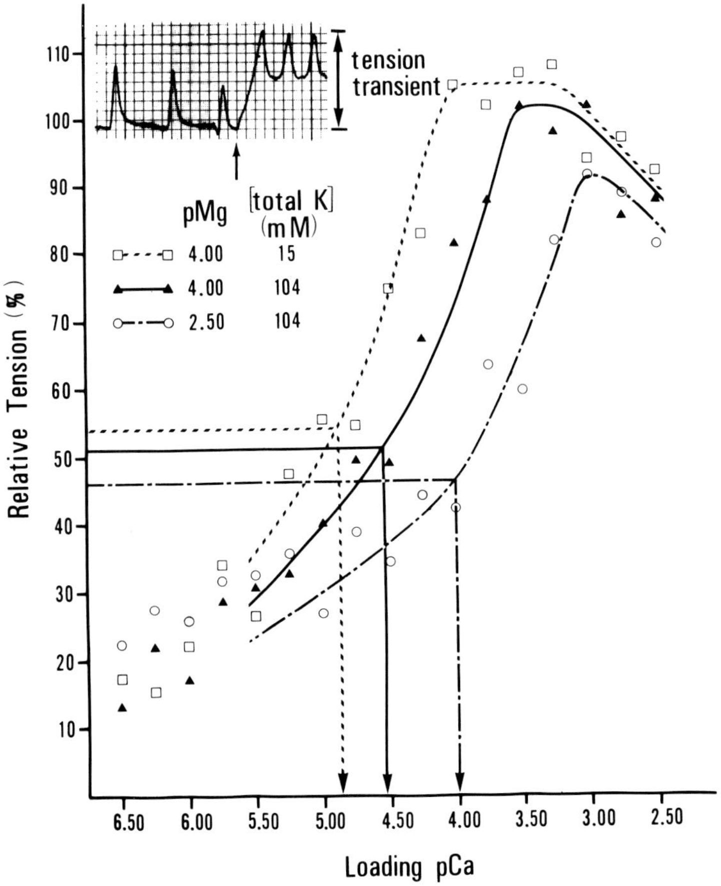

FIGURE 14. *Effect of changing the pMg and the [total K] on the relationship between the pCa used for loading the preparation and the relative tension induced by the replacement of K^+P^- with K^+Cl^- under constant conditions of pMg and [total K] (see text for discussion). The data were pooled from 18 different preparations. Each point corresponds to a single measurement made from a small bundle of about 6 cells with disrupted sarcolemma obtained by homogenization from the rat ventricle according to a protocol similar to that used in Figure 13. In all cases the pMgATP was 2.50, pH 7.10 buffered with 30 mM BES and the temperature 22°C. The solutions contained 12 mM creatine phosphate and 15 U/ml creatine phosphokinase. See text for the normalization procedure used for computing the relative tension.*

10,000 M^{-1}, which is the reciprocal of the value of pCa 4.00 corresponding to 50% of the amplitude of the tension transient shown in Figure 14, was tentatively used as an affinity constant of this relatively high Ca^{2+} affinity site for the computation of the physiological Ca^{2+} buffering of the intact mammalian cardiac cell (27). This seems justified as a first approximation inasmuch as the values derived from these data in cardiac fibers with disrupted sarcolemma in the presence of a low [free Mg^{2+}] with either low or high [total K] are in the same range as the published data for isolated sarcolemma (52-55). There is no steadfast reason against the simplifying assumption that was made (27) that the Ca^{2+} affinity of these sites was the same at the inner and outer surfaces of the sarcolemma. It must be stressed, however, that the relationship between the Ca^{2+} release from sarcolemma and tension is a complex one involving diffusion in the myoplasm where there are other fixed and soluble binding sites, competition with the Ca^{2+} accumulation in the SR and binding to the myofilaments. Thus, the curves shown in Figure 14 cannot be treated through Hill's plots. In addition, the significance of the configuration of the curves is uncertain. The curves shown in Figure 14 have been fitted by eye. No curves have been drawn through the point corresponding to pCa values higher than 5.00 because of the uncertainty in the measurement of the very low level of tension obtained at this low loading [free Ca^{2+}] level.

The sarcolemmal binding site discussed in the interpretation of these experiments corresponds only to the relatively high-affinity binding site described in the isolated sarcolemma (52-55). The isolated sarcolemma also contains a very abundant low-affinity binding site that may play a role in Ca^{2+} binding at the outer surface of the sarcolemma (52-55).

5. Effect of Neuraminidase on Cardiac Fibers with Disrupted Sarcolemma. In cardiac fibers with disrupted sarcolemma cyclic contractions are observed when the [free Ca^{2+}] of the solution is higher than that corresponding to about pCa 7.40 in the presence of pMg 3.50 (5). These cyclic contractions correspond to cycles of Ca^{2+} release and reaccumulation by the SR. They may not have the same mechanism as the tension transients attributed to a perhaps physiologically-relevant Ca^{2+}-induced release of Ca^{2+} from the SR. These cyclic contractions may be merely caused by a Ca^{2+} overload of the SR. Perhaps their mechanism is similar to that responsible for the "Ca^{2+}-induced release of Ca^{2+}" elicited by very slow changes of myoplasmic [free Ca^{2+}] in large skinned fibers of skeletal muscle (see ref. 27 for discussion). In contrast, the Ca^{2+}-induced release of Ca^{2+}

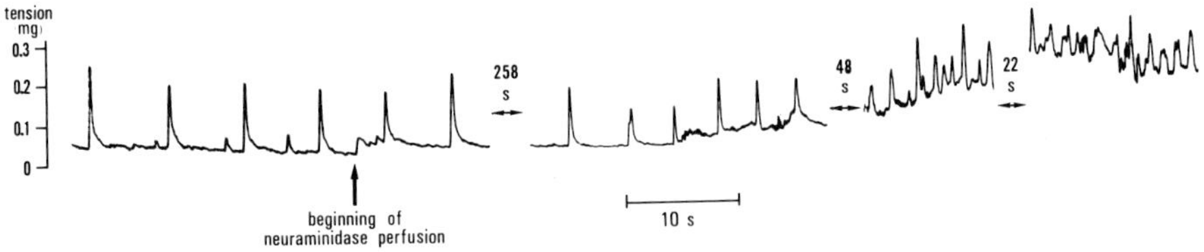

FIGURE 15. *Effect of 0.2 U/ml of neuraminidase in a bundle of about 6 cells with disrupted sarcolemma obtained from the rat ventricle by homogenization. The preparation was about 25 μm wide and 125 μm long. The pCa was 5.00, [total EGTA] 0.050 mM, creatine phosphate 12 mM, creatine phosphokinase 15 U/ml, ionic strength 0.160 M, pMg 3.96, pMgATP 2.52, temperature 35°C with appropriate adjustment of the pH and correction of the apparent stability constant for the Ca-EGTA complex.*

which may be of physiological interest is triggered by a rapid change of [free Ca^{2+}] outside the SR without the requirement of an overloading of the SR. The trigger is not a mere change of [free Ca^{2+}] at the outer surface of the SR but a function of Δ[free Ca^{2+}]/Δt (27, 57). Yet, these cyclic contractions correspond to cyclic releases of Ca^{2+} and their frequency increases when the [free Ca^{2+}] is increased (5). In the presence of a [free Ca^{2+}] of pCa 5.00, these cyclic contractions and relaxations are still observed despite the fact that this [free Ca^{2+}] would be sufficient to almost completely activate the myofilaments in the presence of a strong EGTA buffering. This indicates that the SR effectively competes with the low [total EGTA] buffer (0.050 mM) and causes a much lower [free Ca^{2+}] in the myoplasm than that set in the bulk of the solution (5, 38, 7).

The addition of neuraminidase to this solution produced, after a long delay, an increase of resting tension and an increase of the frequency of the cyclic contractions (Figure 15). Both effects can be attributed to an increase of the [free Ca^{2+}] in the myoplasm since increase of [free Ca^{2+}] increases the tension developed by the myofilaments and increases the frequency of the cyclic contractions (5).

Theoretically, this increase of [free Ca^{2+}] could be due to an effect of neuraminidase on the SR or on the myofilaments. Yet, experiments in skinned single cardiac cells eliminated both possibilities. First, neuraminidase did not modify the frequency of the cyclic contractions or the minimum [free Ca^{2+}] required to trigger Ca^{2+} release from the SR in these skinned single cardiac cells. Secondly, neuraminidase decreased the amplitude of the tonic tension developed by direct activation of the myofilaments in the presence of a high [total EGTA]. This latter effect is opposite to the increase of resting tension shown in Figure 15. Thus, the results observed in Figure 15 must be related to an effect of neuraminidase on the sarcolemma remaining in this preparation, although the nature of this sarcolemmal effect is unclear. It could correspond either to a release of Ca^{2+} bound to the glycocalyx or to a mere increase of the permeability to Ca^{2+} of the remaining sarcolemma facilitating the equilibration of the myoplasmic [free Ca^{2+}] with the [free Ca^{2+}] present in the bulk of the solution.

6. Discussion. These results demonstrate that the abundant sarcolemma remaining in homogenized preparations can indeed represent a binding site in the presence of a high [free Ca^{2+}]. In addition, the replacement of K^+P^- with Ch^+Cl^- can release Ca^{2+} from this binding site. Thus, this type of ionic substitution affects not only the SR but also the sarcolemma and the myofilaments (see preceding section) both in their maximum ability to develop tension (effect on actin and myosin) and in their sensitivity to Ca^{2+} (effect on Troponin C). This renders this ionic switch a very difficult tool to use to demonstrate a "depolarization-induced" release of Ca^{2+} from the SR.

CONCLUSION

These results do not provide any evidence for a "depolarization-induced" release of Ca^{2+} from the SR of skinned cardiac cells. Even if a release of Sr^{2+} (Figure 8), or perhaps Ca^{2+} (Figure 10), can be induced by anionic and cationic substitutions under certain experimental conditions, the present results provide no evidence that this release is caused by a "depolarization" of the SR membrane. The hypothesis of a "depolarization-induced" release of Ca^{2+} has also been challenged for skeletal muscle by the brilliant computations of Hans Oetliker (58) and by the results obtained with electron probe analysis (16).

ACKNOWLEDGMENTS

I thank my wife Dr. Françoise Fabiato for her help in the earliest experiments, Dr. C.M. Baumgarten for reviewing this manuscript, and Drs. E.B. Ridgway and O. Shimomura for the gift of the aequorin. I wish also to express my respectful gratitude to the organizers of the Symposium: Professors S. Fleischer and Y. Tonomura. The data on Ca^{2+}-induced release of Ca^{2+} from the cardiac SR that I presented at this Symposium are reported in another article (27).

REFERENCES

1. Endo, M., Tanaka, M., and Ogawa, Y., Nature 228:34 (1970).
2. Ford, L. E., and Podolsky, R. J., Science 167:58 (1970).
3. Costantin, L. L., and Podolsky, R. J., J. Gen. Physiol. 50:1101 (1967).
4. Nakajima, Y., and Endo, M., Nature, New Biol. 246:216 (1973).
5. Fabiato, A., and Fabiato, F., Circ. Res. 31:293 (1972).
6. Fabiato, A., and Fabiato, F., in "Recent Advances in Studies on Cardiac Structure and Metabolism," Vol. 9, "The Sarcolemma" (P.-E. Roy and N. S. Dhalla, eds.), p. 71, University Park Press, Baltimore, 1976.
7. Fabiato, A., and Fabiato, F., Circ. Res. 40:119 (1977).
8. Fabiato, A., Fabiato, F., and Sonnenblick, E. H., Fed. Proc. 30:668 (1971).
9. Trube, G., Pflügers Arch. 376:217 (1978).
10. Kerrick, W. G. L., and Best, P. M., Science 183:435 (1974).
11. Fabiato, A., and Fabiato, F., J. Gen. Physiol. 70:6a (1977).
12. Endo, M., and Kitazawa, T., in "Biophysical Aspects of Cardiac Muscle" (M. Morad, ed.), p. 307. Academic Press, New York, 1978.
13. Trube, G., Pflügers Arch. 379:121 (1979).
14. Fabiato, A., in "Mechanism of Gated Calcium Transport Across Biological Membranes" (S. T. Ohnishi and M. Endo, eds.), p. 237. Academic Press, New York, 1981.
15. Fabiato, A., Can. J. Physiol. Pharmacol. 60:556 (1982).
16. Somlyo, A. V., Gonzalez-Serratos, H., Shuman, H., McClellan, G., and Somlyo, A. P., J. Cell Biol. 90:577 (1981).
17. Best, P. M., Am. J. Physiol. 244:H167 (1983).
18. Fabiato, A., and Fabiato, F., Eur. J. Cardiol. 1/2:143 (1973).
19. Natori, R., Jikei Med. J. 1:119 (1954).
20. Fabiato, A., J. Gen. Physiol. 78:457 (1981).
21. Fabiato, A., and Fabiato, F., J. Physiol. (Paris) 75: 463 (1979).
22. Natori, R., Jap. J. Physiol. 25:51 (1975).
23. Rougier, O., Vassort, G., and Stämpfli, R., Pflügers Arch. ges. Physiol. 301:91 (1968).
24. Natori, R., Jikei Med. J. 12:214 (1965).
25. Stephenson, D. G., Wendt, I. R., and Forrest, Q. G., Nature 289:690 (1981).
26. Endo, M., Physiol. Rev. 57:71 (1977).
27. Fabiato, A., Am. J. Physiol.: Cell Physiol., in press (July, 1983).

28. Kerrick, W. G. L., and Donaldson, S. K. B., Biochim. Biophys. Acta 275:117 (1972).
29. Kerrick, W. G. L., and Donaldson, S. K. B., Pflügers Arch. 358:195 (1975).
30. McClellan, G. B., and Winegrad, S., J. Gen. Physiol. 72:737 (1978).
31. Chapman, R. A., Prog. Biophys. Mol. Biol. 35:1 (1979).
32. Miller, D. J., Nature 277:142 (1979).
33. Weisberg, A., McClellan, G., Tucker, M., Lin, L-E., and Winegrad, S., J. Gen. Physiol. 81:195 (1983).
34. Winegrad, S., J. Gen. Physiol. 58:71 (1971).
35. Sutherland, P. J., Stephenson, D. G., and Wendt, I. R., Proc. Aust. Physiol. Pharmacol. Soc. 11:160P (1980).
36. Blinks, J. R., Wier, W. G., Hess, P., and Prendergast, F. G., in "Progress in Biophysics & Molecular Biology," Vol. 40 (D. Noble and T. L. Blundell, eds.), p. 1, Pergamon Press, Oxford, 1982.
37. Carmeliet, E. E., Chloride and potassium permeability in cardiac Purkinje fibres. Thesè d'Agrégation, Arsia Société Anonyme, Presses Académiques Europêenes, Bruxelles, Belgium, 1961.
38. Fabiato, A., and Fabiato, F., J. Physiol. (Lond.) 249: 469 (1975).
39. Fabiato, A., and Fabiato, F., J. Physiol. (Lond.) 249: 497 (1975).
40. Meissner, G., and McKinley, D., J. Membr. Biol. 30:79 (1976).
41. Boyle, P. J., and Conway, E. J., J. Physiol. (Lond.) 100:1 (1941).
42. Page, E., and Upshaw-Earley, J., Circ. Res. 40:355 (1977).
43. Ford, L. E., and Surdyk, M. F., J. Gen. Physiol. 72:5a (1978).
44. Donaldson, S., Biophys. J. 37:23a (1982).
45. Donaldson, S., Biophys. J. 41:231a (1983).
46. Stephenson, E. W., Am. J. Physiol. 240:C1 (1981).
47. Chance, B., Legallais, V., Sorge, J., and Graham, N., Anal. Biochem. 66:498 (1975).
48. Kass, R. S., and Tsien, R. W., Biophys. J. 38:259 (1982).
49. Beeler, T. J., Farmen, R. H., and Martonosi, A. N., J. Membr. Biol. 62:113 (1981).
50. Fabiato, A., and Fabiato, F., J. Physiol. (Lond.) 276: 233 (1978).
51. Fabiato, A., and Fabiato, F., Ann. N. Y. Acad. Sci. 307:491 (1978).
52. Bers, D. M., A Correlation of the Effects of Cationic Uncouplers on Intact Cardiac Muscle and on Ca^{2+} bound

to Isolated Cardiac Muscle Plasma Membranes. UCLA Ph.D. Thesis, University Microfilms, Ann Arbor, MI, 1978.

53. Bers, D. M., and Langer, G. A., Am. J. Physiol. 237: H332 (1979).
54. Bers, D. M., Philipson, K. D., and Langer, G. A., Am. J. Physiol. 240:H576 (1981).
55. Philipson, K. D., Bers, D. M., Nishimoto, A. Y., and Langer, G. A., Am. J. Physiol. 238:H373 (1980).
56. Langer, G. A., Frank, J. S., Nudd, L. M., and Seraydarian, K., Science 193:1013 (1976).
57. Fabiato, A., and Fabiato, F., Nature 281:146 (1979).
58. Oetliker, H., J. Muscle Res. Cell Motil. 3:247 (1982).

THE CALCIUM-INDUCED CALCIUM RELEASE FROM THE SARCOPLASMIC RETICULUM[1]

Makoto Endo

Department of Pharmacology
University of Tokyo
Tokyo, Japan

I. INTRODUCTION

Ca release from the sarcoplasmic reticulum (SR) is one of the essential steps in excitation-contraction coupling in skeletal muscle. Physiologically, it is caused by a signal from the T-tubules produced somehow by its depolarization, but the nature of the signal is unknown. The skinned fiber preparation has been used to show that various kinds of stimuli directly applied to the SR can cause Ca release (listed in Table I, see ref. 1). Some stimuli appear to act through a common mechanism or a common "Ca channel" as indicated in the Table. Ca release caused by Ca ion itself and that caused by "depolarization" of the SR membrane have quite different characteristics, which strongly suggests different "channels" operating in each case (2). These two Ca release mechanisms, Ca-induced Ca release and "depolarization"-induced Ca release, in the past have been considered to be the only candidates on the list likely to play a major role in physiological excitation-contraction coupling (1). However, in my present view neither of them are likely, at least as far as amphibian skeletal muscle is concerned. A reason for Ca-induced Ca release being an unlikely candidate was given in my review written several years ago (1), and further evidence will be presented in the present article. The reason for the unlikeliness of the

[1]Supported partly by a grant from the Ministry of Education, Science and Culture, Japan.

ISBN 0-12-260380-X

Table I. Stimuli effective in producing Ca release from SR

Stimulus	Comment
1. Possibly physiological	
a. Ca ion	Ca-induced Ca release
b. "Depolarization of the SR membrane"	
2. Not physiological	
a. Lowering Mg ion concentration	Ca-induced Ca release
b. Hypotonicity	
c. Increase in pH	
d. Rapid cooling	Ca-induced Ca release
e. Drugs i. Caffeine	Ca-induced Ca release
ii. Local anesthetics	
iii. Others	

"depolarization"-induced Ca release to be the physiological mechanism comes from a study of the ionic composition of the lumen of the SR by electron-probe analysis (3). The study showed that there are no significant differences between the ionic composition in the lumen of the SR and that in the cytoplasm, except for Ca ion concentration. Although some reservation must be made because concentrations of organic anions were not determined, the above result indicates that in order to produce a "depolarization" of the SR membrane, either electric current must be forced to flow through the membrane, or permeability of the SR membrane to Ca ion must be raised. It is not impossible for the former to occur during physiological excitation-contraction coupling but some very specific assumptions are necessary (4). If "depolarization" is caused by raised Ca ion permeability, it is no longer the primary cause of Ca release but rather it is a parallel phenomenon although it might enhance the Ca release.

Thus, I question the physiological significance of Ca-induced Ca release in skeletal muscle. However, it is an interesting biological phenomenon, through which a regenerative type of Ca release (5) or a propagation of Ca release (6) can occur under appropriate experimental conditions. The mechanism can also produce repeated spontaneous Ca release if coupled with active Ca uptake by the SR (6). The mechanism also may operate in living skeletal muscle if it is under the influence of some drugs such as caffeine (7) or in a certain pathological state called malignant hyperthermia (8-10). In this article, properties of the Ca-induced Ca release mechanism will be summarized and its significance in living amphibian skeletal muscle will be discussed.

II. METHODS

Intact and skinned fibers were prepared from iliofibularis muscles of African clawed toads, Xenopus laevis(11). The fibers were suspended in a covered trough through which solutions could be perfused rapidly. Isometric tension was recorded with a strain-gauge transducer (U-gage, Shinkoh, Tokyo). Experiments were carried out at 2°C . In some case data on skinned fibers of human muscle (10) were also shown.

The amount of Ca in the SR of a skinned fiber was estimated by releasing it almost completely by applying 25 mM caffeine, and measuring the magnitude of resulting isometric contraction of the fiber (11). The procedure for determining rates of Ca-release induced by Ca ion under various conditions was as follows. After depleting Ca from the SR with 25 mM caffeine, the SR was reloaded with Ca to a specific level, typically by incubating the fiber in a solution containing 5×10^{-7} M free Ca ion for 3 min. ATP, which is known to strongly enhance Ca-induced Ca release (12) as well as to support Ca uptake by the Ca pump, was then removed and various concentrations of Ca ion were applied for various periods of time under variety of conditions. After the Ca ion was completely washed out, ATP was reintroduced and the amount of Ca remaining in the SR was estimated as described above. The amount of Ca released during Ca application in the absence of ATP could be determined from the difference in the remaining Ca with and without the Ca treatment. Remaining Ca after the treatment decreased monotonically with the length of duration of application of Ca ion; in other words, there seems to be no inactivation process in the Ca-induced Ca release mechanism under this experimental condition. The average rate of Ca release was, therefore, calculated from the results of experiments of one or two appropriate durations of Ca application for each Ca ion concentration to cause usually 30 - 70 % reduction in the remaining Ca by assuming a linear release. It was expressed in a unit that was a rate at which all the Ca in the SR is released in exactly 1 min if a linear release persisted to the end.

Compositions of experimental solution were as previously described (11). Free Ca ion concentration was calculated by assuming an apparent association constant of 2×10^{6} M^{-1} for

CaEGTA at pH 7.1 and 5 x 10^5 M^{-1} at pH 6.8, which are close to the calculated values based on recent estimates (13).

III. PROPERTIES OF Ca-INDUCED Ca RELEASE

Ca-induced Ca release is considered to be the result of an increase in Ca permeability of the SR membrane caused by Ca ion rather than a reduction in Ca-retaining capacity of the SR, although the participation of the latter has not been excluded. The evidence for the Ca permeability increase consists of (i) demonstration of an increase rather than decrease in Ca influx into the SR when Ca-induced Ca release mechanism was stimulated under the reversed Ca ion concentration gradient, i.e., when Ca ion concentration outside the SR was higher than that in the SR lumen (14), and (ii) the fact that the rate of Ca release could be very much higher than that expected only from raised Ca ion concentration due to complete unbinding of Ca from calsequestrin in the lumen of the SR (15).

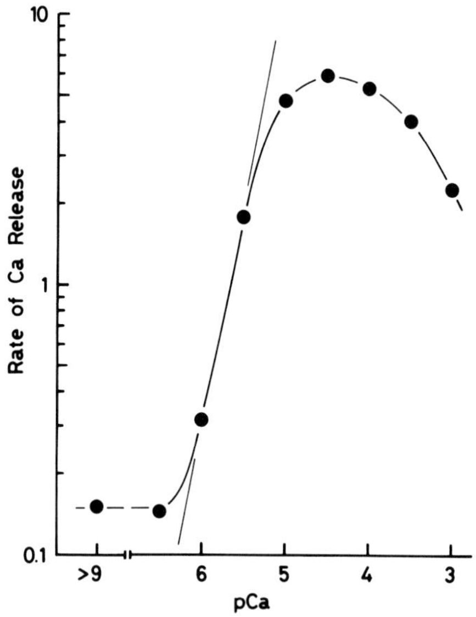

FIGURE 1. Dependence of the rate of Ca release from the SR on pCa. The straight line has a slope of 2. Skinned fibers from human muscles at 20°C, in the absence of Mg ion. Modified from ref. 10. Results of amphibian fibers were similar.

The dependence of the Ca permeability of the SR membrane on the Ca ion concentration is shown in Fig. 1. (As mentioned in METHODS, since ATP is a strong accelerator of Ca-induced Ca release, the rate of Ca release was determined here in the absence of ATP. This condition also allowed us to eliminate the Ca pump activity in the SR membrane, which would have seriously interfered with the accurate measurement of Ca release.) Below a certain level of free Ca ion concentration, the rate of Ca release was not dependent on Ca ion concentration, which suggested that a certain magnitude of Ca release occurred in the absence of Ca through a Ca-insensitive leakage channel independent of that responsible for Ca-induced Ca release. The rate of Ca release starts increasing above about 10^{-6}M, and the increase in the rate was proportional to the square of free Ca ion concentration. However, above 10^{-5}M, this square relationship does not hold and the rate of Ca release tends to saturate. At still higher concentrations of Ca ion, rate of Ca release decreases with an increase in Ca ion concentration.

The apparent inhibition of the rate of Ca release by higher concentrations of Ca might simply be due to the fact that the equilibrium point with higher concentrations of Ca is nearer to the loading level and, therefore, even if "Ca channels" are fully open, not much Ca is released. However, since all the other divalent cations tested so far, including Sr^{2+} and Ba^{2+}, decreased the rate of Ca release at about 1 mM (16), it is quite possible that Ca ion itself as well as other divalent cations may also block the "Ca channel".

Adenine nucleotides (12,15) and xanthine derivatives (7)

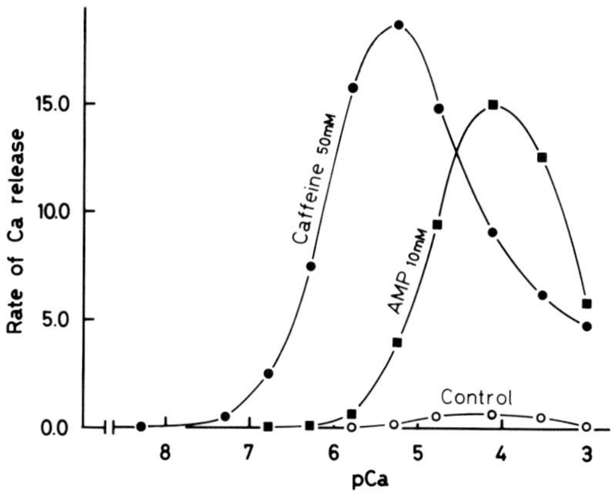

FIGURE 2. Effects of AMP and caffeine on Ca-induced Ca release. Modified from ref. 16.

are known to enhance Ca-induced Ca release very strongly. The mode of action of these two types of purine derivatives, however, is different. As shown in Fig. 2, while adenine nucleotides increase the rate of Ca release at all free Ca ion concentrations more-or-less evenly, xanthine derivatives shift the dependence of rate of Ca release on free Ca ion concentration to a lower concentration range in addition to their effect of increasing the maximum rate of Ca release at optimum free Ca ion concentration (16). Among the adenine nucleotides, ATP is the strongest in accelerating Ca-induced Ca release, but this action does not require the hydrolysis of ATP because a nonhydrolysable analogue of ATP, AMPOPCP, has almost the same magnitude of activity and ADP, AMP, adenosine and adenine have weaker but similar actions (15).

Some local anesthetics such as procaine and tetracaine (2,17) and Mg ion (18) are known to inhibit Ca-induced Ca release. While procaine, and probably other local anesthetics, suppress the rate of Ca release without changing the Ca ion concentration dependence appreciably, Mg ion appears to shift the dependence to a higher free Ca ion concentration range, probably indicating the presence of a component of the competitive inhibition against "gate-opening" action of Ca ion (16). However, since the inhibitory effect of Mg ion does not completely diminish in the presence of a very high free Ca ion concentration, inhibition through some other mode such as "Ca channel"-blocking may also be present (16).

The ease of evoking Ca-induced Ca release is dependent on the level of loading of the SR with Ca (6,18-20). If the SR was too lightly loaded, treatment of the SR with a high concentration of Ca in the presence of MgATP resulted only in a further net uptake of Ca. If the SR was heavily loaded, however, treatment with the same concentration of Ca produced a definite Ca release. Above the "threshold" level that is dependent on conditions, the heavier the loading was, the larger was the amount of release, and the amount of remaining Ca after the Ca-induced Ca release was essentially constant, irrespective of the level of preloading (18). This apparent loading-dependence of Ca-induced Ca release may partly be a natural result of observing net change of two opposing processes, Ca uptake and Ca release; the higher the level of loading, the slower the uptake and the faster the release in the presence of a certain concentration of Ca. Another factor that might contribute to the apparent loading-dependence is a local Ca ion concentration in the vicinity of the SR membrane. The more the SR is loaded, the more calcium tends to leak out from the lumen. Although EGTA-Ca buffer is present in the experimental medium, it is conceivable that due to the leakage, local Ca ion concentration in the vicinity of the SR membrane may be different from that in the bulk solution. If so, the heavier the loading, the higher the Ca ion concentration in the vicinity of the SR membrane, and hence the stronger the release at the same bulk concentration of Ca ion. The loading dependence may in part or in all be due to these factors and therefore, needs to be investigated after excluding them.

IV. EFFECTS OF ADENINE ON Ca-INDUCED Ca RELEASE AND OTHER PROPERTIES OF THE SR AND THE MYOFIBRILS

In order to clarify the physiological significance of Ca-induced Ca release, it would be useful to have a specific inhibitor of Ca-induced Ca release. If this Ca release mechanism plays an important role in normal twitch, then the twitch should be inhibited by such a specific inhibitor provided that the inhibitor penetrates the cell membrane. Results of experiments with an inhibitor of the Ca-induced Ca release mechanism, procaine, strongly suggested that it does not play a major role in the normal twitch. Whereas procaine effectively inhibited caffeine contractures of intact fibers, i.e., contractures due to enhanced Ca-induced Ca release mechanism, it did not inhibit K-contracture (2) or contracture induced by a square wave depolarization applied by a voltage-clamp technique (21). Unfortunately, the effect of procaine is not very specific, and as is well known, it suppresses action potential production. Thus, it has so far been impossible to examine, at least with procaine, whether inhibition of Ca-induced Ca release alters the twitch or not, while no inhibition of twitch is a very likely possibility in view of the results of procaine experiments described above. At this stage therefore, it is most desirable to discover an inhibitor of Ca-induced Ca release that does not have a local anesthetic action. We have recently found that adenine is a useful inhibitor in this sense, although its actions are not highly specific (22).

As described in the previous section, adenine nucleotides are known to enhance Ca-induced Ca release. ATP is the strongest among the group, whereas adenine is weak. If adenine and ATP exert their influence on the Ca-induced Ca release mechanism through their binding to the same site, I would expect adenine to <u>inhibit</u> Ca-induced Ca release in the presence of ATP, although its primary action is the <u>enhancement</u> of the mechanism. It will be shown below that this is indeed the case (22).

As shown in Fig. 3, adenine by itself increased Ca-induced Ca release without significantly changing Ca ion concentration dependence of the release rate. This effect is

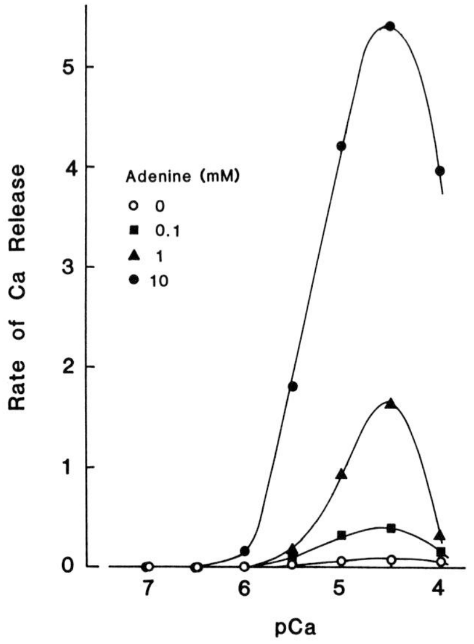

FIGURE 3. Enhancing effect of adenine on Ca-induced Ca release (22).

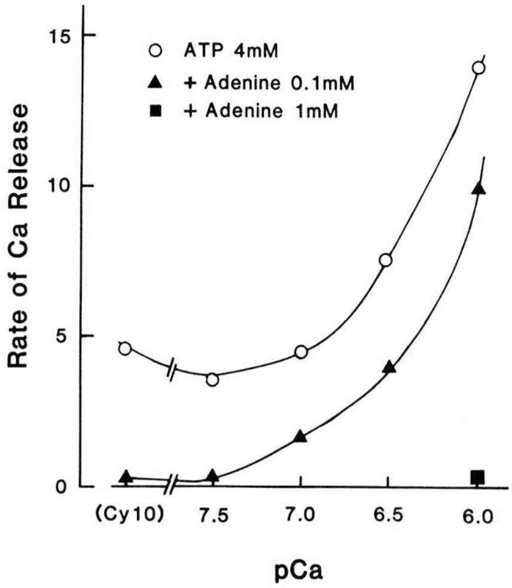

FIGURE 4. Inhibitory effect of adenine on Ca-induced Ca release in the presence of ATP (22).

qualitatively similar to that of adenine nucleotides (Fig. 2), but much smaller in magnitude than that of ATP. In the presence of 4 mM ATP, adenine inhibited rather than enhanced the Ca-induced Ca release mechanism as shown in Fig. 4. Similar results were obtained when Ca-induced Ca release was enhanced by caffeine. In the absence of adenine nucleotides, adenine further enhanced the Ca-induced Ca release that had already been enhanced by caffeine. However, in the presence of ATP, adenine inhibited Ca release by caffeine.

In the absence of Ca ion, 10 mM adenine did not affect the Ca leakage at all, indicating no effect of adenine on the Ca-insensitive leakage channel (22). Ca uptake by the Ca pump in the SR membrane was not affected by 1 mM adenine, but higher concentrations of adenine slightly inhibited Ca uptake, the magnitude of inhibition being some 20 % with 10 mM adenine (22).

The isometric tension-Ca ion concentration relationship in skinned fibers also was affected slightly by adenine. In the presence of mM range of adenine, tension produced by low activating concentrations of Ca ion was increased, but maximum tension at a saturating Ca ion concentration was not altered. In other words, Ca sensitivity of myofibrils was increased by mM range of adenine (22).

Since the inhibition of Ca-induced Ca release by adenine in the presence of ATP is apparent at 0.1 mM (Fig. 4), the results described above taken together indicate that adenine is a fairly specific inhibitor of the Ca-induced Ca release mechanism in the presence of ATP.

V. EFFECTS OF ADENINE ON INTACT AMPHIBIAN MUSCLE FIBERS---THE LACK OF PHYSIOLOGICAL SIGNIFICANCE OF Ca-INDUCED Ca RELEASE

Effects of adenine on intact single muscle fibers of *Xenopus laevis* were examined (23). From the results described in the previous section, we predict that adenine would inhibit the Ca-induced Ca release mechanism in intact fibers, since ATP is present there. To examine this effect we surveyed whether or not caffeine contracture, considered as a result of enhanced Ca-induced Ca release (7), is inhibited by adenine. Fig. 5 shows the time course of development of an inhibitory effect of adenine on caffeine contracture. Caffeine, 4 mM, produced maximum tension reproducibly. If 10 mM adenine was applied at the same time as caffeine application, the development of contracture was delayed and tension developed was 20 % smaller than the control. If adenine was applied before the application of caffeine, the inhibition was more effective with an increase in the duration of pretreatment; the latency was longer and peak tension smaller. With 30 min pretreatment, the caffeine response was completely abolished, This inhibitory effect of adenine was reversible and disappeared with similarly slow time course after washing the inhibitor. The time course probably indicates that of entry of adenine into the fiber. Dose-response curve of caffeine contracture determined in the presence of 10 mM adenine with 30 min pretreatment shifted significantly to the right as shown in Fig. 6. Thus, adenine appears to inhibit Ca-induced Ca release in intact fibers as expected, after it enters the cell.

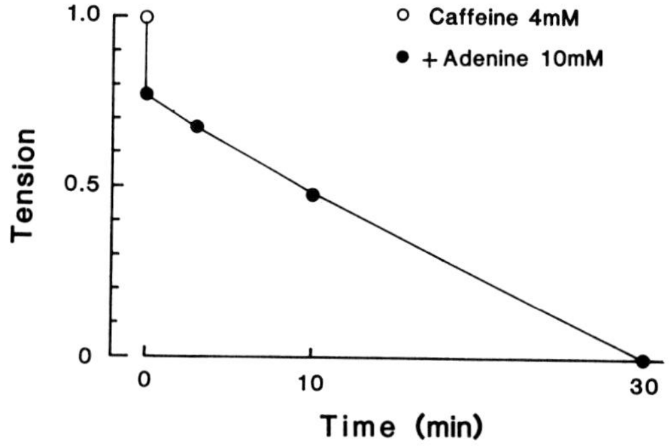

FIGURE 5. Time course of an inhibitory effect of adenine on caffeine contracture of an intact fiber (23).

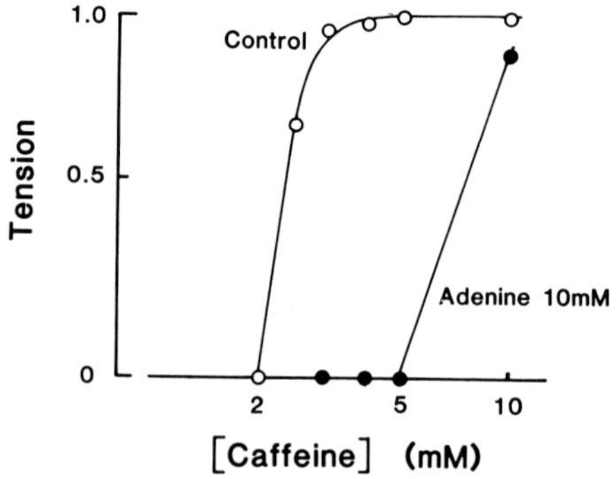

FIGURE 6. Effect of adenine on caffeine contracture of an intact fiber at the steady state (23).

Fig. 7A shows the effect of 10 mM adenine on twitch tension. As is seen, it did not inhibit at all but rather potentiated the twitch tension. The potentiation gradually developed reaching the steady state at 30 min or longer after the start of treatment with adenine. The effect was reversible and disappeared completely with a similarly slow time course on washing. However, if twitch was previously potentiated by caffeine, application of 10 mM adenine now inhibited twitch tension (Fig. 7B). If the twich was potentiated by potentiators other than caffeine such as nitrate replacing chloride (Fig. 7C) or an addition of zinc (not shown), 10 mM adenine exerted no inhibition at all. These results are interpreted as follows. In normal twitch, the Ca-induced Ca release mechanism is not playing a major role, and therefore, even if it is inhibited by adenine, little effect is observed. The same applies for

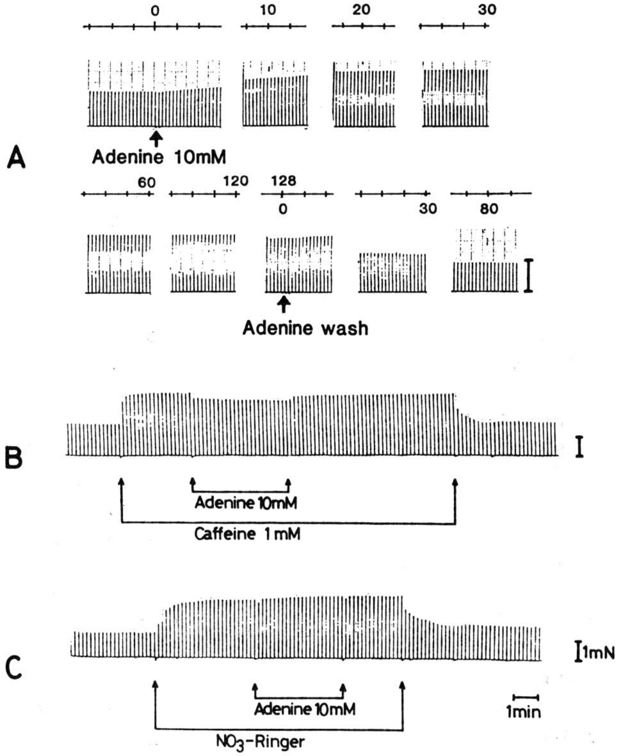

FIGURE 7. Effect of adenine on twitch under various conditions (23). A : Effect on normal twitch. B : Effect on caffeine-potentiated twitch. C : Effect on nitrate-potentiated twitch.

twitches potentiated by nitrate or zinc. However, if twitch was potentiated by caffeine through enhancing the Ca-induced Ca release mechanism, adenine inhibits it by suppressing the enhanced Ca-induced Ca release. The potentiation of twitch by adenine may be due to combination of two effects of adenine described in the previous section, i.e., the increased Ca sensitivity of the contractile system and a small inhibition of Ca uptake by the SR.

The following calculation was made to see whether or not the above interpretation is reasonable. In order to evaluate how the Ca-induced Ca release mechanism affects the net movement of Ca ion across the SR membrane, one must know the exact quantitative dependence of both Ca release and Ca uptake on Ca ion concentration. This is because when cytoplasmic Ca ion concentration was raised, not only the Ca-induced Ca release but also Ca uptake by the Ca pump in the SR membrane are activated. The Ca dependences were assumed to follow equations previously described (10). Briefly, Ca uptake was assumed to be proportional to the square of the fraction of high affinity Ca-binding sites of the Ca pump occupied by Ca^{2+} ion. Ca release from the SR was assumed to occur through two kinds of channels present in parallel, one Ca-independent leakage channel and the other Ca-dependent channel, responsible for the Ca-induced Ca release. Ca release through the latter system was assumed to be proportional to the square of the fraction of Ca-binding sites of channel "gate" occupied by Ca^{2+} ion. Furthermore, Ca release rate was assumed to be proportional to the difference between Ca^{2+} ion concentration in the SR lumen and that in the cytoplasm. It was further assumed that total amount of Ca in the cell is constant. Both of the square relationships described above are actually supported by our skinned fiber experiments (10). The effect of caffeine was incorporated into the equations according to our experimental observation (16), i.e., an increased Ca affinity of the "gate" for Ca-induced Ca release and an increased maximum Ca release rate at optimum Ca ion concentration.

Results of these calculation are shown in Fig. 8. In the resting state, the crossover point between Ca uptake curve and Ca release curve (at pCa 7) is realized since it is a steady-state condition. When the muscle cell is stimulated and Ca ion concentration is raised by Ca release through the third, physiological Ca channel (not incorporated in the calculation) in the SR membrane, Ca ion is rapidly taken up in this model system, since Ca uptake rate greatly exceeds Ca release rate in all the contracting range of Ca ion concentration. In the model shown in Fig. 8, the rate of Ca release in higher Ca ion concentrations is at most 15 % in magnitude of that of Ca uptake in the absence of caffeine. Therefore, if Ca-induced Ca release is inhibited, rate of net Ca uptake would increase and hence decrease twitch tension but only to a small extent. (At most 15% increase in uptake rate with complete inhibition.) The situation does not change when the amount of Ca released is increased as in nitrate or zinc potentiation. However, if the Ca-induced Ca release is mildly enhanced by caffeine, whereas resting Ca ion concentration in the muscle cell (the crossover point) is not appreciably affected, the net rate of

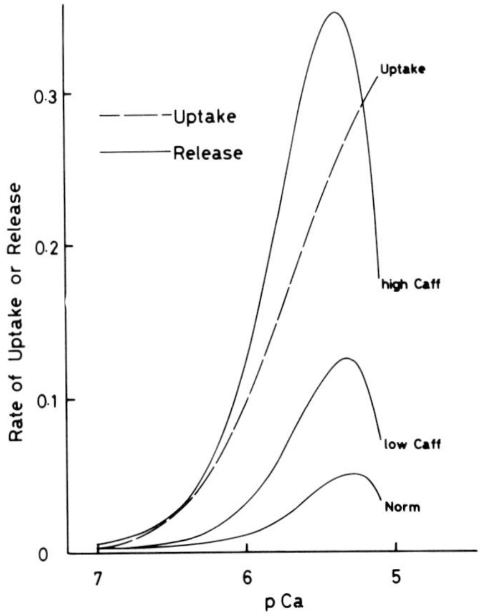

FIGURE 8. Dependence of rate of Ca uptake, U, and that of Ca release, R, in a model system. U and R were assumed to follow Equations (1) and (2), respectively.

$$U = \alpha \cdot \frac{1}{(1 + K_U/[Ca^{2+}])^2} \tag{1}$$

$$R = \left\{ R_0 + \beta \cdot \frac{1}{(1 + K_R/[Ca^{2+}])^2} \right\} \cdot ([Ca^{2+}]_1 - [Ca^{2+}]) \tag{2}$$

Following values were used for calculation. $K_R = 10^{-5.2}$, $10^{-5.4}$, and $10^{-5.65}$ (M) and $\beta/R_0 = 10^{2.2}$, $10^{2.4}$ and $10^{2.65}$ for control (Norm), in the presence of a low concentration of caffeine and that of a high concentration of caffeine, respectively. $K_U = 10^{-6}$ (M). Other parameters were the same as used in ref. 10.

uptake of Ca during a twitch decreases significantly which could result in a prolongation of active state and constitute the basis of twitch potentiation. Under these circumstances, the inhibition of the Ca-induced Ca release reverses the situation produced by caffeine. If the Ca-induced Ca release is further increased by higher concentration of caffeine, the Ca release rate now exceeds the uptake rate in much of the Ca concentration range and net Ca release spontaneously proceeds until the crossover point, which is enough to give strong muscle contracture. This caffeine contracture could of course be inhibited by inhibiting the Ca-induced Ca release.

Whether this is real explanation of the effects of caffeine and adenine or not needs to be examined further, especially because some parameters used in the above calculation were arbitrarily chosen, no experimental data being available at present.

The calculations described above are consistent with our view that the Ca-induced Ca release mechanism does not play an important role in physiological contraction but it does play a role if it is enhanced pharmacologically by caffeine, or, as described elsewhere (10), pathologically in malignant hyperthermia.

ACKNOWLEDGMENT

I thank Dr. E. D. Stevens for helpful comments.

REFERENCES

1. Endo, M., Physiol. Rev. 57:71 (1977).
2. Thorens, S. and Endo, M., Proc. Japan Acad. 51:473 (1975).
3. Somlyo, A. V., Shuman, H. and Somlyo, A. P., J. Cell Biol. 74:828 (1977).
4. Mathias, R. T., Levis, R. A. and Eisenberg, R. S., J. Gen. Physiol. 76:1 (1980).
5. Ford, L. E. and Podolsky, R. J., Science 167:58 (1970).
6. Endo, M., Tanaka, M. and Ogawa, Y., Nature 228:34 (1970).
7. Endo, M., Proc. Japan Acad. 51:479 (1975).
8. Kalow, W., Britt, B. A., Terreau, M. E. and Haist, C., Lancet II:895 (1970).
9. Takagi, A, Sugita, H., Toyokura, Y. and Endo, M., Proc. Japan Acad. 52:603 (1976).
10. Endo, M., Yagi, S., Ishizuka, T., Horiuti, K., Koga, Y. and Amaha, K., Biomed. Res. 4: in press.

11. Endo, M. and Nakajima, Y., Nature (New Biol.) 246:216 (1973).
12. Endo, M. and Kitazawa, T., Proc. Japan Acad., 52:595 (1976).
13. Harafuji, H. and Ogawa, Y., J. Biochem. 87:1305 (1980).
14. Kitazawa, T. and Endo, M., Proc. Japan Acad. 52:599 (1976).
15. Endo, M., Kakuta, Y. and Kitazawa, T., in "The Regulation of Muscle Contraction. Excitation-Contraction Coupling" (A. D. Grinnell and M. A. B. Brazier, ed.), p. 181, Academic Press, New York, 1981.
16. Endo, M., in "The Mechanism of Gated Calcium Transport across Biological Membranes" (S. T. Ohnishi and M. Endo, ed.), p. 257, Academic Press, New York, 1981.
17. Ford, L. E. and Podolsky, R. J., J. Physiol. 223:1 (1972).
18. Endo, M., Proc. Japan Acad. 51:467 (1975).
19. Fabiato, A. and Fabiato, F., Circ. Res. 31:293 (1972).
20. Reuben, J. P., Brandt, P. W. and Grundfest, H., J. Mechanochem. Cell Motil. 2:269 (1974).
21. Heistracker, P. and Hunt, C. C., J. Physiol. 201:627 (1969).
22. Ishizuka, T. and Endo, M., Proc. Japan Acad. 59: in press.
23. Ishizuka, T., Iijima, T. and Endo, M., Proc. Japan Acad. 59: in press.

Section V

Regulation of Ca^{2+} Uptake and Release

ION CHANNELS OF SARCOPLASMIC RETICULUM VESICLES AND CALCIUM RELEASE

Michiki Kasai
Kazuki Nunogaki
Kazuo Nagasaki
Manabu Tanifuji

Department of Biophysical Engineering
Faculty of Engineering Science
Osaka University
Toyonaka, Osaka, Japan

Masahiro Sokabe

Department of Behaviorology
Faculty of Human Science
Osaka University
Suita, Osaka, Japan

I. INTRODUCTION

Sarcoplasmic reticulum (SR) regulates the cytoplasmic Ca^{2+} concentration of skeletal muscle cells and thereby controls muscular contraction and relaxation (1). In connection with these functions, the Ca^{2+} pump mechanism of the SR has been established as the cause of relaxation (1-4). However, the molecular mechanism of Ca^{2+} release from SR is still unknown (4). Knowledge of movement of other ions during the uptake and release of Ca^{2+} is important to understand these functions at the molecular level.

Many researchers have measured the ionic permeability of SR membrane by various methods. However, there are discrepancies among the data reported. For example, as far as the K^+ permeability is concerned, we obtained from the turbidity change a value of 20 sec for permeation time, τ (5.8×10^{-8} cm/sec for permeability coefficient, P, refs. 7,8). McKinkey and Meissner

ISBN 0-12-260380-X

reported that the K^+ permeability is greater than 10^{-6} cm/sec from measurement of membrane potential using a potential sensitive dye (9). Chiu and Haynes obtained a value around 10^{-6} cm/sec by using ANS fluorescence (10).

Recently Miller and his coworkers obtained a channel conductance of 120 pS in 100 mM K-gluconate by using vesicles incorporated into an artificial planar bilayer (11,12). This value corresponds to τ = 1.3 ms (P = 9 x 10^{-4} cm/sec) assuming that one vesicle contains only a single channel (See APPENDIX). This value is much higher than those obtained by other methods (7-10). These discrepancies are mainly attributable to the difference of the method employed. The turbidity method could not follow the faster permeation than water (τ=0.1 sec). It is probable that we measured slower component of the K^+ permeability by the turbidity measurement. The conductance measurement gives the fastest component. However, we do not know how many vesicles in the preparation contain such channels. In order to interpret these data consistently,we needed a new method to follow the rapid permeation of ions.

In this paper, we measured the rapid permeation of ions by a fluorescence quenching technique and got a consistent interpretation about the ion permeation. We could measure the influx of Tl^+ and I^- over a time range from msec to a few hundred sec. It was found there are four components with different permeation times. The permeation times for the fastest components were few msec for Tl^+ and I^-. These ions were used instead of K^+ and Cl^-, respectively. These permeation times are compatible with the value obtained by the single channel conductance measurements (11,12). Through these experiments, we can give a consistent explanation on the reported data about the ionic permeability of SR membrane obtained by various methods.

Further, it was found there exist two types of cation channels (voltage-gated and Ca^{2+}-gated cation channels) and an anion channel. The Ca^{2+}-gated cation channel is suggested to be the Ca^{2+}-gated Ca^{2+} channel for Ca^{2+} release. The physiological significance of other channels is still unknown. Some ideas about the function of these channels are given.

II. EXPERIMENTAL PROCEDURES

A. SR Membrane Fractions

Two types of SR vesicles were isolated as a microsomal fractions by the method of Weber et al. (13) with minor modifications (14,15). Briefly, rabbit dorsal and hind leg muscle were homogenized with 4 volumes of 0.1 M NaCl and 5 mM Tris-Maleate (pH 7.0), using a mixer (Toshiba MX 1300) for 2 min. The homogenate was centrifuged at 4,000 x g for 30 min. The supernatant was centrifuged at 10,000 x g for 30 min and a crude heavy fraction was obtained from the sediment. The supernatant was centrifuged at 44,000 x g for 1 h to yield a crude light fraction. The two sediments were washed with 0.6 M KCl and recovered by centrifugation. The resulting sediments were suspended in an appropriate buffer solution and recentrifuged. The sediments were resuspended in the same buffer and stored at 0°C. These preparations are referred to as HSR and LSR in this paper. The SR vesicles were used within 1 week after isolation. When we compared these preparations with those purified by the sucrose gradient centrifugation, there was no significant difference in permeability properties between our HSR and the purified HSR and between our LSR and the purified LSR, respectively. Our HSR was rich in pure HSR and our LSR was rich in pure LSR. We used the present preparations throughout the experiment unless otherwise stated. In some experiments, Miller's type of preparation was used (16).

B. Miscellaneous Permeability Measurements

Permeability of ions and neutral melecules was measured by three methods.

1. Tracer Assay. The tracer assay was carried out by the Millipore filtration method (17).

2. Turbidity Assay. The osmotic volume change of the vesicles induced by various salts and neutral molecules was followed by measuring the scattered light intensity as described elesewhere (7).

3. CTC Fluorescence Assay. Effluxes of divalent cations were measured by following the change in the fluorescence intensity of chlortetracycline (CTC) as described previously (18). For the measurement of the rapid permeation, some modifications were made (19). Briefly, SR vesicles were incubated in 150 mM KCl, 20 mM Tris-Maleate (pH 6.7), 10 mM $CaCl_2$ and 10 mg SR protein/ml for overnight at 4°C. A 0.1 ml aliquot was diluted in 2 ml of 150 mM KCl, 20 mM Tris-Maleate, 4.5 mM EGTA and 20 μM CTC. Then, 45 sec after the dilution, it was mixed with an equal volume of the same solution as used for the dilution except containing effectors and the fluorescence was recorded with using a cut-off filter at 470 nm on excitation at 390 nm.

C. Permeability Measurement by Fluorescence Quenching Method

Influxes of Tl^+ and I^- were assayed by measuring the quenching rate of the fluorescence of 1,3,6,8-pyrenetetrasulfonate (PTS) by modification (20) of the method described by Moore and Raftery (21). Briefly, Tl^+ influx was measured as follows. SR vesicles were incubated in 0.1 M KNO_3, 5 mM Mes-K buffer (pH 6.5), 1 mM PTS and 9 mg SR protein/ml for overnight at 4°C. Then a 0.35 ml aliquot of the incubated vesicles was passed through a Sephadex column (G-25, coarse, 20 cm x 1 cm) equilibrated with 0.1 M KNO_3 and 5 mM Mes-K buffer to remove the extravesicular PTS. The eluent was mixed with an equal volume of 0.1 M $TlNO_3$ and 5 mM Mes-K using a stopped flow apparatus, and the change in the fluorescence intensity of PTS was followed. PTS fluorescence was excited at 353 nm and recorded with using a cut-off filter at 400 nm. In parallel to the quenching experiment, leakage of PTS from the vesicles was determined as follows. A 0.5 ml aliquot of the eluent from the Sephadex column was diluted with 0.1 M KNO_3 and 5 mM Mes-K buffer to 12 ml and its 3 ml aliquot was passed through a Millipore filter at appropriate time intervals and the fluorescence intensity of the filtrate was measured.

I^- influx was measured by using 0.4 M KNO_3 instead of 0.1 M KNO_3 in the above experiment. The eluent was mixed with 0.4 M KI and the fluorescence quenching rate of PTS was followed. Other procedures were the same as those in the case of the Tl^+ influx experiment.

D. Measurement of Electric Conductivity of SR Membrane

SR vesicles were incorporated into an artificial planar bilayer membrane according to the method described by Miller et al. (11,12) and the conductance was measured. The planar bilayer was formed by asolectin and phosphatidic acid (9:1) in n-decane. The buffer solution used was 5 mM Hepes-Tris (pH 7.2).

III. RESULTS

A. Permeability Measurement by Tracer Method

The simplest method to determine permeability is the tracer assay. Fig. 1 shows an example of the permeability measurement for some ions. The radioactivity of the remaining ions in the vesicles was normalized by the total radioactivity. From this figure, the following facts can be seen. 1)Efflux rates are different from ion to ion. 2) The values extrapolated to zero time are different. Permeability for various ions has been determined from the slope of these curves. However, it is incorrect, because the second point was not taken into account. Since the amount of the permeant in SR vesicles was normalized by the total radioactivity, these values correspond to the apparent intravesicular space (this value is referred to as the

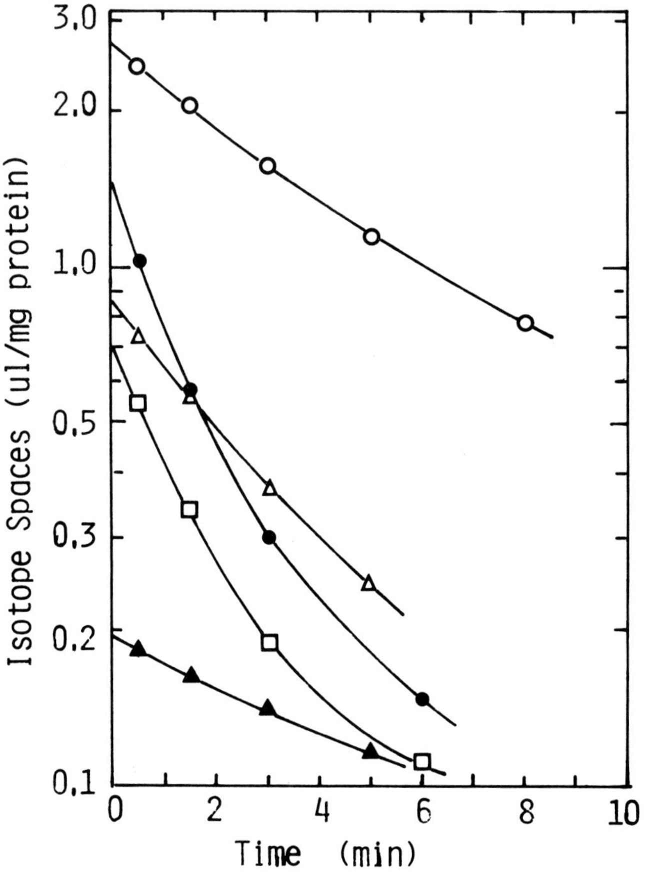

Fig. 1. Permeability measurement by tracer method. SR vesicles were incubated in 0.2 osm of various salts containing radioactive tracers, 5 mM Tris-Maleate (pH 6.5) and 10 mg protein/ml at 0°C for overnight. Efflux of various ions from SR vesicles was followed after the 100-fold dilution in the same solution used for the incubation without the labelled ions. Ordinate represent the relative radioactivity remaining in the vesicles, R/R_oC, where R is the radioactivity in the vesicles, R_o the radioactiviy before the filtration, and C the SR concentration during the incubation. (o) ^{14}C-choline$^+$ (choline Cl), (●) $^{35}SO_4^{2-}$(K_2SO_4), (△) $^{22}Na^+$(NaCl), (□) $^{42}K^+$(KCl), (▲) $^{36}Cl^-$(NaCl)

apparent volume). The apparent volumes for these ions were smaller than the intravesicular water space determined by inulin extrusion volume (about 3-5 μl/mg protein, ref. 22).

The following possible explanations can be considered for the difference of these values from the intravesicular water space. 1) The permeability for the ion was so small that the ion concentration was not equilibrated during the overnight incubation and the amount of the tracer incorporated was small. This case gives a smaller apparent volume. 2) In contrast to the first possibility, the permeability was so high that some ions have already flowed out during the filtration. 3) Some ions can bind to the inside the membrane and give a high apparent volume. 4) Even at the equilibrium condition, the intravesicular ion concentration was different from the extra-vesicular concentration due to the Donnan effect (22).

As it will be made clear in the follwing section, the efflux curve for choline represents the complete time-course since the apparent volume was close to the intravesicular water space determined by the inulin extrusion volume, no specific binding was observed, and the time-courses of choline influx determined by other methods were consistent with the curve in Fig. 1. In the case of other ions, the main effect is attributable to the case 2) although the cases 3) and 4) may have some effect. In other words, the fast components of the efflux could not be followed by the tracer assay. The time-course observed in Fig. 1 corresponds to the slow components of the efflux of these ions. Accordingly, permeability coefficients determined from the slope of these curves give only those of the slow components. In the case of divalent cations such as Ca^{2+}, the effect of binding to the membrane contributes to the apparent volume and a high value of apparent volume was observed (17). In the case of sucrose, permeability was so small that sucrose cannot be equilibrated during the overnight incubation and a small apparent volume was obtained (17,22).

B. Permeability Measurement by Turbidity Change

The tracer method has many advantages, but cannot follow the efflux faster than a few seconds as mentioned above. One method to follow faster flux is to measure turbidity changes (or the changes in scattered light intensity). The turbidity change is caused by the osmotic volume change of the vesicles by an increase in the osmolarity of the solution. Some results have been reported in the previous papers (7,8); Fig. 2 shows an example. When the osmolarity of the solution was increased, the scattered light intensity rapidly increased and then decreased. The rapid increase in the scattered light intensity is due to the shrinkage of the vesicles caused by water efflux.

From this rate, water permeability can be estimated. The latter decrease in the scattered light intensity is due to swelling of the vesicles caused by influx of water accompanied by an influx of solutes. From the rate of decrease in the scattered light intensity, the permeability for the solute can be estimated.

TABLE I summarizes the permeability for various salts and neutral molecules determined accroding to this method. For example, the water permeation time was 0.1 sec, and the permeation time for KCl and choline-Cl were 10 sec and 270 sec. (Permeability for choline depends on the presence of μM Ca^{2+}. See refs. 15,23-25). In the case of KCl, when valinomycin was added to the solution, increase in the permeation rate was observed (5,6). In the case of salt permeation, ions are considered to move as a pair according to the Nernst equation (5,26), then the permeation rate of the slower permeable ion becomes rate determining. This fact shows the permeability for K^+ was smaller than that for Cl^-. According to this method, the permeability for various ions can be determined (7). For example permeation times for K^+ is 20 sec and that for Cl^- is 0.4 sec.

However, many problems arise in the application of this method. The first point is that the extrapolated values of the

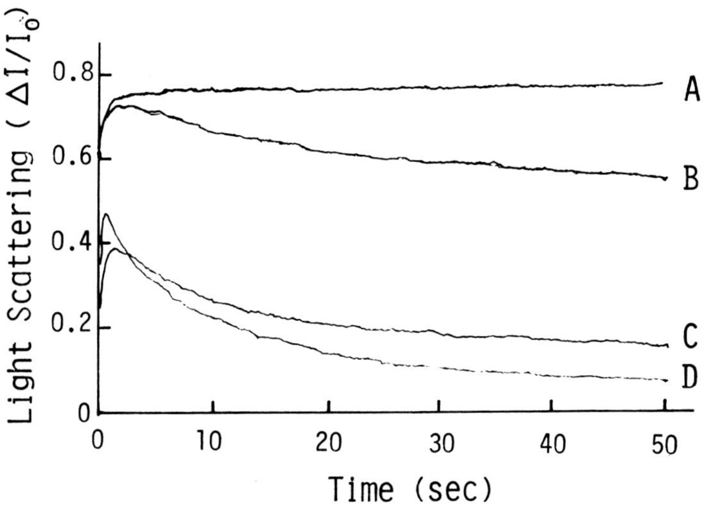

Fig. 2. Permeability measurement by turbidity method. SR vesicles were incubated in 5 mM Tris-Maleate and 0.2 mg protein/ml for 1 h at 23°C. The suspension was rapidly mixed with the solutions containing various salts of 0.2 osm, 5 mM Tris-Maleate (pH 6.5) and change in the scattered light intensity was followed. (A) sucrose, (B) choline-Cl, (C) KCl, (D) glycerol.

falling phase to zero time did not coincide with one another. The extrapolated value corresponds to the volume of the vesicles at the most shrunken states. The extrapolated value for KCl is about a half of that for choline Cl.

In this case, the specific effect of ions can be considered, for example, Ca^{2+} can bind to the membrane and contribute to the turbidity change. However, this effect may be small for other ions. Main reason for this difference is that the permeation of KCl was faster than that of water in some vesicles and we detected the permeation of KCl only through the remaining vesicles. Clearly, this technique can measure only permeation rates that are slower than the water permeation (τ =0.1 sec). Accordingly, in the case of KCl, there exist much

TABLE I. Permeability Properties of the Sarcoplasmic Reticulum Vesicle Membrane Determined by the Turbidity Assay

Species	Permeation time (τ)	Permeability[a] (P)	Increment of light scattering[b] ($\Delta I/I_o$)
	sec	$\times 10^{-7}$ cm/sec	
KCl	10	1.2	0.40
KNO_3	9	1.3	0.30
K-propionate	10	1.2	0.40
KMS	50	0.23	0.57
K-phosphate	100	0.12	0.50
K-gluconate	100	0.12	0.76
K_2-oxalate	90	0.13	0.40
K_2SO_4	80	0.15	0.48
NaMS	50	0.24	0.57
NaCl	13	0.89	0.40
LiCl	18	0.64	0.45
RbCl	8	1.4	0.40
$TlNO_3$	12	1	0.30
Choline-Cl	270	0.044	0.69
Tris-Cl	300	0.04	0.74
Tris-gluconate	1200	0.01	0.83
Tris-maleate	900	0.013	0.71
Ethyleneglycol	0.6	19	0.69
Glycerol	1.8	6.4	0.69
Glucose	500	0.024	0.76
Sucrose	2000	0.0005	0.86
Water	0.1	18000	

[a] Permeability was calculated according to Eq. (12).
[b] Increment of light scattering was defined by the extrapolated value of the falling phase to zero time by the eye.

faster components of the permeation observed here. This shows that our preparation contains at least two types of vesicles: one with channels that are rapidly permeable for K^+ and Cl^- and the other has no such channels. As far as the component observed here is concerned, we can say that the permeability for K^+ is smaller than that for Cl^-. We do not know which ion permeaates faster through the channels involved in the fast component, K^+ or Cl^- (See Section C).

For the solutes such as choline-Cl or glucose, all components of the permeation can be followed by this technique. These curves agree well with the curves of the tracer exeriments. There is no faster component than those observed here.

When we measure the permeability for rapidly permeable neutral molecules such as glycerol, the maximal increment of the scattered light intensity was high. They are considered to permeate through all vesicles. Probably they permeate directly through the membrane by solvation in the lipid phase. It suggests that the mechanisms of permeation of ions and neutral molecules are different. Ions pass the channels.

C. Measurement of Influx of Tl^+ and I^- by Fluorescence Quenching Technique

In order to follow the faster permeation, we developed a fluorescence quenching method. Moore and Raftery measured the Tl^+ influx of acetylcholine receptor rich membrane vesicles by the fluorescence quenching method by using 8-anilino-1,3,6-naphthalenetrisulfonate (ANTS) as a fluorophore (20). We used pyrenetetrasulfonate (PTS) since PTS is a tetravalent anion and less permeable to the SR membrane. SR vesicles were equilibrated with PTS by overnight incubation and free PTS outside the vesicles were removed by a Sephadex column. The eluent was mixed with the solution containing Tl^+ and the quenching process of the PTS fluorescence caused by Tl^+ influx was followed by using a stopped flow apparatus, as shown in Fig. 3. It is known that Tl^+ is able to replace K^+ in the nerve membrane and does not affect its excitability (27). In the case of SR vesicles, Tl^+ permeates as fast as K^+ when measured by the turbidity assay. The ATPase and the Ca^{2+} uptake were little affected. The quenching of fluorescence is caused by the heavy atom effect and follows the Stern-Volmer relationship (28). The quenching process corresponds to the influx of Tl^+. Before the analysis of the quenching curve, a correction was made. Although free PTS outside the vesicles was removed by the Sephadex column, some leakage was observed before the experiment. The amount of the leaked PTS was determined by the Millipore filtration method. The fluorescence quenching was expected to begin from the point indicated by an arrow in the figure.

The initial decreasing phase could not be recorded because of dead time of the stopped flow apparatus.

If all vesicles have a constant permeability, the fluorescence intensity is expected to decrease with time, t, according to the following equation:

$$F = F_o \frac{1}{1 + K\,Q(t)} \quad , \tag{1}$$

where F and F_o are the fluorescence intensity of PTS with and without quencher, respectively, Q(t) the concentration of the quencher (now Tl^+) inside the vesicles, and K is the Stern-Volmer constant. Q(t) is expressed by

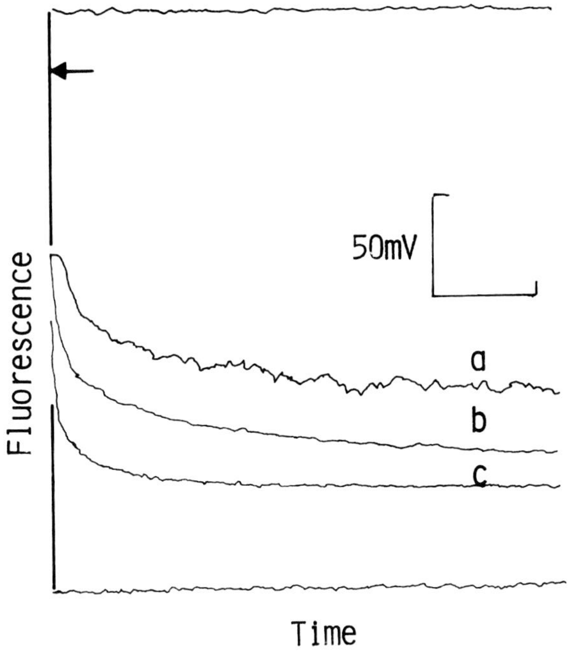

Fig. 3. Measurement of Tl^+ influx by fluorescence quenching method. Tl^+ influx in LSR was measured as described in EXPERIMENTAL PROCEDURES. The lowest trace represents the background of the solvent and the upper trace is the total fluorescence intensity measured by mixing with 0.1 M KNO_3 instead of 0.1 M $TlNO_3$. The arrow shows the level of fluorescence inside the vesicles at zero time. Three traces correspond to the different time scales. The calibration bar represents (a) 20 msec, (b) 0.2 sec, (c) 2 sec.

$$Q(t) = Q_o (1 - \exp(-kt)) , \quad (2)$$

where Q_o is the concentration of the quencher outside the vesicles which is considered to be constant and k is the rate constant of permeation. However, the curve in Fig. 3 did not follow the above equation. Then we assumed the existence of the multicomponent of the permeability. The curve was analyzed according to the following equation:

$$F(t) = F_o \sum_i \frac{c_i}{1 + K Q_i(t)} , \quad (3)$$

where c_i is the fraction of the component i and $\sum_i c_i=1$. $Q_i(t)$ is the concentration of the quencher in component i, and is expressed by

$$Q_i(t) = Q_o (1 - \exp(-k_i t)) , \quad (4)$$

where k_i is the rate constant of the component i.

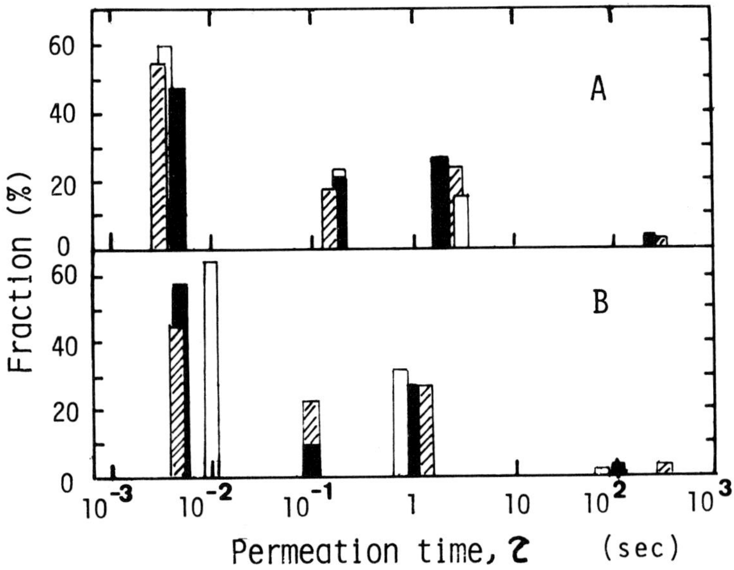

Fig. 4. Distribution of vesicles with different time constants for Tl^+ and I^- influx. The time constants and the fractional amplitudes of each component were obtained from the experiment similar to Figs. 3 and 5 using LSR. (A) Results from Tl^+ influx. (B) Results from I^- influx. Different symbols represent different preparations.

Analysis was made graphically with a microcomputer (Sord 243, Japan) by fitting and subtraction of each component in order from the slowest one. The results obtained are shown in Fig. 4. Four components were obtained. One example of the distribution of the permeation time, τ, is as follows: 3.2 msec 55 %, 0.15 sec 17 %, 2.8 sec 24 %, and 310 sec 4 %, where τ is defined by Eq. 12. The fastest component accounts for 50-60 %. In Fig. 3, the fast decay could not be recorded, but if we extrapolate the fastest component to zero time, we can explain the total intensity of the fluorescence. We can conclude that there is no faster component than those observed here. We made many experiments using different preparations. Some results of

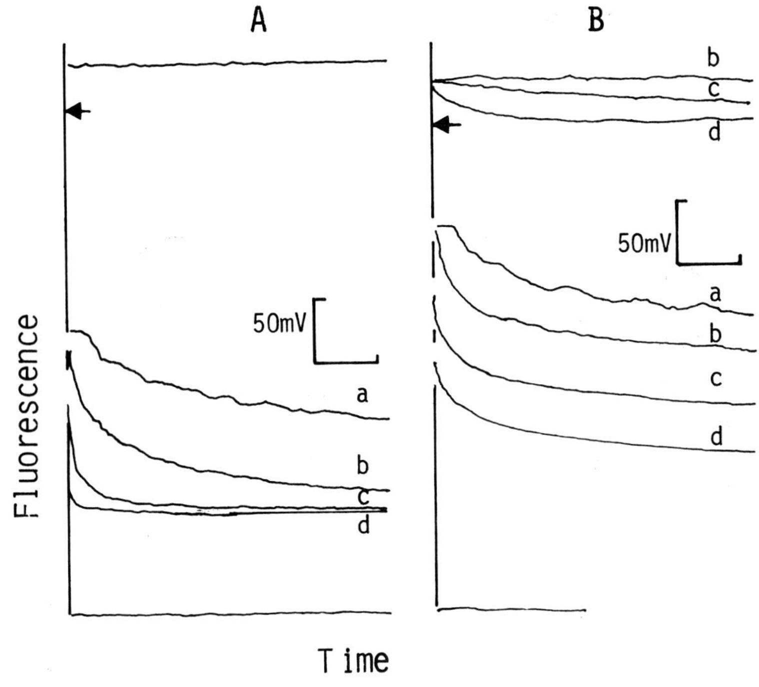

Fig. 5. Measurement of I^- influx by fluorescence quenching method. I^- influx in LSR was measured as described in EXPERIMENTAL PROCEDURES. (A) Control. (B) SR vesicles were treated with 830 µ DIDS at 37°C for 10 min as previously described (31). The upper traces represent the total fluorescence intensity measured by mixing with 0.4 M KNO_3 instead of 0.4 M KI. Decrease in the total fluorescence is due to photobleaching of DIDS. This effect was corrected by subtracting these changes from the records. Other details are the same as in Fig. 3, except that time calibration bar in (d) shows 20 sec.

such experiments are shown in Fig. 4. Some differences were observed in the fractional amplitudes of the components, but the permeation times agree well each other.

Next we carried out similar experiments using HSR. The result shows that the permeation time for each component was approximately the same as that of LSR, but the fractional amplitude was different. The fastest component accounts for 70-80 %. These results indicate that the distribution of the permeation times reflect the permeability properties of the channels involved in each component rather than the difference of the vesicle size, since the change of the size must result in the change of the permeation time.

The anion permeability was also followed by the same technique as that for cation. I^- was used instead of Tl^+. Since the quenching effect of I^- was much smaller than that of Tl^+, 400 mM KI was used as the quencher although this concentration was somewhat higher than physiological. KI permeates as fast as KCl in SR membrane when measured by the turbidity change. Ca^{2+} ATPase and Ca^{2+} uptake was observed in 400 mM KI. A quenching curve is shown in Fig. 5. An analysis similar to that for Tl^+ was obtained. As shown in Fig. 4, four components were obtained as follows: 4.9 msec 57 %, 0.1 sec 11 %, 1.1 sec 27 %, and 120 sec 5 %. For HSR, an increase in the fraction of the fastest component was observed as in the case of Tl^+ influx.

Now we will compare the permeations of Tl^+ and I^-. The fastest component of Tl^+ was slightly faster than that of I^-. The 2nd and the 3rd components of Tl^+ were slower than those of I^-, respectively. As far as the fastest component is concerned, the SR membrane is cation selective although the selectiviey is small. Next we will compare the data of the turbidity change with those of the fluorescence quenching. Since the turbidity change did not follow the permeation faster than 0.1 sec, the observed permeation times correspond to the 2nd to the 4th components. Since the permeation times of these components for I^- is smaller than those for Tl^+, the SR vesicles appeared anion selective. This is the reason why valinomycin accelerates the permeation of KCl in the turbidity experiment.

The tracer experiment shown in Fig. 1 can be expalained as follows. In the case of K^+, we can observe the 3rd and 4th components, while only the 4th component can be followed in Cl^- eflux. This is the reason why the apparent volume for Cl^- was much smaller than that for K^+ and both the apparent volumes were smaller than the intravesicular water space. In this consideration, we did not evaluate the difference between K^+ and Tl^+ and that between Cl^- and I^-. The above explanation is qualitatively correct, but further study is required for the quantitative analysis.

D. Effect of Inhibitors

It is known that the anion permeation was inhibited by stilbene derivatives such as SITS and DIDS (29-31): These inhibitors are known as potent inhibitors for anion transport in red cell membrane (32). When DIDS treated vesicles were used, I^- influx was reduced as shown in Fig. 5. The fractional amplitude of the fastest component decreased from 58 to 28 % without changing the permeation time. The changes in the 2nd and the 3rd components were small, but the 4th component increased. The 4th component corresponds to the practically impermeable vesicles. Now we pay attention to the channel which is responsible for the fastest component. We can conclude that the number of the channels per vesicle is very small. If many channels were involved in a single vesicle, we can expect the increase in the permeation time without changing the amplitude when the anion channel was blocked by DIDS. We analyzed the data under the assumption that the channels were blocked randomly according to the Poisson process. The result shows the average number of the channels per vesicle was about 1.5. From this result, we can calculate the fraction of the vesicles which have no channels. This value was also consistent with the sum of the fractional amplitude which have no fastest channel, i.e., the 2nd to 4th components. This result suggests that other membrane properties of the vesicles which are devoid of the fastest channel are the same as those which have the channels.

As far as the distribution of the cation channels is concerned, the same analysis as for anion channel could not be done since no specific inhibitors have yet been found for the cation channels. However, we can expect that the number of channels for the fastest component must be small, 1-2, since there exist vesicles which are devoid of the fastest component and the permeation time did not change when the fractional amplitude was varied from preparations to preparations. Further analysis is required.

We used the term "channel" for the fastest components for Tl^+ and I^- flux. These results show that the fractional amplitudes of the fastest components represent the fractions of the vesicles which have such channels. It is also suggested that HSR are rich in anion and cation channels and HSR and LSR have common channels.

E. Single Channel Conductance

Miller and his colleagues measured the single channel conductance of the SR vesicles which were incorporated into an artificial planar bilayer and found the existence of a cation and a anion channel (11,12). However, it is not clear how many vesicles in the preparation contain such channels. In order to get the consistent explanation between the fluorescence quenching measurement and the conductance measurement, we followed Miller's experiment (11,12). SR vesicles prepared according to Miller and Rosenberg (16) were compared with our LSR. It was found that the probability of our LSR to fuse with the planar bilayer was smaller than that of Miller's type SR. However, there was no intrinsic difference between the channel properties of the two types of preparations. We confirmed Miller's observations (11,12,16,33,34). An example is shown in Fig. 6. A single channel phenomenon was observed in 0.1 M salts of K-gluconate, KNO_3, and $TlNO_3$. Since the single channel conductance observed in KNO_3 was equivalent to that in K-gluconate, this channel is considered to be the cation channel. The single channel conductances calculated were 160 pS for K^+ and 80 pS for Tl^-. The number of channels per vesicle was about 1-2 in this figure. How many channels are involved in a vesicle is not clear (12). The channel conductance for Cl^- was 120 pS in 0.1 M choline-Cl. Substeps were observed in the anion channel conductance. Details will be reported elsewhere (34).

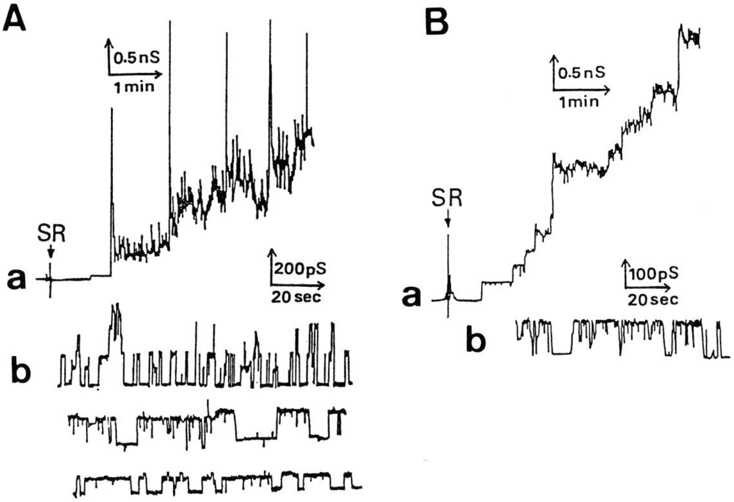

Fig. 6. Single channel conductance of SR vesicles fused to a planar bilayer. (A) Cation channel determined in 0.1 M K-gluconate, KNO_3, and $TlNO_3$. (B) Anion channel determined in 0.1 M choline-Cl. (a) Macroscopic recording of fusion process. (b) Microscopic recordings. From the top to the bottom, K-gluconate, KNO_3, and $TlNO_3$.

F. Comparison between Fluorescence Quenching and Conductance Measurements

Now, we will compare the data obtained from the fluorescence quenching measurements with those of the conductance measurements. If we assume that only one channel is involved in a single vesicle, we can calculate the permeation time of Tl^+ quenching to be 1.9 msec for the single channel conductance of 80 pS and the vesicle size of 50 nm in radius (See APPENDIX). This value is very close to the time constant of the fastest component of the Tl^+ quenching, 3.2 msec. In the fluorescence quenching experiment, the membrane potential was not controlled, and then we do not know the open time of the channel. This fact indicates the fastest component of the fluorescence quenching corresponds to the channel observed by the conductance measurement. For the anion channel, the fastest component of the fluorescence quenching is also considered to correspond to the channel observed by the electric conductance measurement.

As far as the sidedness of the channel is concerned, we measured the choline permeability and concluded from the inhibitory effect of Cs^+ that choline permeates the same voltage-gated cation channel as K^+ and the cis side of the electric measurement corresponds to the outside of the vesicle (33,35) (See next section).

G. Various Kinds of Channels

In order to examine the properties of the cation channels, permeability for choline was followed by the tracer assay and the turbidity assay (15,23-25,35). Permeation of physiological-

TABLE II. Distribution of the Two Types of Channels in SR Vesicles.

Type	Ca^{2+}-gated cation channel	Voltage-gated cation channel	Fraction (%) LSR	Fraction (%) HSR
I	+	+	16	42
II	+	-	4	28
III	-	+	64	18
IV	-	-	15	12

" + ", the vesicle contains the corresponding channel.
" - ", the vesicle do not contain the corresponding channel.

ly important K^+ was so fast that the permeation could not be followed by the usual method, but the permeation of choline can be followed over the whole time-course. As a result, it was found that SR membrane has two types of cation channels: voltage-gated cation channel which was blocked by extravesicular Cs^+ and Ca^{2+}-gated cation channel which was activated by extravesicular Ca^{2+} (15,24,25,35). Further analysis showed that LSR is rich in the voltage-gated cation channel and HSR is rich in the Ca^{2+}-gated cation channel (15).

If we assume that the channels are randomly distributed in membrane vesicles we can calculate the distribution of these channels as shown in TABLE II. This predicts the existence of 4 types of vesicles: type I vesicle having both the voltage-gated and the Ca^{2+}-gated cation channels, type II vesicle having only the Ca^{2+}-gated cation channel, type III vesicle having only the voltage-gated cation channel, and type IV vesicle having neither of the cation channels. If the anion channels are distributed independently to the cation channels, 8 types of vesicles can be expected depending on involvement of 3 types of channels. We do not know how these channels correspond to the channel determined by the fluorescence quenching measurement.

H. Cation Channel and Ca^{2+} Release

In a previous paper (36), we showed from the tracer assay that μM concentration of Ca^{2+} increased the Ca^{2+} permeability and mM concentration of Ca^{2+} reduced the Ca^{2+} permeability. Further analysis showed that this Ca^{2+} release was enhanced by caffeine and reduced by procaine and Mg^{2+}. These properties are very similar to the Ca^{2+}-induced Ca^{2+} release phenomenon observed in skinned muscle fiber (4,37,38). Accordingly, the isolated SR vesicles have such Ca^{2+} releasing ability. The Ca^{2+} release was observed more highly in HSR than in LSR. In this paper Ca^{2+} release rate from HSR was measured by fluorescence assay using CTC. In Fig. 7, Ca^{2+} release from CTC loaded vesicles was followed by using a stopped flow apparatus. When the permeability for Ca^{2+} is low, fluorescence of CTC follows the Ca^{2+} movement directly (18). When Ca^{2+} efflux is very fast, however, the fluorescence change becomes slower than the Ca^{2+} efflux since CTC remains in SR; a correction was made. The Ca^{2+} release rate determined by this procedure is increased by caffeine and ATP (19). The dissociation constant of Ca^{2+} for the channel opening was about 34 μM in the absence of caffeine and 0.1 μM in the presence of 25 mM caffeine. Caffeine did not affect the maximal rate of Ca^{2+} release. The dissociation constant of Ca^{2+} for the channel closing was 120-150 μM irrespective of the presence and absence of caffeine. ATP did not affect the affinity for Ca^{2+}, but increases the channel permeability 20-40 times. The similar phenomenon of ATP was observed in skinned muscle fiber (38). The maximal Ca^{2+} permeation rate in the presence of caffeine and ATP was about 50 sec^{-1} (0.02 sec for τ). This value is high enough to cause muscular contraction. It is not clear, however, whether this phenomenon is operative in living muscle.

Further, we got an interesting result about the channel for Ca^{2+} release. The Ca^{2+}-gated cation channel reported in the previous section was further examined by measuring the choline influx (15,24,25). There was a good correlation between the choline flux through the Ca^{2+}-gated cation channel and the Ca^{2+}-induced Ca^{2+} release. Micromolar concentrations of Ca^{2+} activated choline influx and mM concentration of Ca^{2+} inhibited it. Divalent cations such as Mg^{2+}, Co^{2+}, Mn^{2+} and etc. inhibited it and calcium antagonists such as nicardipine inhibited it.

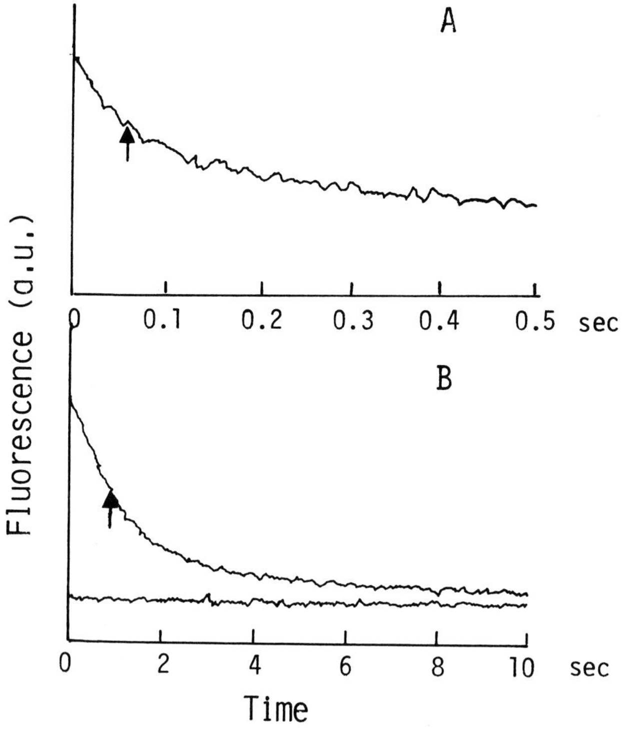

Fig. 7. Ca^{2+} efflux measured by chlortetracycline fluorescence. Efflux of Ca^{2+} was measured as described in EXPERIMENTAL PROCEDURES. Efflux was measured in (A) 12.5 mM caffeine, 2.5 μM Ca^{2+} and 12.5 mM ATP, (B) 12.5 mM caffeine and 2.5 μM Ca^{2+}. Arrow show the permeation times.

Further, DCCD blocked the choline influx. It is known that DCCD is a blocker of the Ca^{2+}-ATPase. Although this blockade is protected in the presence of μM concentration of Ca^{2+}, the effect on the Ca^{2+}-gated choline influx and the Ca^{2+}-induced Ca^{2+} release could not be protected in the presence of μM Ca^{2+}, but required mM Ca^{2+}. These facts show that the Ca^{2+}-gated cation channel is the same channel for the Ca^{2+}-induced Ca^{2+} release.

IV. DISCUSSION AND CONCLUSION

In this paper, we showed that permeability of sarcoplasmic reticulum membrane for physiologically important ions such as K^+ and Cl^- are very high. Permeability for Tl^+ instead of K^+ and permeability for I^- instead of Cl^- were measured by the fluorescence quenching technique. All components of influx of these ions were determined over a time range from msec to hundred seconds. The fastest permeation time is consistent with the channel conductance obtained from the electric measurement of the SR vesicles fused into a planar bilayer. The number of the channels involved in the vesicle is very small, about 1-2, for the anion channel and also suggested to be small for the cation channel. This value is consistent with the value suggested by McKinley and Meissner (9). HSR contained a little larger number of channels than LSR and the properties of the channels are not different.

There are various methods to measure the permeability of the vesicles. We will mention about two methods other than those reported here: the fluorescence measurement of ANS developed by Haynes and his coworkers (10,39,40) and the fluorescence measurement of the potential sensitive dyes used by many researchers (9,41). The former method measures the surface potential of the membrane, then the permeability can be estimated from the rate of the surface potential change due to the shielding or binding effect on the surface charge in the inner membrane. The latter method measures the diffusion potential of the vesicles, then the permeability ratio of the permeant ions can be estimated from the absolute value of the membrane potential and the permeation rate can also be determined from the membrane potential change caused by the change in the ionic concentrations due to the ion influx. The data obtained by these methods are also consistent with our interpretation (9, 10).

We showed that there exist two types of cation channels and an anion channel. We do not know exactly the physiological role of these channels. The possible explanation is as follows. In the previous paper (25), it was shown that choline permeation through the Ca^{2+}-gated cation channel is similar to the Ca^{2+}-induced Ca^{2+} release. Therefore, it is probable that this channel works as a Ca^{2+} channel in Ca^{2+} release. However, as far as the other channels are concerned, there is no direct evidence that they are related to Ca^{2+} uptake or Ca^{2+} release. One possible function is as follows. When Ca^{2+} is accumulated or released, it must move against osmotic and electric resistance. If the permeability for cations and for anions is very high, Ca^{2+} can move readily without osmotic and electric resistance. This idea is consistent with that of McKinley and Meissner (9).

As mentioned in the RESULTS, it is not clear which channels are contributing to the components observed in the fluorescence quenching measurement among various kinds of channels. For the further analysis of the channel properties, it is desired to develop the technique for separating vesicles which have only one type of channel, isolation of the channels, and reconstitution of the vesicles which have one type of the channels.

Finally, Meissner and Young measured the proton permeability by means of membrane potential measurement by using the voltage-sensitive dye and found that all vesicles are permeable to proton and permeability coefficient is greater than 10^{-3} cm/sec (42). This fact suggests that the proton channel is independent of the cation channel and anion channel reported here and the number of the proton channel is greater than those of the cation and anion channels. We did not study on the proton transport here.

ACKNOWLEDGEMENTS

The authors wish to thank Professor Fumio Oosawa for his valuable suggestions and discussion and also to the members of his laboratory for their support. The authors also thank Professor Sidney Fleischer for reading the manuscript. Part of this investigation was supported by Grant-in-Aid for Scientific Research (5644085) from the Ministry of Education, Science and Culture of Japan.

REFERENCES

1. Ebashi, S. and Endo, M., Prog. Biophys. Mol. Biol. 18:123 (1968).
2. Hasselbach, W., Prog. Biophys. Mol. Biol. 14:167 (1964).
3. Tonomura, Y., "Muscle Protein, Muscle Contraction and Cation Transport", p. 305. University of Tokyo Press,

Tokyo, 1972.
4. Endo, M., Physiol. Rev. 57:71 (1977).
5. Kometani, T. and Kasai, M., J. Membrane Biol. 56:159 (1980).
6. Haynes, D. H., Arch. Biochem. Biophys. 215:444 (1982).
7. Kometani, T. and Kasai, M., J. Membrane Biol. 41:295 (1978).
8. Kasai, M., Kanemasa, T., and Fukumoto, S., J. Membrane Biol. 51:311 (1979).
9. McKinley, D., and Meissner, G., J. Membrane Biol. 44:1138 (1979).
10. Chiu, V. C. K., and Haynes, D. H., J. Membrane Biol. 56: 203 (1980).
11. Miller, C., J. Membrane Biol. 40:1 (1978).
12. Labarca, P., Coronado, R., and Miller, C., J. Gen. Physiol. 76:397 (1980).
13. Weber, A., Herz, R., and Reiss, I., Biochem. Z. 345:329 (1966).
14. Kasai, M., and Miyamoto, H., J. Biochem. 79:1053 (1976).
15. Yamamoto, N., and Kasai, M., J. Biochem. 92:465 (1982).
16. Miller, C., and Rosenberg, R. L., Biochemistry 18:1138 (1979).
17. Kasai, M., and Miyamoto, H., J. Biochem. 79:1067 (1976).
18. Nagasaki, K., and Kasai, M., J. Biochem. 90:749 (1980).
19. Nagasaki, K., and Kasai, M., Abstract of the 20th Annual Meeting of the Biophysical Society of Japan, p.69. Osaka, 1982.
20. Nunogaki, K., and Kasai, M., Abstract of the 20th Annual Meeting of the Biophysical Society of Japan, p. 70, Osaka, 1982.
21. Moore, H.-P. H., and Raftery, M. A., Proc. Natl. Acad. Sci. 77:4509 (1980).
22. Kasai, M., J. Biochem. 88:1081 (1980).
23. Yamamoto, N., and Kasai, M., J. Biochem. 90:1351 (1981).
24. Yamamoto, N., and Kasai, M., J. Biochem. 92:477 (1982).
25. Yamamoto, N., and Kasai, M., J. Biochem. 92:485 (1982).
26. Schultz, S. G., "Basic Principles of Membrane Transport", p. 21. Cambridge University Press, Cambridge, 1980.
27. Landowne, D., J. Physiol. 252:79 (1975).
28. Eftink, M. R., and Ghiron, C. A., J. Phys. Chem. 80:486 (1976).
29. Kasai, M., and Kometani, T., Biochim. Biophys. Acta 557: 243 (1979).
30. Kasai, M., J. Biochem. 89:943 (1981).
31. Kasai, M., and Taguchi, T., Biochim. Biophys. Acta 643: 213 (1981).
32. Cabanchik, Z. I., Knauf, P. A., and Rothstein, A., Biochim. Biophys. Acta 515:239 (1978).
33. Coronado, R., and Miller, C., Nature 280:807 (1979).

34. Sokabe, M., Kometani, T., Tanifuji, M., and Kasai, M., Abstract of the 20th Annual Meeting of the Biophysical Society of Japan, p. 72, Osaka, 1982.
35. Yamamoto, N., and Kasai, M., Biochim. Biophys. Acta 692: 98 (1982).
36. Nagasaki, K., and Kasai, M., J. Biochem. 90:749 (1981).
37. Endo, M., Tanaka, M., and Ogawa, Y., Nature 228:34 (1970).
38. Endo, M., in "The Mechanism of Gated Calcium Transport across Biological Membranes" (S. T. Ohnishi and M. Endo, eds.), p.257. Academic Press, New York, 1981.
39. Chiu, V. K., Mouring, D., Watson, B. D., and Haynes, D. H., J. Membrane Biol. 56:121 (1980).
40. Chiu, V. K., and Haynes, D. H., J. Membrane Biol. 56:219 (1980).
41. Russell, J. T., Beeler, T., and Martonosi, A., J. Biol. Chem. 254:2047 (1979).
42. Meissner, G., and Young, R. C., J. Biol. Chem. 255:6814 (1980).

APPENDIX

For the interconversion of two among flux ϕ, permeability coefficient P, conductance g, permeation time τ, and rate constant k, the following relations were used. Under the assumption of constant field (26), ion flux toward the inside of the vesicles is given by

$$\phi = - P x \frac{C^o - C^i e^x}{1 - e^x} , \tag{5}$$

$$x \equiv \frac{z F E}{R T} ,$$

where C^o and C^i are the concentrations of the ions outside and inside the vesicle, z the valence of the ion, F the Faraday constant, E the given membrane potential of the membrane, R the gas constant and T the absolute temperature.

Conductance g is given by

$$g = PC^o \frac{z^2 F^2}{RT} \frac{x}{x - x_o} \frac{1 - e^{(x-x_o)}}{1 - e^x} , \tag{6}$$

where x_o is a parameter related to the equilibrium potential for the considered ions and is given by $x_o = \ln(C^o/C^i)$.

If we assume the independent principle (26), unidirectional flux is given by

$$\vec{\phi}_i = P\,C^o \frac{x}{1 - e^{-x}} \qquad (7)$$

$$\vec{\phi}_o = P\,C^i \frac{x}{e^{x} - 1} \qquad (8)$$

where $\vec{\phi}_i$ and $\vec{\phi}_o$ are, respectively, the unidirectional influx and efflux rates of permeant ions.

Assuming that $E = E_o$ and set $C^i = C^o = C$, we get

$$\vec{\phi} = \vec{\phi}_i = \vec{\phi}_o = P\,C \qquad (9)$$

$$g = P\,C \frac{z^2F^2}{RT} = \vec{\phi}\frac{z^2F^2}{RT} \qquad (10)$$

Efflux from the vesicle can be equated by

$$C_i = C_{io}\, e^{-kt} \qquad (11)$$

where C_{io} and C_i are the concentrations of tracer at t=o and t, respectively, and k the rate constant. If we assume a spherical shape of the vesicle, k is given by k=3P/r where r is the radius of the vesicle neglecting the effect of the thickness of the membrane. The permeatiom time, which is defined by the time when C_i becomes a half of the initial value C_{io}, is given by

$$\tau = \frac{0.693}{k} = \frac{0.231\,r}{P} \qquad (12)$$

If we assume r=50 nm, we get the following values from Eqs. 9, 10, and 12 for the permeation time τ=1 sec at room temperature.

τ = 1 sec, P = 1.16 x 10^{-6} cm/sec, $\vec{\phi}$ = 116 pmol/cm^2/s,
$\vec{\phi}$ = 6.96 x 10^{13} ions/cm^2/s, g = 436 μS/cm^2.

The values in TABLE I were calculated from these relations.

Now we apply the relations to SR vesicles. If we assume the radius of the vesicles r = 50 nm, membrane width d = 6 nm, the protein to lipid wt ratio of 1, and density 1, we get the intravesicular water space of 5 μl/mg protein, the vesicular weight 2 x 10^{-16} g protein, and the surface area 3.5 x 10^{-10} cm^2/vesicle in average. If ATPase pump protein accounts for 60 % of the membrane proteins, 400 molecules are present in a vesicle.

If one vesicle has one channel, a conductance of 160 pS, 80 pS, and 120 pS for K^+, Tl^+ and Cl^- corresponds to a permeation time of 0.96 msec, 1.92 msec, and 1.28 msec, respectively.

REGULATION OF Ca^{2+} PUMPING IN HEART SARCOPLASMIC RETICULUM

Michihiko Tada[1]
Masaaki Kadoma
Makoto Inui
Hiroshi Abe

First Department of Medicine
Osaka University School of Medicine
Osaka, Japan

I. INTRODUCTION

Calcium fluxes across the membrane of cardiac sarcoplasmic reticulum (SR) largely determine the intracellular Ca^{2+} levels, mediating the myofibrillar contraction and relaxation (1, 2). Calcium release from the SR increases the intracellular Ca^{2+} concentration ($\sim 10^{-5}$ M) inducing contraction, while calcium accumulation, often termed as calcium uptake, by the SR reduces the Ca^{2+} concentration ($\sim 10^{-7}$ M), producing relaxation (2, 3). Calcium uptake by the SR represents the ATP-supported active calcium transport, which is carried out by the Ca^{2+}-dependent ATPase, the major component of approximately 100,000 daltons within the membrane of cardiac SR (1).

The β-adrenergic action of catecholamines is known to exhibit remarkable influence on the excitation-contraction coupling of the myocardium to alter the contractility of heart muscle (inotropic effect). Cyclic AMP (cAMP) is considered to be a second messenger of such catecholamine actions, activating cAMP-dependent protein kinase (4, 5). In view of the central role of calcium in the excitation-contraction coupling of the myocardium, it is of importance to elucidate the processes through which cAMP exerts its action on calcium movement across cardiac membranes, notably SR and sarcolemma (6). A line of evidence indicate that a specific protein named phospholamban with the molecular weight of 22,000 daltons is the major substrate of cAMP-dependent protein kinase in cardiac SR ; phosphorylation of phospholamban significantly

[1] Supported by research grants from the Ministry of Education, Science and Culture of Japan, and a Grant-in-Aid from Muscular Dystrophy Association of America.

ISBN 0-12-260380-X

increases calcium uptake by augmenting the turnover rate of the Ca^{2+}-dependent ATPase (6-9). Such an interaction between calcium and cAMP within the myocardial cell may represent a key reaction step that mediates the inotropic effect of catecholamines on the myocardium. A working hypothesis was proposed, in which phospholamban is assumed to serve as a regulatory factor of the ATPase in cardiac SR.

This paper gives an outline of the present consensus on the characteristics of phospholamban that serves to regulate Ca^{2+}-dependent ATPase.

II. PHOSPHORYLATION OF 22,000-DALTON PROTEIN (PHOSPHOLAMBAN)

A. Cyclic AMP-Dependent Phosphorylation of Phospholamban

When cardiac SR was incubated with cAMP in the absence (10, 11) and presence (12) of cAMP-dependent protein kinase, a phosphoester phosphoprotein was formed. Formation of this phosphoprotein was independent of Ca^{2+} up to 0.1 mM, but was markedly dependent on cAMP between 0.1 μM and 10 μM with the half-maximal activation occurring at 0.2 μM (12). Tada et al. (9) found that a single microsomal protein of approximately 22,000 daltons incorporated most of the ^{32}P label when cardiac

TABLE I. Molecular Weights of Proteins Phosphorylated by cAMP-Dependent Protein Kinase in Cardiac SR

Investigators	Year	Animals	Phospholamban	Other phosphoprotein
			Molecular weight (1000 daltons)	Molecular weight (1000 daltons)
LaRaia & Morkin (10)	1974	Rabbit	20K	
Tada et al. (9)	1975	Dog	22K	11K
Kirchberger & Tada (14)	1976	Dog, cat, guinea pig rabbit	22K	11K
Schwartz et al. (15)	1976	Dog	20K	
Wray & Gray (16)	1977	Dog	20K	
Will et al. (17)	1978	Pigeon	22K	15K
St. Louis & Sulakhe (18)	1979	Pig	22K	56K 16K
Jones et al. (19)	1979	Dog	20K	7K
Lamers & Stinis (20)	1980	Rat	24K	9K
Bidlack & Shamoo (21)	1980	Dog	22K	6K
Le Peuch et al. (22)	1980	Dog	22K	11K 7K
Tada et al. (23)	1983	Dog	22K	11K

(From Tada and Katz (6).)

microsomes incubated with [γ-^{32}P]ATP and cAMP-dependent protein kinase were solubilized in sodium dodecyl sulfate (SDS) and electrophoresed on SDS-polyacrylamide gel. Since the 22,000-dalton protein was associated with functional alteration in SR (see below), it was termed as "phospholamban", i.e. "phosphate receptor" (6, 9, 13), ($\lambda\alpha\mu\beta\alpha\nu\epsilon\iota\nu$ = receive). In the presence of NaF, inhibitor of phosphoprotein phosphatase, phosphorylation of this protein increased several fold, and the other two proteins (55,000 and 11,000 daltons) were also phosphorylated to a lesser extent. The 55,000-dalton phosphoprotein was attributed to autophosphorylation of the regulatory subunits of cAMP-dependent protein kinase. These findings were virtually in agreement with those by LaRaia and Morkin (10) who also found phosphorylation of 20,000-dalton microsomal protein, catalyzed by an endogenous protein kinase. These reports were subsequently followed by a number of evidence (Table I), supporting that phospholamban is the major site at which cAMP-dependent protein kinase could react. Existence of phospholamban phosphorylation was confirmed in SR preparations virtually freed from sarcolemmal vesicles by means of density gradient centrifugations (17, 19, 20) (Table I).

Structural Characteristics of Phospholamban. Phospholamban is tightly associated with SR membrane, and is an acidic proteolipid that could be extracted by acidified chloroform-methanol (21, 22). A part of phospholamban was suggested to be exposed at the outer surface of SR membranes, the remainder of the molecule being buried in the membrane interior, forming a tight complex with the ATPase. The lack of iodination of phospholamban (24) probably reflects the location of the tyrosine residues (22) within the membrane interior.

Phospholamban was originally thought to represent a single peptide. However, addition of Triton X-100 and boiling of the phosphorylated SR membranes in SDS resulted in replacement of the 22,000-dalton phosphoprotein by an 11,000-dalton and even lower, 5,000- to 7,000 dalton, phosphoprotein (23, 25, 26). From the analysis of the amino acid composition, the possible minimum molecular weight is estimated to be approximately 5,500-daltons (22, 27, 28). 22,000-dalton phospholamban would probably represent an oligomer. Possible existence of the oligomer-monomer transition should be further sought.

The content of phospholamban in SR is estimated to be about 4% (29) or 6% (26) of the total protien. A 1 : 1 stoichiometry between phospholamban and the ATPase is probably acceptable since protein contents of these two components and amounts of phosphoprotein formed within the membrane exhibited roughly a 1 : 1 molar ratio (23, 26, 29 –31). At the present, however, it is unknown how oligomeric form of phospholamban is related to the ATPase protein which also functions as an oligomer (1).

B. Calmodulin-Dependent Phosphorylation of Phospholamban

In the presence of Ca^{2+} and calmodulin, phosphate could be incorporated into a 22,000-dalton protein, presumably phospholamban (23, 26, 32). The formation of this phosphoprotein with stability characteristics of a phosphoester was dependent on calmodulin but independent of cAMP. It was postulated that phospholamban could be phosphorylated by an endogenous protein kinase (calmodulin-dependent protein kinase) associated with SR which was activated by calmodulin and Ca^{2+}. The maximal amount of phospholamban phosphorylation catalyzed by calmodulin-dependent protein kinase is about the same as that found in cAMP-dependent protein kinase (23). Phosphorylation of phospholamban catalyzed by calmodulin-dependent protein kinase and cAMP-dependent protein kinase occurs independently and additively, suggesting that phosphorylation catalyzed by two kinases occurs at different sites within phospholamban (23, 32).

It was reported that preincubation of cardiac SR with calmodulin enhances the rate of calcium uptake (33) and that calmodulin-dependent phosphorylation of phospholamban is accompanied by an enhanced rate of calcium uptake (26, 32, 34). cAMP-mediated enhancement of calcium transport could be seen even when phosphorylation was carried out under Ca^{2+}-free conditions ($pCa > 8$) (23). The stimulatory effects of calmodulin-dependent protein kinase and cAMP-dependent protein kinase on calcium transport can be additive (23, 33, 34), in accord with the finding that phosphorylation by the two kinases occurred independently and additively. The ATPase activity is also enhanced by calmodulin and physiological concentrations of Ca^{2+} (0.1-50 μM) (23, 27, 34). Similar to calcium uptake, the effects of calmodulin-dependent protein kinase and cAMP-dependent protein kinase on the ATPase activity occurred in independent and additive manners (23). In contrast, one group (26) claims that stimulation of calcium uptake by cAMP-dependent phosphorylation is not observed unless calmodulin-dependent phosphorylation of phospholamban occurs and that calmodulin-dependent phosphorylation cannot affect the ATPase activity. These discrepancies may be derived from the difference of Ca^{2+} concentrations employed in the calcium uptake and ATPase assays. The higher Ca^{2+} concentration (0.1mM) used by the latter group (26) might mask the cAMP-mediated augmentation of calcium uptake by the SR and calmodulin-mediated enhancement of ATPase activity, which are seen at the lower Ca^{2+} concentrations ($< 10\ \mu$M) (6, 35). Two different kinases can catalyze phosphorylation of phospholamban, suggesting the possible existence of heterogenous population of phospholamban. It should be further sought how the two different phosphorylation sites are associated with molecular structure of phospholamban.

III. ROLE OF PHOSPHOLAMBAN IN THE REGULATION OF Ca^{2+}-DEPENDENT ATPase

A. Active Calcium Transport

ATP-supported active calcium transport by SR vesicles was found to be markedly augmented when membrane vesicles were previously phosphorylated by cAMP-dependent protein kinase (12, 13, 35, 36). Under these conditions, where phospholamban was specifically phosphorylated, the rate of oxalate-facilitated calcium uptake was more than doubled. Calcium uptake by SR from fast-contracting skeletal muscle did not exhibit such stimulation, nor was there significant phosphorylation of similar protein (14). The stimulation of calcium uptake by cardiac SR was well correlated with the increase in phospholamban phosphorylation (12, 37). Other reports also indicated the existence of such stimulation caused by endogenous and exogenous protein kinases (10, 15, 38). When the rate of calcium uptake by cardiac SR was examined at different Ca^{2+} concentrations in the presence and absence of cAMP-dependent phosphorylation, the rate of calcium uptake by cardiac SR in the absence of cAMP-dependent phosphorylation exhibited half-maximal activation at around 1 μM Ca^{2+}, whereas those in the presence of cAMP-dependent phosphorylation of phospholamban exhibited half-maximal activation at lower Ca^{2+} concentrations (0.3 μM) (35). These results suggest that cAMP-dependent phosphorylation of phospholamban may increase an apparent affinity of the ATPase of cardiac SR for Ca^{2+}. These effects of protein kinase on calcium transport of cardiac SR were confirmed by treatment of cardiac SR with phosphoprotein phosphatase (39, 40) or protein kinase modulator (41, 42), which can antagonize the effect of cAMP-dependent protein kinase. When phosphorylated SR was incubated with phosphoprotein phosphatase obtained from bovine heart, most of the phosphorylated protein (phospholamban) was dephosphorylated (39). Resultant dephosphorylation of the membrane led to complete reversal of the effects produced by protein kinase (40). When calcium uptake and phospholamban phosphorylation of caridac SR were compared in the presence of a fixed amount of protein kinase and different amounts of protein kinase inhibitor (modulator), a linear relationship was observed between the decrease in calcium uptake and the inhibition of phosphorylation of the membrane (41). Phosphorylase kinase was found to stimulate the rate of calcium uptake, although no underlying mechanism was clarified. Employing a rapid quenching apparatus, Will et al. (43) indicated that protein kinase-catalyzed phosphorylation was capable of stimulating calcium transport even in the absence of oxalate. Under these conditions, the effect was apparent only at the initial phase (up to 300 ms) of the reaction. In more recent experiments performed by us (44), in which the rapid initial phase (5-200 ms) of calcium transport in the absence of oxalate was followed by high-performance quenching device, the enhancement of the calcium uptake rate was evident at the initial phase (5-50 ms) of the reaction (see below).

B. Ca^{2+}-Dependent ATPase

Since hydrolysis of ATP by Ca^{2+}-dependent ATPase is tightly coupled to active calcium transport in that two moles of Ca^{2+} being taken up for each mole of ATP hydrolyzed (1), the observed stimulation of calcium uptake by phosphorylation of phospholamban could reflect either enhanced turnover of the ATPase or increased efficiency of calcium transport. Within a wide range of Ca^{2+} concentrations, the ATPase activity was markedly enhanced when phospholamban was phosphorylated (35). Ca^{2+} concentrations at which half-maximal activation was attained effectively decreased (30, 35) (Fig. 1). Stoichiometric coupling between the ATPase activity and calcium uptake was maintained at a ratio of about 2 after phosphorylation of phospholamban (35).

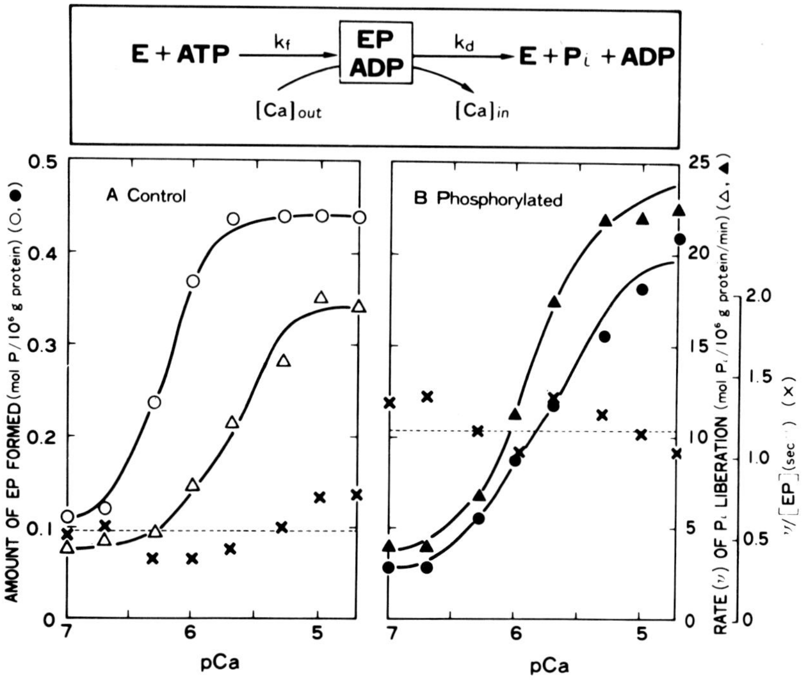

FIGURE 1. Effect of phospholamban phosphorylation on the Ca^{2+} dependence profiles of EP formation and the rate of Pi liberation. (From Tada et al. (30).)

Under these conditions, the maximal velocity of the ATPase was found by Tada et al. (30) to be markedly increased. Wray and Gray (16) also found augmentation of Ca^{2+}-dependent ATP hydrolysis by endogenous protein kinase. It is likely, therefore, that the turnover rate of normally coupled transport is enhanced when phospholamban is at phosphorylated state.

C. Elementary Steps of ATPase

During translocation of Ca^{2+} across the membrane of SR, the ATPase undergoes a complex series of reactions in which a number of phosphorylated intermediates (EP) are sequentially formed and degraded (see Equation 1). Steady-state EP levels were not altered by phospholamban phosphorylation at saturating concentrations of Ca^{2+} and ATP (30, 45). However, phosphorylation of phospholamban reduced EP levels at Ca^{2+} concentrations below 10 μM while it was accompanied by the increased rate of Pi liberation (ν) (30) (Fig. 1). Thus, the ratio ν/[EP] is augmented by phospholamban phosphorylation (30), indicating that the rate-limiting step of EP decomposition (1) is

TABLE II. Comparison of Enzymatic Parameters of Ca^{2+}-Dependent ATPase of Unphosphorylated (Control) and Phosphorylated Cardiac SR

Cardiac SR	Maximal velocity V_{max}	EP decomposition $\nu/[EP]$	k_d	EP formation $t_{1/2}$
	nmol Pi/mg·min	sec^{-1}	sec^{-1}	*msec*
Control	26.3	0.54	0.55	42.7
Phosphorylated	54.3	1.14	1.03	21.7

$\nu/[EP]$: The rate of ATPase activity per unit of concentration of EP at steady state.
k_d : Estimated from the rate of decay in the amount of EP after EP formation is terminated by excess EGTA.
$t_{1/2}$: Time at which a half of maximal EP is attained when Ca^{2+}-free enzyme.

markedly enhanced (Table II). These results are consistent with the finding that the rate constant, K_d, of EP decay was enhanced (30, 46) (Table II). Arrhenius plots for K_d indicated that the activation energy remained constant, with a marked increase in free energy when phospholamban is phosphorylated (30). At ATP concentrations below 5 μM, where EP formation was slowed, EP levels were also lowered by phospholamban phosphorylation (44). These results indicate that phospholamban phosphorylation can enhance the rate of EP decomposition relative to that of EP formation when the latter is inhibited at low ATP concentrations.

The EP formation step is extremely rapid, taking place within tens of milliseconds (1). Employing a rapid mixing device (47), the rate of EP formation is determined for reactions initiated at two different states of the ATPase. E_1 represents the state of enzyme that binds Ca^{2+} (Equation 1), produced when SR vesicles are incubated at micromolar concentrations of Ca^{2+}, while E_2 represents the state of the enzyme that does not bind Ca^{2+} (Equation 1) when the SR is incubated with EGTA (44). In reactions initiated at the E_1 state, the initial rates of EP formation were virtually unaltered by phospholamban phosphorylation, whereas EP levels were slightly augmented (44, 46) (Fig. 2). In reactions initiated at the E_2 state, the initial rate of EP formation is much lower than that initiated at the E_1 state since the EP formation requires the initial conversion of E_2 to E_1 (see Equation 1), which is rate-determining. Under these conditions, phosphorylation of phospholamban resulted in a marked increase in the initial rate of EP formation (Fig. 2) (44). Thus, the value of $t_{1/2}$, which represents the reciprocal of the initial rate and is virtually independent of Ca^{2+} concentrations, was much shortened after phosphorylation of phospholamban; phosphorylated SR exhibited $t_{1/2}$ of 22 msec, compared with control value of 43 msec for determinations made at 6 different Ca^{2+} concentrations (Table II).

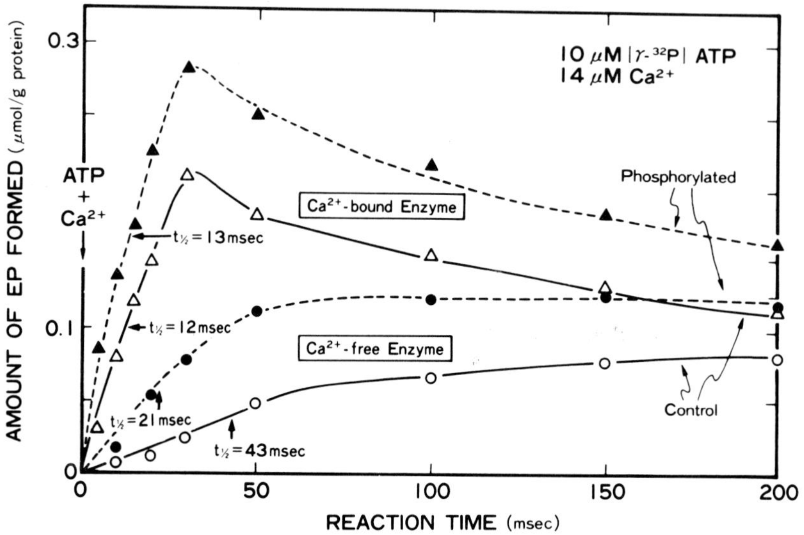

FIGURE 2. Effect of treatment with cAMP-dependent protein kinase on the initial rate of Ca^{2+}-dependent formation of EP of ATPase of Ca^{2+}-free and Ca^{2+}-bound cardiac SR. (From Tada et al. (44).)

IV. MECHANISM BY WHICH Ca^{2+}-DEPENDENT ATPase IS REGULATED

A. Actions of Phospholamban on Elementary Steps of ATPase

The observed alterations in kinetic properties of the ATPase reaction can be interpreted in light of the following equation of the Ca^{2+}-dependent ATPase reaction (1, 48):

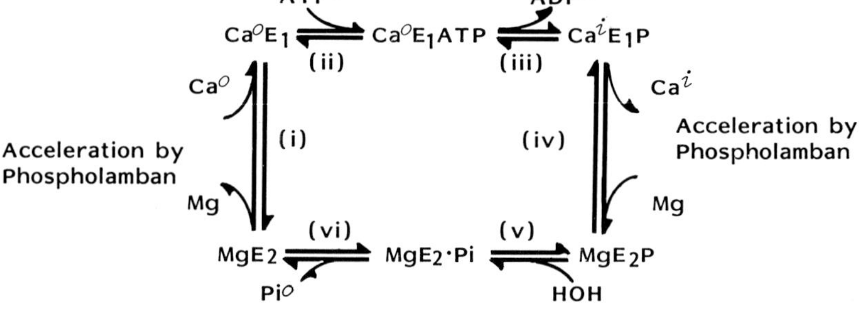

Equation 1

where E_1 and E_2 represent two different states of the ATPase (see above); *i* and *o* indicate the inside and outside of SR membranes, respectively. E_1P is the phosphorylated intermediate which has high affinity for Ca^{2+}, while E_2P has low affinity for Ca^{2+}. E_1P can react with ADP to form ATP, and is thus termed as ADP-sensitive phosphoenzyme, while E_2P could not form ATP upon addition of ADP, thus termed as ADP-insensitive phosphoenzyme. As usually measured, however, EP represents the sum of E_1P and E_2P. Under conditions where concentrations of Ca^{2+}, Mg^{2+} and KCl are saturating, the decomposition of E_2P (Step v) is accelerated and conversion from E_1P to E_2P (Step iv) is rate-determining (8).

As pointed out above, the rate of conversion from E_2 to E_1 (Step i), the rate-limiting step during EP formation, is enhanced when phospholamban is phosphorylated (44). The reversal rate of Step i ($E_1 \rightarrow E_2$) is also augmented by phospholamban phosphorylation (7). No change was found when the reaction was initiated at E_1 (44, 46).

The finding that the rate of EP decomposition is largely accelerated by phospholamban phosphorylation (30, 44) can be derived from stimulation of any of the Steps iv, v and vi. However, among these steps, Step iv is probably accelerated since it is rate-limiting during EP decomposition and E_1P usually predominates under the conditions where saturating concentration of Ca^{2+}, Mg^{2+}, and K^+ are present. Indeed, transient state measurement of EP decomposition indicated that EP formed under these conditions largely represented ADP-sensitive E_1P whose decay was accelerated by phospholamban phosphorylation (7).

It is of prime significance, therefore, that both of the two key rate-determining steps (Steps i and iv) during the turnover of ATPase are enhanced when phospholamban is at phosphorylated state. These are the major steps at which the affinity of the enzyme for divalent cations Ca^{2+} and Mg^{2+} is greatly altered, suggesting that phospholamban would probably exert its action by regulating the cation-mediated conformational change of the ATPase enzyme.

B. Molecular Model

cAMP-dependent phosphorylation of phospholamban profoundly alters calcium transport by cardiac SR. Since such an alteration is accompanied by significant changes in kinetic (9, 30, 44, 46) properties of the Ca^{2+}-dependent ATPase, it may be assumed that phospholamban is closely associated with Ca^{2+}-dependent ATPase protein. Thus, as working hypotheses, molecular models have been presented in which phosphorylated phospholamban was suggested to serve as an activator (45) or a derepressor (38, 49) of this ATPase. Fig. 3 represents a diagrammatical illustration of the present consensus views on the interplay between phospholamban and Ca^{2+}-dependent ATPase.

The Ca^{2+}-dependent ATPase is an amphiphathic single polypeptide of 100,000 daltons that functions either as a monomer or an oligomer, during ATP-dependent translocation of Ca^{2+} across the SR membrane (1, 50). A protein-protein interaction has been suggested to exist between phospholamban and the ATPase (30, 38, 44), although an alternative mode of interaction such as that involving the membrane lipid is also possible. A direct molecular interaction between these proteins is supported by the evidence that the ATPase enzyme and phospholamban remain associated with each other after solubilization of SR with detergents (21, 22, 29). This association appears to be tighter when phospholamban is at its phosphorylated state (21, 29). The action of phospholamban to augment the ATPase activity remains operational in the presence of saturating amount of dodecyloctaoxyethylene glycol monoether ($C_{12}E_8$), suggesting that phospholamban remains associated with the ATPase enzyme under the condition where ATPase is assumed to be monomeric (Table III). However, it should be further sought whether the sequence of events leading to increased ATPase activity in the solubilized SR is identical to that of the intact SR.

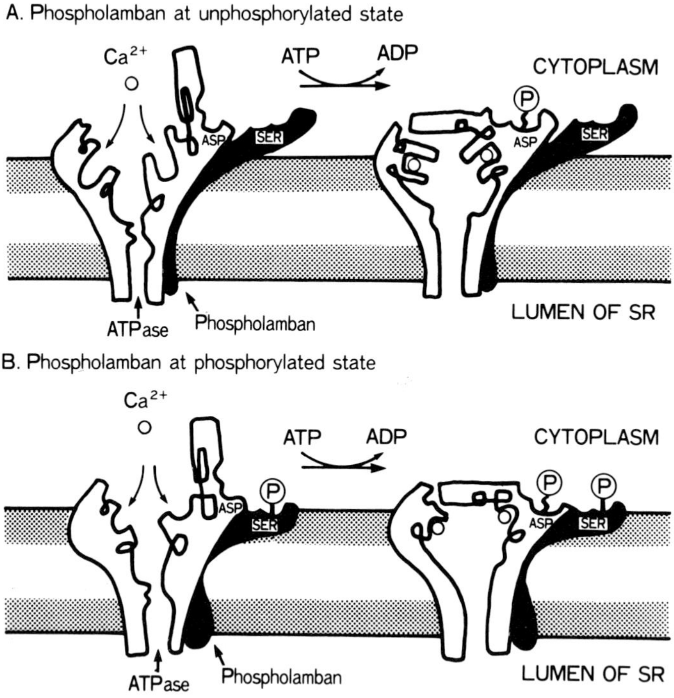

FIGURE 3. Diagrammatic representation of the action of phospholamban on the calcium pump of cardiac SR.
ASP represents an asparatic acid residue in the ATPase and SER a serine residue in phospholamban. The phosphates attached to these amino acids are ~ P, the acyl phosphate ATPase intermediate, and –P, the regulatory phosphoester (From Tada and Katz (6).)

TABLE III. Effect of $C_{12}E_8$ Treatment on Phospholamban Control of Overall Turnover Rate of Ca^{2+}-Dependent ATPase

Cardiac SR	ν/[EP] at $C_{12}E_8$/cardiac SR ratio (mg/mg) of		
	0	1	10
		sec^{-1}	
Control	0.8	1.2	1.5
Phosphorylated	1.2	1.2	2.5

After pretreatment of cardiac SR with cAMP-dependent protein kinase, cardiac SR vesicles were solubilized by various concentrations of dodecyloctaoxyethylene glycol monoether ($C_{12}E_8$) in the presence of 0.5 M KCl, 20% glycerol and 5 mM $CaCl_2$ at pH 7.5 and 22°C for 5 min. The ATPase reaction was subsequently started by adding [γ-^{32}P] ATP and terminated by the addition of 5% TCA. ν/[EP] is calculated from the rate of inorganic phosphate ($^{32}P_i$) liberation and the amount of ATPase intermediate ($E^{32}P$).

Although part of phospholamban is located at the cytoplasmic surface of SR membrane (9, 22), a large part of the molecule appears to be imbedded within the membrane (21, 22, 24). Upon phosphorylation, the phospholamban molecule may undergo a conformational change towards the interior of the membrane (9, 21), so that tryptic digestion becomes less effective. A 1 : 1 stoichiometry between phospholamban and the ATPase is proposed by the approximately 1 : 1 ratio between the amounts of the two phosphoproteins formed within the membrane (26, 29 – 31).

During Ca^{2+} translocation, the moiety of the ATPase molecule containing the active ATPase site (30,000 daltons) may undergo a conformational change that causes the Ca-binding moiety (20,000 daltons) to translocate from outer to inner position (1, 50). Because both formation (44) and decomposition (30, 46) of EP are enhanced by phospholamban phosphorylation, the conformational state of the ATPase appears to be under the direct control of phospholamban. Thus, phosphorylated phospholamban would enhance the rate of the cation-mediated conformational changes of the ATPase region, resulting in an increased rate of translocation of Ca^{2+}.

V. PURIFICATION OF PHOSPHOLAMBAN

Several attempts have been made to purify phospholamban from cardiac SR. However, due to hydrophobic nature of phospholamban, which was assumed to be tightly associated with the membrane ATPase, previous procedures employed SDS and organic solvent (22, 51), the drastic treatment that could have deteriorated the inherent properties of phospholamban. Therefore, it is of importance to purify phospholamban from cardiac SR that largely retains its inherent properties. Fractionation by deoxycholate (DOC) and ammonium sulfate gave a clue to purify phospholamban. The procedures are briefly described below. Cardiac SR (10 mg/ml) was solubilized by DOC (10 mg/ml) in the presence of 0.6 M KC1 and 0.3 M sucrose. When 50% saturated ammonium sulfate was added in a final proportion of 0.3 ml/ml, phospholamban and a significant amount of the ATPase and the 50,000-dalton protein precipitated. Several other proteins could be removed during this procedure. The pellet was dissolved with 2 mg/ml $C_{12}E_8$ and 2 M KI. Phospholamban was further purified from the ATPase and the 50,000-dalton protein by two successive chromatographic procedures on Sephacryl S-200 in the presence of 20 mM Tris/HC1, pH 7.5, 1 mg/ml $C_{12}E_8$ and 1 M KI. The latter conditions gave separation of co-purified proteins into the ATPase enriched fractions and the purified phospholamban. According to these procedures, 10 mg/ml of cardiac SR, which was estimated to contain about 500 μg of phospholamban by densitometric scanning of Coomassie blue-stained SDS-polyacrylamide gel, yielded about 50 μg of phospholamban. The obtained phospholamban was more than 95% purified, judging from the densitometric tracing.

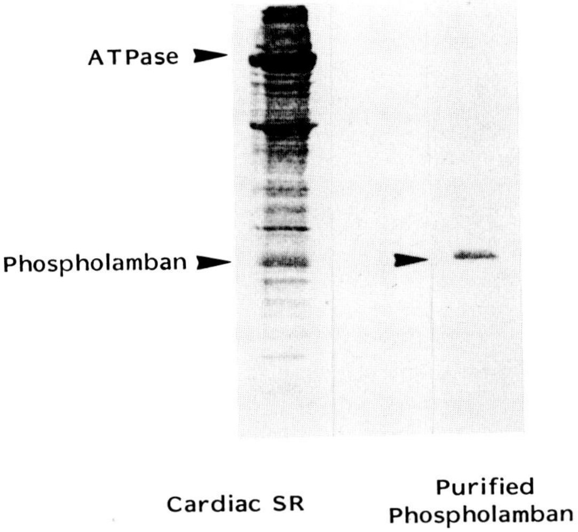

FIGURE 4. SDS-polyacrylamide (12.5%) gel electrophoretic pattern of cardiac SR proteins (left) and the purified phospholamban (right).

The purified phospholamban showed a single band of approximately 22,000 daltons on SDS-polyacrylamide gel (Fig. 4). However, the boiling of phospholamban in the presence of SDS resulted in a dissociation of the 22,000-dalton protein into a smaller component (5,000 to 7,000 daltons). This finding is in accord with the fact that a 22,000-dalton phosphoprotein in cardiac SR vesicles was dissociated into 5,500-dalton phosphoprotein by boiling of the phosphorylated SR vesicles in the presence of SDS (25). The purified phospholamban could be phosphorylated by cAMP-dependent protein kinase, incorporating 38.8 nmol of phosphate per mg protein or 0.85 mol of phosphate per mol of 22,000-dalton phospholamban. The corresponding value in intact SR vesicles was found to be 0.65 mol of phosphate per mol of 22,000-dalton phospholamban, taking 4% for the content of phospholamban. In several preparations, we found that the extent of phosphate incorporation into the purified phospholamban significantly exceeded that into phospholamban of intact SR. It is probably due to the existence of phosphatase in the SR preparations or due to the limited accessibility of the protein kinase to phospholamban in the SR membrane.

VI. SUMMARY

In the excitation-contraction coupling of the myocardium, active calcium transport by cardiac SR assumes a central role. Ca^{2+}-dependent ATPase (mol. wt. 100,000) of cardiac SR serves as an energy transducer and a translocator of Ca^{2+} across the membrane. A line of evidence indicated that a specific protein named phospholamban (mol. wt. 22,000) regulates Ca^{2+}-dependent ATPase of cardiac SR. Phosphorylation of phospholamban by cAMP-dependent protein kinase resulted in a marked increase in the rate of turnover of the ATPase, through the enhancement of the key elementary steps, i.e. the steps at which the intermediate EP is formed and decomposed. Thus, phospholamban is putatively thought to serve as a modulator of Ca^{2+}-dependent ATPase of cardiac SR. To interpret the mechanism, a working model was proposed. Also documented is a possibility that another protein kinase activatable by Ca^{2+} and calmodulin is functional in regulating the phospholamban-ATPase system. Thus, the existence of a dual control system of phospholamban is suggested in which both cAMP- and calmodulin-dependent phosphorylation are in control of the Ca^{2+}-dependent ATPase. The effect of phospholamban on Ca^{2+}-dependent ATPase remains operational in the solubilized cardiac SR, in which the ATPase protein is assumed to be monomeric. Phospholamban was more than 95% purified.

It should be further sought whether the molecular structure of phospholamban, that is assumed to consist of heterogeneous substructures, is associated with functional properties of the protein kinase-phospholamban-ATPase system.

REFERENCES

1. Tada, M., Yamamoto, T., and Tonomura, Y., Physiol. Rev. 58 : 1 (1978).
2. Endo. M., Physiol. Rev. 57 : 71 (1977).
3. Fabiato, A., and Fabiato, F., Ann. Rev. Physiol. 41 : 473 (1979).
4. Sutherland, E. W., and Rall, T. W., Pharmacol. Rev. 12 : 265 (1960).
5. Tsien, R. W., Adv. Cyclic Nuc. Res. 8 : 363 (1977).
6. Tada, M., and Katz, A. M., Ann. Rev. Physiol. 44 : 401 (1982).

7. Tada, M., Yamada, M., Kadoma, M., Inui, M., and Ohmori, F. Mol. Cell. Biochem. 46 : 73 (1982).
8. Tada, M., Shigekawa, M., and Nimura, Y., in "Function of the Heart in Normal and Pathological States" (N. Sperelakis, ed.), in press, Martinus Nijhoff Publishers, The Hague, 1983.
9. Tada, M., Kirchberger, M. A., and Katz, A. M., J. Biol. Chem. 250 : 2640 (1975).
10. LaRaia, P. J., and Morkin, E., Circ. Res. 35 : 298 (1974).
11. Wray, H. L., Gray, R. R., and Olsson, R. A., J. Biol. Chem. 248 : 1496 (1973).
12. Kirchberger, M. A., Tada, M., and Katz, A. M., J. Biol. Chem. 249 : 6166 (1974).
13. Katz, A. M., Tada, M., and Kirchberger, M. A., Adv. Cyclic Nuc. Res. 5 : 453 (1975).
14. Kirchberger, M. A., and Tada, M., J. Biol. Chem. 251 : 725 (1976).
15. Schwartz, A., Entman, M. L., Kaniike, K., Lane, L. K., Van Winkle, W. B., and Bornet, E. P., Biochim. Biophys. Acta 426 : 57 (1976).
16. Wray, H. L., and Gray, R. R., Biochim. Biophys. Acta 461 : 441 (1977).
17. Will, H., Levchenko, T. S., Levitsky, D. O., Smirnov, V. N., and Wollenberger, A., Biochim. Biophys. Acta 543 : 175 (1978).
18. St. Louis, P. J., and Sulakhe, P. V., Arch. Biochem. Biophys. 198 : 227 (1979).
19. Jones, L. R., Besch, H. R., Jr., Fleming, J. W., McConnaughey, M. M., and Watanabe, A. M., J. Biol. Chem. 254 : 530 (1979).
20. Lamers, J. M. J., and Stinis, J. T., Biochim. Biophys. Acta 624 : 443 (1980).
21. Bidlack, J. M., and Shamoo, A. E., Biochim. Biophys. Acta 632 : 310 (1980).
22. Le Peuch, C. J., Le Peuch, D. A. M., and Demaille, J. G., Biochemistry 19 : 3368 (1980).
23. Tada, M., Inui, M., Yamada, M., Kadoma, M., Kuzuya, T., Abe, H., and Kakiuchi, S., J. Mol. Cell Cardiol. in press (1983).
24. Louis, C. F., and Katz, A. M., Biochim. Biophys. Acta 494 : 255 (1977).
25. Kirchberger, M. A., and Antonetz, T., Biochem. Biophys. Res. Commun. 105 : 152 (1982).
26. Le Peuch, C. J., Haiech, J., and Demaille, J. G., Biochemistry 18 : 5150 (1979).
27. Kirchberger, M. A., and Antonetz, T., J. Biol. Chem. 257 : 5685 (1982).
28. Bidlack, J. M., Ambudkar, I. S., and Shamoo, A. E., J. Biol. Chem. 257 : 4501 (1982).
29. Tada, M., Yamada, M., Ohmori, F., Kuzuya, T., and Abe, H., in "Cation Flux Across Biomembranes" (Y. Mukohata, and L. Packer, eds.), p. 179 Academic Press, New York, 1979.
30. Tada, M., Ohmori, F., Yamada, M., and Abe, H., J. Biol. Chem. 254 : 319 (1979).
31. Katz, A. M., Adv. Cyclic Nuc. Res. 11 : 303 (1979).
32. Bilezikjian, L. M., Kranias, E. G., Potter, J. D., and Schwartz, A., Circ. Res. 49 : 1356 (1981).

33. Katz, S., and Remtulla, M. A., Biochem. Biophys. Res. Commun. 83 : 1373 (1978).
34. Lopaschuk, G., Richter, B., and Katz, S., Biochemistry 19 : 5603 (1980).
35. Tada, M., Kirchberger, M. A., Repke, D. I., and Katz, A. M., J. Biol. Chem. 249 : 6174 (1974).
36. Kirchberger, M. A., Tada, M., Repke, D. I., and Katz, A. M., J. Mol. Cell. Cardiol. 4 : 673 (1972).
37. Kirchberger, M. A., and Chu, G., Biochim. Biophys. Acta 419 : 559 (1976).
38. Hicks, M. J., Shigekawa, M., and Katz, A. M., Circ. Res. 44 : 384 (1979).
39. Tada, M., Kirchberger, M. A., and Li, H.-C., J. Cyclic Nuc. Res. 1 : 329 (1975).
40. Kirchberger, M. A., and Raffo, A., J. Cyclic Nuc. Res. 3 : 45 (1977).
41. Tada, M., Ohmori, F., Nimura, Y., and Abe, H., J. Biochem. Tokyo 82 : 885 (1977).
42. Ohmori, F., Tada, M., Kinoshita, N., Matsuo, H., Sakakibara H., Nimura, Y., and Abe, H., "Recent Advances in Studies on Cardiac Structure and Metabolism", Vol. 11 (T. Kobayashi, T. Sano, N. S. Dhalla, eds.) p. 279 University Park Press, Baltimore, 1978.
43. Will, H., Blanck, J., Smettan, G., and Wollenberger, A., Biochim. Biophys. Acta 449 : 295 (1976).
44. Tada, M., Yamada, M., Ohmori, F., Kuzuya, T., Inui, M., and Abe, H., J. Biol. Chem. 255 : 1985 (1980).
45. Tada, M., Ohmori, F., Kinoshita, N., and Abe, H., Adv. Cyclic Nuc. Res. 9 : 355 (1978).
46. Kranias, E. G., Mandel, F., Wang, T., and Schwartz, A., Biochemistry 19 : 5434 (1980).
47. Froehlich, J. P., Sullivan, J. V., and Berger, R. L., Anal. Biochem. 73 : 331 (1976).
48. De Meis, L., and Vianna, A. L., Ann. Rev. Biochem. 48 : 275 (1975).
49. Katz, A. M., Trends Pharmacol. Sci. 1 : 434 (1980).
50. Yamamoto, T., Takisawa, H., and Tonomura., Y., Curr. Top. Bioenerg. 9 179. (1979).
51. Collins, J. H., Kranias, E. G., Reeves, A. S., Bilezikjian, L. M., and Schwartz, A., Biochem. Biophys. Res. Commun. 99 : 796 (1981).

RESOLUTION OF THE REGULATORY SYSTEMS OF THE CARDIAC SARCOPLASMIC RETICULUM $Ca^{2+} + Mg^{2+}$ -ATPase

Adil E. Shamoo
Indu S. Ambudkar[1]

Membrane Biochemistry Research Laboratory
Department of Biological Chemistry
University of Maryland School of Medicine
Baltimore, Maryland

I. INTRODUCTION

Muscle contraction is induced by the depolarization of sarcolemma. This depolarization is propagated to the transverse tubule which causes the release of calcium that is stored in sarcoplasmic reticulum. In cardiac muscle, calcium also enters the cell through the slow calcium channels in the sarcolemma. The increased cytosolic calcium concentration leads to contraction. Relaxation occurs when calcium is removed from the cytoplasm by the sarcoplasmic reticulum and sarcolemmal membrane systems.

The sarcoplasmic reticulum membrane is involved in active transport of calcium, mediated by the activity of the $Ca^{2+} + Mg^{2+}$-ATPase, a 105K protein associated with it. The mechanism of coupling between ATP hydrolysis and calcium transport has been an issue of great interest, though not clearly defined as yet. The system exists both in the cardiac and skeletal muscles and considerable work has been accomplished in characterizing it from both tissue sources (1-4). The molecular mechanisms involved in the calcium transport process have not been established as yet. It has been

[1]This work was done while the author was a recipient of a Post-doctoral fellowship from the American Heart Association, Maryland Affiliate, Inc.

ISBN 0-12-260380-X

reported by our laboratory (5) for the skeletal enzyme, that distinct molecular domains (sites) of the protein are involved in the two functions. The results suggest an ATP hydrolytic domain and a calcium binding/ transport (i.e. ionophoric) domain, which are juxtapositioned such that the energy of ATP cleavage is transferred from the ATP hydrolytic site to the calcium transport site.

The cardiac calcium transporting mechanism is more complex and relatively less understood. The complexity is attributed to the regulatory systems that are associated with it. β-adrenergic amines such as norepinephrine, cause ionotropic and chronotropic effects in the heart (6,7). Catecholamines thus cause an increase in peak tension; shortening of the time to reach peak tension and shortening of the refractory period. The catecholamine action is attributed to a cascade reaction which is initiated by hormone-receptor interaction, leading to an increase in the intracellular c-AMP. This, in turn, activates a c-AMP-dependent protein kinase that phosphorylates the regulatory protein, phospholamban (8,9), which then stimulates Ca^{2+}-uptake into sarcoplasmic reticulum by activating the Ca^{2+} + Mg^{2+}-ATPase. The Ca^{2+}/ATP stoichiometry is not altered by this phosphorylation-induced activation. The molecular interactions involved in these covalent modifications of sarcoplasmic reticulum have not yet been fully explained. The phosphorylation has been reported to alter the rates of E-P formation and the intermediate steps involved in the ATPase activity (10). Kirchberger (11) has also reported an effect of the phosphorylation on calcium release.

Recent reports (12-14) show a role for calmodulin in regulation of the ATPase. This regulation has also been suggested to be mediated by phospholamban phosphorylation, at a site distinct from the c-AMP dependent one. Again the exact mechanism involved in this regulation still remains to be elucidated.

We had earlier reported a very simple method for the purification of the 22K protein (15), and its biochemical characterization. In this paper we present our studies to correlate the membrane phosphorylation status and the level of the ATPase function that is regulated. The results presented give evidence of a regulatory role for phospholamban in the calcium transport activity. In cardiac membranes this regulation is achieved by the phosphorylation and probable de-phosphorylation of the 22K protein. Our results suggest that when the protein is unphosphorylated the transport function is uncoupled from the ATPase function. Phosphorylation at one site couples the calcium transport and ATP-hydrolytic functions, while a second phosphorylation activates this function, expressing the β-adrenergic hormone-dependent levels of uptake. The results discussed in this paper provide an insight into the possible mechanism of phosphorylation-dependent regulation of calcium transport, as has been reported in the case of a number of other membrane systems (16,17).

II. METHODS

A. Isolation of Cardiac Microsomes

Cardiac microsomes were prepared from canine heart ventricle according to the method of Harigaya and Schwartz (18) with the following modifications (15). Cardiac ventricular muscle was homogenized in 10 mM $NaHCO_3$ and centrifuged at 8,7000 xg for 20 min. The supernatant was then centrifuged at 10,000 xg for 20 min. The supernatant was strained through four layers of cheesecloth, followed by centrifugation at 37,000 xg for 60 min. The pellet was suspended in 20 mM Tris/maleate, 0.6 M KCl, pH 6.8, and centrifuged at 100,000 xg for 20 min. The pellet, containing microsomes, was suspended in 0.25 M sucrose. Protein was estimated by the method of Lowry (19) or from the absorption at 280 nm, in the presence of 1% SDS.

B. Phosphorylation Assay

cAMP dependent phosphorylation was accomplished essentially by the method of Kirchberger et. al. (20). The protein fraction (0.25 to 0.50 mg/ml) was suspended in 40 mM histidine (pH 6.8), 0.12 M KCl, 0.5 mM $MgCl_2$, 5 mM NaN_3, 5 mM EGTA, 25 mM NaF buffer at 25,C. Cyclic AMP was added to a final concentration of 1 μM. Protein kinase was added at the concentration of 0.125 mg/ml. The reaction was started by the addition of 0.20 mM [γ-^{32}P]ATP. The final volume was 0.2 ml. Calmodulin dependent phosphorylation was carried out according to LePeuch et. al. (12) in the presence of 0.5 μM CAM and 0.5 mM Ca^{2+}. After 10 min, the reaction was stopped by adding 2 ml of cold 10% trichloroacetic acid, 0.1 mM KH_2PO_4. 0.2 ml of 0.63% bovine serum albumin was added as a carrier protein. After remaining on ice for 15 min, samples were centrifuged at 2,000 xg for 10 min. The pellets were suspended in 0.2 ml of 0.6 N NaOH. The protein was reprecipitated by the addition of 2 ml of 10% trichloroacetic acid, centrifuged, resuspended in NaOH, and washed a couple of times. The pellet was finally suspended in 0.2 ml of 0.6 N NaOH and transferred to scintillation vials containing 10 ml of Instagel (Packard) for counting.

C. DOC Treatment of Microsomes

DOC treatment was carried out as described by Bidlack et. al. (15). Cardiac microsomes suspended at a concentration of 7 mg/ml in 0.25 M sucrose. A small aliquot of deoxycholate was added at the ratio of 2 μg of dexoycholate/7 mg of microsomal protein (0.286 μg of deoxycholate/mg of protein). The suspension was incubated with stirring on ice for 15 min. After incubation, the suspension was centrifuged at 100,000 xg for 30 min. The pellet was resuspended in 0.25 M sucrose.

D. KCl Treatment of Microsomes

This was carried out as reported by LePeuch et. al. (12). Microsomes were suspended in 0.6 M KCl, 10 mM Tris-HCl, pH 6.8 and incubated for 24 hours, centrifuged at 100,000 xg for 30 min, washed and suspended in 0.25 M sucrose.

E. ATPase Activity

The Ca^{++}-dependent ATPase activity was measured by a method similar to that used by Tada et al. (21). Membrane vesicles were suspended at a concentration of 30-50 μg/ml in 40 mM in 40 mM Hist. (pH 6.8), 0.125 M KCl, 5 mM NaN_3, 5 mM $MgCl_2$, 30 μM $CaCl_2$. The reaction was initiated by the addition of 5 mM Tris-ATP. The Ca^{2+}-independent ATPase activity was measured by including 0.5 mM EGTA, instead of calcium. The reaction was carried out at 25°C. At the end of the incubation period, the reaction was quenched by adding 0.2 ml of 10% TCA. The samples were centrifuged and the Pi in the supernatant estimated by the method of LeBel et al. (22).

F. Calcium Uptake

^{45}Calcium uptake was assayed following the method of LePeuch et al. (12) in the presence of 150 mM KCl, 0.1 mM $^{45}CaCl_2$, 5 mM ATP, 5 mM Mg^{2+}, 10 mM Hepes buffer, pH 7.2, containing 50 to 100 μg of protein/ ml. The oxalate concentration is specified wherever used.

The method using Arsenzo III, a calcium specific dye, was carried out according to Van Winkle et al. (23) in the presence of 10 mM Tris-HCl or Mops, pH 7.0, 60 μM Az III, 40 μM $CaCl_2$, 5 mM ATP, 5 mM $MgCl_2$, 150 mM KCl, 10 mM oxalate containing 100 μg protein per ml.

G. Passive Permeability of Vesicles

The passive permeability of membrane vesicles was determined according to Campbell et. al. (24). Membrane vesicles (20 mg/ml) were incubated with 1 mM $^{45}CaCl_2$ in 100 mM KCl, 5 mM $MgCl_2$, 25 mM Hepes, pH 7.0 at 4°C for the mentioned time intervals and then diluted 40-fold with the same buffer containing 5 mM EGTA. Aliquots were removed at 0.5, 1.0, 1.5, 2.0 minutes after dilution and filtered through 0.33μ millipore filters. The radioactivity associated with the filters was determined. The counts at time 0 (Co) and rate constants were calculated as described by Campbell et al.

III. RESULTS

A. Effect of Phosphorylation on ATPase function

Table I shows the ATP-dependent calcium uptake activity associated with the sarcoplasmic reticulum. Phosphorylation of the membrane with c-AMP and c-AMP dependent protein kinase induces a 40% increase in this activity. This activation of Ca^{2+}-uptake involves the 22K protein as the substrate and has been previously reported by this laboratory (25) and others (8).

The effect of phosphorylation on the calcium-dependent ATP hydrolytic activity is also shown in Table I. The phosphorylation results in a 40-50% stimulation in the calcium dependent rates of hydrolysis. The Ca^{2+}/ATP stoichiometry is approximately 0.9 in our system, which agrees with other reports (8,14). This ratio is retained after phosphorylation with c-AMP-dependent protein kinase.

B. Effect of Alterations in the Phosphorylation Status of Cardiac SR

In view of the effect of phosphorylation, as evidenced by the data presented above, we altered the phosphorylation status of the SR vesicles by three different techniques. The consequent effects on membrane phosphorylation, ATP hydrolysis, and calcium uptake are presented.

<u>1. DOC Treatment of Cardiac Sarcoplasmic Reticulum</u>. We have previously reported that 0.0002% DOC solubilizes up to 90% of the phosphorylatable substrate from the membranes (15). This technique has been used to purify the 22K protein and characterize it biochemically. The isolated protein shows c-AMP dependent phosphorylation and a requirement for

TABLE I.

	Calcium uptake µmoles/mg pr/ min[1]	ATP hydrolysis µmoles Pi/mg pr/ min[2]	Ca^{2+}/ATP
SR-control	0.12	0.132	0.91
SR-phosphorylated	0.17	0.198	0.85
% of control	142	150	-

[1,2] The values given represent the rates of ATP dependent uptake and calcium-dependent hydrolysis respectively.

exogenous c-AMP and c-AMP-dependent protein kinase. The membrane fraction remaining after DOC treatment retains vesicle morphology as seen by freeze fracture electron microscopy (data not shown). The permeability to calcium ions is not altered as shown in Table II. The rates of EGTA-induced calcium release is not altered by DOC even under various extents of calcium loading.

The ATPase enzyme (100K dalton) remains associated with the DOC treated membrane fraction and can be visualized by SDS-gel electrophoresis. Table III shows the ATP hydrolytic rates of the DOC treated membranes. The data clearly reveal that the DOC treatment does not alter ATP hydrolysis, both in terms of rates and calcium sensitivity.

The levels of c-AMP dependent phosphorylation are also given in Table III. The DOC treated membrane fraction shows significantly lower levels of phosphate label incorporation. The phosphorylated substrate was localized in the fraction solubilized by DOC, as reported by us (15). The effect of the DOC treatment on calcium uptake is shown in Table III. A 90% decrease in the rates of oxalate supported, ATP-dependent calcium transport is seen in DOC treated membranes. These membranes do not regain the initial levels, as in control membranes, even after phosphorylation with c-AMP dependent protein kinase. A slight stimulation in the activity is seen after phosphorylation, which is probably due to residual levels of the protein left in the SR after DOC treatment. This decrease in calcium uptake is significant in terms of the ATP hydrolytic activity, which remains unaltered even after the DOC treatment. These results suggest an uncoupling of the transport function from ATP hydrolysis in the absence of the 22K protein.

TABLE II. Calcium Release in SR and DOC-SR

Membrane	1.75h	3h	4h	24h	Mean*
	Rate of Ca^{2+} release (nmoles Ca^{2+}/mg pr./min)				
SR (Control)	0.3	0.455	0.394	0.394	0.383 ± 0.061
DOC-SR	0.38	0.32	0.394	0.321	0.363 ± 0.049
% of Control	126.7	72	100	82	95 ± 13

*Mean rate of all four time points

The rates were calculated from the slopes of plots of Ct/Co vs. Time for each time point as described by Campbell et al. (24).

TABLE III. Comparison of c-AMP-Dependent Phosphorylation, ATP Hydrolysis and Calcium Uptake in Cardiac SR.

Membrane	% ^{32}P in 22K	% Ca^{2+} ATPase	% Ca^{2+} Uptake
SR	100	100	100
DOC-SR	10.3 ± 2.6	98 ± 9.2	16.7 ± 8.3
Phosphorylated* before DOC	94.8 ± 7.6	115 ± 15.6	128 ± 4.2
Phosphorylated* after DOC	10.3 ± 2.6	95 ± 11.0	37.9 ± 6.7

The values represent ± SE of five experiments.

*Phosphorylation was carried out in the presence of c-AMP and protein kinase as described in methods.

We have shown previously (25), that phosphorylation prior to DOC treatment allows retention of the 22K in the membrane. The data presented in Table III show that under these conditions, the calcium uptake function of these membranes is also retained. The function also exhibits the c-AMP dependent stimulated levels.

2. Dephosphorylation of Cardiac Sarcoplasmic Reticulum. Cardiac sarcoplasmic vesicles were treated with alkaline phosphatase to cleave any endogenous phosphate bonds. This technique has been used earlier by Demaille (12) to characterize the ester serine bond in the 22K protein. The phosphatase treated membranes have drastically decreased levels of calcium uptake, comparable to that of DOC treated membranes shown in Figure 1. The phosphorylation efficiency of these membranes is shown in Table IV. Phosphatase treated membranes exhibit c-AMP dependent phosphorylation, to levels equal to control membranes. A slight stimulation (30%) is observed which could be due to removal of the phosphoryl group from an endogenously phosphorylated c-AMP dependent site.

SR vesicles isolated by the method described here have been reported to contain a site that can be phosphorylated by calmodulin. This phosphorylation can be detected only after the calmodulin associated with the membrane has been removed, by treatment with 0.6 M KCl. Since all phosphorylatable sites are probably saturated in the isolated membranes (12). Figure 2 shows the calmodulin dependent phosphorylation of KCl treated sarcoplasmic reticulum. A maximal phosphorylation is observed around 0.5 μM calmodulin. This

TABLE IV. c-AMP and Protein Kinase Dependent Phosphorylation in Cardiac SR

Membrane	nmoles ^{32}P/mg Protein*	%
SR	0.99	100
24 h. KCl treated SR	1.35	136
Phosphatase treated SR	1.26	127

All values are means of three experiments.

*This value is calculated from the difference between the levels obtained in controls or those in the presence of c-AMP and protein kinase.

agrees rather well with the results reported by LePeuch et al., (12), in a similar experiment. These membranes also retain maximal levels of c-AMP dependent phosphorylation efficiency as evidenced by the data given in Table III. The membrane phosphorylation level is 1.35 nmoles of ^{32}Pi per milligram protein. This level is about 30% more than levels obtained for control membranes. The increased level of phosphorylation, could be a result of inhibition of phosphodiesterase, which is activated by calmodulin.

The calcium uptake activity of the KCl-treated membranes is shown in Fig. 1. There is a significant decrease in the rates of uptake, reaching about 90% of the intial levels after 24h of KCl treatment. DOC treatment of these membranes further decrease the level to that in the case of direct DOC treatment.

We have reported previously that KCl-treated membranes, when phosphorylated, regain more than 70% of the activity associated with control membranes phosphorylated under the same conditions; in the presence of calmodulin, c-AMP and c-AMP dependent protein kinase. DOC treated membranes, under the same conditions, recover less than 25% of the transport activity (data not shown) (26). These results suggest that the presence of phospholamban in the membrane and its phosphorylation status contribute to the expression of the calcium transport function.

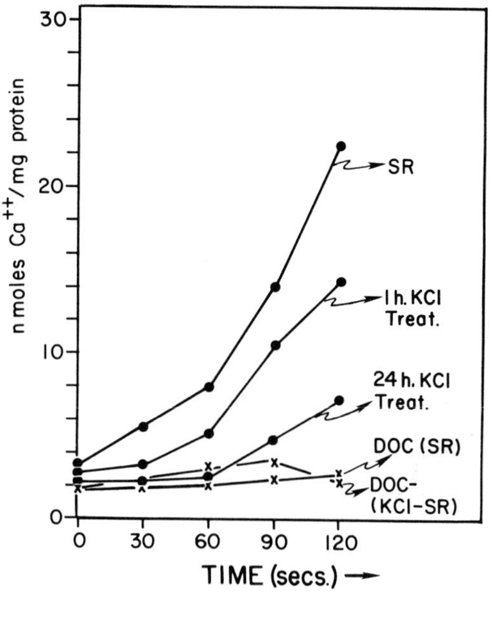

FIGURE 1. Effect of KCl and DOC treatments on cardiac SR calcium uptake.

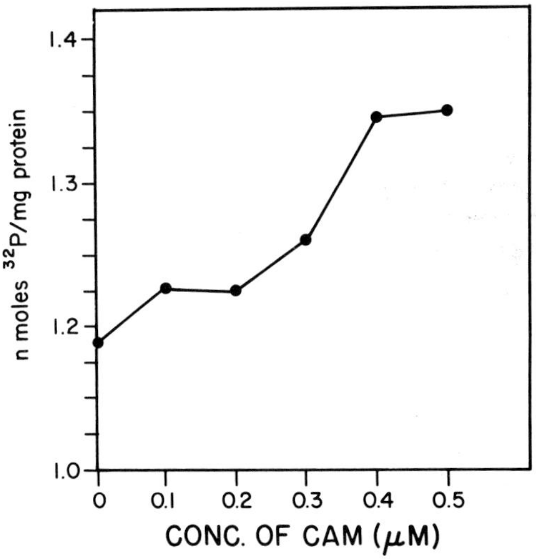

FIGURE 2. CAM-dependent phosphorylation of KCl-treated cardiac SR.

IV. DISCUSSION

Several calcium transport systems have been reported to be regulated by membrane phosphorylation (16,17). Specific membrane proteins have been implicated in some instances to be the regulatory proteins, the regulatory mechanism being achieved by their phosphorylation and dephosphorylation. In most cases the protein phosphorylation has been identified and the resultant change in the activity of the system has been characterized. However, the inter-molecular interactions involved in these regulatory mechanisms have not been characterized or established.

Sarcoplasmic reticulum contains a Ca^{2+} + Mg^{2+} -ATPase which is involved in the calcium transport activity exhibited by this membrane. Membranes from both the skeletal as well the cardiac tissue contain this system, though some basic differences have been reported for the two systems (23,27, 28). The differences are mainly in the membrane association of the ATPase with the various proteins associated with it. The skeletal enzyme has been isolated and characterized. No direct hormonal regulation of its functions has been reported yet. The activity of the calcium transport system is considerably higher in skeletal SR as compared to the cardiac membranes (29), and these are correlated with higher rates of

ATP hydrolysis. The cardiac SR contains several more proteins as compared to its skeletal counterpart. At least two, phospholamban and calmodulin, are associated with the Ca^{2+}-transport system and thus associated with the ATPase. Phosphorylation of the former by the c-AMP dependent system and the calcium-calmodulin system is postulated to be involved in the activation of the ATPase function by β-adrenergic hormones and during beat-to-beat activity. No studies have been reported to describe the molecular events associated with the hormonal control.

We had reported earlier that the 22K protein can be solubilized from the SR at very low DOC concentrations if it is unphosphorylated (25). Phosphorylation renders phospholamban (22K) insoluble to DOC (up to 0.1%), and also resistant to cleavage by trypsin. These results give evidence for an alteration in the membrane interaction of the protein as a result of phosphorylation, which probably renders it less exposed to the cytosolic side and also results in the activation of the ATPase function.

The results presented in this paper directly correlate the presence of the 22K protein in the membrane and its phosphorylation status to the level of calcium transport expressed by the ATPase. The results give evidence for a differential role of the regulatory system. Only the calcium transport function is affected by alterations in the calmodulin dependent phosphorylation levels of the membrane. The ATP hydrolytic function remains unaffected and retains normal levels of activity, except after c-AMP-dependent phosphorylation, when it is activated. A differential effect on the two ATPase functions can be visualized, taking into account the involvement of distinct molecular domains in these activities. Differential effects at the sub-molecular level, even of the same amino acid species localized differently, has been proposed earlier in synaptosomal membranes (30) and in the sodium channel (31).

The results illustrate the levels of the calcium uptake activity that are regulated by the two systems. The normal levels of transport, associated with the membranes as isolated, are related to the endogenously phosphorylated levels of the membrane. This level of phosphorylation is evidenced by the effect of dephosphorylation by phosphatase on the transport. Since KCl-treated membranes show the same effects, the phosphorylation could be attributed to calmodulin, which is removed by KCl. C-AMP-dependent phosphorylation would then be associated with calcium transport levels higher than the normal levels. The fact that in DOC-treated membranes, which do not contain the phosphorylatable moiety, calcium transport cannot be restored even in the presence of both the calmodulin and the c-AMP dependent systems, implicates the requirement of phospholamban in these phosphorylation dependent effects.

From the above discussion, it can be concluded that in cardiac sarcoplasmic reticulum, the Ca^{2+}+Mg^{2+}-ATPase requires the presence of phosphorylated phospholamban in order to express its calcium transport function, the levels of transport being correlated to the level of calmodulin dependent phosphorylation of the protein. Furthermore, the stimulation of calcium transport occurs when the c-AMP-dependent protein kinase system results in an additional phosphorylation of phospholamban.

REFERENCES

1. Berman, M. C. (1982). Biochem. Biophys. Acta. 694:95-121.
2. Tada, M., Yamamoto, T., Tohomura, Y. (1978). Phys. Rev. 58:1-79.
3. Dhalla, N. S., Das, P. K. Sharma, G. P. (1978). Cardiology. 10:363-385.
4. Fabiato, A. and Fabiato, F. (1977). Ann. Rev. Physiol. 40:119-129.
5. Scott, T. K. and Shamoo, A. E. (1982). J. Membr. Biol. 64:137-144.
6. Stull, J. T. and Mayer, S. E. (1979). In Handbook of Physiology, Sec. 2, Vol. 1, (R. M. Berne, ed), 741-774.
7. Hoffman, B. B. and Lefkowitz, R. J. (1982). Ann. Rev. Physiol. 44:475-484.
8. Tada, M., Yamada, M., Kadoma, M., Inui, M., and Ohmori, F. (1982). Mol. and Cell. Biochem. 46:74-95.
9. Cavadore, J. C., LePeuch, D. J., Walsh, M. P., Vallet, B., Molla, A. and Demaille, J.O. (1981). Biochimie. 63:301-306.
10. Rada, M., Yamada, M., Ohmori, F., Kuzuya, T., Inui, M. and Abe, H. (1980). J. Biol. Chem. 255:1985-1992.
11. Kirchberger, M. A. and Wong, D. (1978). J. Biol. Chem. 253:6941-6945.
12. LePeuch, C. J., Haiech, J., Demaille, J. G. (1979). Biochemistry 18:5150-5157.
13. Lopaschuk, G., Richter, B., Katz, S. (1980). Biochemistry 19:5603-5607.
14. Kirchberger, M. A. and Antonetz, T. (1982). J. Biol. Chem. 257:5685-5691.
15. Bidlack, J. M. Ambudkar, I. S., and Shamoo, A. E. (1982). J. Biol. Chem. 257:4501-4506.
16. Brockbank, K. J. and England P. J. (1980). Febs. Letters 122:67-71.

17. Kometiani, P., Kometiani, Z., and Mikeladze, D. (1978). Progress in Neurobiology 11:223-247.
18. Harigaya, S. and Schwartz, A. (1969). Circ. Res. 25:781-794.
19. Lowry, L. H., Rosebrough, N. J., Farr, A. L., and Randall, R. J. (1951). J. Biol. Chem. 193:265-275.
20. Kirchberger, M. A., Tada, M. and Katz, A. M. (1974). J. Biol. Chem. 249:6166-6173.
21. Tada, M., Kirchberger, M. A., Repke, D. L. and Katz, A. M. (1974). J. Biol. Chem. 249:6174-6180.
22. LeBel, D., Poirier, G. G., Beaudoin, A. R. (1978). Anal. Biochem. 85:86-89.
23. VanWinkle, W. B., Tate, C. A., Bick, R. J., and Entman, M. L. (1981). J. Biol. Chem. 256:2268-2274.
24. Campbell, K. P., Armstrong, C. F., and Shamoo, A. E. (1980). Biochem. Biophys. Acta. 602:97-116.
25. Bidlack, J. M. and Shamoo, A. E. (1980). Biochem. Biophys. Acta. 632:310-325.
26. Shamoo, A. E. and Ambudkar, I. S. (1982). EBEC Reports (L.B.T.M., C.N.R.S. Ed.) 43-44.
27. DeFoor, P. H., Levitsky, D., Biryubona, T., and Fleischer, S. (1980). Arch. Biochem. Biophys. 200:196-205.
28. Chiesi, M. (1979). J. Mol. and Cell. Cardiol. 11:245-259.
29. Sumida, M., Wang, T., Mandel, F., Froehlich, J. P., and Schwartz, A. (1978). J. Biol. Chem. 253:8772-8777.
30. Huttner, W. B., DeGennaro, L. J., and Greengard, P. (1981). J. Biol. Chem. 256:1482-1488.
31. Costa, M. R. C., Casnellie, J. E., and Catterall, W. A. (1982). J. Biol. Chem. 257:7918-7921.

ACKNOWLEDGMENT

This work was supported in part by a Program Project of the National Institute of Environmental Health Sciences (ES-1248), the Department of Energy (DE-AS0580EV10329) and the Naval Research (N00014-80-C-0030).

Index

Q

R

S